ATF Federal Explosives Law and Regulations

Plus

ATF Federal Firearms Regulations Reference Guide

ATF Federal Explosives Law and Regulations

Table of Contents

Organized Crime Control Act of 1970, Title XI

Regulation of Explosives

Public Law 91-452, Approved October 15, 1970 (as Amended)

[**NOTE:** Any reference to the Internal Revenue Code of 1954 refers to the Internal Revenue Code of 1986 (§2, Public Law 99-514, 100 Stat. 2085, October 22, 1986.)]

Purpose

SEC. 1101. The Congress hereby declares that the purpose of this title is to protect interstate and foreign commerce against interference and interruption by reducing the hazard to persons and property arising from misuse and unsafe or insecure storage of explosive materials. It is not the purpose of this title to place any undue or unnecessary Federal restrictions or burdens on law-abiding citizens with respect to the acquisition, possession, storage, or use of explosive materials for industrial, mining, agricultural, or other lawful purposes, or to provide for the imposition by Federal regulations of any procedures or requirements other than those reasonably necessary to implement and effectuate the provisions of this title.

Sec. 1102. Title 18, United States Code, is amended by adding after chapter 39 the following chapter:

Chapter 40. Importation, Manufacture, Distribution and Storage of Explosive Materials

Editor's Note: The sections of law set out herein were added by Public Law 91-452, Title XI, §1102(a), Oct. 15, 1970, 84 Stat. 952-959, and remain un-changed unless otherwise footnoted.

Sec.

§ 841. Definitions

As used in this chapter

(a) "Person" means any individual, corporation, company, association, firm, partnership, society, or joint stock company.

(b) "Interstate or foreign commerce" means commerce between any place in a State and any place outside of that State, or within any possession of the United States (not including the Canal Zone) or the District of Columbia, and commerce between places within the same State but through any place outside of that State. "State" includes the District of Columbia, the Commonwealth of Puerto Rico, and the possessions of the United States (not including the Canal Zone).

(c) "Explosive materials" means explosives, blasting agents, and detonators.

(d) Except for the purposes of subsections (d), (e), (f), (g), (h), (i), and (j) of section 844 of this title, **"explosives"** means any chemical compound mixture, or device, the primary or common purpose of which is to function by explosion; the term includes, but is not limited to, dynamite and other high explosives, black powder, pellet powder, initiating explosives, detonators, safety fuses, squibs, detonating cord, igniter cord, and igniters. The Attorney General shall publish and revise at least annually in the Federal Register a list of these and any additional explosives which he determines to be within the coverage of this chapter. For the purposes of subsections (d), (e), (f), (g), (h), and (i) of section 844 of this title, the term "explosive" is defined in subsection (j) of such section 844.

(e) "Blasting agent" means any material or mixture, consisting of fuel and oxidizer, intended for blasting, not otherwise defined as an explosive: Provided, That the finished product, as mixed for use or shipment, cannot be detonated by means of a numbered 8 test blasting cap when unconfined.

(f) "Detonator" means any device containing a detonating charge that is used for initiating detonation in an explosive; the term includes, but is not limited to, electric blasting caps of instantaneous and delay types, blasting caps for use with safety fuses and detonating-cord delay connectors.

(g) "Importer" means any person engaged in the business of importing or bringing explosive materials into the United States for purposes of sale or distribution.

(h) "Manufacturer" means any person engaged in the business of manufacturing explosive materials for purposes of sale or distribution or for his own use.

(i) "Dealer" means any person engaged in the business of distributing explosive materials at wholesale or retail.

(j) "Permittee" means any user of explosives for a lawful purpose, who has obtained either a user permit or a limited permit under the provisions of this chapter.

(k) "Attorney General" means the Attorney General of the United States.

(l) "Crime punishable by imprisonment for a term exceeding one year" shall not mean (1) any Federal or State offenses pertaining to antitrust violations, unfair trade practices, restraints of trade, or other similar offenses relating to the regulation of business practices as the Attorney General may be regulation designate, or (2) any State offense (other than one involving a firearm or explosive) classified by the laws of the State as a misdemeanor and punishable by a term of imprisonment of two years or less.

(m) "Licensee" means any importer, manufacturer, or dealer licensed under the provisions of this chapter.

(n) "Distribute" means sell, issue, give, transfer, or otherwise dispose of.

(o) "Convention on the Marking of Plastic Explosives" means the Convention on the Marking of Plastic Explosives for the Purpose of Detection, Done at Montreal on 1 March 1991.

(p) "Detection agent" means any one of the substances specified in this subsection when introduced into a plastic explosive or formulated in such explosive as a part of the manufacturing process in such a manner as to achieve homogeneous distribution in the finished explosive, including—

(1) Ethylene glycol dinitrate (EGDN), C[2]H[4](NO[3])[2], molecular weight 152, when the minimum concentration in the finished explosive is 0.2 percent by mass;

(2) 2,3-Dimethyl-2,3-dinitrobutane (DMNB), C[6]H[12](NO[2])[2], molecular weight 176, when the minimum concentration in the finished explosive is 0.1 percent by mass;

(3) Para-Mononitrotoluene (p-MNT), C[7]H[7]NO[2], molecular weight 137, when the minimum concentration in the finished explosive is 0.5 percent by mass;

(4) Ortho-Mononitrotoluene (o-MNT), C[7]H[7]NO[2], molecular weight 137, when the minimum concentration in the finished explosive is 0.5 percent by mass; and

(5) any other substance in the concentration specified by the Attorney General, after consultation with the Secretary of State and the Secretary of Defense, that has been added to the table in part 2 of the Technical Annex to the Convention on the Marking of Plastic Explosives.

(q) "Plastic explosive" means an explosive material in flexible or elastic sheet form formulated with one or more high explosives which in their pure form has a vapor pressure less than 10^{-}<4> Pa at a temperature of 25 degrees C., is formulated with a binder material, and is as a mixture malleable or flexible at normal room temperature.

(r) "Alien" means any person who is not a citizen or national of the United States.

(s) "Responsible person" means an individual who has the power to direct the management and policies of the applicant pertaining to explosive materials.

(Added Pub. L. 91-452, title XI, Sec. 1102(a), Oct. 15, 1970, 84 Stat. 952; amended Pub. L. 104-132, title VI, Sec. 602, Apr. 24, 1996, 110 Stat. 1288; Pub. L. 107-296, title XI, Sec. 1112(e)(1), (3), 1122(a), Nov. 25, 2002, 116 Stat. 2276, 2280.)

§ 842. Unlawful acts

(a) It shall be unlawful for any person—

(1) to engage in the business of importing, manufacturing, or dealing in explosive materials without a license issued under this chapter;

(2) knowingly to withhold information or to make any false or fictitious oral or written statement or to furnish or exhibit any false, fictitious, or misrepresented identification, intended or likely to deceive for the purpose of obtaining explosive materials, or a license, permit, exemption, or relief from disability under the provisions of this chapter;

(3) other than a licensee or permittee knowingly–

(A) to transport, ship, cause to be transported, or receive any explosive materials; or

(B) to distribute explosive materials to any person other than a licensee or permittee; or

(4) who is a holder of a limited permit—

(A) to transport, ship, cause to be transported, or receive in interstate or foreign commerce any explosive materials; or

(B) to receive explosive materials from a licensee or permittee, whose premises are located outside the State of residence of the limited permit holder, or on more than 6 separate occasions, during the period of the permit, to receive explosive materials from 1 or more licensees or permittees whose premises are located within the State of residence of the limited permit holder.

(b) It shall be unlawful for any licensee or permittee to knowingly distribute any explosive materials to any person other than—

(1) a licensee;

(2) a holder of a user permit; or

(3) a holder of a limited permit who is a resident of the State where distribution is made and in which the premises of the transferor are located.

(c) It shall be unlawful for any licensee to distribute explosive materials to any person who the licensee has reason to believe intends to transport such explosive materials into a State where the purchase, possession, or use of explosive materials is prohibited or which does not permit its residents to transport or ship explosive materials into it or to receive explosive materials in it.

(d) It shall be unlawful for any person knowingly to distribute explosive materials to any individual who:

(1) is under twenty-one years of age;

(2) has been convicted in any court of a crime punishable by imprisonment for a term exceeding one year;

(3) is under indictment for a crime punishable by imprisonment for a term exceeding one year;

(4) is a fugitive from justice;

(5) is an unlawful user of or addicted to any controlled substance (as defined in section 102 of the Controlled Substances Act);

(6) has been adjudicated a mental defective or who has been committed to a mental institution;

(7) is an alien, other than an alien who—

(A) is lawfully admitted for permanent residence (as defined in section 101 (a)(20) of the Immigration and Nationality Act);

(B) is in lawful nonimmigrant status, is a refugee admitted under section 207 of the Immigration and Nationality Act or is in asylum status under section 208 of the Immigration and Nationality Act and—

(i) is a foreign law enforcement officer of a friendly foreign government, as determined by the Secretary in consultation with the Secretary of State, entering the United States on official law enforcement business, and the shipping, transporting, possession, or receipt of explosive materials is in furtherance of this official law enforcement business; or

(ii) is a person having the power to direct or cause the direction of the management and policies of a corporation, partnership, or association licensed pursuant to section 843(a), and the shipping, transporting, possession, or receipt of explosive materials is in furtherance of such power;

(C) is a member of a North Atlantic Treaty Organization (NATO) or other friendly foreign military force, as determined by the Attorney General in consultation with the Secretary of Defense, who is present in the United States under military orders for training or other military purpose authorized by the United States and the shipping, transporting, possession, or receipt of explosive materials is in furtherance of the authorized military purpose; or

(D) is lawfully present in the United States in cooperation with the Director of Central Intelligence, and the shipment, transportation, receipt, or possession of the explosive materials is in furtherance of such cooperation;

(8) has been discharged from the armed forces under dishonorable conditions;

(9) having been a citizen of the United States, has renounced the citizenship of that person.

(e) It shall be unlawful for any licensee knowingly to distribute any explosive materials to any person in any State where the purchase, possession, or use by such person of such explosive materials would be in violation of any State law or any published ordinance applicable at the place of distribution.

(f) It shall be unlawful for any licensee or permittee willfully to manufacture, import, purchase, distribute, or receive explosive materials without making such records as the Attorney General may by regulation require, including, but not limited to, a statement of intended use, the name, date, place of birth, social security number or taxpayer identification number, and place of residence of any natural person to whom explosive materials are distributed. If explosive materials are distributed to a corporation or other business entity, such records shall include the identity and principal and local places of business and the name, date, place of birth, and place of residence of the natural person acting as agent of the corporation or other business entity in arranging the distribution.

(g) It shall be unlawful for any licensee or permittee knowingly to make any false entry in any record which he is required to keep pursuant to this section or regulations promulgated under section 847 of this title.

(h) It shall be unlawful for any person to receive, possess, transport, ship, conceal, store, barter, sell, dispose of, or pledge or accept as security for a loan, any stolen explosive materials which are moving as, which are part of, which constitute, or which have been shipped or transported in, interstate or foreign commerce, either before or after such materials were stolen, knowing or having reasonable cause to believe that the explosive materials were stolen.

(i) It shall be unlawful for any person—

(1) who is under indictment for, or who has been convicted in any court of, a crime punishable by imprisonment for a term exceeding one year;

(2) who is a fugitive from justice;

(3) who is an unlawful user of or addicted to any controlled substance (as defined in section 102 of the Controlled Substances Act);

(4) who has been adjudicated as a mental defective or who has been committed to a mental institution;

(5) who is an alien, other than an alien who—

(A) is lawfully admitted for permanent residence (as that term is defined in section 101(a)(20) of the Immigration and Nationality Act);

(B) is in lawful nonimmigrant status, is a refugee admitted under section 207 of the Immigration and Nationality Act or is in asylum status under section 208 of the Immigration and Nationality Act and--

(i) is a foreign law enforcement officer of a friendly foreign government, as determined by the Secretary in consultation with the Secretary of State, entering the United States on official law enforcement business, and the shipping, transporting, possession, or receipt of explosive materials is in furtherance of this official law enforcement business; or

(ii) is a person having the power to direct or cause the direction of the management and policies of a corporation, partnership, or association licensed pursuant to section 843(a), and the shipping, transporting, possession, or receipt of explosive materials is in furtherance of such power;

(C) is a member of a North Atlantic Treaty Organization (NATO) or other friendly foreign military force, as determined by the Attorney General in consultation with the Secretary of Defense, who is present in the United States under military orders for training or other military purpose authorized by the United States and the shipping, transporting, possession, or receipt of explosive materials is in furtherance of the authorized military purpose; or

(D) is lawfully present in the United States in cooperation with the Director of Central Intelligence, and the shipment, transportation, receipt, or possession of the explosive materials is in furtherance of such cooperation;

(6) who has been discharged from the armed forces under dishonorable conditions;

(7) who, having been a citizen of the United States, has renounced the citizenship of that person; to ship or transport any explosive in or affecting interstate or foreign commerce or to receive or possess any explosive which has been shipped or transported in or affecting interstate or foreign commerce.

(j) It shall be unlawful for any person to store any explosive material in a manner not in conformity with regulations promulgated by the Attorney General. In promulgating such regulations, the Attorney General shall take into consideration the class, type, and quantity of explosive materials to be stored, as well as the standards of safety and security recognized in the explosives industry.

(k) It shall be unlawful for any person who has knowledge of the theft or loss of any explosive materials from his stock, to fail to report such theft or loss within twenty-four hours of discovery thereof, to the Attorney General and to appropriate local authorities.

(l) It shall be unlawful for any person to manufacture any plastic explosive that does not contain a detection agent.

(m) **(1)** It shall be unlawful for any person to import or bring into the United States, or export from the United States, any plastic explosive that does not contain a detection agent.

(2) This subsection does not apply to the importation or bringing into the United States, or the exportation from the United States, of any plastic explosive that was imported or brought into, or manufactured in the United States prior to the date of enactment of this subsection [enacted April 24, 1996] by or on behalf of any agency of the United States performing military or police functions (including any military reserve component) or by or on behalf of the National Guard of any State, not later than 15 years after the date of entry into force of the Convention on the Marking of Plastic Explosives, with respect to the United States.

(n) **(1)** It shall be unlawful for any person to ship, transport, transfer, receive, or possess any plastic explosive that does not contain a detection agent.

(2) This subsection does not apply to—

(A) the shipment, transportation, transfer, receipt, or possession of any plastic explosive that was imported or brought into, or manufactured in the United States prior to the date of enactment of this subsection [enacted April 24, 1996] by any person during the period beginning on that date and ending 3 years after that date of enactment; or

(B) the shipment, transportation, transfer, receipt, or possession of any plastic explosive that was imported or brought into, or manufactured in the United States prior to the date of enactment of this subsection [enacted April 24, 1996] by or on behalf of any agency of the United States performing a military or police function (including any military reserve component) or by or on behalf of the National Guard of any State, not later than 15 years after the date of entry into force of the Convention on the Marking of Plastic Explosives, with respect to the United States.

(o) It shall be unlawful for any person, other than an agency of the United States (including any military reserve component) or the National Guard of any State, possessing any plastic explosive on the date of enactment of this subsection [enacted April 24, 1996], to fail to report to the Attorney General within 120 days after such date of enactment the quantity of such explosives possessed, the manufacturer or importer, any marks of identification on such explosives, and such other information as the Attorney General may prescribe by regulation.

(p) Distribution of information relating to explosives, destructive devices, and weapons of mass destruction.

(1) Definitions. In this subsection—

(A) the term "destructive device" has the same meaning as in section 921(a)(4);

(B) the term "explosive" has the same meaning as in section 844(j); and

(C) the term "weapon of mass destruction" has the same meaning as in section 2332a(c)(2).

(2) Prohibition. It shall be unlawful for any person—

(A) to teach or demonstrate the making or use of an explosive, a destructive device, or a weapon of mass destruction, or to distribute by any means information pertaining to, in whole or in part, the manufacture or use of an explosive, destructive device, or weapon of mass destruction, with the intent that the teaching, demonstration, or information be used for, or in furtherance of, an activity that constitutes a Federal crime of violence; or

(B) to teach or demonstrate to any person the making or use of an explosive, a destructive device, or a weapon of mass destruction, or to distribute to any person, by any means, information pertaining to, in whole or in part, the manufacture or use of an explosive, destructive device, or weapon of mass destruction, knowing that such person intends to use the teaching, demonstration, or information for, or in furtherance of, an activity that constitutes a Federal crime of violence.

(Added Pub. L. 91-452, title XI, Sec. 1102(a), Oct. 15, 1970, 84 Stat. 953; amended Pub. L. 100-690, title VI, Sec. 6474(c), (d), Nov. 18, 1988, 102 Stat. 4380; Pub. L. 101-647, title XXXV, Sec. 3521, Nov. 29, 1990, 104 Stat. 4923; Pub. L. 103-322, title XI, Sec. 110508, 10516, Sept. 13, 1994, 108 Stat. 2018, 2020; Pub. L. 104-132, title VI, Sec. 603, title VII, Sec. 707, Apr. 24, 1996, 110 Stat. 1289, 1296; Pub. L. 106-54, Sec. 2(a), Aug. 17, 1999, 113 Stat. 398; Pub. L. 107-296, title XI, Sec. 1112(e)(3), 1122(b), 1123, Nov. 25, 2002, 116 Stat. 2276, 2280, 2283.)

§ 843. Licenses and user permits

(a) An application for a user permit or limited permit or a license to import, manufacture, or deal in explosive materials shall be in such form and contain such information as the Attorney General shall by regulation prescribe, including the names of and appropriate identifying information regarding all employees who will be authorized by the applicant to possess explosive materials, as well as fingerprints and a photograph of each responsible person. Each applicant for a license or permit shall pay a fee to be charged as set by the Attorney General, said fee not to exceed $ 50 for a limited permit and $ 200 for any other license or permit. Each license or user permit shall be valid for not longer than 3 years from the date of issuance and each limited permit shall be valid for not longer than 1 year from the date of issuance. Each license or permit shall be renewable upon the same conditions and subject to the same restrictions as the original license or permit, and upon payment of a renewal fee not to exceed one-half of the original fee.

(b) Upon the filing of a proper application and payment of the prescribed fee, and subject to the provisions of this chapter and other applicable laws, the Attorney General shall issue to such applicant the appropriate license or permit if--

(1) the applicant (or, if the applicant is a corporation, partnership, or association, each responsible person with respect to the applicant) is not a person described in section 842(i);

(2) the applicant has not willfully violated any of the provisions of this chapter or regulations issued hereunder;

(3) the applicant has in a State premises from which he conducts or intends to conduct business;

(4) (A) the Secretary verifies by inspection or, if the application is for an original limited permit or the first or second renewal of such a permit, by such other means as the Secretary determines appropriate, that the applicant has a place of storage for explosive materials which meets such standards of public safety and security against theft as the Attorney General by regulations shall prescribe; and

(B) subparagraph (A) shall not apply to an applicant for the renewal of a limited permit if the Secretary has verified, by inspection within the preceding 3 years, the matters described in subparagraph (A) with respect to the applicant; and

(5) the applicant has demonstrated and certified in writing that he is familiar with all published State laws and local ordinances relating to explosive materials for the location in which he intends to do business;

(6) none of the employees of the applicant who will be authorized by the applicant to possess explosive materials is any person described in section 842(i); and

(7) in the case of a limited permit, the applicant has certified in writing that the applicant will not receive explosive materials on more than 6 separate occasions during the 12-month period for which the limited permit is valid.

(c) The Attorney General shall approve or deny an application within a period of 90 days for licenses and permits beginning on the date such application is received by the Attorney General.

(d) The Attorney General may revoke any license or permit issued under this section if in the opinion of the Attorney General the holder thereof has violated any provision of this chapter or any rule or regulation prescribed by the Attorney General under this chapter, or has become ineligible to acquire explosive materials under section 842(d). The Attorney General's action under this subsection may be reviewed only as provided in subsection (e)(2) of this section.

(e) (1) Any person whose application is denied or whose license or permit is revoked shall receive a written notice from the Attorney General stating the specific grounds upon which such denial or revocation is based. Any notice of a revocation of a license or permit shall be given to the holder of such license or permit prior to or concurrently with the effective date of the revocation.

(2) If the Attorney General denies an application for, or revokes a license, or permit, he shall, upon request by the aggrieved party, promptly hold a hearing to review his denial or revocation. In the case of a revocation, the Attorney General may upon a request of the holder stay the effective date of the revocation. A hearing under this section shall be at a location convenient to the aggrieved party. The Attorney General shall give written notice of his decision to the aggrieved party within a reasonable time after the hearing. The aggrieved party may, within sixty days

after receipt of the Secretary's written decision, file a petition with the United States court of appeals for the district in which he resides or has his principal place of business for a judicial review of such denial or revocation, pursuant to sections 701-706 of title 5, United States Code.

(f) Licensees and holders of user permits shall make available for inspection at all reasonable times their records kept pursuant to this chapter or the regulations issued hereunder, and licensees and permittees shall submit to the Attorney General such reports and information with respect to such records and the contents thereof as he shall by regulations prescribe. The Attorney General may enter during business hours the premises (including places of storage) of any licensee or holder of a user permit, for the purpose of inspecting or examining (1) any records or documents required to be kept by such licensee or permittee, under the provisions of this chapter or regulations issued hereunder, and (2) any explosive materials kept or stored by such licensee or permittee at such premises. Upon the request of any State or any political subdivision thereof, the Attorney General may make available to such State or any political subdivision thereof, any information which he may obtain by reason of the provisions of this chapter with respect to the identification of persons within such State or political subdivision thereof, who have purchased or received explosive materials, together with a description of such explosive materials. The Secretary may inspect the places of storage for explosive materials of an applicant for a limited permit or, at the time of renewal of such permit, a holder of a limited permit, only as provided in subsection (b)(4).

(g) Licenses and user permits issued under the provisions of subsection (b) of this section shall be kept posted and kept available for inspection on the premises covered by the license and permit.

(h) **(1)** If the Secretary receives, from an employer, the name and other identifying information of a responsible person or an employee who will be authorized by the employer to possess explosive materials in the course of employment with the employer, the Secretary shall determine whether the responsible person or employee is one of the persons described in any paragraph of section 842(i). In making the determination, the Secretary may take into account a letter or document issued under paragraph (2).

(2) **(A)** If the Secretary determines that the responsible person or the employee is not one of the persons described in any paragraph of section 842(i), the Secretary shall notify the employer in writing or electronically of the determination and issue, to the responsible person or employee, a letter of clearance, which confirms the determination.

(B) If the Secretary determines that the responsible person or employee is one of the persons described in any paragraph of section 842(i), the Secretary shall notify the employer in writing or electronically of the determination and issue to the responsible person or the employee, as the case may be, a document that—

(i) confirms the determination;

(ii) explains the grounds for the determination;

(iii) provides information on how the disability may be relieved; and

(iv) explains how the determination may be appealed.

(i) Furnishing of samples.

(1) In general. Licensed manufacturers and licensed importers and persons who manufacture or import explosive materials or ammonium nitrate shall, when required by letter issued by the Secretary, furnish—

(A) samples of such explosive materials or ammonium nitrate;

(B) information on chemical composition of those products; and

(C) any other information that the Secretary determines is relevant to the identification of the explosive materials or to identification of the ammonium nitrate.

(2) Reimbursement. The Secretary shall, by regulation, authorize reimbursement of the fair market value of samples furnished pursuant to this subsection, as well as the reasonable costs of shipment.

(Added Pub. L. 91-452, title XI, Sec. 1102(a), Oct. 15, 1970, 84 Stat. 955; amended Pub. L. 107-296, title XI, Sec. 1112(e)(3), 1122(c)-(h), 1124, Nov. 25, 2002, 116 Stat. 2276, 2281, 2282, 2285.)

§ 844. Penalties

(a) Any person who—

(1) violates any of subsections (a) through (i) or (l) through (o) of section 842 shall be fined under this title, imprisoned for not more than 10 years, or both; and

(2) violates subsection (p)(2) of section 842, shall be fined under this title, imprisoned not more than 20 years, or both.

(b) Any person who violates any other provision of section 842 of this chapter shall be fined under this title or imprisoned not more than one year, or both.

(c) **(1)** Any explosive materials involved or used or intended to be used in any violation of the provisions of this chapter or any other rule or regulation promulgated thereunder or any violation of any criminal law of the United States shall be subject to seizure and forfeiture, and all provisions of the Internal Revenue Code of 1986 relating to the seizure, forfeiture, and disposition of firearms, as defined in section 5845(a) of that Code, shall, so far as applicable, extend to seizures and forfeitures under the provisions of this chapter.

(2) Notwithstanding paragraph (1), in the case of the seizure of any explosive materials for any offense for which the materials would be subject to forfeiture in which it would be impracticable or unsafe to remove the materials to a place of storage or would be unsafe to store them, the seizing officer may destroy the explosive materials forthwith. Any destruction under this paragraph shall be in the presence of at least 1 credible witness. The seizing officer shall make a report of the seizure and take samples as the Attorney General may by regulation prescribe.

(3) Within 60 days after any destruction made pursuant to paragraph (2), the owner of (including any person having an interest in) the property so destroyed may make application to the Attorney General for reimbursement of the value of the property. If the claimant establishes to the satisfaction of the Attorney General that—

(A) the property has not been used or involved in a violation of law; or

(B) any unlawful involvement or use of the property was without the claimant's knowledge, consent, or willful blindness, the Attorney General shall make an allowance to the claimant not exceeding the value of the property destroyed.

(d) Whoever transports or receives, or attempts to transport or receive, in interstate or foreign commerce any explosive with the knowledge or intent that it will be used to kill, injure, or intimidate any individual or unlawfully to damage or destroy any building, vehicle, or other real or personal property, shall be imprisoned for not more than ten years, or fined under this title, or both; and if personal injury results to any person, including any public safety officer performing duties as a direct or proximate result of conduct prohibited by this subsection, shall be imprisoned for not more than twenty years or fined under this title, or both; and if death results to any person, including any public safety officer performing duties as a direct or proximate result of conduct prohibited by this subsection, shall be subject to imprisonment for any term of years, or to the death penalty or to life imprisonment.

(e) Whoever, through the use of the mail, telephone, telegraph, or other instrument of interstate or foreign commerce, or in or affecting interstate or foreign commerce, willfully makes any threat, or maliciously conveys false information knowing the same to be false, concerning an attempt or alleged attempt being made, or to be made, to kill, injure, or intimidate any individual or unlawfully to damage or destroy any building, vehicle, or other real or personal property by means of fire or an explosive shall be imprisoned for not more than 10 years or fined under this title, or both.

(f) **(1)** Whoever maliciously damages or destroys, or attempts to damage or destroy, by means of fire or an explosive, any building, vehicle, or other personal or real property in whole or in part owned or possessed by, or leased to, the United States, or any department or agency thereof, or any institution or organization receiving Federal financial assistance, shall be imprisoned for not less than 5 years and not more than 20 years, fined under this title, or both.

(2) Whoever engages in conduct prohibited by this subsection, and as a result of such conduct, directly or proximately causes personal injury or creates a substantial risk of injury to any person, including any public safety officer performing duties, shall be imprisoned for not less than 7 years and not more than 40 years, fined under this title, or both.

(3) Whoever engages in conduct prohibited by this subsection, and as a result of such conduct directly or proximately causes the death of any person, including any public safety officer performing duties, shall be subject to the death penalty, or imprisoned for not less than 20 years or for life, fined under this title, or both.

(g) **(1)** Except as provided in paragraph (2), whoever possesses an explosive in an airport that is subject to the regulatory authority of the Federal Aviation Administration, or in any building in whole or in part owned, possessed, or used by, or leased to, the United States or any department or agency thereof, except with the written consent of the agency, department, or other person responsible for the management of such building or airport, shall be imprisoned for not more than five years, or fined under this title, or both.

(2) The provisions of this subsection shall not be applicable to—

(A) the possession of ammunition (as that term is defined in regulations issued pursuant to this chapter) in an airport that is subject to the regulatory authority of the Federal Aviation Administration if such ammunition is either in checked baggage or in a closed container; or

(B) the possession of an explosive in an airport if the packaging and transportation of such explosive is exempt from, or subject to and in accordance with, regulations of the Pipeline and Hazardous Materials Safety Administration for the handling of hazardous materials pursuant to chapter 51 of title 49.

(h) Whoever—

(1) uses fire or an explosive to commit any felony which may be prosecuted in a court of the United States, or

(2) carries an explosive during the commission of any felony which may be prosecuted in a court of the United States, including a felony which provides for an enhanced punishment if committed by the use of a deadly or dangerous weapon or device shall, in addition to the punishment provided for such felony, be sentenced to imprisonment for 10 years. In the case of a second or subsequent conviction under this subsection, such person shall be sentenced to imprisonment for 20 years. Notwithstanding any other provision of law, the court shall not place on probation or suspend the sentence of any person convicted of a violation of this subsection, nor shall the term of imprisonment imposed under this subsection run concurrently with any other term of imprisonment including that imposed for the felony in which the explosive was used or carried.

(i) Whoever maliciously damages or destroys, or attempts to damage or destroy, by means of fire or an explosive, any building, vehicle, or other real or personal property used in interstate or foreign commerce or in any activity affecting interstate or foreign commerce shall be imprisoned for not less than 5 years and not more than 20 years, fined under this title, or both; and if personal injury results to any person, including any public safety officer performing duties as a direct or proximate result of conduct prohibited by this subsection, shall be imprisoned for not less than 7 years and not more than 40 years, fined under this title, or both; and if death results to any person, including any public safety officer performing duties as a direct or proximate result of conduct prohibited by this subsection, shall also be subject to imprisonment for any term of years, or to the death penalty or to life imprisonment.

(j) For the purposes of subsections (d), (e), (f), (g), (h), and (i) of this section and section 842(p), the term "explosive" means gunpowders, powders used for blasting, all forms of high explosives, blasting materials, fuses (other than electric circuit breakers), detonators, and other detonating agents, smokeless powders, other explosive or incendiary devices within the meaning of paragraph (5) of section 232 of this title, and any chemical compounds, mechanical mixture, or device that contains any oxidizing and combustible units, or other ingredients, in such proportions, quantities, or packing that ignition by fire, by friction, by concussion, by percussion, or by detonation of the compound, mixture, or device or any part thereof may cause an explosion.

(k) A person who steals any explosives materials which are moving as, or are a part of, or which have moved in, interstate or foreign commerce shall be imprisoned for not more than 10 years, fined under this title, or both.

(l) A person who steals any explosive material from a licensed importer, licensed manufacturer, or licensed dealer, or from any permittee shall be fined under this title, imprisoned not more than 10 years, or both.

(m) A person who conspires to commit an offense under subsection (h) shall be imprisoned for any term of years not exceeding 20, fined under this title, or both.

(n) Except as otherwise provided in this section, a person who conspires to commit any offense defined in this chapter shall be subject to the same penalties (other than the penalty of death) as the penalties prescribed for the offense the commission of which was the object of the conspiracy.

(o) Whoever knowingly transfers any explosive materials, knowing or having reasonable cause to believe that such explosive materials will be used to commit a crime of violence (as defined in section 924(c)(3)) or drug trafficking crime (as defined in section 924(c)(2)) shall be subject to the same penalties as may be imposed under subsection (h) for a first conviction for the use or carrying of an explosive material.

(p) Theft reporting requirement.

(1) In general. A holder of a license or permit who knows that explosive materials have been stolen from that licensee or permittee, shall report the theft to the Attorney General not later than 24 hours after the discovery of the theft.

(2) Penalty. A holder of a license or permit who does not report a theft in accordance with paragraph (1), shall be fined not more than $ 10,000, imprisoned not more than 5 years, or both.

(Added Pub. L. 91-452, title XI, Sec. 1102(a), Oct. 15, 1970, 84 Stat. 956; amended Pub. L. 97-298, Sec. 2, Oct. 12, 1982, 96 Stat. 1319; Pub. L. 98-473, title II, Sec. 1014, Oct. 12, 1984, 98 Stat. 2142; Pub. L. 99-514, Sec. 2, Oct. 22, 1986, 100 Stat. 2095; Pub. L. 100-690, title VI, Sec. 6474(a), (b), Nov. 18, 1988, 102 Stat. 4379; Pub. L. 101-647, title XXXV, Sec. 3522, Nov. 29, 1990, 104 Stat. 4924; Pub. L. 103-272, Sec. 5(e)(7), July 5, 1994, 108 Stat. 1374; Pub. L. 103-322, title VI, Sec. 60003(a)(3), title XI, Sec. 110504(b), 110509, 110515(b), 110518(b), title XXXII, Sec. 320106, 320917(a), title XXXIII, Sec. 330016(1)(H), (K), (L), (N), Sept. 13, 1994, 108 Stat. 1969, 2016, 2018, 2020, 2111, 2129, 2147, 2148; Pub. L. 104-132, title VI, Sec. 604, title VII, Sec. 701, 706, 708(a), (c)(3), 724, Apr. 24, 1996, 110 Stat. 1289, 1291, 1295-1297, 1300; Pub. L. 104-294, title VI, Sec. 603(a), Oct. 11, 1996, 110 Stat. 3503; Pub. L. 106-54, Sec. 2(b), Aug. 17, 1999, 113 Stat. 399; Pub. L. 107-296, title XI, Sec. 1112(e)(3), 1125, 1127, Nov. 25, 2002, 116 Stat. 2276, 2285.)

§ 845. Exceptions; relief from disabilities

(a) Except in the case of subsections (l), (m), (n), or (o) of section 842 and subsections (d), (e), (f), (g), (h), and (i) of section 844 of this title, this chapter shall not apply to:

(1) aspects of the transportation of explosive materials via railroad, water, highway, or air that pertain to safety, including security, and are regulated by the United States Department of Transportation or the Department of Homeland Security;

(2) the use of explosive materials in medicines and medicinal agents in the forms prescribed by the official United States Pharmacopeia, or the National Formulary;

(3) the transportation, shipment, receipt, or importation of explosive materials for delivery to any agency of the United States or to any State or political subdivision thereof;

(4) small arms ammunition and components thereof;

(5) commercially manufactured black powder in quantities not to exceed fifty pounds, percussion caps, safety and pyrotechnic fuses, quills, quick and slow matches, and friction primers, intended to be used solely for sporting, recreational, or cultural purposes in antique firearms as defined in section 921(a)(16) of title 18 of the United States Code, or in antique devices as exempted from the term "destructive device" in section 921(a)(4) of title 18 of the United States Code; and

(6) the manufacture under the regulation of the military department of the United States of explosive materials for, or their distribution to or storage or possession by the military or naval services or other agencies of the United States; or to arsenals, navy yards, depots, or other establishments owned by, or operated by or on behalf of, the United States.

(b) **(1)** A person who is prohibited from shipping, transporting, receiving, or possessing any explosive under section 842(i) may apply to the Secretary for relief from such prohibition.

(2) The Secretary may grant the relief requested under paragraph (1) if the Secretary determines that the circumstances regarding the applicability of section 842(i), and the applicant's record and reputation, are such that the applicant will not be likely to act in a manner dangerous to public safety and that the granting of such relief is not contrary to the public interest.

(3) A licensee or permittee who applies for relief, under this subsection, from the disabilities incurred under this chapter as a result of an indictment for or conviction of a crime punishable by imprisonment for a term exceeding 1 year shall not be barred by such disability from further operations under the license or permit pending final action on an application for relief filed pursuant to this section.

(c) It is an affirmative defense against any proceeding involving subsections (l) through (o) of section 842 if the proponent proves by a preponderance of the evidence that the plastic explosive—

(1) consisted of a small amount of plastic explosive intended for and utilized solely in lawful—

(A) research, development, or testing of new or modified explosive materials;

(B) training in explosives detection or development or testing of explosives detection equipment; or

(C) forensic science purposes; or

(2) was plastic explosive that, within 3 years after the date of enactment of the Antiterrorism and Effective Death Penalty Act of 1996 [enacted April 24, 1996], will be or is incorporated in a military device within the territory of the United States and remains an integral part of such military device, or is intended to be, or is incorporated in, and remains an integral part of a military device that is intended to become, or has become, the property of any agency of the United States performing military or police functions (including any military reserve component) or the National Guard of any State, wherever such device is located.

(3) For purposes of this subsection, the term "military device" includes, but is not restricted to, shells, bombs, projectiles, mines, missiles, rockets, shaped charges, grenades, perforators, and similar devices lawfully manufactured exclusively for military or police purposes.

(Added Pub. L. 91-452, title XI, Sec. 1102(a), Oct. 15. 1970, 84 Stat. 958; amended Pub. L. 93-639, Sec. 101, Jan. 4, 1975, 88 Stat. 2217; Pub. L. 104-132, title VI, Sec. 605, Apr. 24, 1996, 110 Stat. 1289; Pub. L. 107-296, title XI, Sec. 1126, Nov. 25, 2002, 116 Stat. 2285; Pub. L. 109-59, title VII, Sec. 7127, Aug. 10, 2005, 119 Stat. 1909.)

§ 846. Additional powers of the Attorney General

(a) The Attorney General is authorized to inspect the site of any accident, or fire, in which there is reason to believe that explosive materials were involved, in order that if any such incident has been brought about by accidental means, precautions may be taken to prevent similar accidents from occurring. In order to carry out the purpose of this subsection, the Attorney General is authorized to enter into or upon any property where explosive materials have been used, are suspected of having been used, or have been found in an otherwise unauthorized location. Nothing in this chapter shall be construed as modifying or otherwise affecting in any way the investigative authority of any other Federal agency. In addition to any other investigatory authority they have with respect to violations of provisions of this, the Federal Bureau of Investigation, together with the Bureau of Alcohol, Tobacco, Firearms, and Explosives, shall have authority to conduct investigations with respect to violations of subsection (d), (e), (f), (g), (h), or (i) of section 844 of this title.

(b) The Attorney General is authorized to establish a national repository of information on incidents involving arson and the suspected criminal misuse of explosives. All Federal agencies having information concerning such incidents shall report the information to the Attorney General pursuant to such regulations as deemed necessary to carry out the provisions of this subsection. The repository shall also contain information on incidents voluntarily reported to the Attorney General by State and local authorities.

(Added Pub. L. 91-452, title XI, Sec. 1102(a), Oct. 15, 1970, 84 Stat. 959; amended Pub. L. 104-208, div. A, title I, Sec. 101(f) (title VI, Sec. 654(a)), Sept. 30, 1996, 110 Stat. 3009-314, 3009-369; Pub. L. 107-296, title XI, Sec. 1112(e)(2), (3), Nov. 25, 2002, 116 Stat. 2276.)

§ 847. Rules and regulations

The administration of this chapter shall be vested in the Attorney General. The Attorney General may prescribe such rules and regulations as he deems reasonably necessary to carry out the provisions of this chapter. The Attorney General shall give reasonable public notice, and afford to interested parties opportunity for hearing, prior to prescribing such rules and regulations.

(Added Pub. L. 91-452, title XI, Sec. 1102(a), Oct. 15, 1970, 84 Stat. 959; amended Pub. L. 107-296, title XI, Sec. 1112(e)(3), Nov. 25, 2002, 116 Stat. 2276.)

§ 848. Effect on State Law

No provision of this chapter shall be construed as indicating an intent on the part of the Congress to occupy the field in which such provision operates to the exclusion of the law of any State on the same subject matter, unless there is a direct and positive conflict between such provision and the law of the State so that the two cannot be reconciled or consistently stand together.

(Added Pub. L. 91-452, title XI, Sec. 1102(a), Oct. 15, 1970, 84 Stat. 959.)

Title 27, Code of Federal Regulations
Part 555—Commerce in Explosives

Section Contents

Subpart H—Exemptions

Subpart I—Unlawful Acts, Penalties, Seizures and Forfeitures

Subpart J—Marking of Plastic Explosives

Subpart K—Storage

Authority: 18 U.S.C. 847.

Source: T.D. ATF–87, 46 FR 40384, Aug. 7, 1981, unless otherwise noted.

Editor's Note: There are several name changes for titles referenced in the regulations, 27 CFR Part 555. The title "Chief, Firearms and Explosives Licensing Center" has been changed to "Chief, National Explosives Licensing Center". Also, the title "Regional director (compliance)" has been changed to "Director, Industry Operations" within the respective ATF Field Divisions. All regulations and rulings contained within this publication were current as of 1/26/07. For more recent information, please refer to the ATF website at www.atf.gov.

Subpart A—Introduction

§ 555.1 Scope of regulations.

(a) In general. The regulations contained in this part relate to commerce in explosives and implement Title XI, Regulation of Explosives (18 U.S.C. Chapter 40; 84 Stat. 952), of the Organized Crime Control Act of 1970 (84 Stat. 922), Pub. L. 103–322 (108 Stat. 1796), Pub. L. 104–132 (110 Stat. 1214), and Pub. L. 107–296 (116 Stat. 2135).

(b) Procedural and substantive requirements. This part contains the procedural and substantive requirements relative to:

(1) The interstate or foreign commerce in explosive materials;

(2) The licensing of manufacturers and importers of, and dealers in, explosive materials;

(3) The issuance of permits;

(4) The conduct of business by licensees and operations by permittees;

(5) The storage of explosive materials;

(6) The records and reports required of licensees and permittees;

(7) Relief from disabilities under this part;

(8) Exemptions, unlawful acts, penalties, seizures, and forfeitures; and

(9) The marking of plastic explosives.

[T.D. ATF–87, 46 FR 40384, Aug. 7, 1981, as amended by T.D. ATF–363, 60 FR 17449, Apr. 6, 1995; T.D. ATF–387, 62 FR 8376, Feb. 25, 1997; ATF No. 1, 68 FR 13780, Mar. 20, 2003]

§ 555.2 Relation to other provisions of law.

The provisions in this part are in addition to, and are not in lieu of, any other provision of law, or regulations, respecting commerce in explosive materials. For regulations applicable to commerce in firearms and ammunition, see Part 478 of this chapter. For regulations applicable to traffic in machine guns, destructive devices, and certain other firearms, see Part 479 of this chapter. For statutes applicable to the registration and licensing of persons engaged in the business of manufacturing, importing or exporting arms, ammunition, or implements of war, see section 38 of the Arms Export Control Act (22 U.S.C. 2778), and regulations of Part 447 of this chapter and in Parts 121 through 128 of Title 22, Code of Federal Regulations. For statutes applicable to nonmailable materials, see 18 U.S.C. 1716 and implementing regulations. For statutes applicable to water quality standards, see 33 U.S.C. 1341.

Subpart B—Definitions

§ 555.11 Meaning of terms.

When used in this part, terms are defined as follows in this section. Words in the plural form include the singular, and vice versa, and words indicating the masculine gender include the feminine. The terms "includes" and "including" do not exclude other things not named which are in the same general class or are otherwise within the scope of the term defined.

Act. 18 U.S.C. Chapter 40.

Adjudicated as a mental defective. (a) A determination by a court, board, commission, or other lawful authority that a person, as a result of marked subnormal intelligence, or mental illness, incompetency, condition, or disease:

(1) Is a danger to himself or to others; or

(2) Lacks the mental capacity to contract or manage his own affairs.

(b) The term will include—

(1) A finding of insanity by a court in a criminal case; and

(2) Those persons found incompetent to stand trial or found not guilty by reason of lack of mental responsibility by any court or pursuant to articles 50a and 76b of the Uniform Code of Military Justice, 10 U.S.C. 850a, 876b.

Alien. Any person who is not a citizen or national of the United States.

Ammunition. Small arms ammunition or cartridge cases, primers, bullets, or smokeless propellants designed for use in small arms, including percussion caps, and 3/32 inch and other external burning pyrotechnic hobby fuses. The term does not include black powder.

Appropriate identifying information. The term means, in relation to an individual:

(a) The full name, date of birth, place of birth, sex, race, street address, State of residence, telephone numbers (home and work), country or countries of citizenship, and position at the employer's business or operations of responsible persons and employees authorized to possess explosive materials;

(b) The business name, address, and license or permit number with which the responsible person or employee is affiliated;

(c) If an alien, INS-issued alien number or admission number; and

(d) Social security number, as optional information (this information is not required but is helpful in avoiding misidentification when a background check is conducted).

Approved storage facility. A place where explosive materials are stored, consisting of one or more approved magazines, conforming to the requirements of this part and covered by a license or permit issued under this part.

Articles pyrotechnic. Pyrotechnic devices for professional use similar to consumer fireworks in chemical composition and construction but not intended for consumer use. Such articles meeting the weight limits for consumer fireworks but not labeled as such and classified by U.S. Department of Transportation regulations in 49 CFR 172.101 as UN0431 or UN0432.

Artificial barricade. An artificial mound or revetted wall of earth of a minimum thickness of three feet, or any other approved barricade that offers equivalent protection.

ATF. (a) Prior to January 24, 2003. The Bureau of Alcohol, Tobacco and Firearms, Department of the Treasury, Washington, DC.

(b) On and after January 24, 2003. The Bureau of Alcohol, Tobacco, Firearms and Explosives, Department of Justice, Washington, DC.

ATF officer. (a) Prior to January 24, 2003. An officer or employee of the Bureau of Alcohol, Tobacco and Firearms (ATF) authorized to perform any function relating to the administration or enforcement of this part.

(b) On and after January 24, 2003. An officer or employee of the Bureau of Alcohol, Tobacco, Firearms and Explosives (ATF) authorized to perform any function relating to the administration or enforcement of this part.

Authority having jurisdiction for fire safety. The fire department having jurisdiction over sites where explosives are manufactured or stored.

Barricaded. The effective screening of a magazine containing explosive materials from another magazine, a building, a railway, or a highway, either by a natural barricade or by an artificial barricade. To be properly barricaded, a straight line from the top of any sidewall of the magazine containing explosive materials to the eave line of any other magazine or building, or to a point 12 feet above the center of a railway or highway, will pass through the natural or artificial barricade.

Blasting agent. Any material or mixture, consisting of fuel and oxidizer, that is intended for blasting and not otherwise defined as an explosive; if the finished product, as mixed for use or shipment, cannot be detonated by means of a number 8 test blasting cap when unconfined. A number 8 test blasting cap is one containing 2 grams of a mixture of 80 percent mercury fulminate and 20 percent potassium chlorate, or a blasting cap of equivalent strength. An equivalent strength cap comprises 0.40–0.45 grams of PETN base charge pressed in an aluminum shell with bottom thickness not to exceed to 0.03 of an inch, to a specific gravity of not less than 1.4 g/cc., and primed with standard weights of primer depending on the manufacturer.

Bulk salutes. Salute components prior to final assembly into aerial shells, and finished salute shells held separately prior to being packed with other types of display fireworks.

Bullet-sensitive explosive materials. Explosive materials that can be exploded by 150-grain M2 ball ammunition having a nominal muzzle velocity of 2700 fps (824 mps) when fired from a .30 caliber rifle at a distance of 100 ft (30.5 m), measured perpendicular. The test material is at a temperature of 70 to 75 degrees F (21 to 24 degrees C) and is placed against a ½ inch (12.4 mm) steel backing plate.

Bureau. (a) Prior to January 24, 2003. The Bureau of Alcohol, Tobacco and Firearms, Department of the Treasury, Washington, DC.

(b) On and after January 24, 2003. The Bureau of Alcohol, Tobacco, Firearms and Explosives, Department of Justice, Washington, DC.

Business premises. When used with respect to a manufacturer, importer, or dealer, the property on which explosive materials are manufactured, imported, stored or distributed. The premises include the property where the records of a manufacturer, importer, or dealer are kept if different than the premises where explosive materials are manufactured, imported, stored or distributed. When used with respect to a user of explosive materials, the property on which the explosive materials are received or stored. The premises includes the property where the records of the users are kept if different than the premises where explosive materials are received or stored.

Chief, Firearms and Explosives Licensing Center. The ATF official responsible for the issuance and renewal of licenses and permits under this part.

Committed to a mental institution. A formal commitment of a person to a mental institution by a court, board, commission, or other lawful authority. The term includes a commitment to a mental institution involuntarily. The term includes commitment for mental defectiveness or mental illness. It also includes commitments for other reasons, such as for drug use. The term does not include a person in a mental institution for observation or a voluntary admission to a mental institution.

Common or contract carrier. Any individual or organization engaged in the business of transporting passengers or goods.

Consumer fireworks. Any small firework device designed to produce visible effects by combustion and which must comply with the construction, chemical composition, and labeling regulations of the U.S. Consumer Product Safety Commission, as set forth in title 16, Code of Federal Regulations, parts 1500 and 1507. Some small devices designed to produce audible effects are included, such as whistling devices, ground devices containing 50 mg or less of explosive materials, and aerial devices containing 130 mg or less of explosive materials. Consumer fireworks are classified as fireworks UN0336, and UN0337 by the U.S. Department of Transportation at 49 CFR 172.101. This term does not include fused setpieces containing components which together exceed 50 mg of salute powder.

Controlled substance. A drug or other substance, or immediate precursor, as defined in section 102 of the Controlled Substances Act, 21 U.S.C. 802. The term includes, but is not limited to, marijuana, depressants, stimulants, and narcotic drugs. The term does not include distilled spirits, wine, malt beverages, or tobacco, as those terms are defined or used in Subtitle E of the Internal Revenue Code of 1986, as amended.

Crime punishable by imprisonment for a term exceeding one year. Any offense for which the maximum penalty, whether or not imposed, is capital punishment or imprisonment in excess of one year. The term does not include (a) any Federal or State offenses pertaining to antitrust violations, unfair trade practices, restraints of trade, or (b) any State offense (other than one involving a firearm or explosive) classified by the laws of the State as a misdemeanor and punishable by a term of imprisonment of two years or less.

Customs officer. Any officer of the Customs Service or any commissioned, warrant, or petty officer of the Coast Guard, or any agent or other person authorized to perform the duties of an officer of the Customs Service.

Dealer. Any person engaged in the business of distributing explosive materials at wholesale or retail.

Detonator. Any device containing a detonating charge that is used for initiating detonation in an explosive. The term includes, but is not limited to, electric blasting caps of instantaneous and delay types, blasting caps for use with safety fuses, detonating-cord delay connectors, and nonelectric instantaneous and delay blasting caps.

Director. (a) Prior to January 24, 2003. The Director, Bureau of Alcohol, Tobacco and Firearms, Department of the Treasury, Washington, DC. (b) On and after January 24, 2003. The Director, Bureau of Alcohol, Tobacco, Firearms and Explosives, Department of Justice, Washington, DC.

Discharged under dishonorable conditions. Separation from the U.S. Armed Forces resulting from a dishonorable discharge or dismissal adjudged by general court-martial. The term does not include any separation from the Armed Forces resulting from any other discharge, e.g., a bad conduct discharge.

Display fireworks. Large fireworks designed primarily to produce visible or audible effects by combustion, deflagration, or detonation. This term includes, but is not limited to, salutes containing more than 2 grains (130 mg) of explosive materials, aerial shells containing more than 40 grams of pyrotechnic compositions, and other display pieces which exceed the limits of explosive materials for classification as "consumer fireworks." Display fireworks are classified as fireworks UN0333, UN0334 or UN0335 by the U.S. Department of Transportation at 49 CFR 172.101. This term also includes fused setpieces containing components which together exceed 50 mg of salute powder.

Distribute. To sell, issue, give, transfer, or otherwise dispose of. The term does not include a mere change of possession from a person to his agent or employee in connection with the agency or employment.

Executed under penalties of perjury. Signed with the required declaration under the penalties of perjury as provided on or with respect to the return, form, or other document or, where no form of declaration is required, with the declaration:

"I declare under the penalties of perjury that this —(insert type of document, such as, statement, application, request, certificate), including the documents submitted in support thereof, has been examined by me and, to the best of my knowledge and belief, is true, correct, and complete".

Explosive actuated device. Any tool or special mechanized device which is actuated by explosives, but not a propellent actuated device.

Explosive materials. Explosives, blasting agents, water gels and detonators. Explosive materials include, but are not limited to, all items in the "List of Explosive Materials" provided for in §555.23.

Explosives. Any chemical compound, mixture, or device, the primary or common purpose of which is to function by explosion. The term includes, but is not limited to, dynamite and other high explosives, black powder, pellet powder, initiating explosives, detonators, safety fuses, squibs, detonating cord, igniter cord, and igniters.

Fireworks. Any composition or device designed to produce a visible or an audible effect by combustion, deflagration, or detonation, and which meets the definition of "consumer fireworks" or "display fireworks" as defined by this section.

Fireworks mixing building. Any building or area used for mixing and blending pyrotechnic compositions except wet sparkler mix.

Fireworks nonprocess building. Any office building or other building or area in a fireworks plant where no fireworks, pyrotechnic compositions or explosive materials are processed or stored.

Fireworks plant. All land and buildings thereon used for or in connection with the assembly or processing of fireworks, including warehouses used with or in connection with fireworks plant operations.

Fireworks plant warehouse. Any building or structure used exclusively for the storage of materials which are neither explosive materials nor pyrotechnic compositions used to manufacture or assemble fireworks.

Fireworks process building. Any mixing building; any building in which pyrotechnic compositions or explosive materials is pressed or otherwise prepared for finished and assembly; or any finishing or assembly building.

Fireworks shipping building. A building used for the packing of assorted display fireworks into shipping cartons for individual public displays and for the loading of packaged displays for shipment to purchasers.

Flash powder. An explosive material intended to produce an audible report and a flash of light when ignited which includes but is not limited to oxidizers such as potassium chlorate or potassium perchlorate, and fuels such as sulfur or aluminum powder.

Fugitive from justice. Any person who has fled from the jurisdiction of any court of record to avoid prosecution for any crime or to avoid giving testimony in any criminal proceeding. The term also includes any person who has been convicted of any crime and has fled to avoid imprisonment.

Hardwood. Oak, maple, ash, hickory, or other hard wood, free from loose knots, spaces, or similar defects.

Highway. Any public street, public alley, or public road, including a privately financed, constructed, or maintained road that is regularly and openly traveled by the general public.

Identification document. A document containing the name, residence address, date of birth, and photograph of the holder and which was made or issued by or under the authority of the United States Government, a State, political subdivision of a State, a foreign government, a political subdivision of a foreign government, an international governmental or an international quasi-governmental organization which, when completed with information concerning a particular individual, is of a type intended or commonly accepted for the purpose of identification of individuals.

Importer. Any person engaged in the business of importing or bringing explosive materials into the United States for purposes of sale or distribution.

Indictment. Includes an indictment or information in any court under which a crime punishable by imprisonment for a term exceeding one year may be prosecuted.

Inhabited building. Any building regularly occupied in whole or in part as a habitation for human beings, or any church, schoolhouse, railroad station, store, or other structure where people are accustomed to assemble, except any building occupied in connection with the manufacture, transportation, storage, or use of explosive materials.

Interstate or foreign commerce. Commerce between any place in a State and any place outside of that State, or within any possession of the United States or the District of Columbia, and commerce between places within the same State but through any place outside of that State.

Licensed dealer. A dealer licensed under this part.

Licensed importer. An importer licensed under this part.

Licensed manufacturer. A manufacturer licensed under this part to engage in the business of manufacturing explosive materials for purposes of sale or distribution or for his own use.

Licensee. Any importer, manufacturer, or dealer licensed under this part.

Limited permit. A permit issued to a person authorizing him to receive for his use explosive materials from a licensee or permittee in his state of residence on no more than 6 occasions during the 12-month period in which the permit is valid. A limited permit does not authorize the receipt or transportation of explosive materials in interstate or foreign commerce.

Magazine. Any building or structure, other than an explosives manufacturing building, used for storage of explosive materials.

Manufacturer. Any person engaged in the business of manufacturing explosive materials for purposes of sale or distribution or for his own use.

Mass detonation (mass explosion). Explosive materials mass detonate (mass explode) when a unit or any part of a larger quantity of explosive material explodes and causes all or a substantial part of the remaining material to detonate or explode.

Mental institution. Includes mental health facilities, mental hospitals, sanitariums, psychiatric facilities, and other facilities that provide diagnoses by licensed professionals of mental retardation or mental illness, including a psychiatric ward in a general hospital.

Natural barricade. Natural features of the ground, such as hills, or timber of sufficient density that the surrounding exposures which require protection cannot be seen from the magazine when the trees are bare of leaves.

Number 8 test blasting cap. (See definition of "blasting agent.")

Permittee. Any user of explosives for a lawful purpose who has obtained either a user permit or a limited permit under this part.

Person. Any individual, corporation, company, association, firm, partnership, society, or joint stock company.

Plywood. Exterior, construction grade (laminated wood) plywood.

Propellant actuated device. Any tool or special mechanized device or gas generator system which is actuated by a propellant or which releases and directs work through a propellant charge.

Pyrotechnic compositions. A chemical mixture which, upon burning and without explosion, produces visible, brilliant displays, bright lights, or sounds.

Railway. Any steam, electric, or other railroad or railway which carries passengers for hire.

Region. A geographical region of the Bureau of Alcohol, Tobacco and Firearms.

Regional director (compliance). The principal regional official responsible for administering regulations in this part.

Renounced U.S. citizenship. (a) A person has renounced his U.S. citizenship if the person, having been a citizen of the United States, has renounced citizenship either—

(1) Before a diplomatic or consular officer of the United States in a foreign state pursuant to 8 U.S.C. 1481(a)(5); or

(2) Before an officer designated by the Attorney General when the United States is in a state of war pursuant to 8 U.S.C. 1481(a)(6).

(b) The term will not include any renunciation of citizenship that has been reversed as a result of administrative or judicial appeal.

Responsible person. An individual who has the power to direct the management and policies of the applicant pertaining to explosive materials. Generally, the term includes partners, sole proprietors, site managers, corporate officers and directors, and majority shareholders.

Salute. An aerial shell, classified as a display firework, that contains a charge of flash powder and is designed to produce a flash of light and a loud report as the pyrotechnic effect.

Screen barricade. Any barrier that will contain the embers and debris from a fire or deflagration in a process building, thus preventing propagation of fire to other buildings or areas. Such barriers shall be constructed of metal roofing, ¼ to ½ inch (6 to 13 mm) mesh screen, or equivalent material. The barrier extends from floor level to a height such that a straight line from the top of any side wall of the donor building to the eave line of any exposed building intercepts the screen at a point not less than 5 feet (1.5 m) from the top of the screen. The top 5 feet (1.5 m) of the screen is inclined towards the donor building at an angle of 30 to 45 degrees.

Softwood. Fir, pine, or other soft wood, free from loose knots, spaces, or similar defects.

State. A State of the United States. The term includes the District of Columbia, the Commonwealth of Puerto Rico, and the possessions of the United States.

State of residence. The State in which an individual regularly resides or maintains his home. Temporary stay in a State does not make the State of temporary stay the State of residence.

Theatrical flash powder. Flash powder commercially manufactured in premeasured kits not exceeding 1 ounce and mixed immediately prior to use and intended for use in theatrical shows, stage plays, band concerts, magic acts, thrill shows, and clown acts in circuses.

Unlawful user of or addicted to any controlled substance. A person who uses a controlled substance and has lost the power of self-control with reference to the use of a controlled substance; and any person who is a current user of a controlled substance in a manner other than as prescribed by a licensed physician. Such use is not limited to the use of drugs on a particular day, or within a matter of days or weeks before possession of the explosive materials, but rather that the unlawful use has occurred recently enough to indicate that the individual is actively engaged in such conduct. A person may be an unlawful current user of a controlled substance even though the substance is not being used at the precise time the person seeks to acquire explosive materials or receives or possesses explosive materials. An inference of current use may be drawn from evidence of a recent use or possession of a controlled substance or a pattern of use or possession that reasonably covers the present time, e.g., a conviction for use or possession of a controlled substance within the past year; multiple arrests for such offenses within the past 5 years if the most recent arrest occurred within the past year; or persons found through a drug test to use a controlled substance unlawfully, provided that the test was administered within the past year. For a current or former member of the Armed Forces, an inference of current use may be drawn from recent disciplinary or other administrative action based on confirmed drug use, e.g., court-martial conviction, nonjudicial punishment, or an administrative discharge based on drug use or drug rehabilitation failure.

U.S.C. The United States Code.

User-limited permit. A user permit valid only for a single purchase transaction, a new permit being required for a subsequent purchase transaction.

User permit. A permit issued to a person authorizing him (a) to acquire for his own use explosive materials from a licensee in a State other than the State in which he resides or from a foreign country, and (b) to transport explosive materials in interstate or foreign commerce.

Water gels. Explosives or blasting agents that contain a substantial proportion of water.

(18 U.S.C. 847 (84 Stat. 959); 18 U.S.C. 926 (82 Stat. 1226)
[T.D. ATF–87, 46 FR 40384, Aug. 7, 1981, as amended by T.D. ATF–92, 46 FR 46916, Sept. 23, 1981; T.D. ATF–200, 50 FR 10497, Mar. 15, 1985; T.D. ATF–290, 54 FR 53053, Dec. 27, 1989; T.D. ATF–293, 55 FR 3720, Feb. 5, 1990; T.D. ATF–314, 56 FR 49140, Sept. 27, 1991; T.D. ATF–382, 61 FR 38084, July 23, 1996; T.D. ATF–400, 63 FR 45001, Aug. 24, 1998; T.D. ATF–446, 66 FR 16602, Mar. 27, 2001; ATF No. 1, 68 FR 13780, Mar. 20, 2003]

Subpart C—Administrative and Miscellaneous Provisions

§ 555.21 Forms prescribed.

(a) The Director is authorized to prescribe all forms required by this part. All of the information called for in each form shall be furnished as indicated by the headings on the form and the instructions on or pertaining to the form. In addition, information called for in each form shall be furnished as required by this part.

(b) Requests for forms should be mailed to the ATF Distribution Center, 7943 Angus Court, Springfield, Virginia 22153.

[T.D. ATF–92, 46 FR 46916, Sept. 23, 1981, as amended by T.D. ATF–249, 52 FR 5961, Feb. 27, 1987; T.D. 372, 61 FR 20724, May 8, 1996]

§ 555.22 Alternate methods or procedures; emergency variations from requirements.

(a) Alternate methods or procedures. The permittee or licensee, on specific approval by the Director as provided by this paragraph, may use an alternate method or procedure in lieu of a method or procedure specifically prescribed in this part. The Director may approve an alternate method or procedure, subject to stated conditions, when he finds that:

(1) Good cause is shown for the use of the alternate method or procedure;

(2) The alternate method or procedure is within the purpose of, and consistent with the effect intended by, the specifically prescribed method or procedure and that the alternate method or procedure is substantially equivalent to that specifically prescribed method or procedure; and

(3) The alternate method or procedure will not be contrary to any provision of law and will not result in an increase in cost to the Government or hinder the effective administration of this part. Where the permittee or licensee desires to employ an alternate method or procedure, he shall submit a written application to the regional director (compliance), for transmittal to the Director. The application shall specifically describe the proposed alternate method or procedure and shall set forth the reasons for it. Alternate methods or procedures may not be employed until the application is approved by the Director. The permittee or licensee shall, during the period of authorization of an alternate method or procedure, comply with the terms of the approved application. Authorization of any alternate method or procedure may be withdrawn whenever, in the judgment of the Director, the effective administration of this part is hindered by the continuation of the authorization. As used in this paragraph, alternate methods or procedures include alternate construction or equipment.

(b) Emergency variations from requirements. The Director may approve construction, equipment, and methods of operation other than as specified in this part, where he finds that an emergency exists and the proposed variations from the specified requirements are necessary and the proposed variations:

(1) Will afford security and protection that are substantially equivalent to those prescribed in this part;

(2) Will not hinder the effective administration of this part; and

(3) Will not be contrary to any provisions of law.

Variations from requirements granted under this paragraph are conditioned on compliance with the procedures, conditions, and limitations set forth in the approval of the application. Failure to comply in good faith with the procedures, conditions, and limitations shall automatically terminate the authority for the variations and the licensee or permittee shall fully comply with the prescribed requirements of regulations from which the variations were authorized. Authority for any variation may be withdrawn whenever, in the judgment of the Director, the effective administration of this part is hindered by the continuation of the variation. Where the licensee or permittee desires to employ an emergency variation, he shall submit a written application to the regional director (compliance) for transmittal to the Director. The application shall describe the proposed variation and set forth the reasons for it. Variations may not be employed until the application is approved, except when the emergency requires immediate action to correct a situation that is threatening to life or property. Corrective action may then be taken concurrent with the filing of the application and notification of the Director via telephone.

(c) Retention of approved variations. The licensee or permittee shall retain, as part of his records available for examination by ATF officers, any application approved by the Director under this section.

§ 555.23 List of explosive materials.

The Director shall compile a list of explosive materials, which shall be published and revised at least annually in the Federal Register. The "List of Explosive Materials" (ATF Publication 5400.8) is available at no cost upon request from the ATF Distribution Center, 7943 Angus Court, Springfield, Virginia 22153.

[T.D. ATF–290, 54 FR 53054, Dec. 27, 1989, as amended by T.D. ATF–446, 66 FR 16602, Mar. 27, 2001]

§ 555.24 Right of entry and examination.

(a) Any ATF officer may enter during business hours the premises, including places of storage, of any licensee or holder of a user permit for the purpose of inspecting or examining any records or documents required to be kept under this part, and any facilities in which explosive materials are kept or stored.

(b) Any ATF officer may inspect the places of storage for explosive materials of an applicant for a limited permit or, in the case of a holder of a limited permit, at the time of renewal of such permit.

(c) The provisions of paragraph (b) of this section do not apply to an applicant for the renewal of a limited permit if an ATF officer has, within the preceding 3 years, verified by inspection that the applicant's place of storage for explosive materials meets the requirements of subpart K of this part.
[ATF No. 1, 68 FR 13781, Mar. 20, 2003]

§ 555.25 Disclosure of information.

Upon receipt of written request from any State or any political subdivision of a State, the regional director (compliance) may make available to the State or political subdivision any information which the regional director (compliance) may obtain under the Act with respect to the identification of persons within the State or political subdivision, who have purchased or received explosive materials, together with a description of the explosive materials.

§ 555.26 Prohibited shipment, transportation, receipt, possession, or distribution of explosive materials.

(a) General. No person, other than a licensee or permittee knowingly may transport, ship, cause to be transported, or receive any explosive materials: Provided, That the provisions of this paragraph (a) do not apply to the lawful purchase by a nonlicensee or nonpermittee of commercially manufactured black powder in quantities not to exceed 50 pounds, if the black powder is intended to be used solely for sporting, recreational, or cultural purposes in antique firearms as defined in 18 U.S.C. 921(a)(16), or in antique devices as exempted from the term "destructive device" in 18 U.S.C. 921(a)(4).

(b) Holders of a limited permit. No person who is a holder of a limited permit may—

(1) Transport, ship, cause to be transported, or receive in interstate or foreign commerce any explosive materials;

(2) Receive explosive materials from a licensee or permittee, whose premises are located outside the State of residence of the limited permit holder; or

(3) Receive explosive materials on more than 6 separate occasions, during the period of the permit, from one or more licensees or permittees whose premises are located within the State of residence of the limited permit holder. (See §555.105(b) for the definition of "6 separate occasions.")

(c) Possession by prohibited persons. No person may ship or transport any explosive material in or affecting interstate or foreign commerce or receive or possess any explosive materials which have been shipped or transported in or affecting interstate or foreign commerce who:

(1) Is under indictment or information for, or who has been convicted in any court of, a crime punishable by imprisonment for a term exceeding one year;

(2) Is a fugitive from justice;

(3) Is an unlawful user of or addicted to any controlled substance (as defined in section 102 of the Controlled Substances Act (21 U.S.C. 802) and §555.11);

(4) Has been adjudicated as a mental defective or has been committed to a mental institution;

(5) Is an alien, other than an alien who—

(i) Is lawfully admitted for permanent residence (as that term is defined in section 101(a)(20) of the Immigration and Nationality Act (8 U.S.C. 1101)); or

(ii) Is in lawful nonimmigrant status, is a refugee admitted under section 207 of the Immigration and Nationality Act (8 U.S.C. 1157), or is in asylum status under section 208 of the Immigration and Nationality Act (8 U.S.C. 1158), and—

(A) Is a foreign law enforcement officer of a friendly foreign government, as determined by the Attorney General in consultation with the Secretary of State, entering the United States on official law enforcement business, and the shipping, transporting, possession, or receipt of explosive materials is in furtherance of this official law enforcement business;

(B) Is a person having the power to direct or cause the direction of the management and policies of a corporation, partnership, or association licensed pursuant to section 843(a) of the Act, and the shipping, transporting, possession, or receipt of explosive materials is in furtherance of such power;

(C) Is a member of a North Atlantic Treaty Organization (NATO) or other friendly foreign military force, as determined by the Attorney General in consultation with the Secretary of Defense, (whether or not admitted in a nonimmigrant status) who is present in the United States under military orders for training or other military purpose authorized by the United States, and the shipping, transporting, possession, or receipt of explosive materials is in furtherance of the military purpose; or

(D) Is lawfully present in the United States in cooperation with the Director of Central Intelligence, and the shipment, transportation, receipt, or possession of the explosive materials is in furtherance of such cooperation;

(6) Has been discharged from the armed forces under dishonorable conditions; or

(7) Having been a citizen of the United States, has renounced citizenship.

(d) Distribution to prohibited persons. No person may knowingly distribute explosive materials to any individual who:

(1) Is under twenty-one years of age;

(2) Is under indictment or information for, or who has been convicted in any court of, a crime punishable by imprisonment for a term exceeding one year;

(3) Is a fugitive from justice;

(4) Is an unlawful user of or addicted to any controlled substance (as defined in section 102 of the Controlled Substances Act (21 U.S.C. 802) and §555.11);

(5) Has been adjudicated as a mental defective or has been committed to a mental institution;

(6) Is an alien, other than an alien who—

(i) Is lawfully admitted for permanent residence (as that term is defined in section 101(a)(20) of the Immigration and Nationality Act (8 U.S.C. 1101)); or

(ii) Is in lawful nonimmigrant status, is a refugee admitted under section 207 of the Immigration and Nationality Act (8 U.S.C. 1157), or is in asylum status under section 208 of the Immigration and Nationality Act (8 U.S.C. 1158), and—

(A) Is a foreign law enforcement officer of a friendly foreign government, as determined by the Attorney General in consultation with the Secretary of State, entering the United States on official law enforcement business, and the shipping, transporting, possession, or receipt of explosive materials is in furtherance of this official law enforcement business;

(B) Is a person having the power to direct or cause the direction of the management and policies of a corporation, partnership, or association licensed pursuant to section 843(a) of the Act, and the shipping, transporting, possession, or receipt of explosive materials is in furtherance of such power;

(C) Is a member of a North Atlantic Treaty Organization (NATO) or other friendly foreign military force, as determined by the Attorney General in consultation with the Secretary of Defense, (whether or not admitted in a nonimmigrant status) who is present in the United States under military orders for training or other military purpose authorized by the United States, and the shipping, transporting, possession, or receipt of explosive materials is in furtherance of the military purpose; or

(D) Is lawfully present in the United States in cooperation with the Director of Central Intelligence, and the shipment, transportation, receipt, or possession of the explosive materials is in furtherance of such cooperation;

(7) Has been discharged from the armed forces under dishonorable conditions; or

(8) Having been a citizen of the United States, has renounced citizenship.

(e) See §555.180 for regulations concerning the prohibited manufacture, importation, exportation, shipment, transportation, receipt, transfer, or possession of plastic explosives that do not contain a detection agent.

[ATF No. 1, 68 FR 13781, Mar. 20, 2003]

§ 555.27 Out-of-State disposition of explosive materials.

(a) No nonlicensee or nonpermittee may distribute any explosive materials to any other nonlicensee or nonpermittee who the distributor knows or who has reasonable cause to believe does not reside in the State in which the distributor resides.

(b) The provisions of this section do not apply on and after May 24, 2003.

[ATF No. 1, 68 FR 13782, Mar. 20, 2003]

§ 555.28 Stolen explosive materials.

No person shall receive, conceal, transport, ship, store, barter, sell, or dispose of any stolen explosive materials knowing or having reasonable cause to believe that the explosive materials were stolen.

§ 555.29 Unlawful storage.

No person shall store any explosive materials in a manner not in conformity with this part.

§ 555.30 Reporting theft or loss of explosive materials.

(a) Any licensee or permittee who has knowledge of the theft or loss of any explosive materials from his stock shall, within 24 hours of discovery, report the theft or loss by telephoning 1–800–800–3855 (nationwide toll free number) and on ATF F 5400.5 (formerly Form 4712) in accordance with the instructions on the form. Theft or loss of any explosive materials shall also be reported to appropriate local authorities.

(b) Any other person, except a carrier of explosive materials, who has knowledge of the theft or loss of any explosive materials from his stock shall, within 24 hours of discovery, report the theft or loss by telephoning 1–800–800–3855 (nationwide toll free number) and in writing to the nearest ATF office. Theft or loss shall be reported to appropriate local authorities.

(c) Reports of theft or loss of explosive materials under paragraphs (a) and (b) of this section must include the following information, if known:

(1) The manufacturer or brand name.

(2) The manufacturer's marks of identification (date and shift code).

(3) Quantity (applicable quantity units, such as pounds of explosives, number of detonators, etc.).

(4) Description (dynamite, blasting agents, detonators, etc.) and United Nations (UN) identification number, hazard division number, and classification letter, e.g., 1.1D, as classified by the U.S. Department of Transportation at 49 CFR 172.101 and 173.52.

(5) Size (length and diameter).

(d) A carrier of explosive materials who has knowledge of the theft or loss of any explosive materials shall, within 24 hours of discovery, report the theft or loss by telephoning 1–800–800–3855 (nationwide toll free number). Theft or loss shall also be reported to appropriate local authorities. Reports of theft or loss of explosive materials by carriers shall include the following information, if known:

(1) The manufacturer or brand name.

(2) Quantity (applicable quantity units, such as pounds of explosives, number of detonators, etc.).

(3) Description (United Nations (UN) identification number, hazard division number, and classification letter, e.g., 1.1D) as classified by the U.S. Department of Transportation at 49 CFR 172.101 and 173.52.

[T.D. ATF–87, 46 FR 40384, Aug. 7, 1981, as amended by T.D. ATF–400, 63 FR 45002, Aug. 24, 1998]

§ 555.31 Inspection of site accidents or fires; right of entry.

Any ATF officer may inspect the site of any accident or fire in which there is reason to believe that explosive materials were involved. Any ATF officer may enter into or upon any property where explosive materials have been used, are suspected of having been used, or have been found in an otherwise unauthorized location.

§ 555.32 Special explosive devices.

The Director may exempt certain explosive actuated devices, explosive actuated tools, or similar devices from the requirements of this part. A person who desires to obtain an exemption under this section for any special explosive device, which as designed does not constitute a public safety or security hazard, shall submit a written request to the Director. Each request shall be executed under the penalties of perjury and contain a complete and accurate description of the device, the name and address of the manufacturer or importer, the purpose of and use for which it is intended, and any photographs, diagrams, or drawings as may be necessary to enable the Director to make a determination. The Director may require that a sample of the device be submitted for examination and evaluation. If it is not possible to submit the device, the person requesting the exemption shall advise the Director and designate the place where the device will be available for examination and evaluation.

§ 555.33 Background checks and clearances (effective May 24, 2003).

(a) Background checks.

(1) If the Director receives from a licensee or permittee the names and appropriate identifying information of responsible persons and employees who will be authorized by the employer to possess explosive materials in the course of employment with the employer, the Director will conduct a background check in accordance with this section.

(2) The Director will determine whether the responsible person or employee is one of the persons described in any paragraph of section 842(i) of the Act (see §555.26). In making such determination, the Director may take into account a letter or document issued under paragraph (a)(3) of this section.

(3)(i) If the Director determines that the responsible person or the employee is not one of the persons described in any paragraph of section 842(i) of the Act (see §555.26), the Director will notify the employer in writing or electronically of the determination and issue, to the responsible person or employee, as the case may be, a letter of clearance which confirms the determination.

(ii) If the Director determines that the responsible person or employee is one of the persons described in any paragraph of section 842(i) of the Act (see §555.26), ATF will notify the employer in writing or electronically of the determination and issue to the responsible person or the employee, as the case may be, a document that confirms the determination; explains the grounds for the determination; provides information on how the disability may be relieved; and explains how the determination may be appealed. The employer will retain the notification as part of his permanent records in accordance with §555.121. The employer will take immediate steps to remove the responsible person from his position directing the management or policies of the business or operations as they relate to explosive materials or, as the case may be, to remove the employee from a position requiring the possession of explosive materials. Also, if the employer has listed the employee as a person authorized to accept delivery of explosive materials, as specified in §555.103 or §555.105, the employer must remove the employee from such list and immediately, and in no event later than the second business day after such change, notify distributors of such change.

(b) Appeals and correction of erroneous system information—

(1) In general. A responsible person or employee may challenge the adverse determination set out in the letter of denial, in writing and within 45 days of issuance of the determination, by directing his or her challenge to the basis for the adverse determination, or to the accuracy of the record upon which the adverse determination is based, to the Director. The appeal request must include appropriate documentation or record(s) establishing the legal and/or factual basis for the challenge. Any record or document of a court or other government entity or official furnished in support of an appeal must be certified by the court or other government entity or official as a true copy. In the case of an employee, or responsible person who did not submit fingerprints, such appeal must be accompanied by two properly completed FBI Forms FD–258 (fingerprint card). The Director will advise the individual in writing of his decision and the reasons for the decision.

(2) Employees. The letter of denial, among other things, will advise an employee who elects to challenge an adverse determination to submit the fingerprint cards as described above. The employee also will be advised of the agency name and address that originated the record containing the information causing the adverse determination ("originating agency"). At that time, and where appropriate, an employee is encouraged to apply to the originating agency to challenge the accuracy of the record(s) upon which the denial is based. The originating agency may respond to the individual's application by addressing the individual's specific reasons for the challenge, and by indicating whether additional information or documents are required. If the record is corrected as a result of the application to the originating agency, the individual may so notify ATF which will, in turn, verify the record correction with the originating agency and take all necessary steps to contact the agency responsible for the record system and correct the record. The employee may provide to ATF additional and appropriate documentation or record(s) establishing the legal and/or factual basis for the challenge to ATF's decision to uphold the initial denial. If ATF does not receive such additional documentation or record(s) within 45 days of the date of the decision upholding the initial denial, ATF will close the appeal.

(3) Responsible persons. The letter of denial, among other things, will advise a responsible person of the agency name and address which originated the record containing the information causing the adverse determination ("originating agency"). A responsible person who elects to challenge the adverse determination, where appropriate, is encouraged to apply to the originating agency to challenge the accuracy of the record(s) upon which the denial is based. The originating agency may respond to the individual's application by addressing the individual's specific reasons for the challenge, and by indicating whether additional information or documents are required. If the record is corrected as a result of the application to the originating agency, the individual may so notify ATF which will, in turn, verify the record correction with the originating agency and take all necessary steps to contact the agency responsible for the record system and correct the record. A responsible person may provide additional documentation or records as specified for employees in paragraph (b)(2) of this section.

(Approved by the Office of Management and Budget under control number 1140–0081)
[ATF No. 1, 68 FR 13783, Mar. 20, 2003]

§ 555.34 Replacement of stolen or lost ATF Form 5400.30 (Intrastate Purchase of Explosives Coupon (IPEC)).

When any Form 5400.30 is stolen, lost, or destroyed, the person losing possession will, upon discovery of the theft, loss, or destruction, immediately, but in all cases before 24 hours have elapsed since discovery, report the matter to the Director by telephoning 1–888–ATF-BOMB (nationwide toll free number). The report will explain in detail the circumstances of the theft, loss, or destruction and will include all known facts that may serve to identify the document. Upon receipt of the report, the Director will make such investigation as appears appropriate and may issue a duplicate document upon such conditions as the circumstances warrant.

(Approved by the Office of Management and Budget under control number 1140–0077)
[ATF No. 1, 68 FR 13783, Mar. 20, 2003]

Subpart D—Licenses and Permits

§ 555.41 General.

(a) Licenses and permits issued prior to May 24, 2003.

(1) Each person intending to engage in business as an importer or manufacturer of, or a dealer in, explosive materials, including black powder, must, before commencing business, obtain the license required by this subpart for the business to be operated. Each person who intends to acquire for use explosive materials from a licensee in a State other than the State in which he resides, or from a foreign country, or who intends to transport explosive materials in interstate or foreign commerce, must obtain a permit under this subpart; except that it is not necessary to obtain a permit if the user intends to lawfully purchase:

(i) Explosive materials from a licensee in a State contiguous to the user's State of residence and the user's State of residence has enacted legislation, currently in force, specifically authorizing a resident of that State to purchase explosive materials in a contiguous State; or

(ii) Commercially manufactured black powder in quantities not to exceed 50 pounds, intended to be used solely for sporting, recreational, or cultural purposes in antique firearms or in antique devices.

(2) Each person intending to engage in business as an explosive materials importer, manufacturer, or dealer must file an application, with the required fee (see §555.42), with ATF in accordance with the instructions on the form (see §555.45). A license will, subject to law, entitle the licensee to transport, ship, and receive explosive materials in interstate or foreign commerce, and to engage in the business specified by the license, at the location described on the license. A separate license must be obtained for each business premises at which the applicant is to manufacture, import, or distribute explosive materials except under the following circumstances:

(i) A separate license will not be required for storage facilities operated by the licensee as an integral part of one business premises or to cover a location used by the licensee solely for maintaining the records required by this part.

(ii) A separate license will not be required of a licensed manufacturer with respect to his on-site manufacturing.

(iii) It will not be necessary for a licensed importer or a licensed manufacturer (for purposes of sale or distribution) to also obtain a dealer's license in order to engage in business on his licensed premises as a dealer in explosive materials.

(iv) A separate license will not be required of licensed manufacturers with respect to their on-site manufacture of theatrical flash powder.

(3) Except as provided in paragraph (a)(1) of this section, each person intending to acquire explosive materials from a licensee in a State other than a State in which he resides, or from a foreign country, or who intends to transport explosive materials in interstate or foreign commerce, must file an application, with the required fee (see §555.43), with ATF in accordance with the instructions on the form (see §555.45). A permit will, subject to law, entitle the permittee to acquire, transport, ship, and receive in interstate or foreign commerce explosive materials. Only one permit is required under this part.

(b) Licenses and permits issued on and after May 24, 2003—

(1) In general.

(i) Each person intending to engage in business as an importer or manufacturer of, or a dealer in, explosive materials, including black powder, must, before commencing business, obtain the license required by this subpart for the business to be operated.

(ii) Each person who intends to acquire for use explosive materials within the State in which he resides on no more than 6 separate occasions during the 12-month period in which the permit is valid must obtain a limited permit under this subpart. (See §555.105(b) for definition of "6 separate occasions.")

(iii) Each person who intends to acquire for use explosive materials from a licensee or permittee in a State other than the State in which he resides, or from a foreign country, or who intends to transport explosive materials in interstate or foreign commerce, or who intends to acquire for use explosive materials within the State in which he resides on more than 6 separate occasions during a 12-month period, must obtain a user permit under this subpart.

(iv) It is not necessary to obtain a permit if the user intends only to lawfully purchase commercially manufactured black powder in quantities not to exceed 50 pounds, intended to be used solely for sporting, recreational, or cultural purposes in antique firearms or in antique devices.

(2) Importers, manufacturers, and dealers. Each person intending to engage in business as an explosive materials importer, manufacturer, or dealer must file an application, with the required fee (see §555.42), with ATF in accordance with the instructions on the form (see §555.45). A license will, subject to law, entitle the licensee to transport, ship, and receive explosive materials in interstate or foreign commerce, and to engage in the business specified by the license, at the location described on the license. A separate license must be obtained for each business premises at which the applicant is to manufacture, import, or distribute explosive materials except under the following circumstances:

(i) A separate license will not be required for storage facilities operated by the licensee as an integral part of one business premises or to cover a location used by the licensee solely for maintaining the records required by this part.

(ii) A separate license will not be required of a licensed manufacturer with respect to his on-site manufacturing.

(iii) It will not be necessary for a licensed importer or a licensed manufacturer (for purposes of sale or distribution) to also obtain a dealer's license in order to engage in business on his licensed premises as a dealer in explosive materials. No licensee will be required to obtain a user permit to lawfully transport, ship, or receive explosive materials in interstate or foreign commerce.

(iv) A separate license will not be required of licensed manufacturers with respect to their on-site manufacture of theatrical flash powder.

(3) Users of explosive materials.

(i) A limited permit will, subject to law, entitle the holder of such permit to receive for his use explosive materials from a licensee or permittee in his state of residence on no more than 6 separate occasions during the 12-month period in which the permit is valid. A limited permit does not authorize the receipt or transportation of explosive materials in interstate or foreign commerce. Holders of limited permits who need to receive explosive materials on more than 6 separate occasions during a 12-month period must obtain a user permit in accordance with this subpart.

(ii) Each person intending to acquire explosive materials from a licensee in a State other than a State in which he resides, or from a foreign country, or who intends to transport explosive materials in interstate or foreign commerce, must file an application for a user permit, with the required fee (see §555.43), with ATF in accordance with the instructions on the form (see §555.45). A user permit will, subject to law, entitle the permittee to transport, ship, and receive in interstate or foreign commerce explosive materials. Only one user permit per person is required under this part, irrespective of the number of locations relating to explosive materials operated by the holder of the user permit.

(Approved by the Office of Management and Budget under control number 1140–0083)

[ATF No. 1, 68 FR 13783, Mar. 20, 2003, as amended by ATF 5F, 70 FR 30633, May 27, 2005]

§ 555.42 License fees.

(a) Each applicant shall pay a fee for obtaining a three year license, a separate fee being required for each business premises, as follows:

(1) Manufacturer—$200.

(2) Importer—$200.

(3) Dealer—$200.

(b) Each applicant for a renewal of a license shall pay a fee for a three year license as follows:

(1) Manufacturer—$100.

(2) Importer—$100.

(3) Dealer—$100.

[T.D. ATF–400, 63 FR 45002, Aug. 24, 1998]

§ 555.43 Permit fees.

(a) Each applicant must pay a fee for obtaining a permit as follows:

(1) User—$100 for a three-year period.

(2) User-limited (nonrenewable)—$75.

(3) Limited—$25 for a one-year period.

(b) **(1)** Each applicant for renewal of a user permit must pay a fee of $50 for a three-year period.

(2) Each applicant for renewal of a limited permit must pay a fee of $12 for a one-year period.

[ATF No. 1, 68 FR 13785, Mar. 20, 2003]

§ 555.44 License or permit fee not refundable.

No refund of any part of the amount paid as a license or permit fee will be made where the operations of the licensee or permittee are, for any reason, discontinued during the period of an issued license or permit. However, the license or permit fee submitted with an application for a license or permit will be refunded if that application is denied, withdrawn, or abandoned, or if a license is cancelled subsequent to having been issued through administrative error.

§ 555.45 Original license or permit.

(a) Licenses issued prior to May 24, 2003. Any person who intends to engage in business as an explosive materials importer, manufacturer, or dealer, or who has not timely submitted application for renewal of a previous license issued under this part, shall file with ATF an application for License, Explosives, ATF F 5400.13 with ATF in accordance with the instructions on the form. The application must be executed under the penalties of perjury and the penalties imposed by 18 U.S.C. 844(a). The application is to be accompanied by the appropriate fee in the form of a money order or check made payable to the Bureau of Alcohol, Tobacco and Firearms. ATF F 5400.13 may be obtained from any ATF office. The Chief, Firearms and Explosives Licensing Center, will not approve an application postmarked on or after March 20, 2003, unless it is submitted with a Responsible Person Questionnaire, ATF Form 5400.28. Form 5400.28 must be completed in accordance with the instructions on the form.

(b) Permits issued prior to May 24, 2003. Any person, except as provided in §555.41(a), who intends to acquire explosive materials from a licensee in a state other than the State in which that person resides, or from a foreign country, or who intends to transport explosive materials in interstate or foreign commerce, or who has not timely submitted application for renewal of a previous permit issued under this part, shall file an application for Permit, Explosives, ATF F 5400.16 or Permit, User Limited Special Fireworks, ATF F 5400.21 with ATF in accordance with the instructions on the form. The application must be executed under the penalties of perjury and the penalties imposed by 18 U.S.C. 844(a). The application is to be accompanied by the appropriate fee in the form of a money order or check made payable to the Bureau of Alcohol, Tobacco and Firearms. ATF F 5400.16 and ATF F 5400.21 may be obtained from any ATF office. The Chief, Firearms and Explosives Licensing Center, will not approve an application postmarked on or after March 20, 2003, unless it is submitted with a Responsible Person Questionnaire, ATF Form 5400.28. Form 5400.28 must be completed in accordance with the instructions on the form.

(c) Licenses and permits issued on and after May 24, 2003—

(1) License. Any person who intends to engage in the business as an importer of, manufacturer of, or dealer in explosive materials, or who has not timely submitted an application for renewal of a previous license issued under this part, must file an application for License, Explosives, ATF F 5400.13, with ATF in accordance with the instructions on the form. ATF Form 5400.13 may be obtained by contacting any ATF office. The application must:

(i) Be executed under the penalties of perjury and the penalties imposed by 18 U.S.C. 844(a);

(ii) Include appropriate identifying information concerning each responsible person;

(iii) Include a photograph and fingerprints for each responsible person;

(iv) Include the names of and appropriate identifying information regarding all employees who will be authorized by the applicant to possess explosive materials by submitting ATF F 5400.28 for each employee; and

(v) Include the appropriate fee in the form of money order or check made payable to the Bureau of Alcohol, Tobacco, Firearms and Explosives.

(2) User permit and limited permit. Except as provided in 555.41(b)(1)(iv), any person who intends to acquire explosive materials in the State in which that person resides or acquire explosive materials from a licensee or holder of a user permit in a State other than the State in which that person resides, or from a foreign country, or who intends to transport explosive materials in interstate or foreign commerce, or who has not timely submitted an application for renewal of a previous permit issued under this part, must file an application for Permit, Explosives, ATF F 5400.16 or Permit, User Limited Display Fireworks, ATF F 5400.21 with ATF in accordance with the instructions on the form. ATF Form 5400.16 and ATF Form 5400.21 may be obtained by contacting any ATF office. The application must:

(i) Be executed under the penalties of perjury and the penalties imposed by 18 U.S.C. 844(a);

(ii) Include a photograph, fingerprints, and appropriate identifying information for each responsible person;

(iii) Include the names of and appropriate identifying information regarding all employees who will be authorized by the applicant to possess explosive materials by submitting ATF F 5400.28 for each employee; and

(iv) Include the appropriate fee in the form of money order or check made payable to the Bureau of Alcohol, Tobacco, Firearms and Explosives.

(3) The Chief, Firearms and Explosives Licensing Center, will conduct background checks on responsible persons and employees authorized by the applicant to possess explosive materials in accordance with §555.33. If it is determined that any responsible person or employee is described in any paragraph of section 842(i) of the Act, the applicant must submit an amended application indicating removal or reassignment of that person before the license or permit will be issued.

(Approved by the Office of Management and Budget under control number 1140–0083)

(18 U.S.C. 847 (84 Stat. 959); 18 U.S.C. 926 (82 Stat. 1226))
[T.D. ATF–200, 50 FR 10497, Mar. 15, 1985, as amended by T.D. ATF–400, 63 FR 45002, Aug. 24, 1998; ATF No. 1, 68 FR 13785, Mar. 20, 2003]

§ 555.46 Renewal of license or permit.

(a) If a licensee or permittee intends to continue the business or operation described on a license or permit issued under this part after the expiration date of the license or permit, he shall, unless otherwise notified in writing by the Chief, Firearms and Explosives Licensing Center, execute and file prior to the expiration of his license or permit an application for license renewal, ATF F 5400.14 (Part III), or an application for permit renewal, ATF F 5400.15 (Part III), accompanied by the required fee, with ATF in accordance with the instructions on the form. In the event the licensee or permittee does not timely file a renewal application, he shall file an original application as required by §555.45, and obtain the required license or permit in order to continue business or operations.

(b) A user-limited permit is not renewable and is valid for a single purchase transaction. Applications for all user-limited permits must be filed on ATF F 5400.16 or ATF F 5400.21, as required by §555.45.

(18 U.S.C. 847 (84 Stat. 959); 18 U.S.C. 926 (82 Stat. 1226))
[T.D. ATF–87, 46 FR 40384, Aug. 7, 1981, as amended by T.D. ATF–200, 50 FR 10497, Mar. 15, 1985; T.D. ATF–290, 54 FR 53054, Dec. 27, 1989; T.D. ATF–400, 63 FR 45002, Aug. 24, 1998]

§ 555.47 Insufficient fee.

If an application is filed with an insufficient fee, the application and fee submitted will be returned to the applicant.

(18 U.S.C. 847 (84 Stat. 959); 18 U.S.C. 926 (82 Stat. 1226))
[T.D. ATF–200, 50 FR 10498, Mar. 15, 1985]

§ 555.48 Abandoned application.

Upon receipt of an incomplete or improperly executed application, the applicant will be notified of the deficiency in the application. If the application is not corrected and returned within 30 days following the date of notification, the application will be considered as having been abandoned and the license or permit fee returned.

§ 555.49 Issuance of license or permit.

(a) Issuance of license or permit prior to May 24, 2003. (1) The Chief, Firearms and Explosives Licensing Center, will issue a license or permit if—

(i) A properly executed application for the license or permit is received; and

(ii) Through further inquiry or investigation, or otherwise, it is found that the applicant is entitled to the license or permit.

(2) The Chief, Firearms and Explosives Licensing Center, will approve a properly executed application for a license or permit, if:

(i) The applicant is 21 years of age or over;

(ii) The applicant (including, in the case of a corporation, partnership, or association, any individual possessing, directly or indirectly, the power to direct or cause the direction of the management and policies of the corporation, partnership, or association) is not a person to whom distribution of explosive materials is prohibited under the Act;

(iii) The applicant has not willfully violated any provisions of the Act or this part;

(iv) The applicant has not knowingly withheld information or has not made any false or fictitious statement intended or likely to deceive, in connection with his application;

(v) The applicant has in a State, premises from which he conducts business or operations subject to license or permit under the Act or from which he intends to conduct business or operations;

(vi) The applicant has storage for the class (as described in §555.202) of explosive materials described on the application, unless he establishes to the satisfaction of the Chief, Firearms and Explosives Licensing Center, that the business or operations to be conducted will not require the storage of explosive materials;

(vii) The applicant has certified in writing that he is familiar with and understands all published State laws and local ordinances relating to explosive materials for the location in which he intends to do business; and

(viii) The applicant for a license has submitted the certificate required by section 21 of the Federal Water Pollution Control Act, as amended (33 U.S.C. 1341).

(3) The Chief, Firearms and Explosives Licensing Center, will approve or the regional director (compliance) will deny any application for a license or permit within the 45-day period beginning on the date a properly executed application was received. However, when an applicant for license or permit renewal is a person who is, under the provisions of §555.83 or §555.142, conducting business or operations under a previously issued license or permit, action regarding the application will be held in abeyance pending the completion of the proceedings against the

applicant's existing license or permit, or renewal application, or final action by the Director on an application for relief submitted under §555.142, as the case may be.

(4) The license or permit and one copy will be forwarded to the applicant, except that in the case of a user-limited permit, the original only will be issued.

(5) Each license or permit will bear a serial number and this number may be assigned to the licensee or permittee to whom issued for as long as he maintains continuity of renewal in the same region.

(b) Issuance of license or permit on and after May 24, 2003.

(1) The Chief, Firearms and Explosives Licensing Center, will issue a license or permit if:

(i) A properly executed application for the license or permit is received; and

(ii) Through further inquiry or investigation, or otherwise, it is found that the applicant is entitled to the license or permit.

(2) The Chief, Firearms and Explosives Licensing Center, will approve a properly executed application for a license or permit, if:

(i) The applicant (or, if the applicant is a corporation, partnership, or association, each responsible person with respect to the applicant) is not a person described in any paragraph of section 842(i) of the Act;

(ii) The applicant has not willfully violated any provisions of the Act or this part;

(iii) The applicant has not knowingly withheld information or has not made any false or fictitious statement intended or likely to deceive, in connection with his application;

(iv) The applicant has in a State, premises from which he conducts business or operations subject to license or permit under the Act or from which he intends to conduct business or operations;

(v) The applicant has storage for the class (as described in §555.202) of explosive materials described on the application;

(vi) The applicant has certified in writing that he is familiar with and understands all published State laws and local ordinances relating to explosive materials for the location in which he intends to do business;

(vii) The applicant for a license has submitted the certificate required by section 21 of the Federal Water Pollution Control Act, as amended (33 U.S.C. 1341);

(viii) None of the employees of the applicant who will be authorized by the applicant to possess explosive materials is a person described in any paragraph of section 842(i) of the Act; and

(ix) In the case of an applicant for a limited permit, the applicant has certified in writing that the applicant will not receive explosive materials on more than 6 separate occasions during the 12-month period for which the limited permit is valid.

(3) The Chief, Firearms and Explosives Licensing Center, will approve or the regional director (compliance) will deny any application for a license or permit within the 90-day period beginning on the date a properly executed application was received. However, when an applicant for license or permit renewal is a person who is, under the provisions of §555.83 or §555.142, conducting business or operations under a previously issued license or permit, action regarding the application will be held in abeyance pending the completion of the proceedings against the applicant's existing license or permit, or renewal application, or final action by the Director on an application for relief submitted under §555.142, as the case may be.

(4) The license or permit and one copy will be forwarded to the applicant, except that in the case of a user-limited permit, the original only will be issued.

(5) Each license or permit will bear a serial number and this number may be assigned to the licensee or permittee to whom issued for as long as he maintains continuity of renewal in the same region.

(Approved by the Office of Management and Budget under control number 1140–0082)

[ATF No. 1, 68 FR 13785, Mar. 20, 2003]

§ 555.50 Correction of error on license or permit.

(a) Upon receipt of a license or permit issued under this part, each licensee or permittee shall examine the license or permit to insure that the information on it is accurate. If the license or permit is incorrect, the licensee or permittee shall return the license or permit to the Chief, Firearms and Explosives Licensing Center, with a statement showing the nature of the error. The Chief, Firearms and Explosives Licensing Center, shall correct the error, if the error was made in his office, and return the license or permit. However, if the error resulted from information contained in the licensee's or permittee's application for the license or permit, the Chief, Firearms and Explosives Licensing Center, shall require the licensee or permittee to file an amended application setting forth the correct information and a statement explaining the error contained in the application. Upon receipt of the amended application and a satisfactory explanation of the error, the Chief, Firearms and Explosives Licensing Center, shall make the correction on the license or permit and return it to the licensee or permittee.

(b) When the Chief, Firearms and Explosives Licensing Center, finds through any means other than notice from the licensee or permittee that an incorrect license or permit has been issued, (1) the Chief, Firearms and Explosives Licensing Center, may require the holder of the incorrect license or permit to return the license or permit for correction, and (2) if the error resulted from information contained in the licensee's or permittee's application for the license or permit, the Chief, Firearms and Explosives Licensing Center, shall require the licensee or permittee to file an amended application setting forth the correct information, and a

statement satisfactorily explaining the error contained in the application. The Chief, Firearms and Explosives Licensing Center, then shall make the correction on the license or permit and return it to the licensee or permittee.
[T.D. ATF–87, 46 FR 40384, Aug. 7, 1981, as amended by T.D. ATF–290, 54 FR 53054, Dec. 27, 1989]

§ 555.51 Duration of license or permit.

(a) Prior to May 24, 2003. An original license or permit is issued for a period of three years. A renewal license or permit is issued for a period of three years. However, a user-limited permit is valid only for a single purchase transaction.

(b) On and after May 24, 2003.

(1) An original license or user permit is issued for a period of three years. A renewal license or user permit is also issued for a period of three years. However, a user-limited permit is valid only for a single purchase transaction.

(2) A limited permit is issued for a period of one year. A renewal limited permit is also issued for a period of one year.
[ATF No. 1, 68 FR 13786, Mar. 20, 2003]

§ 555.52 Limitations on license or permit.

(a) The license covers the business of explosive materials specified in the license at the licensee's business premises (see §555.41(b)).

(b) The permit is valid with respect to the type of operations of explosive materials specified in the permit.
[T.D. ATF–87, 46 FR 40384, Aug. 7, 1981, as amended by T.D. ATF–387, 62 FR 8376, Feb. 25, 1997; ATF 5F, 70 FR 30633, May 27, 2005]

§ 555.53 License and permit not transferable.

Licenses and permits issued under this part are not transferable to another person. In the event of the lease, sale, or other transfer of the business or operations covered by the license or permit, the successor must obtain the license or permit required by this part before commencing business or operations. However, for rules on right of succession, see §555.59.

§ 555.54 Change of address.

(a) During the term of a license or permit, a licensee or permittee may move his business or operations to a new address at which he intends to regularly carry on his business or operations, without procuring a new license or permit. However, in every case, the licensee or permittee shall—

(1) Give notification of the new location of the business or operations to the Chief, Firearms and Explosives Licensing Center at least 10 days before the move; and

(2) Submit the license or permit to the Chief, Firearms and Explosives Licensing Center. The Chief, Firearms and Explosives Licensing Center will issue an amended license or permit, which will contain the new address (and new license or permit number, if any).

(b) Licensees and permittees whose mailing address will change must notify the Chief, Firearms and Explosives Licensing Center, at least 10 days before the change.

(Paragraph (b) approved by the Office of Management and Budget under control number 1140–0080)
[T.D. ATF–87, 46 FR 40384, Aug. 7, 1981, as amended by T.D. ATF–290, 54 FR 53054, Dec. 27, 1989; ATF No. 1, 68 FR 13786, Mar. 20, 2003]

§ 555.56 Change in trade name.

A licensee or permittee continuing to conduct business or operations at the location shown on his license or permit is not required to obtain a new license or permit by reason of a mere change in trade name under which he conducts his business or operations. However, the licensee or permittee shall furnish his license or permit and any copies furnished with the license or permit for endorsement of the change to the Chief, Firearms and Explosives Licensing Center, within 30 days from the date the licensee or permittee begins his business or operations under the new trade name.
[T.D. ATF–87, 46 FR 40384, Aug. 7, 1981, as amended by T.D. ATF–290, 54 FR 53054, Dec. 27, 1989]

§ 555.57 Change of control, change in responsible persons, and change of employees.

(a) In the case of a corporation or association holding a license or permit under this part, if actual or legal control of the corporation or association changes, directly or indirectly, whether by reason of change in stock ownership or control (in the corporation holding a license or permit or in any other corporation), by operation of law, or in any other manner, the licensee or permittee shall, within 30 days of the change, give written notification executed under the penalties of perjury, to the Chief, Firearms and Explosives Licensing Center. Upon expiration of the license or permit, the corporation or association shall file an ATF F 5400.13 or an ATF F 5400.16 as required by §555.45, and pay the fee prescribed in §555.42(b) or §555.43(b).

(b) For all licenses or permits issued on and after May 24, 2003, each person holding the license or permit must report to the Chief, Firearms and Explosives Licensing Center, any change in responsible persons or employees authorized to possess explosive materials. Such report must be submitted within 30 days of the change and must include appropriate identifying information for each responsible person. Reports relating to newly hired employees authorized to possess explosive materials must be submitted on ATF F 5400.28 for each employee.

(c) Upon receipt of a report, the Chief, Firearms and Explosives Licensing Center, will conduct a background check, if appropriate, in accordance with §555.33.

(d) The reports required by paragraph (b) of this section must be retained as part of a licensee's or permittee's permanent records for the period specified in §555.121.

(Approved by the Office of Management and Budget under control number 1140–0074)
[T.D. ATF–87, 46 FR 40384, Aug. 7, 1981, as amended by T.D. ATF–290, 54 FR 53054, Dec. 27, 1989; ATF No. 1, 68 FR 13786, Mar. 20, 2003]

§ 555.58 Continuing partnerships.

Where, under the laws of the particular State, the partnership is not terminated on death or insolvency of a partner, but continues until the winding up of the partnership affairs is completed, and the surviving partner has the exclusive right to the control and possession of the partnership assets for the purpose of liquidation and settlement, the surviving partner may continue to conduct the business or operations under the license or permit of the partnership. If the surviving partner acquires the business or operations on completion of settlement of the partnership, he shall obtain a license or permit in his own name from the date of acquisition, as provided in §555.45. The rule set forth in this section will also apply where there is more than one surviving partner.

§ 555.59 Right of succession by certain persons.

(a) Certain persons other than the licensee or permittee may secure the right to carry on the same explosive materials business or operations at the same business premises for the remainder of the term of license or permit. These persons are:

(1) The surviving spouse or child, or executor, administrator, or other legal representative of a deceased licensee or permittee; and

(2) A receiver or trustee in bankruptcy, or an assignee for benefit of creditors.

(b) In order to secure the right of succession, the person or persons continuing the business or operations shall submit the license or permit and all copies furnished with the license or permit for endorsement of the succession to the Chief, Firearms and Explosives Licensing Center, within 30 days from the date on which the successor begins to carry on the business or operations.
[T.D. ATF–87, 46 FR 40384, Aug. 7, 1981, as amended by T.D. ATF–290, 54 FR 53054, Dec. 27, 1989]

§ 555.60 Certain continuances of business or operations.

A licensee or permittee who furnishes his license or permit to the Chief, Firearms and Explosives Licensing Center, for correction, amendment, or endorsement, as provided in this subpart, may continue his business or operations while awaiting its return.
[T.D. ATF–87, 46 FR 40384, Aug. 7, 1981, as amended by T.D. ATF–290, 54 FR 53054, Dec. 27, 1989]

§ 555.61 Discontinuance of business or operations.

Where an explosive materials business or operations is either discontinued or succeeded by a new owner, the owner of the business or operations discontinued or succeeded shall, within 30 days, furnish notification of the discontinuance or succession and submit his license or permit and any copies furnished with the license or permit to the Chief, Firearms and Explosives Licensing Center. (See also §555.128.)
[T.D. ATF–87, 46 FR 40384, Aug. 7, 1981, as amended by T.D. ATF–290, 54 FR 53054, Dec. 27, 1989]

§ 555.62 State or other law.

A license or permit issued under this part confers no right or privilege to conduct business or operations, including storage, contrary to State or other law. The holder of a license or permit issued under this part is not, by reason of the rights and privileges granted by that license or permit, immune from punishment for conducting an explosive materials business or operations in violation of the provisions of any State or other law. Similarly, compliance with the provisions of any State or other law affords no immunity under Federal law or regulations.

§ 555.63 Explosives magazine changes.

(a) General.

(1) The requirements of this section are applicable to magazines used for other than temporary (under 24 hours) storage of explosives.

(2) A magazine is considered suitable for the storage of explosives if the construction requirements of this part are met during the time explosives are stored in the magazine.

(3) A magazine is considered suitable for the storage of explosives if positioned in accordance with the applicable table of distances as specified in this part during the time explosives are stored in the magazine.

(4) For the purposes of this section, notification of the regional director (compliance) may be by telephone or in writing. However, if notification of the regional director (compliance) is in writing it must be at least three business days in advance of making changes in construction to an existing magazine or constructing a new magazine, and at least five business days in advance of using any reconstructed magazine or added magazine for the storage of explosives.

(b) Exception. Mobile or portable type 5 magazines are exempt from the requirements of paragraphs (c) and (d) of this section, but must otherwise be in compliance with paragraphs (a) (2) and (3) of this section during the time explosives are stored in such magazines.

(c) Changes in magazine construction. A licensee or permittee who intends to make changes in construction of an existing magazine shall notify the regional director (compliance) describing the proposed changes prior to making any changes. Unless otherwise advised by the regional director (compliance), changes in construction may commence after explosives are removed from the magazine. Explosives may not be stored in a reconstructed magazine before the regional director (compliance) has been notified in accordance with paragraph (a)(4) of this section that the changes have been completed.

(d) Magazines acquired or constructed after permit or license is issued. A licensee or permittee who intends to construct or acquire additional magazines shall notify the regional director (compliance) in accordance with paragraph (a)(4) of this section describing the additional magazines and the class and quantity of explosives to be stored in the magazine. Unless otherwise advised by the regional director (compliance), additional magazines may be constructed, or acquired magazines may be used for the storage of explosives. Explosives must not be stored in a magazine under construction. The regional director (compliance) must be notified that construction has been completed.

[T.D. ATF–87, 46 FR 40384, Aug. 7, 1981, as amended by T.D. ATF–400, 63 FR 45002, Aug. 24, 1998]

Subpart E—License and Permit Proceedings

§ 555.71 Opportunity for compliance.

Except in cases of willfulness or those in which the public interest requires otherwise, and the regional director (compliance) so alleges in the notice of denial of an application or revocation of a license or permit, no license or permit will be revoked or renewal application denied without first calling to the attention of the licensee or permittee the reasons for the contemplated action and affording him an opportunity to demonstrate or achieve compliance with all lawful requirements and to submit facts, arguments, or proposals of adjustment. The notice of contemplated action, ATF F 5400.12, will afford the licensee or permittee 15 days from the date of receipt of the notice to respond. If no response is received within the 15 days, or if after consideration of relevant matters presented by the licensee or permittee, the regional director (compliance) finds that the licensee or permittee is not likely to abide by the law and regulations, he will proceed as provided in §555.74.

[T.D. ATF–87, 46 FR 40384, Aug. 7 1981, as amended by T.D. ATF–446, 66 FR 16602, Mar. 27, 2001]

§ 555.72 Denial of initial application.

Whenever the regional director (compliance) has reason to believe that an applicant for an original license or permit is not eligible to receive a license or permit under the provisions of §555.49, he shall issue a notice of denial on ATF F 5400.11. The notice will set forth the matters of fact and law relied upon in determining that the application should be denied, and will afford the applicant 15 days from the date of receipt of the notice in which to request a hearing to review the denial. If no request for a hearing is filed within that time, a copy of the application, marked *"Disapproved"*, will be returned to the applicant.

§ 555.73 Hearing after initial application is denied.

If the applicant for an original license or permit desires a hearing, he shall file a request with the regional director (compliance) within 15 days after receipt of the notice of denial. The request should include a statement of the reasons for a hearing. On receipt of the request, the regional director (compliance) shall refer the matter to an administrative law judge who shall set a time and place (see §555.77) for a hearing and shall serve notice of the hearing upon the applicant and the regional director (compliance) at least 10 days in advance of the hearing date. The hearing will be conducted in accordance with the hearing procedures prescribed in part 71 of this chapter (see §555.82). Within a reasonable time after the conclusion of the hearing, and as expeditiously as possible, the administrative law judge shall render his recommended decision. He shall certify to the complete record of the proceedings before him and shall immediately forward the complete certified record, together with four copies of his recommended decision, to the regional director (compliance) for decision.

§ 555.74 Denial of renewal application or revocation of license or permit.

If following the opportunity for compliance under §555.71, or without opportunity for compliance under §555.71, as circumstances warrant, the regional director (compliance) finds that the licensee or permittee is not likely to comply with the law or regulations or is otherwise not eligible to continue operations authorized under his license or permit, the regional director (compliance) shall issue a notice of denial of the renewal application or revocation of the license or permit, ATF F 5400.11 or ATF F 5400.10, as appropriate. In either case, the notice will set forth the matters of fact constituting the violations specified, dates, places, and the sections of law and regulations violated. The notice will, in the case of revocation of a license or permit, specify the date on which the action is effective, which date will be on or after the date the notice is served on the licensee or permittee. The notice will also advise the licensee or permittee that he may, within 15 days after receipt of the notice, request a hearing and, if applicable, a stay of the effective date of the revocation of his license or permit.

§ 555.75 Hearing after denial of renewal application or revocation of license or permit.

If a licensee or permittee whose renewal application has been denied or whose license or permit has been revoked desires a hearing, he shall file a request for a hearing with the regional director (compliance). In the case of the revocation of a license or permit, he may include a request for a stay of the effective date of the revocation. On receipt of the request the regional director (compliance) shall advise the licensee or permittee whether the stay of the effective date of the revocation is granted. If the stay of the effective date of the revocation is granted, the regional director (compliance) shall refer the matter to an administrative law judge who shall set a time and place (see §555.77) for a hearing and shall serve notice of the hearing upon the licensee or permittee and the regional director (compliance) at least 10 days in advance of the hearing date. If the stay of the effective date of the revocation is denied, the licensee or permittee may request an immediate hearing. In this event, the regional director (compliance) shall immediately refer the matter to an administrative law judge who shall set a date and place for a hearing, which date shall be no later than 10 days from the date the licensee or permittee requested an immediate hearing. The hearing will be held in accordance with the applicable provisions of part 71 of this chapter. Within a reasonable time after the conclusion of the hearing, and as expeditiously as possible, the administrative law judge shall render his decision. He shall certify to the complete record of the proceeding before him and shall immediately forward the complete certified record, together with two copies of his decision, to the regional director (compliance), serve one copy of his decision on the licensee or permittee or his counsel, and transmit a copy to the attorney for the Government.

§ 555.76 Action by regional director (compliance).

(a) Initial application proceedings. If, upon receipt of the record and the recommended decision of the administrative law judge, the regional director (compliance) decides that the license or permit should be issued, the regional director (compliance) shall cause the application to be approved, briefly stating, for the record, his reasons. If he contemplates that the denial should stand, he shall serve a copy of the administrative law judge's recommended decision on the applicant, informing the applicant of his contemplated action and affording the applicant not more than 10 days in which to submit proposed findings and conclusions or exceptions to the recommended decision with supporting reasons. If the regional director (compliance), after consideration of the record of the hearing and of any proposed findings, conclusions, or exceptions filed with him by the applicant, approves the findings, conclusions and recommended decision of the administrative law judge, the regional director (compliance) shall cause the license or permit to be issued or disapproved the application accordingly. If he disapproves the findings, conclusions, and recommendation of the administrative law judge, in whole or in part, he shall by order make such findings and conclusions as in his opinion are warranted by the law and the facts in the record. Any decision of the regional director (compliance) ordering the disapproval of an initial application for a license or permit shall state the findings and conclusions upon which it is based, including his ruling upon each proposed finding, conclusion, and exception to the administrative law judge's recommended decision, together with a statement of his findings and conclusions, and reasons or basis for his findings and conclusions, upon all material issues of fact, law or discretion presented on the record. A signed duplicate original of the decision will be served upon the applicant and the original copy containing certificate of service will be placed in the official record of the proceedings. If the decision of the regional director (compliance) is in favor of the applicant, he shall issue the license or permit, to be effective on issuance.

(b) Renewal application and revocation proceedings. Upon receipt of the complete certified records of the hearing, the regional director (compliance) shall enter an order confirming the revocation of the license or permit, or disapproving the application, in accordance with the administrative law judge's findings and decision, unless he disagrees with the findings and decision. A signed duplicate original of the order, ATF F 5400.9, will be served upon the licensee or permittee and the original copy containing certificate of service will be placed in the official record of the proceedings. If the regional director (compliance) disagrees with the findings and decision of the administrative law judge, he shall file a petition with the Director for review of the findings and decision, as provided in §555.79. In either case, if the renewal application denial is sustained, a copy of the application marked "Disapproved" will be returned to the applicant. If the renewal application denial is reversed, a license or permit will be issued to become effective on expiration of the license or permit being renewed, or on the date of issuance, whichever is later. If the proceedings involve the revocation of a license or permit which expired before a decision is in favor of the licensee or permittee, the regional director (compliance) shall:

(1) If renewal application was timely filed and a stay of the effective date of the revocation was granted, cause to be issued a license or permit effective on the date of issuance;

(2) If renewal application was not timely filed but a stay of the effective date of the revocation had been granted, request that a renewal application be filed and, following that, cause to be issued a license or permit to be effective on issuance; or

(3) If a stay of the effective date of the revocation had not been granted, request that an application be filed as provided in §555.45, and process it in the same manner as for an application for an original license or permit.

(c) Curtailment of stay of revocation effective date. If, after approval of a request for a stay of the effective date of an order revoking a license or permit but before actions are completed under this subpart, the regional director (compliance) finds that it is contrary to the public interest for the licensee or permittee to continue the operations or activities covered by his license or permit, the regional director (compliance) may issue a notice of withdrawal of the approval, effective on the date of issuance. Notice of withdrawal will be served upon the licensee or permittee in the manner provided in §555.81.

[T.D. ATF-87, 46 FR 40384, Aug. 7, 1981, as amended by T.D. ATF-290, 54 FR 53054, Dec. 27, 1989]

§ 555.77 Designated place of hearing.

The designated place of hearing set as provided in §555.73 or §555.75, will be at the location convenient to the aggrieved party.

§ 555.78 Representation at a hearing.

An applicant, licensee, or permittee may be represented by an attorney, certified public accountant, or other person recognized to practice before the Bureau of Alcohol, Tobacco and Firearms as provided in 31 CFR Part 8, if he has otherwise complied with the applicable requirements of 26 CFR 601.521 through 601.527. The regional director (compliance) shall be represented in proceedings under §§555.73 and 555.75 by an attorney in the office of the chief counsel or regional counsel who is authorized to execute and file motions, briefs, and other papers in the proceedings, on behalf of the regional director (compliance), in his own name as *"Attorney for the Government"*.

[T.D. ATF–87, 46 FR 40384, Aug. 7, 1981, as amended by T.D. ATF–92, 46 FR 46916, Sept. 23, 1981]

§ 555.79 Appeal on petition to the Director.

An appeal to the Director is not required prior to filing an appeal with the U.S. Court of Appeals for judicial review. An appeal may be taken by the applicant, licensee, or permittee to the Director from a decision resulting from a hearing under §555.73 or §555.75. An appeal may also be taken by a regional director (compliance) from a decision resulting from a hearing under §555.75 as provided in §555.76(b). The appeal shall be taken by filing a petition for review on appeal with the Director within 15 days of the service of an administrative law judge's decision or an order. The petition will set forth facts tending to show (a) action of an arbitrary nature, (b) action without reasonable warrant in fact, or (c) action contrary to law and regulations. A copy of the petition will be filed with the regional director (compliance) or served on the applicant, licensee, or permittee, as the case may be. In the event of appeal, the regional director (compliance) shall immediately forward the complete original record, by certified mail, to the Director for his consideration, review, and disposition as provided in subpart I of part 71 of this chapter. When, on appeal, the Director affirms the initial decision of the regional director (compliance) or the administrative law judge, as the case may be, the initial decision will be final.

§ 555.80 Court review.

An applicant, licensee, or permittee may, within 60 days after receipt of the decision of the administrative law judge or the final order of the regional director (compliance) or the Director, file a petition for a judicial review of the decision, with the U.S. Court of Appeals for the district in which he resides or has his principal place of business. The Director, upon notification that a petition has been filed, shall have prepared a complete transcript of the record of the proceedings. The regional director (compliance) or the Director, as the case may be, shall certify to the correctness of the transcript of the record, forward one copy to the attorney for the Government in the review of the case, and file the original record of the proceedings with the original certificate in the U.S. Court of Appeals.

§ 555.81 Service on applicant, licensee, or permittee.

All notices and other formal documents required to be served on an applicant, licensee, or permittee under this subpart will be served by certified mail or by personal delivery. Where service is by personal delivery, the signed duplicate original copy of the formal document will be delivered to the applicant, licensee, or permittee, or, in the case of a corporation, partnership, or association, by delivering it to an officer, manager, or general agent, or to its attorney of record.

§ 555.82 Provisions of part 200 made applicable.

The provisions of subpart G of part 200 of this chapter, as well as those provisions of part 71 relative to failure to appear, withdrawal of an application or surrender of a permit, the conduct of hearings before an administrative law judge, and record of testimony, are hereby made applicable to application, license, and permit proceedings under this subpart to the extent that they are not contrary to or incompatible with this subpart.

§ 555.83 Operations by licensees or permittees after notice of denial or revocation.

In any case where a notice of revocation has been issued and a request for a stay of the effective date of the revocation has not been granted, the licensee or permittee shall not engage in the activities covered by the license or permit pending the outcome of proceedings under this subpart. In any case where notice of revocation has been issued but a stay of the effective date of the revocation has been granted, the licensee or permittee may continue to engage in the activities covered by his license or permit unless, or until, formally notified to the contrary: Provided, That in the event the license or permit would have expired before proceedings under this subpart are completed, timely renewal application must have been filed to continue the license or permit beyond its expiration date. In any case where a notice of denial of a renewal application has been issued, the licensee or permittee may continue to engage in the activities covered by the existing license or permit after the date of expiration of the license or permit until proceedings under this subpart are completed.

Subpart F—Conduct of Business or Operations

§ 555.101 Posting of license or user permit.

A license or user permit issued under this part, or a copy of a license or user permit, will be posted and available for inspection on the business premises at each place where explosive materials are manufactured, imported, or distributed.

[T.D. ATF–87, 46 FR 40384, Aug. 7, 1981. Redesignated by T.D. ATF–487, 68 FR 3748, Jan. 24, 2003. ATF No. 1, 68 FR 13786, Mar. 20, 2003]

§ 555.102 Authorized operations by permittees.

(a) In general. A permit issued under this part does not authorize the permittee to engage in the business of manufacturing, importing, or dealing in explosive materials. Accordingly, if a permittee's operations bring him within the definition of manufacturer, importer, or dealer under this part, he shall qualify for the appropriate license.

(b) Distributions of surplus stocks—

(1) Distributions of surplus stocks prior to May 24, 2003. Permittees are not authorized to engage in the business of sale or distribution of explosive materials. However, permittees may dispose of surplus stocks of explosive materials to other licensees or permittees in accordance with §555.103, and to nonlicensees or to nonpermittees in accordance with §555.105(a)(4).

(2) Distributions of surplus stocks on and after May 24, 2003. Permittees are not authorized to engage in the business of sale or distribution of explosive materials. However, permittees may dispose of surplus stocks of explosive materials to other licensees or permittees in accordance with §555.103 and §555.105.

[T.D. ATF–400, 63 FR 45002, Aug. 24, 1998, as amended by ATF No. 1, 68 FR 13787, Mar. 20, 2003]

§ 555.103 Transactions among licensees/permittees and transactions among licensees and holders of user permits.

(a) Transactions among licensees/permittees prior to May 24, 2003—

(1) General.

(i) A licensed importer, licensed manufacturer or licensed dealer selling or otherwise distributing explosive materials (or a permittee disposing of surplus stock to a licensee or another permittee) who has the certified information required by this section may sell or distribute explosive materials to a licensee or permittee for not more than 45 days following the expiration date of the distributee's license or permit, unless the distributor knows or has reason to believe that the distributee's authority to continue business or operations under this part has been terminated.

(ii) A licensed importer, licensed manufacturer or licensed dealer selling or otherwise distributing explosive materials (or a permittee disposing of surplus stock to another licensee or permittee) must verify the license or permit status of the distributee prior to the release of explosive materials ordered, as required by this section.

(iii) Licensees or permittees desiring to return explosive materials to a licensed manufacturer may do so without obtaining a certified copy of the manufacturer's license.

(iv) Where possession of explosive materials is transferred at the distributor's premises, the distributor must in all instances verify the identity of the person accepting possession on behalf of the distributee before relinquishing possession. Before the delivery at the distributor's premises of explosive materials to an employee of a licensee or permittee, or to an employee of a common or contract carrier transporting explosive materials to a licensee or permittee, the distributor delivering explosive materials must obtain an executed ATF F 5400.8, Explosives Delivery Record, from the employee before releasing the explosive materials. The ATF F 5400.8 must contain all of the information required on the form and required by this part.

Example 1. An ATF F 5400.8 is required when:

a. An employee of the purchaser takes possession at the distributor's premises.

b. An employee of a common or contract carrier hired by the purchaser takes possession at the distributor's premises.

Example 2. An ATF F 5400.8 is not required when:

a. An employee of the distributor takes possession of the explosives for the purpose of transport to the purchaser.

b. An employee of a common or contract carrier hired by the distributor takes possession of the explosives for the purpose of transport to the purchaser.

(2) License/permit verification of individuals.

(i) The distributee must furnish a certified copy (or, in the case of a user-limited, the original) of the license or permit. The certified copy need be furnished only once during the current term of the license or permit. Also, a licensee need not furnish certified copies of licenses to other licensed locations operated by such licensee.

(ii) The distributor may obtain any additional verification as the distributor deems necessary.

(3) License/permit verification of business organizations.

(i) A business organization may (in lieu of furnishing a certified copy of a license) furnish the distributor a certified list which contains the name, address, license number and date of license expiration of each licensed location. The certified list need be furnished only once during the current term of the license or permit. Also, a business organization need not furnish a certified list to other licensed locations operated by such business organization.

(ii) A business organization must, prior to ordering explosive materials, furnish the licensee or permittee a current certified list of the representatives or agents authorized to order explosive materials on behalf of the business organization showing the name, address, and date and place of birth of each representative or agent. A licensee or permittee may not distribute explosive materials to a business organization on the order of a person who does not appear on the certified list of representatives or agents and, if the person does appear on the certified list, the licensee or permittee must verify the identity of such person.

(4) Licensee/permittee certified statement.

(i) A licensee or permittee ordering explosive materials from another licensee or permittee must furnish a current, certified statement of the intended use of the explosive materials, e.g., resale, mining, quarrying, agriculture, construction, sport rocketry, road building, oil well drilling, seismographic research, to the distributor.

(ii) For individuals, the certified statement of intended use must specify the name, address, date and place of birth, and social security number of the distributee.

(iii) For business organizations, the certified statement of intended use must specify the taxpayer identification number, the identity and the principal and local places of **business.**

(iv) The licensee or permittee purchasing explosive materials must revise the furnished copy of the certified statement only when the information is no longer current.

(5) User-limited permit transactions. A user-limited permit issued under the provisions of this part is valid for only a single purchase transaction and is not renewable (see §555.51). Accordingly, at the time a user-limited permittee orders explosive materials, the licensed distributor must write on the front of the user-limited permit the transaction date, his signature, and the distributor's license number prior to returning the permit to the user-limited permittee.

(b) Transactions among licensees/permittees on and after May 24, 2003—

(1) General.

(i) A licensed importer, licensed manufacturer or licensed dealer selling or otherwise distributing explosive materials (or a holder of a user permit disposing of surplus stock to a licensee; a holder of a user permit; or a holder of a limited permit who is within the same State as the distributor) who has the certified information required by this section may sell or distribute explosive materials to a licensee or permittee for not more than 45 days following the expiration date of the distributee's license or permit, unless the distributor knows or has reason to believe that the distributee's authority to continue business or operations under this part has been terminated.

(ii) A licensed importer, licensed manufacturer or licensed dealer selling or otherwise distributing explosive materials (or a holder of a user permit disposing of surplus stock to another licensee or permittee) must verify the license or permit status of the distributee prior to the release of explosive materials ordered, as required by this section.

(iii) Licensees or permittees desiring to return explosive materials to a licensed manufacturer may do so without obtaining a certified copy of the manufacturer's license.

(2) Verification of license/user permit.

(i) Prior to or with the first order of explosive materials, the distributee must provide the distributor certified copy (or, in the case of a user-limited, the original) of the distributee's license or user permit. However, licensees or holders of user permits that are business organizations may (in lieu of a certified copy of a license or user permit) provide the distributor with a certified list that contains the name, address, license or user permit number, and date of the license or user permit expiration of each location.

(ii) The distributee must also provide the distributor with a current list of the names of persons authorized to accept delivery of explosive materials on behalf of the distributee. The distributee ordering explosive materials must keep the list current and provide updated lists to licensees and holders of user permits on a timely basis. A distributor may not transfer possession of explosive materials to any person whose name does not appear on the current list of names of persons authorized to accept delivery of explosive materials on behalf of the distributee. Except as provided in paragraph (b)(3) of this section, in all instances the distributor must verify the identity of the distributee, or the employee of the distributee accepting possession of explosive materials on behalf of the distributee, by examining an identification document (as defined in §555.11) before relinquishing possession.

(iii) A licensee or holder of a user permit ordering explosive materials from another licensee or permittee must provide to the distributor a current, certified statement of the intended use of the explosive materials, e.g., resale, mining, quarrying, agriculture, construction, sport rocketry, road building, oil well drilling, seismographic research, etc.

(A) For individuals, the certified statement of intended use must specify the name, address, date and place of birth, and social security number of the distributee.

(B) For business organizations, the certified statement of intended use must specify the taxpayer identification number, the identity and the principal and local places of business.

(C) The licensee or holder of a user permit purchasing explosive materials must revise the furnished copy of the certified statement only when the information is no longer current.

(3) Delivery of explosive materials by common or contract carrier. When a common or contract carrier will transport explosive materials from a distributor to a distributee who is a licensee or holder of a user permit, the distributor must take the following actions before relinquishing possession of the explosive materials:

(i) Verify the identity of the person accepting possession for the common or contract carrier by examining such person's valid, unexpired driver's license issued by any State, Canada, or Mexico; and

(ii) Record the name of the common or contract carrier (i.e., the name of the driver's employer) and the full name of the driver. This information must be maintained in the distributor's permanent records in accordance with §555.121.

(4) User-limited permit transactions. A user-limited permit issued under the provisions of this part is valid for only a single purchase transaction and is not renewable (see §555.51). Accordingly, at the time a user-limited permittee orders explosive materials, the licensed distributor must write on the front of the user-limited permit the transaction date, his signature, and the distributor's license number prior to returning the permit to the user-limited permittee.

(Approved by the Office of Management and Budget under control number 1140–0079)

[ATF No. 1, 68 FR 13787, Mar. 20, 2003, as amended by ATF No. 2, 68 FR 53512, Sept. 11, 2003]

§ 555.104 Certified copy of license or permit.

Except as provided in §555.49(a), each person issued a license or permit under this part shall be furnished together with his license or permit a copy for his certification. If a person desires an additional copy of his license or permit for certification and for use under §555.103, he shall:

(a) Make a reproduction of the copy of his license or permit and execute the certification on it;

(b) Make a reproduction of his license or permit, enter on the reproduction the statement: "I certify that this is a true copy of a (insert the word license or permit) issued to me to engage in the specified business or operations", and sign his name next to the statement; or

(c) Submit a request, in writing, for certified copies of his license or permit to the Chief, Firearms and Explosives Licensing Center. The request will show the name, trade name (if any), and address of the licensee or permittee and the number of copies of the license or permit desired. There is a fee of $1 for each copy of a license or permit issued by the Chief, Firearms and Explosives Licensing Center under this paragraph. Fee payment must accompany each request for additional copies of a license or permit. The fee must be paid by (1) cash, or (2) money order or check made payable to the Bureau of Alcohol, Tobacco and Firearms.

[T.D. ATF–87, 46 FR 40384, Aug. 7, 1981, as amended by T.D. ATF–290, 54 FR 53054, Dec. 27, 1989]

§ 555.105 Distributions to nonlicensees, nonpermittees, and limited permittees.

(a) Distributions to nonlicensees and nonpermittees prior to May 24, 2003.

(1) This section will apply in any case where distribution of explosive materials to the distributee is not otherwise prohibited by the Act or this part.

(2) Except as provided in paragraph (a)(3) of this section, a licensed importer, licensed manufacturer, or licensed dealer may distribute explosive materials to a nonlicensee or nonpermittee if the nonlicensee or nonpermittee is a resident of the same State in which the licensee's business premises are located, and the nonlicensee or nonpermittee furnishes to the licensee the explosives transaction record, ATF F 5400.4, required by §555.126. Disposition of ATF F 5400.4 will be made in accordance with §555.126.

(3) A licensed importer, licensed manufacturer, or licensed dealer may sell or distribute explosive materials to a resident of a State contiguous to the State in which the licensee's place of business is located if the purchaser's State of residence has enacted legislation, currently in force, specifically authorizing a resident of that State to purchase explosive materials in a contiguous State and the purchaser and the licensee have, prior to the distribution of the explosive materials, complied with all the requirements of paragraphs (a)(2), (a)(5), and (a)(6) of this section applicable to intrastate transactions occurring on the licensee's business premises.

(4) A permittee may dispose of surplus stocks of explosive materials to a nonlicensee or nonpermittee if the nonlicensee or nonpermittee is a resident of the same State in which the permittee's business premises or operations are located, or is a resident of a State contiguous to the State in which the permittee's place of business or operations are located, and if the requirements of paragraphs (a)(2), (a)(3), (a)(5), and (a)(6) of this section are fully met.

(5) A licensed importer, licensed manufacturer, or licensed dealer selling or otherwise distributing explosive materials to a business entity must verify the identity of the representative or agent of the business entity who is authorized to order explosive materials on behalf of the business entity. Each business entity ordering explosive materials must furnish the distributing licensee prior to or with the first order of explosive materials a current certified list of the names of representatives or agents authorized to order explosive materials on behalf of the business entity. The business entity ordering explosive materials is responsible for keeping the certified list current. A licensee may not distribute explosive materials to a business entity on the order of a person whose name does not appear on the certified list.

(6) Where the possession of explosive materials is transferred at the distributor's premises, the distributor must in all instances verify the identity of the person accepting possession on behalf of the distributee before relinquishing possession. Before the delivery at the distributor's premises of explosive materials to an employee of a nonlicensee or nonpermittee, or to an employee of a common or contract carrier transporting explosive materials to a nonlicensee or nonpermittee, the distributor delivering explosive materials must obtain an executed ATF F 5400.8 from the employee before releasing the explosive materials. The ATF F 5400.8 must contain all of the information required on the form and by this part. (See examples in **§555.103(a)).**

(7) A licensee or permittee disposing of surplus stock may sell or distribute commercially manufactured black powder in quantities of 50 pounds or less to a nonlicensee or nonpermittee if the black powder is intended to be used solely for sporting, recreational, or cultural purposes in antique firearms as defined in 18 U.S.C. 921(a)(16), or in antique devices as exempted from the term "destructive device" in 18 U.S.C. 921(a)(4).

(b) Distributions to holders of limited permits on and after May 24, 2003.

(1) This section will apply in any case where distribution of explosive materials to the distributee is not otherwise prohibited by the Act or this part.

(2) A licensed importer, licensed manufacturer or a licensed dealer may distribute explosive materials to a holder of a limited permit if such permittee is a resident of the same State in which the licensee's business premises are located, the holder of the limited permit presents in person or by mail ATF Form 5400.4, Limited Permittee Transaction Report (LPTR), and the licensee completes Form 5400.4 in accordance with §555.126(b). In no event will a licensee distribute explosive materials to a holder of a limited permit unless the holder presents a Form 5400.4 with an original unaltered and unexpired Intrastate Purchase of Explosives Coupon (IPEC), ATF Form 5400.30, affixed. The coupon must bear the name, address, permit number, and the coupon number of the limited permittee seeking distribution of the explosives.

(3) A holder of a limited permit is authorized to receive explosive materials from a licensee or permittee whose premises are located in the same State of residence in which the premises of the holder of the limited permit are located on no more than 6 separate occasions during the one-year period of the permit. For purposes of this section, the term "6 separate occasions" means six deliveries of explosive materials. Each delivery must—

(i) Relate to a single purchase transaction made on one ATF F 5400.4;

(ii) Be referenced on one commercial invoice or purchase order; and

(iii) Be delivered to the holder of the limited permit in one shipment delivered at the same time.

(4) A holder of a user permit may dispose of surplus stocks of explosive materials to a licensee or holder of a user permit, or a holder of a limited permit who is a resident of the same State in which the premises of the holder of the user permit are located. A holder of a limited permit may dispose of surplus stocks of explosive materials to another holder of a limited permit who is a resident of the same State in which the premises of the distributor are located, if the transaction complies with the requirements of paragraph (b)(2) of this section and §555.126(b). A holder of a limited permit may also dispose of surplus stocks of explosive materials to a licensee or holder of a user permit if the disposition occurs in the State of residence of the holder of the limited permit. (See §555.103.)

(5) Each holder of a limited permit ordering explosive materials must furnish the distributing licensee prior to or with the first order of the explosive materials a current list of the names of employees authorized to accept delivery of explosive materials on behalf of the limited permittee. The distributee ordering explosive materials must keep the list current and provide updated lists to licensees and holders of user permits on a timely basis. A licensed importer, licensed manufacturer, licensed dealer, or permittee, selling or otherwise distributing explosive materials to a holder of a limited permit must, prior to delivering the explosive materials, obtain from the limited permittee a current list of persons who are authorized to accept deliveries of explosive materials on behalf of the limited permittee. A licensee or permittee may not deliver explosive materials to a person whose name does not appear on the list.

(6)(i) Delivery at the distributor's premises. Where possession of explosive materials is transferred directly to the distributee at the distributor's premises, the distributor must obtain an executed Form 5400.4 in accordance with §555.126(b) and must in all instances verify the identity of the person accepting possession on behalf of the distributee by examining an identification document (as defined in §555.11) before relinquishing possession.

(ii) Delivery by distributor. Where possession of explosive materials is transferred by the distributor to the distributee away from the distributor's premises, the distributor must obtain an executed Form 5400.4 in accordance with §555.126(b) and must in all instances verify the identity of the person accepting possession on behalf of the distributee by examining an identification document (as defined in §555.11) before relinquishing possession.

(iii) Delivery by common or contract carrier hired by the distributor. Where a common or contract carrier hired by the distributor will transport explosive materials from the distributor to a holder of a limited permit:

(A) The limited permittee must, prior to delivery of the explosive materials, complete the appropriate section on Form 5400.4, affix to the Form 5400.4 one of the six IPECs he has been issued, and provide the form to the distributor in person or by mail.

(B) The distributor must, before relinquishing possession of the explosive materials to the common or contract carrier:

(1) Verify the identity of the person accepting possession for the common or contract carrier by examining such person's valid, unexpired driver's license issued by any State, Canada, or Mexico; and

(2) Record the name of the common or contract carrier (i.e., the name of the driver's employer) and the full name of the driver. This information must be maintained in the distributor's permanent records in accordance with §555.121.

(C) At the time of delivery of the explosive materials, the common or contract carrier, as agent for the distributor, must verify the identity of the person accepting delivery on behalf of the distributee, note the type and number of the identification document (as defined in §555.11) and provide this information to the distributor. The distributor must enter this information in the appropriate section on Form 5400.4.

(iv) Delivery by common or contract carrier hired by the distributee. Where a common or contract carrier hired by the distributee will transport explosive materials from the distributor to a holder of a limited permit:

(A) The limited permittee must, prior to delivery of the explosive materials, complete the appropriate section on Form 5400.4, affix to the Form 5400.4 one of the six IPECs he has been issued, and provide the form to the distributor in person or by mail.

(B) Before the delivery at the distributor's premises to the common or contract carrier who will transport explosive materials to the holder of a limited permit, the distributor must:

(1) Verify the identity of the person accepting possession for the common or contract carrier by examining such person's valid, unexpired driver's license issued by any State, Canada, or Mexico; and

(2) Record the name of the common or contract carrier (i.e., the name of the driver's employer) and the full name of the driver. This information must be maintained in the distributor's permanent records in accordance with §555.121.

(7) A licensee or permittee disposing of surplus stock may sell or distribute commercially manufactured black powder in quantities of 50 pounds or less to a holder of a limited permit, nonlicensee, or nonpermittee if the black powder is intended to be used solely for sporting, recreational, or cultural purposes in antique firearms as defined in 18 U.S.C. 921(a)(16), or in antique devices as exempted from the term "destructive device" in 18 U.S.C. 921(a)(4).

(Approved by the Office of Management and Budget under control number 1140–0075)

[ATF No. 1, 68 FR 13788, Mar. 20, 2003, as amended by ATF No. 2, 68 FR 53513, Sept. 11, 2003]

§ 555.106 Certain prohibited distributions.

(a) A licensee or permittee may not distribute explosive materials to any person except—

(1) A licensee;

(2) A holder of a user permit; or

(3) A holder of a limited permit who is a resident of the State where distribution is made and in which the premises of the transferor are located.

(b) A licensee shall not distribute any explosive materials to any person:

(1) Who the licensee knows is less than 21 years of age;

(2) In any State where the purchase, possession, or use by a person of explosive materials would be in violation of any State law or any published ordinance applicable at the place of distribution;

(3) Who the licensee has reason to believe intends to transport the explosive materials into a State where the purchase, possession, or use of explosive materials is prohibited or which does not permit its residents to transport or ship explosive materials into the State or to receive explosive materials in the State; or

(4) Who the licensee has reasonable cause to believe intends to use the explosive materials for other than a lawful purpose.

(c) A licensee shall not distribute any explosive materials to any person knowing or having reason to believe that the person:

(1) Is, except as provided under §555.142 (d) and (e), under indictment or information for, or was convicted in any court of, a crime punishable by imprisonment for a term exceeding 1 year;

(2) Is a fugitive from justice;

(3) Is an unlawful user of marijuana, or any depressant or stimulant drug, or narcotic drug (as these terms are defined in the Controlled Substances Act, 21 U.S.C. 802);

(4) Was adjudicated as a mental defective or was committed to a mental institution;

(5) Is an alien, other than an alien who—

(i) Is lawfully admitted for permanent residence (as that term is defined in section 101(a)(20) of the Immigration and Nationality Act (8 U.S.C. 1101));

(ii) Is in lawful nonimmigrant status, is a refugee admitted under section 207 of the Immigration and Nationality Act (8 U.S.C. 1157), or is in asylum status under section 208 of the Immigration and Nationality Act (8 U.S.C. 1158), and—

(A) Is a foreign law enforcement officer of a friendly foreign government, as determined by the Attorney General in consultation with the Secretary of State, entering the United States on official law enforcement business, and the shipping, transporting, possession, or receipt of explosive materials is in furtherance of this official law enforcement business;

(B) Is a person having the power to direct or cause the direction of the management and policies of a corporation, partnership, or association licensed pursuant to section 843(a), and the shipping, transporting, possession, or receipt of explosive materials is in furtherance of such power;

(C) Is a member of a North Atlantic Treaty Organization (NATO) or other friendly foreign military force, as determined by the Attorney General in consultation with the Secretary of Defense, (whether or not admitted in a nonimmigrant status) who is present in the United States under military orders for training or other military purpose authorized by the United States, and the shipping, transporting, possession, or receipt of explosive materials is in furtherance of the military purpose; or

(D) Is lawfully present in the United States in cooperation with the Director of Central Intelligence, and the shipment, transportation, receipt, or possession of the explosive materials is in furtherance of such cooperation;

(6) Has been discharged from the armed forces under dishonorable conditions; or

(7) Having been a citizen of the United States, has renounced citizenship.

(d) The provisions of this section do not apply to the purchase of commercially manufactured black powder in quantities not to exceed 50 pounds, intended to be used solely for sporting, recreational, or cultural purposes in antique firearms or in antique devices, if the requirements of §555.105(a)(7) or (b)(7) are fully met.

[T.D. ATF–87, 46 FR 40384, Aug. 7, 1981. Redesignated by T.D. ATF–487, 68 FR 3748, Jan. 24, 2003. ATF No. 1, 68 FR 13790, Mar. 20, 2003]

§ 555.107 Record of transactions.

Each licensee and permittee shall keep records of explosive materials as required by subpart G of this part.

§ 555.108 Importation.

(a) Explosive materials imported or brought into the United States by a licensed importer or holder of a user permit may be released from customs custody to the licensed importer or holder of a user permit upon proof of his status as a licensed importer or holder of a user permit. Proof of status must be made by the licensed importer or holder of a user permit furnishing to the customs officer a certified copy of his license or permit (see §555.103).

(b) A nonlicensee or nonpermittee may import or bring into the United States commercially manufactured black powder in quantities not to exceed 50 pounds. Upon submitting to the customs officer completed ATF F 5400.3, certifying that the black powder is intended to be used solely for sporting, recreational, or cultural purposes in antique firearms or in antique devices, black powder may be released from customs custody. The disposition of the executed ATF F 5400.3 will be in accordance with the instructions on the form.

(c) The provisions of this section are in addition to, and are not in lieu of, any applicable requirement under 27 CFR Part 447.

(d) For additional requirements relating to the importation of plastic explosives into the United States on or after April 24, 1997, see §555.183.

(e) For requirements relating to the marking of imported explosive materials, see §555.109.

[T.D. ATF–87, 46 FR 40384, Aug. 7, 1981, as amended by T.D. ATF–387, 62 FR 8376, Feb. 25, 1997; ATF No. 1, 68 FR 13790, Mar. 20, 2003; ATF 5F, 70 FR 30633, May 27, 2005]

§ 555.109 Identification of explosive materials.

(a) General. Explosive materials, whether manufactured in the United States or imported, must contain certain marks of identification.

(b) Required marks

(1) Licensed manufacturers. Licensed manufacturers who manufacture explosive materials for sale or distribution must place the following marks of identification on explosive materials at the time of manufacture:

(i) The name of the manufacturer; and

(ii) The location, date, and shift of manufacture. Where a manufacturer operates his plant for only one shift during the day, he does not need to show the shift of manufacture.

(2) Licensed importers.

(i) Licensed importers who import explosive materials for sale or distribution must place the following marks of identification on the explosive materials they import:

(A) The name and address (city and state) of the importer; and

(B) The location (city and country) where the explosive materials were manufactured, date, and shift of manufacture. Where the foreign manufacturer operates his plant for only one shift during the day, he does not need to show the shift of manufacture.

(ii) Licensed importers must place the required marks on all explosive materials imported prior to distribution or shipment for use, and in no event later than 15 days after the date of release from Customs custody.

(c) General requirements.

(1) The required marks prescribed in this section must be permanent and legible.

(2) The required marks prescribed in this section must be in the English language, using Roman letters and Arabic numerals.

(3) Licensed manufacturers and licensed importers must place the required marks on each cartridge, bag, or other immediate container of explosive materials that they manufacture or import, as well as on any outside container used for the packaging of such explosive materials.

(4) Licensed manufacturers and licensed importers may use any method, or combination of methods, to affix the required marks to the immediate container of explosive materials, or outside containers used for the packaging thereof, provided the identifying marks are legible, permanent, show all the required information, and are not rendered unreadable by extended periods of storage.

(5) If licensed manufacturers or licensed importers desire to use a coding system and omit printed markings on the container that show all the required information specified in paragraphs (b)(1) and (2) of this section, they must file with ATF a letterhead application displaying the coding that they plan to use and explaining the manner of its application. The Director must approve the application before the proposed coding can be used.

(d) Exceptions —

(1) Blasting caps. Licensed manufacturers or licensed importers are only required to place the identification marks prescribed in this section on the containers used for the packaging of blasting caps.

(2) Alternate means of identification. The Director may authorize other means of identifying explosive materials, including fireworks, upon receipt of a letter application from the licensed manufacturer or licensed importer showing that such other identification is reasonable and will not hinder the effective administration of this part.

(Paragraph (b)(2) approved by the Office of Management and Budget under control number 1140–0055)
[ATF 5F, 70 FR 30633, May 27, 2005]

§ 555.110 Furnishing of samples (Effective on and after January 24, 2003).

(a) In general. Licensed manufacturers and licensed importers and persons who manufacture or import explosive materials or ammonium nitrate must, when required by letter issued by the Director, furnish—

(1) Samples of such explosive materials or ammonium nitrate;

(2) Information on chemical composition of those products; and

(3) Any other information that the Director determines is relevant to the identification of the explosive materials or to identification of the ammonium nitrate.

(b) Reimbursement. The Director will reimburse the fair market value of samples furnished pursuant to paragraph (a) of this section, as well as reasonable costs of shipment.

(Approved by the Office of Management and Budget under control number 1140–0073)
[ATF No. 1, 68 FR 13790, Mar. 20, 2003]

Subpart G—Records and Reports

§ 555.121 General.

(a) **(1)** Licensees and permittees shall keep records pertaining to explosive materials in permanent form (i.e., commercial invoices, record books) and in the manner required in this subpart.

(2) Licensees and permittees shall keep records required by this part on the business premises for five years from the date a transaction occurs or until discontinuance of business or operations by the licensee or permittee. (See also §555.128 for discontinuance of business or operations.)

(b) ATF officers may enter the premises of any licensee or holder of a user permit for the purpose of examining or inspecting any record or document required by or obtained under this part (see §555.24). Section 843(f) of the Act requires licensees and holders of user permits to make all required records available for examination or inspection at all reasonable times. Section 843(f) of the Act also requires licensees and permittees (including holders of limited permits) to submit all reports and information relating to all required records and their contents, as the regulations in this part prescribe.

(c) Each licensee and permittee shall maintain all records of importation, production, shipment, receipt, sale, or other disposition, whether temporary or permanent, of explosive materials as the regulations in this part prescribe. Sections 842(f) and 842(g) of the Act make it unlawful for any licensee or permittee knowingly to make any false entry in, or fail to make entry in, any record required to be kept under the Act and the regulations in this part.

(Approved by the Office of Management and Budget under control number 1512–0373)
[T.D. ATF–87, 46 FR 40384, Aug. 7, 1981, as amended by T.D. ATF–172, 49 FR 14941, Apr. 16, 1984; ATF No. 1, 68 FR 13790, Mar. 20, 2003]

§ 555.122 Records maintained by licensed importers.

(a) Each licensed importer shall take true and accurate physical inventories which will include all explosive materials on hand required to be accounted for in the records kept under this part. The licensed importer shall take a special inventory

(1) At the time of commencing business, which is the effective date of the license issued upon original qualification under this part;

(2) At the time of changing the location of his business to another region;

(3) At the time of discontinuing business; and

(4) At any time the regional director (compliance) may in writing require. Each special inventory is to be prepared in duplicate, the original of which is submitted to the regional director (compliance), and the duplicate retained by the licensed importer. If a special inventory specified by paragraphs (a) (1) through (4) of this section has not been taken during the calendar year, at least one physical inventory will be taken. However, the record of the yearly inventory, other than a special inventory required by paragraphs (a) (1) through (4) of this section, will remain on file for inspection instead of being sent to the regional director (compliance). (See also §555.127.)

(b) Each licensed importer shall, not later than the close of the next business day following the date of importation or other acquisition of explosive materials, enter the following information in a separate record:

(1) Date of importation or other acquisition.

(2) Name or brand name of manufacturer and country of manufacture.

(3) Manufacturer's marks of identification.

(4) Quantity (applicable quantity units, such as pounds of explosives, number of detonators, number of display fireworks, etc.).

(5) Description (dynamite (dyn), blasting agents (ba), detonators (det), display fireworks (df), etc.) and size (length and diameter or diameter only of display fireworks).

(c) Each licensed importer shall, not later than the close of the next business day following the date of distribution of any explosive materials to another licensee or a permittee, enter in a separate record the following information:

(1) Date of disposition.

(2) Name or brand name of manufacturer and country of manufacture.

(3) Manufacturer's marks of identification.

(4) Quantity (applicable quantity units, such as pounds of explosives, number of detonators, number of display fireworks, etc.).

(5) Description (dynamite (dyn), blasting agents (ba), detonators (det), display fireworks (df), etc.) and size (length and diameter or diameter only of display fireworks).

(6) License or permit number of licensee or permittee to whom the explosive materials are distributed.

(d) The regional director (compliance) may authorize alternate records to be maintained by a licensed importer to record his distribution of explosive materials when it is shown by the licensed importer that alternate records will accurately and readily disclose the information required by paragraph (c) of this section. A licensed importer who proposes to use alternate records shall submit a letter application to the regional director (compliance) and shall describe the proposed alternate records and the need for them. Alternate records are not to be employed by the licensed importer until approval is received from the regional director (compliance).

(e) Each licensed importer shall maintain separate records of the sales or other distribution made of explosive materials to nonlicensees or nonpermittees. These records are maintained as prescribed by §555.126.

(Approved by the Office of Management and Budget under control number 1512–0373)

[T.D. ATF–87, 46 FR 40384, Aug. 7, 1981, as amended by T.D. ATF–172, 49 FR 14941, Apr. 16, 1984; T.D. ATF–293, 55 FR 3721, Feb. 5, 1990; T.D. ATF–400, 63 FR 45003, Aug. 24, 1998]

§ 555.123 Records maintained by licensed manufacturers.

(a) Each licensed manufacturer shall take true and accurate physical inventories which will include all explosive materials on hand required to be accounted for in the records kept under this part. The licensed manufacturer shall take a special inventory

(1) At the time of commencing business, which is the effective date of the license issued upon original qualification under this part;

(2) At the time of changing the location of his premises to another region;

(3) At the time of discontinuing business; and

(4) At any other time the regional director (compliance) may in writing require. Each special inventory is to be prepared in duplicate, the original of which is submitted to the regional director (compliance), and the duplicate retained by the licensed manufacturer. If a special inventory required by paragraphs (a) (1) through (4) of this section has not been taken during the calendar year, at least one physical inventory will be taken. However, the record of the yearly inventory, other than a special inventory required by paragraphs (a) (1) through (4) of this section, will remain on file for inspection instead of being sent to the regional director (compliance). (See also §555.127.)

(b) Each licensed manufacturer shall not later than the close of the next business day following the date of manufacture or other acquisition of explosive materials, enter the following information in a separate record:

(1) Date of manufacture or other acquisition.

(2) Manufacturer's marks of identification.

(3) Quantity (applicable quantity units, such as pounds of explosives, number of detonators, number of display fireworks, etc.).

(4) Name, brand name or description (dynamite (dyn), blasting agents (ba), detonators (det), display fireworks (df), etc.) and size (length and diameter or diameter only of display fireworks).

(c) Each licensed manufacturer shall, not later than the close of the next business day following the date of distribution of any explosive materials to another licensee or a permittee, enter in a separate record the following information:

(1) Date of disposition.

(2) Name or brand name of manufacturer or name of importer, as applicable, if acquired other than by his own manufacture.

(3) Manufacturer's marks of identification.

(4) Quantity (applicable quantity units, such as pounds of explosives, number of detonators, number of display fireworks, etc.).

(5) Description (dynamite (dyn), blasting agents (ba), detonators (det), display fireworks (df), etc.) and size (length and diameter or diameter only of display fireworks).

(6) License or permit number of licensee or permittee to whom the explosive materials are distributed.

(d) Each licensed manufacturer who manufactures explosive materials for his own use shall, not later than the close of the next business day following the date of use, enter in a separate record the following information:

(1) Date of use.

(2) Quantity (applicable quantity units, such as pounds of explosives, number of detonators, number of special fireworks, etc.).

(3) Description (dynamite (dyn), blasting agents (ba), detonators (det), display fireworks (df), etc.) and size (length and diameter or diameter only of display fireworks).

Exception: A licensed manufacturer is exempt from the recordkeeping requirements of this subsection if the explosive materials are manufactured for his own use and used within a 24 hour period at the same site.

(e) The regional director (compliance) may authorize alternate records to be maintained by a licensed manufacturer to record his distribution or use of explosive materials when it is shown by the licensed manufacturer that alternate records will accurately and readily disclose the information required by paragraph (c) of this section. A licensed manufacturer who proposes to use alternate records shall submit a letter application to the regional director (compliance) and shall describe the proposed alternate records and the need for them. Alternate records are not to be employed by the licensed manufacturer until approval is received from the regional director (compliance).

(f) Each licensed manufacturer shall maintain separate records of the sales or other distribution made of explosive materials to nonlicensees or nonpermittees. These records are maintained as prescribed by §555.126.

(Approved by the Office of Management and Budget under control number 1512–0373)

[T.D. ATF–87, 46 FR 40384, Aug. 7, 1981, as amended by T.D. ATF–172, 49 FR 14941, Apr. 16, 1984; T.D. ATF–293, 55 FR 3721, Feb. 5, 1990; T.D. ATF–400, 63 FR 45003, Aug. 24, 1998]

§ 555.124 Records maintained by licensed dealers.

(a) Each licensed dealer shall take true and accurate physical inventories which will include all explosive materials on hand required to be accounted for in the records kept under this part. The licensed dealer shall take a special inventory

(1) At the time of commencing business, which is the effective date of the license issued upon original qualification under this part;

(2) At the time of changing the location of his premises to another region;

(3) At the time of discontinuing business; and

(4) At any other time the regional director (compliance) may in writing require. Each special inventory is to be prepared in duplicate, the original of which is submitted to the regional director (compliance), and the duplicate retained by the licensed dealer. If a special inventory required by paragraphs (a)(1) through (4) of this section has not been taken during the calendar year, at least one physical inventory will be taken. However, the record of the yearly inventory, other than a special inventory required by paragraphs (a)(1) through (4) of this section, will remain on file for inspection instead of being sent to the regional director (compliance). (See also §555.127.)

(b) Each licensed dealer shall, not later than the close of the next business day following the date of purchase or other acquisition of explosive materials (except as provided in paragraph (d) of this section), enter the following information in a separate record:

(1) Date of acquisition.

(2) Name or brand name of manufacturer and name of importer (if any).

(3) Manufacturer's marks of identification.

(4) Quantity (applicable quantity units, such as pounds of explosives, number of detonators, number of display fireworks, etc.).

(5) Description (dynamite (dyn), blasting agents (ba), detonators (det), display fireworks (df), etc.) and size (length and diameter or diameter only of display fireworks).

(6) Name, address, and license or permit number of the person from whom the explosive materials are received.

(c) Each licensed dealer shall, not later than the close of the next business day following the date of use (if the explosives are used by the dealer) or the date of distribution of any explosive materials to another licensee or a permittee (except as provided in paragraph (d) of this section), enter in a separate record the following information:

(1) Date of disposition.

(2) Name or brand name of manufacturer and name of importer (if any).

(3) Manufacturer's marks of identification.

(4) Quantity (applicable quantity units, such as pounds of explosives, number of detonators, number of display fireworks, etc.).

(5) Description (dynamite (dyn), blasting agents (ba), detonators (det), display fireworks (df), etc.) and size (length and diameter or diameter only of display fireworks).

(6) License or permit number of licensee or permittee to whom the explosive materials are distributed.

(d) When a commercial record is kept by a licensed dealer showing the purchase or other acquisition information required for the permanent record prescribed by paragraph (b) of this section, or showing the distribution information required for the permanent record prescribed by paragraph (c) of this section, the licensed dealer acquiring or distributing the explosive materials may, for a period not exceeding seven days following the date of acquisition of distribution of the explosive materials, delay making the required entry into the permanent record of acquisition or distribution. However, until the required entry of acquisition or disposition is made in the permanent record, the commercial record must be (1) kept by the licensed dealer separate from other commercial documents kept by the licensee, and (2) readily available for inspection on the licensed premises.

(e) The regional director (compliance) may authorize alternate records to be maintained by a licensed dealer to record his acquisition or disposition of explosive materials, when it is shown by the licensed dealer that alternate records will accurately and readily disclose the required information. A licensed dealer who proposes to use alternate records shall submit a letter application to the regional director (compliance) and shall describe the proposed alternate records and the need for them. Alternate records are not to be employed by the licensed dealer until approval is received from the regional director (compliance).

(f) Each licensed dealer shall maintain separate records of the sales or other distribution made of explosive materials to nonlicensees or nonpermittees. These records are maintained as prescribed by §555.126.

(Approved by the Office of Management and Budget under control number 1512–0373)

[T.D. ATF–87, 46 FR 40384, Aug. 7, 1981, as amended by T.D. ATF–172, 49 FR 14941, Apr. 16, 1984; T.D. ATF–293, 55 FR 3721, Feb. 5, 1990; T.D. ATF–400, 63 FR 45003, Aug. 24, 1998]

§ 555.125 Records maintained by permittees.

(a) Records maintained by permittees prior to May 24, 2003.

(1) Each permittee must take true and accurate physical inventories that will include all explosive materials on hand required to be accounted for in the records kept under this part. The permittee must take a special inventory—

(i) At the time of commencing business, which is the effective date of the permit issued upon original qualification under this part;

(ii) At the time of changing the location of his premises to another region;

(iii) At the time of discontinuing business; and

(iv) At any other time the regional director (compliance) may in writing require. Each special inventory is to be prepared in duplicate, the original of which is submitted to the regional director (compliance) and the duplicate retained by the permittee. If a special inventory required by paragraphs (a)(1)(i) through (iv) of this section has not been taken during the calendar year, a permittee is required to take at least one physical inventory. However, the record of the yearly inventory, other than a special inventory required by paragraphs (a)(1)(i) through (iv) of this section, will remain on file for inspection instead of being sent to the regional director (compliance). (See also §555.127.)

(2) Each permittee must, not later than the close of the next business day following the date of acquisition of explosive materials, enter the following information in a separate record:

(i) Date of acquisition;

(ii) Name or brand name of manufacturer;

(iii) Manufacturer's marks of identification;

(iv) Quantity (applicable quantity units, such as pounds of explosives, number of detonators, number of display fireworks, etc.);

(v) Description (dynamite (dyn), blasting agents (ba), detonators (det), display fireworks (df), etc., and size (length and diameter or diameter only of display fireworks)); and

(vi) Name, address, and license number of the persons from whom the explosive materials are received.

(3) Each permittee must, not later than the close of the next business day following the date of disposition of surplus explosive materials to another permittee or a licensee, enter in a separate record the information prescribed in §555.124(c).

(4) Each permittee must maintain separate records of disposition of surplus stocks of explosive materials to nonlicensees or nonpermittees as prescribed in §555.126.

(5) The regional director (compliance) may authorize alternate records to be maintained by a permittee to record his acquisition of explosive materials, when it is shown by the permittee that alternate records will accurately and readily disclose the required information. A permittee who proposes to use alternate records must submit a letter application to the regional director (compliance) and must describe the proposed alternate records and the need for them. Alternate records are not to be employed by the permittee until approval is received from the regional director (compliance).

(b) Records maintained by permittees on and after May 24, 2003.

(1) Each holder of a user permit must take true and accurate physical inventories that will include all explosive materials on hand required to be accounted for in the records kept under this part. The permittee must take a special inventory—

(i) At the time of commencing business, which is the effective date of the permit issued upon original qualification under this part;

(ii) At the time of changing the location of his premises;

(iii) At the time of discontinuing business; and

(iv) At any other time the regional director (compliance) may in writing require. Each special inventory is to be prepared in duplicate, the original of which is submitted to the regional director (compliance) and the duplicate retained by the permittee. If a special inventory required by paragraphs (b)(1)(i) through (iv) of this section has not been taken during the calendar year, a permittee is required to take at least one physical inventory.

The record of the yearly inventory, other than a special inventory required by paragraphs (b)(1)(i) through (iv) of this section, will remain on file for inspection instead of being sent to the regional director (compliance). (See also §555.127.)

(2) Each holder of a limited permit must take true and accurate physical inventories, at least annually, that will include all explosive materials on hand required to be accounted for in the records kept under this part.

(3) Each holder of a user permit or a limited permit must, not later than the close of the next business day following the date of acquisition of explosive materials, enter the following information in a separate record:

(i) Date of acquisition;

(ii) Name or brand name of manufacturer;

(iii) Manufacturer's marks of identification;

(iv) Quantity (applicable quantity units, such as pounds of explosives, number of detonators, number of display fireworks, etc.);

(v) Description (dynamite (dyn), blasting agents (ba), detonators (det), display fireworks (df), etc., and size (length and diameter or diameter only of display fireworks)); and

(vi) Name, address, and license number of the persons from whom the explosive materials are received.

(4) Each holder of a user permit or a limited permit must, not later than the close of the next business day following the date of disposition of surplus explosive materials to another permittee or a licensee, enter in a separate record the information prescribed in §555.124(c).

(5) When a record book is used as a permittee's permanent record the permittee may delay entry of the required information for a period not to exceed seven days if the commercial record contains all of the required information prescribed by paragraphs (b)(3) and (b)(4) of this section. However, the commercial record may be used instead of a record book as a permanent record provided that the record contains all of the required information prescribed by paragraphs (b)(3) and (b)(4) of this section.

(6) Each holder of a user permit or a limited permit must maintain separate records of disposition of surplus stocks of explosive materials to holders of a limited permit as prescribed in §555.126.

(7) The regional director (compliance) may authorize alternate records to be maintained by a holder of a user permit or a limited permit to record his acquisition of explosive materials, when it is shown by the permittee that alternate records will accurately and readily disclose the required information. A permittee who proposes to use alternate records must submit a letter application to the regional director (compliance) and must describe the proposed alternate records and the need for them. Alternate records are not to be employed by the permittee until approval is received from the regional director (compliance).

(Approved by the Office of Management and Budget under control number 1140–0030)

[ATF No. 1, 68 FR 13790, Mar. 20, 2003]

§ 555.126 Explosives transaction record for distribution of explosive materials prior to May 24, 2003 and Limited Permittee Transaction Report for distribution of explosive materials on and after May 24, 2003.

(a) Explosives transaction record for distribution of explosive materials prior to May 24, 2003.

(1) A licensee or permittee shall not temporarily or permanently distribute explosive materials to any person, other than another licensee or permittee, unless he records the transaction on an explosives transaction record, ATF F 5400.4.

(2) Before the distribution of explosive materials to a nonlicensee or nonpermittee who is a resident of the State in which the licensee or permittee maintains his business premises, or to a nonlicensee or nonpermittee who is not a resident of the State in which the licensee or permittee maintains his business premises and is acquiring explosive materials under §555.105(a)(3), the licensee or permittee distributing the explosive materials shall obtain an executed ATF F 5400.4 from the distributee which contains all of the information required on the form and by the regulations in this part.

(3) Completed ATF F 5400.4 is to be retained by the licensee or permittee as part of his permanent records in accordance with paragraph (a)(4) of this section.

(4) Each ATF F 5400.4 is retained in numerical (by transaction serial number) order commencing with "1" and continuing in regular sequence. When the numbering of any series reaches "1,000,000," the licensee or permittee may recommence the series. The recommenced series is to be given an alphabetical prefix or suffix. Where there is a change in proprietorship, or in the individual, firm, corporate name or trade name, the series in use at the time of the change may be continued.

(5) The requirements of this section are in addition to any other recordkeeping requirement contained in this part.

(6) A licensee or permittee may obtain, upon request, a supply of ATF F 5400.4 from the Director.

(b) Limited Permittee Transaction Report for distribution of explosive materials on and after May 24, 2003.

(1) A licensee or permittee may not distribute explosive materials to any person who is not a licensee or permittee. A licensee or permittee may not distribute explosive materials to a limited permittee unless the distributor records the transaction on ATF Form 5400.4, Limited Permittee Transaction Report.

(2) Before distributing explosive materials to a limited permittee, the licensee or permittee must obtain an executed Form 5400.4 from the limited permittee with an original unaltered and unexpired Intrastate Purchase of Explosives Coupon (IPEC) affixed. Except when delivery of explosive materials is made by a common or contract carrier who is an agent of the limited permittee, the licensee, permittee, or an agent of the licensee or permittee, must verify the identity of the of the holder of the limited permit by examining an identification document (as defined in §555.11) and noting on the Form 5400.4 the type of document presented. The licensee or permittee must complete the appropriate section on Form 5400.4 to indicate the type and quantity of explosive materials distributed, the license or permit number of the seller, and the date of the transaction. The licensee or permittee must sign and date the form and include any other information required by the instructions on the form and the regulations in this part.

(3) One copy of Form 5400.4 must be retained by the distributor as part of his permanent records in accordance with paragraph (b)(4) of this section and for the period specified in §555.121. The distributor must mail the other copy of Form 5400.4 to the Bureau of Alcohol, Tobacco, Firearms and Explosives in accordance with the instructions on the form.

(4) Each Form 5400.4 must be retained in chronological order by date of disposition, or in alphabetical order by name of limited permittee. A licensee may not, however, use both methods in a single recordkeeping system. Where there is a change in proprietorship by a limited permittee, the forms may continue to be filed together after such change.

(5) The requirements of this section are in addition to any other recordkeeping requirement contained in this part.

(Approved by the Office of Management and Budget under control number 1140–0078)

[T.D. ATF–87, 46 FR 40384, Aug. 7, 1981, as amended by T.D. ATF–93, 46 FR 50787, Oct. 15, 1981; T.D. ATF–172, 49 FR 14941, Apr. 16, 1984; T.D. ATF–446, 66 FR 16602, Mar. 27, 2001; ATF No. 1, 68 FR 13791, Mar. 20, 2003]

§ 555.127 Daily summary of magazine transactions.

In taking the inventory required by §§555.122, 555.123, 555.124, and 555.125, a licensee or permittee shall enter the inventory in a record of daily summary transactions to be kept at each magazine of an approved storage facility; however, these records may be kept at one central location on the business premises if separate records of daily transactions are kept for each magazine. Not later than the close of the next business day, each licensee and permittee shall record by manufacturer's name or brand name, the total quantity received in and removed from each magazine during the day, and the total remaining on hand at the end of the day. Quantity entries for display fireworks may be expressed as the number and size of individual display fireworks in a finished state or as the number of packaged display segments or packaged displays. Information as to the number and size of display fireworks contained in any one packaged display segment or packaged display shall be provided to any ATF officer on request. Any discrepancy which might indicate a theft or loss of explosive materials is to be reported in accordance with §555.30.

[T.D. ATF–293, 55 FR 3722, Feb. 5, 1990, as amended by T.D. ATF–400, 63 FR 45003, Aug. 24, 1998]

§ 555.128 Discontinuance of business.

Where an explosive materials business or operations is discontinued and succeeded by a new licensee or new permittee, the records prescribed by this subpart shall appropriately reflect such facts and shall be delivered to the successor. Where discontinuance of the business or operations is absolute, the records required by this subpart must be delivered within 30 days following the business or operations discontinuance to any ATF office located in the region in which the business was located, or to the ATF Out-of-Business Records Center, Spring Mills Office Park, 882 T.J. Jackson Drive, Falling Waters, West Virginia 25419. Where State law or local ordinance requires the delivery of records to other responsible authority, the Chief, Firearms and Explosives Licensing Center may arrange for the delivery of the records required by this subpart to such authority. (See also, §555.61.)

[T.D. ATF–290, 54 FR 53054, Dec. 27, 1989, as amended by T.D. ATF–446a, 66 FR 19089, Apr. 13, 2001; ATF No. 1, 68 FR 13792, Mar. 20, 2003]

§ 555.129 Exportation.

Exportation of explosive materials is to be in accordance with the applicable provisions of section 38 of the Arms Export Control Act (22 U.S.C. 2778) and implementing regulations. However, a licensed importer, licensed manufacturer, or licensed dealer exporting explosive materials shall maintain records showing the manufacture or acquisition of explosive materials as required by this part and records showing the quantity, the manufacturer's name or brand name of explosive materials, the name and address of the foreign consignee of the explosive materials, and the date the explosive materials were exported. See §555.180 for regulations concerning the exportation of plastic explosives.

[T.D. ATF–87, 46 FR 40384, Aug. 7, 1981, as amended by T.D. ATF–387, 62 FR 8377, Feb. 25, 1997]

§ 555.130 [Reserved]

Subpart H—Exemptions

§ 555.141 Exemptions.

(a) General. Except for the provisions of §§555.180 and 555.181, this part does not apply to:

(1) Any aspect of the transportation of explosive materials via railroad, water, highway, or air which is regulated by the U.S. Department of Transportation and its agencies, and which pertains to safety. For example, regulations issued by the Department of Transportation addressing the security risk of aliens transporting explosives by commercial motor or railroad carrier from Canada preclude the enforcement of 18 U.S.C. 842(i)(5) against persons shipping, transporting, receiving, or possessing explosives incident to and in connection with the commercial transportation of explosives by truck or rail from Canada into the United States. Questions concerning this exception should be directed to ATF's Public Safety Branch in Washington, DC.

(2) The use of explosive materials in medicines and medicinal agents in the forms prescribed by the official United States Pharmacopeia or the National Formulary. "The United States Pharmacopeia and The National Formulary," USP and NF Compendia, are available from the United States Pharmacopeial Convention, Inc., 12601 Twinbrook Parkway, Rockville, Maryland 20852.

(3) The transportation, shipment, receipt, or importation of explosive materials for delivery to any agency of the United States or to any State or its political subdivision.

(4) Small arms ammunition and components of small arms ammunition.

(5) The manufacture under the regulation of the military department of the United States of explosive materials for, or their distribution to or storage or possession by, the military or naval services or other agencies of the United States.

(6) Arsenals, navy yards, depots, or other establishments owned by, or operated by or on behalf of, the United States.

(7) The importation, distribution, and storage of fireworks classified as UN0336, UN0337, UN0431, or UN0432 explosives by the U.S. Department of Transportation at 49 CFR 172.101 and generally known as "consumer fireworks" or "articles pyrotechnic."

(8) Gasoline, fertilizers, propellant actuated devices, or propellant actuated industrial tools manufactured, imported, or distributed for their intended purposes.

(9) Industrial and laboratory chemicals which are intended for use as reagents and which are packaged and shipped pursuant to U.S. Department of Transportation regulations, 49 CFR Parts 100 to 177, which do not require explosives hazard warning labels.

(10) Model rocket motors that meet all of the following criteria—

(i) Consist of ammonium perchlorate composite propellant, black powder, or other similar low explosives;

(ii) Contain no more than 62.5 grams of total propellant weight; and

(iii) Are designed as single-use motors or as reload kits capable of reloading no more than 62.5 grams of propellant into a reusable motor casing.

(b) Black powder. Except for the provisions applicable to persons required to be licensed under subpart D, this part does not apply with respect to commercially manufactured black powder in quantities not to exceed 50 pounds, percussion caps, safety and pyrotechnic fuses, quills, quick and slow matches, and friction primers, if the black powder is intended to be used solely for sporting, recreational, or cultural purposes in antique firearms, as defined in 18 U.S.C. 921(a)(16) or antique devices, as exempted from the term "destructive devices" in 18 U.S.C. 921(a)(4).

[T.D. ATF–87, 46 FR 40384, Aug. 7, 1981 as amended by T.D. ATF–87, 46 FR 46916, Sept. 23, 1981; T.D. ATF–293, 55 FR 3722, Feb. 5, 1990; T.D. ATF–387, 62 FR 8377, Feb. 25, 1997; T.D. ATF–400, 63 FR 45003, Aug. 24, 1998; ATF No. 1, 68 FR 13792, Mar. 20, 2003; ATF 6F, 71 FR 46101, Aug. 11, 2006]

§ 555.142 Relief from disabilities (effective January 24, 2003).

(a) Any person prohibited from shipping or transporting any explosive in or affecting interstate or foreign commerce or from receiving or possessing any explosive which has been shipped or transported in or affecting interstate or foreign commerce may make application for relief from disabilities under section 845(b) of the Act.

(b) An application for relief from disabilities must be filed with the Director by submitting ATF Form 5400.29, Application for Restoration of Explosives Privileges, in accordance with the instructions on the form. The application must be supported by appropriate data, including the information specified in paragraph (f) of this section. Upon receipt of an incomplete or improperly executed application for relief, the applicant will be notified of the deficiency in the application. If the application is not corrected and returned within 30 days following the date of notification, the application will be considered abandoned.

(c) **(1)** The Director may grant relief to an applicant if it is established to the satisfaction of the Director that the circumstances regarding the disability and the applicant's record and reputation are such that the applicant will not be likely to act in a manner dangerous to public safety and that the granting of such relief is not contrary to the public interest.

(2) Except as provided in paragraph (c)(3) of this section, the Director will not grant relief if the applicant—

(i) Has not been discharged from parole or probation for a period of at least 2 years;

(ii) Is a fugitive from justice;

(iii) Is a prohibited alien;

(iv) Is an unlawful user of or addicted to any controlled substance;

(v) Has been adjudicated a mental defective or committed to a mental institution, unless the applicant was subsequently determined by a court, board, commission, or other lawful authority to have been restored to mental competency, to be no longer suffering from a mental disorder, and to have had all rights restored; or

(vi) Is prohibited by the law of the State where the applicant resides from receiving or possessing explosive materials.

(3)(i) The Director may grant relief to aliens who have been lawfully admitted to the United States or to persons who have not been discharged from parole or probation for a period of at least 2 years if he determines that the applicant has a compelling need to possess explosives, such as for purposes of employment.

(ii) The Director may grant relief to the persons identified in paragraph (c)(2) of this section in extraordinary circumstances where the granting of such relief is consistent with the public interest.

(d) A person who has been granted relief under this section is relieved of all disabilities imposed by the Act for the disabilities disclosed in the application. The granting of relief will not affect any disabilities incurred subsequent to the date the application was filed. Relief from disabilities granted to aliens will be effective only so long as the alien retains his or her lawful immigration status.

(e)(1) A licensee or permittee who is under indictment or information for, or convicted of, a crime punishable by imprisonment for a term exceeding one year during the term of a current license or permit, or while he has pending a license or permit renewal application, shall not be barred from licensed or permit operations for 30 days after the date of indictment or information or 30 days after the date upon which his conviction becomes final. Also, if he files his application for relief under this section within such 30 day period, he may further continue licensed or permit operations while his application is pending. A licensee or permittee who does not file an application within 30 days from the date of his indictment or information, or within 30 days from the date his conviction becomes final, shall not continue licensed or permit operations beyond 30 days from the date of his indictment or information or beyond 30 days from the date his conviction becomes final.

(2) In the event the term of a license or permit of a person expires during the 30 day period following the date of indictment of information of during the 30 day period after the date upon which his conviction becomes final or while his application for relief is pending, he shall file a timely application for renewal of his license or permit in order to continue licensed or permit operations. The license or permit application is to show that the applicant has been indicted or under information for, or convicted of, a crime punishable by imprisonment for a term exceeding one year.

(3) A licensee or permittee shall not continue licensed or permit operations beyond 30 days following the date the Director issues notification that the licensee's or permittee's application for removal of the disabilities resulting from an indictment, information or conviction has been denied.

(4) When a licensee or permittee may no longer continue licensed or permit operations under this section, any application for renewal of license of permit filed by the licensee or permittee while his application for removal of disabilities resulting from an indictment, information or conviction is pending, will be denied by the regional director (compliance).

(f)(1) Applications for relief from disabilities must include the following information:

(i) In the case of a corporation, or of any person having the power to direct or control the management of the corporation, information as to the absence of culpability in the offense for which the corporation, or any such person, was indicted, formally accused or convicted;

(ii) In the case of an applicant who is an individual, two properly completed FBI Forms FD–258 (fingerprint card), and a written statement from each of three references who are not related to the applicant by blood or marriage and have known the applicant for at least 3 years, recommending the granting of relief;

(iii) Written consent to examine and obtain copies of records and to receive statements and information regarding the applicant's background, including records, statements and other information concerning employment, medical history, military service, immigration status, and criminal record;

(iv) In the case of an applicant having been convicted of a crime punishable by imprisonment for a term exceeding one year, a copy of the indictment or information on which the applicant was convicted, the judgment of conviction or record of any plea of nolo contendere or plea of guilty or finding of guilt by the court;

(v) In the case of an applicant under indictment, a copy of the indictment or information;

(vi) In the case of an applicant who has been adjudicated a mental defective or committed to a mental institution, a copy of the order of a court, board, commission, or other lawful authority that made the adjudication or ordered the commitment, any petition that sought to have the applicant so adjudicated or committed, any medical records reflecting the reasons for commitment and diagnoses of the applicant, and any court order or finding of a court, board, commission, or other lawful authority showing the applicant's discharge from commitment, restoration of mental competency and the restoration of rights;

(vii) In the case of an applicant who has been discharged from the Armed Forces under dishonorable conditions, a copy of the applicant's Certificate of Release or Discharge from Active Duty (Department of Defense Form 214), Charge Sheet (Department of Defense Form 458), and final court martial order;

(viii) In the case of an applicant who, having been a citizen of the United States, has renounced his or her citizenship, a copy of the formal renunciation of nationality before a diplomatic or consular officer of the United States in a foreign state or before an officer designated by the Attorney General when the United States was in a state of war (see 8 U.S.C. 1481(a)(5) and (6)); and

(ix) In the case of an applicant who is an alien, documentation that the applicant is an alien who has been lawfully admitted to the United States; certification from the applicant including the applicant's INS-issued alien number or admission number, country/countries of citizenship, and immigration status, and certifying that the applicant is legally authorized to work in the United States, or other purposes for which possession of explosives is required; certification from an appropriate law enforcement agency of the applicant's country of citizenship stating that the applicant does not have a criminal record; and, if applicable, certification from a Federal explosives licensee or permittee or other employer stating that the applicant is employed by the employer and must possess explosive materials for purposes of employment. These certifications must be submitted in English.

(2) Any record or document of a court or other government entity or official required by paragraph (f)(1) of this section must be certified by the court or other government entity or official as a true copy.

(Approved by the Office of Management and Budget under control number 1140–0076)

[T.D. ATF–87, 46 FR 40384, Aug. 7, 1981. Redesignated by T.D. ATF–487, 68 FR 3748, Jan. 24, 2003. ATF No. 1, 68 FR 13792, Mar. 20, 2003]

Subpart I—Unlawful Acts, Penalties, Seizures and Forfeitures

§ 555.161 Engaging in business without a license.

Any person engaging in the business of importing, manufacturing, or dealing in explosive materials without a license issued under the Act, shall be fined not more than $10,000 or imprisoned not more than 10 years, or both.

§ 555.162 False statement or representation.

Any person who knowingly withholds information or makes any false or fictitious oral or written statement or furnishes or exhibits any false, fictitious, or misrepresented identification, intended or likely to deceive for the purpose of obtaining explosive materials, or a license, permit, exemption, or relief from disability under the Act, shall be fined not more than $10,000 or imprisoned not more than 10 years, or both.

§ 555.163 False entry in record.

Any licensed importer, licensed manufacturer, licensed dealer, or permittee who knowingly makes any false entry in any record required to be kept under subpart G of this part, shall be fined not more than $10,000 or imprisoned not more than 10 years, or both.

[T.D. ATF–87, 46 FR 40384, Aug. 7, 1981, as amended by T.D. ATF–400, 63 FR 45003, Aug. 24, 1998]

§ 555.164 Unlawful storage.

Any person who stores any explosive material in a manner not in conformity with this part, shall be fined not more than $1,000 or imprisoned not more than one year, or both.

§ 555.165 Failure to report theft or loss.

(a) Any person who has knowledge of the theft or loss of any explosive materials from his stock and fails to report the theft or loss within 24 hours of discovery in accordance with §555.30, shall be fined not more than $1,000 or imprisoned not more than one year, or both.

(b) On and after January 24, 2003, any licensee or permittee who fails to report a theft of explosive materials in accordance with §555.30 will be fined under title 18 U.S.C., imprisoned not more than 5 years, or both.

[T.D. ATF–87, 46 FR 40384, Aug. 7, 1981, as amended by ATF No. 1, 68 FR 13793, Mar. 20, 2003]

§ 555.166 Seizure or forfeiture.

Any explosive materials involved or used or intended to be used in any violation of the Act or of this part or in any violation of any criminal law of the United States are subject to seizure and forfeiture, and all provisions of title 26, U.S.C. relating to the seizure, forfeiture, and disposition of firearms, as defined in 26 U.S.C. 5845(a), will, so far as applicable, extend to seizures and forfeitures under the Act. (See §72.27 of this title for regulations on summary destruction of explosive materials which are impracticable or unsafe to remove to a place of storage.)

[T.D. ATF–87, 46 FR 40384, Aug. 7, 1981, as amended by T.D. ATF–363, 60 FR 17449, Apr. 6, 1995]

Subpart J—Marking of Plastic Explosives

§ 555.180 Prohibitions relating to unmarked plastic explosives.

(a) No person shall manufacture any plastic explosive that does not contain a detection agent.

(b) No person shall import or bring into the United States, or export from the United States, any plastic explosive that does not contain a detection agent. This paragraph does not apply to the importation or bringing into the United States, or the exportation from the United States, of any plastic explosive that was imported or brought into, or manufactured in the United States prior to April 24, 1996, by or on behalf of any agency of the United States performing military or police functions (including any military reserve component) or by or on behalf of the National Guard of any State, not later than 15 years after the date of entry into force of the Convention on the Marking of Plastic Explosives with respect to the United States, i.e., not later than June 21, 2013.

(c) No person shall ship, transport, transfer, receive, or possess any plastic explosive that does not contain a detection agent. This paragraph does not apply to:

(1) The shipment, transportation, transfer, receipt, or possession of any plastic explosive that was imported or brought into, or manufactured in the United States prior to April 24, 1996, by any person during the period beginning on that date and ending on April 24, 1999; or

(2) The shipment, transportation, transfer, receipt, or possession of any plastic explosive that was imported or brought into, or manufactured in the United States prior to April 24, 1996, by or on behalf of any agency of the United States performing a military or police function (including any military reserve component) or by or on behalf of the National Guard of any State, not later than 15 years after the date of entry into force of the Convention on the Marking of Plastic Explosives with respect to the United States, i.e., not later than June 21, 2013.

(d) When used in this subpart, terms are defined as follows:

(1) Convention on the Marking of Plastic Explosives means the Convention on the Marking of Plastic Explosives for the Purposes of Detection, Done at Montreal on 1 March 1991.

(2) "Date of entry into force" of the Convention on the Marking of Plastic Explosives means that date on which the Convention enters into force with respect to the U.S. in accordance with the provisions of Article XIII of the Convention on the Marking of Plastic Explosives. The Convention entered into force on June 21, 1998.

(3) Detection agent means any one of the substances specified in this paragraph when introduced into a plastic explosive or formulated in such explosive as a part of the manufacturing process in such a manner as to achieve homogeneous distribution in the finished explosive, including—

(i) Ethylene glycol dinitrate (EGDN), C[2]H[4](NO[3])[2], molecular weight 152, when the minimum concentration in the finished explosive is 0.2 percent by mass;

(ii) 2,3-Dimethyl-2,3-dinitrobutane (DMNB), C[6]H[12](NO[2])[2], molecular weight 176, when the minimum concentration in the finished explosive is 0.1 percent by mass;

(iii) Para-Mononitrotoluene (p-MNT), C[7]H[7]NO[2], molecular weight 137, when the minimum concentration in the finished explosive is 0.5 percent by mass;

(iv) Ortho-Mononitrotoluene (o-MNT), C[7]H[7]NO[2], molecular weight 137, when the minimum concentration in the finished explosive is 0.5 percent by mass; and

(v) Any other substance in the concentration specified by the Director, after consultation with the Secretary of State and Secretary of Defense, that has been added to the table in Part 2 of the Technical Annex to the Convention on the Marking of Plastic Explosives.

(4) Plastic explosive means an explosive material in flexible or elastic sheet form formulated with one or more high explosives which in their pure form has a vapor pressure less than 10<4>Pa at a temperature of 25 degrees C, is formulated with a binder material, and is as a mixture malleable or flexible at normal room temperature. High explosives, as defined in §555.202(a), are explosive materials which can be caused to detonate by means of a blasting cap when unconfined.

[T.D. ATF–387, 62 FR 8376, Feb. 25, 1997, as amended by T.D. ATF–419, 64 FR 55628, Oct. 14, 1999]

§ 555.181 Reporting of plastic explosives.

All persons, other than an agency of the United States (including any military reserve component) or the National Guard of any State, possessing any plastic explosive on April 24, 1996, shall submit a report to the Director no later than August 22, 1996. The report shall be in writing and mailed by certified mail (return receipt requested) to the Director at P.O. Box 50204, Washington, DC 20091–0204. The report shall include the quantity of plastic explosives possessed on April 24, 1996; any marks of identification on such explosives; the name and address of the manufacturer or importer; the storage location of such explosives, including the city and State; and the name and address of the person possessing the plastic explosives.

(Approved by the Office of Management and Budget under control number 1512–0535)

[T.D. ATF–382, 61 FR 38085, July 23, 1996, as amended by T.D. ATF–387, 62 FR 8377, Feb. 25, 1997]

§ 555.182 Exceptions.

It is an affirmative defense against any proceeding involving §§555.180 and 555.181 if the proponent proves by a preponderance of the evidence that the plastic explosive—

(a) Consisted of a small amount of plastic explosive intended for and utilized solely in lawful—

(1) Research, development, or testing of new or modified explosive materials;

(2) Training in explosives detection or development or testing of explosives detection equipment; or

(3) Forensic science purposes; or

(b) Was plastic explosive that, by April 24, 1999, will be or is incorporated in a military device within the territory of the United States and remains an integral part of such military device, or is intended to be, or is incorporated in, and remains an integral part of a military device that is intended to become, or has become, the property of any agency of the United States performing military or police functions (including any military reserve component) or the National Guard of any State, wherever such device is located. For purposes of this paragraph, the term "military device" includes, but is not restricted to, shells, bombs, projectiles, mines, missiles, rockets, shaped charges, grenades, perforators, and similar devices lawfully manufactured exclusively for military or police purposes.
[T.D. ATF–387, 62 FR 8377, Feb. 25, 1997]

§ 555.183 Importation of plastic explosives on or after April 24, 1997.

Persons filing Form 6 applications for the importation of plastic explosives on or after April 24, 1997, shall attach to the application the following written statement, prepared in triplicate, executed under the penalties of perjury:

(a) "I declare under the penalties of perjury that the plastic explosive to be imported contains a detection agent as required by 27 CFR 555.180(b)"; or

(b) "I declare under the penalties of perjury that the plastic explosive to be imported is a "small amount" to be used for research, training, or testing purposes and is exempt from the detection agent requirement pursuant to 27 CFR 555.182."
[T.D. ATF–387, 62 FR 8377, Feb. 25, 1997]

§ 555.184 Statements of process and samples.

(a) A complete and accurate statement of process with regard to any plastic explosive or to any detection agent that is to be introduced into a plastic explosive or formulated in such plastic explosive shall be submitted by a licensed manufacturer or licensed importer, upon request, to the Director.

(b) Samples of any plastic explosive or detection agent shall be submitted by a licensed manufacturer or licensed importer, upon request, to the Director.

(Paragraph (a) approved by the Office of Management and Budget under control number 1512–0539)
[T.D. ATF–387, 62 FR 8378, Feb. 25, 1997]

§ 555.185 Criminal sanctions.

Any person who violates the provisions of 18 U.S.C. 842(l)–(o) shall be fined under title 18, U.S.C., imprisoned for not more than 10 years, or both.
[T.D. ATF–387, 62 FR 8378, Feb. 25, 1997]

§ 555.186 Seizure or forfeiture.

Any plastic explosive that does not contain a detection agent in violation of 18 U.S.C. 842(l)–(n) is subject to seizure and forfeiture, and all provisions of 19 U.S.C. 1595a, relating to seizure, forfeiture, and disposition of merchandise introduced or attempted to be introduced into the U.S. contrary to law, shall extend to seizures and forfeitures under this subpart. See §72.27 of this chapter for regulations on summary destruction of plastic explosives that do not contain a detection agent.
[T.D. ATF–387, 62 FR 8378, Feb. 25, 1997]

Subpart K—Storage

§ 555.201 General.

(a) Section 842(j) of the Act and §555.29 of this part require that the storage of explosive materials by any person must be in accordance with the regulations in this part. Further, section 846 of this Act authorizes regulations to prevent the recurrence of accidental explosions in which explosive materials were involved. The storage standards prescribed by this subpart confer no right or privileges to store explosive materials in a manner contrary to State or local law.

(b) The Director may authorize alternate construction for explosives storage magazines when it is shown that the alternate magazine construction is substantially equivalent to the standards of safety and security contained in this subpart. Any alternate explosive magazine construction approved by the Director prior to August 9, 1982, will continue as approved unless notified in writing by the Director. Any person intending to use alternate magazine construction shall submit a letter application to the regional director (compliance) for transmittal to the Director, specifically describing the proposed magazine. Explosive materials may not be stored in alternate magazines before the applicant has been notified that the application has been approved.

(c) A licensee or permittee who intends to make changes in his magazines, or who intends to construct or acquire additional magazines, shall comply with §555.63.

(d) The regulations set forth in §§555.221 through 555.224 pertain to the storage of display fireworks, pyrotechnic compositions, and explosive materials used in assembling fireworks and articles pyrotechnic.

(e) The provisions of §555.202(a) classifying flash powder and bulk salutes as high explosives are mandatory after March 7, 1990: Provided, that those persons who hold licenses or permits under this part on that date shall, with respect to the premises covered by such licenses or permits, comply with the high explosives storage requirements for flash powder and bulk salutes by March 7, 1991.

(f) Any person who stores explosive materials shall notify the authority having jurisdiction for fire safety in the locality in which the explosive materials are being stored of the type, magazine capacity, and location of each site where such explosive materials are stored. Such notification shall be made orally before the end of the day on which storage of the explosive materials commenced and in writing within 48 hours from the time such storage commenced.

(Paragraph (f) approved by the Office of Management and Budget under control number 1512–0536)

[T.D. ATF–87, 46 FR 40384, Aug. 7, 1981, as amended by T.D. ATF–293, 55 FR 3722, Feb. 5, 1990; T.D. ATF–400, 63 FR 45003, Aug. 24, 1998]

§ 555.202 Classes of explosive materials.

For purposes of this part, there are three classes of explosive materials. These classes, together with the description of explosive materials comprising each class, are as follows:

(a) High explosives. Explosive materials which can be caused to detonate by means of a blasting cap when unconfined, (for example, dynamite, flash powders, and bulk salutes). See also §555.201(e).

(b) Low explosives. Explosive materials which can be caused to deflagrate when confined (for example, black powder, safety fuses, igniters, igniter cords, fuse lighters, and "display fireworks" classified as UN0333, UN0334, or UN0335 by the U.S. Department of Transportation regulations at 49 CFR 172.101, except for bulk salutes).

(c) Blasting agents. (For example, ammonium nitrate-fuel oil and certain water-gels (see also §555.11).

[T.D. ATF–87, 46 FR 40384, Aug. 7, 1981, as amended by T.D. ATF–293, 55 FR 3722, Feb. 5, 1990; T.D. ATF–400, 63 FR 45003, Aug. 24, 1998]

§ 555.203 Types of magazines.

For purposes of this part, there are five types of magazines. These types, together with the classes of explosive materials, as defined in §555.202, which will be stored in them, are as follows:

(a) Type 1 magazines. Permanent magazines for the storage of high explosives, subject to the limitations prescribed by §§555.206 and 555.213. Other classes of explosive materials may also be stored in type 1 magazines.

(b) Type 2 magazines. Mobile and portable indoor and outdoor magazines for the storage of high explosives, subject to the limitations prescribed by §§555.206, 555.208(b), and 555.213. Other classes of explosive materials may also be stored in type 2 magazines.

(c) Type 3 magazines. Portable outdoor magazines for the temporary storage of high explosives while attended (for example, a "day-box"), subject to the limitations prescribed by §§555.206 and 555.213. Other classes of explosives materials may also be stored in type 3 magazines.

(d) Type 4 magazines. Magazines for the storage of low explosives, subject to the limitations prescribed by §§555.206(b), 555.210(b), and 555.213. Blasting agents may be stored in type 4 magazines, subject to the limitations prescribed by §§555.206(c), 555.211(b), and 555.213. Detonators that will not mass detonate may also be stored in type 4 magazines, subject to the limitations prescribed by §§555.206(a), 555.210(b), and 555.213.

(e) Type 5 magazines. Magazines for the storage of blasting agents, subject to the limitations prescribed by §§555.206(c), 555.211(b), and 555.213.

§ 555.204 Inspection of magazines.

Any person storing explosive materials shall inspect his magazines at least every seven days. This inspection need not be an inventory, but must be sufficient to determine whether there has been unauthorized entry or attempted entry into the magazines, or unauthorized removal of the contents of the magazines.

§ 555.205 Movement of explosive materials.

All explosive materials must be kept in locked magazines meeting the standards in this subpart unless they are:

(a) In the process of manufacture;

(b) Being physically handled in the operating process of a licensee or user;

(c) Being used; or

(d) Being transported to a place of storage or use by a licensee or permittee or by a person who has lawfully acquired explosive materials under §555.106.

§ 555.206 Location of magazines.

(a) Outdoor magazines in which high explosives are stored must be located no closer to inhabited buildings, passenger railways, public highways, or other magazines in which high explosives are stored, than the minimum distances specified in the table of distances for storage of explosive materials in §555.218.

(b) Outdoor magazines in which low explosives are stored must be located no closer to inhibited buildings, passenger railways, public highways, or other magazines in which explosive materials are stored, than the minimum distances specified in the table of distances for storage of low explosives in §555.219, except that the table of distances in §555.224 shall apply to the storage of display fireworks. The distances shown in §555.219 may not be reduced by the presence of barricades.

(c)(1) Outdoor magazines in which blasting agents in quantities of more than 50 pounds are stored must be located no closer to inhabited buildings, passenger railways, or public highways than the minimum distances specified in the table of distances for storage of explosive materials in §555.218.

(2) Ammonium nitrate and magazines in which blasting agents are stored must be located no closer to magazines in which high explosives or other blasting agents are stored than the minimum distances specified in the table of distances for the separation of ammonium nitrate and blasting agents in §555.220. However, the minimum distances for magazines in which explosives and blasting agents are stored from inhabited buildings, etc., may not be less than the distances specified in the table of distances for storage of explosives materials in §555.218.

[T.D. ATF–87, 46 FR 40384, Aug. 7, 1981, as amended by T.D. ATF–293, 55 FR 3722, Feb. 5, 1990; T.D. ATF–400, 63 FR 45003, Aug. 24, 1998]

§ 555.207 Construction of type 1 magazines.

A type 1 magazine is a permanent structure: a building, an igloo or "Army-type structure", a tunnel, or a dugout. It is to be bullet-resistant, fire-resistant, weather-resistant, theft-resistant, and ventilated.

(a) Buildings. All building type magazines are to be constructed of masonry, wood, metal, or a combination of these materials, and have no openings except for entrances and ventilation. The ground around building magazines must slope away for drainage or other adequate drainage provided.

(1) Masonry wall construction. Masonry wall construction is to consist of brick, concrete, tile, cement block, or cinder block and be not less than 6 inches in thickness. Hollow masonry units used in construction must have all hollow spaces filled with well-tamped, coarse, dry sand or weak concrete (at least a mixture of one part cement and eight parts of sand with enough water to dampen the mixture while tamping in place). Interior walls are to be constructed of, or covered with, a nonsparking material.

(2) Fabricated metal wall construction. Metal wall construction is to consist of sectional sheets of steel or aluminum not less than number 14-gauge, securely fastened to a metal framework. Metal wall construction is either lined inside with brick, solid cement blocks, hardwood not less than four inches thick, or will have at least a six inch sand fill between interior and exterior walls. Interior walls are to be constructed of, or covered with, a nonsparking material.

(3) Wood frame wall construction. The exterior of outer wood walls is to be covered with iron or aluminum not less than number 26-gauge. An inner wall of, or covered with nonsparking material will be constructed so as to provide a space of not less than six inches between the outer and inner walls. The space is to be filled with coarse, dry sand or weak concrete.

(4) Floors. Floors are to be constructed of, or covered with, a nonsparking material and shall be strong enough to bear the weight of the maximum quantity to be stored. Use of pallets covered with a nonsparking material is considered equivalent to a floor constructed of or covered with a nonsparking material.

(5) Foundations. Foundations are to be constructed of brick, concrete, cement block, stone, or wood posts. If piers or posts are used, in lieu of a continuous foundation, the space under the buildings is to be enclosed with metal.

(6) Roof. Except for buildings with fabricated metal roofs, the outer roof is to be covered with no less than number 26-guage iron or aluminum, fastened to at least ⅞ inch sheathing.

(7) Bullet-resistant ceilings or roofs. Where it is possible for a bullet to be fired directly through the roof and into the magazine at such an angle that the bullet would strike the explosives within, the magazine is to be protected by one of the following methods:

(i) A sand tray lined with a layer of building paper, plastic, or other nonporous material, and filled with not less than four inches of coarse, dry sand, and located at the tops of inner walls covering the entire ceiling area, except that portion necessary for ventilation.

(ii) A fabricated metal roof constructed of 3⁄16-inch plate steel lined with four inches of hardwood. (For each additional ⅙ inch of plate steel, the hardwood lining may be decreased one inch.)

(8) Doors. All doors are to be constructed of not less than ¼ inch plate steel and lined with at least two inches of hardwood. Hinges and hasps are to be attached to the doors by welding, riveting or bolting (nuts on inside of door). They are to be installed in such a manner that the hinges and hasps cannot be removed when the doors are closed and locked.

(9) Locks. Each door is to be equipped with (i) two mortise locks; (ii) two padlock fastened in separate hasps and staples; (iii) a combination of a mortise lock and a padlock; (iv) a mortise lock that requires two keys to open; or (v) a three-point lock. Padlocks must have at least five tumblers and a casehardened shackle of at least ⅜ inch diameter. Padlocks must be protected with not less than ¼ inch steel hoods constructed so as to prevent sawing or lever action on the locks, hasps, and staples. These requirements do not apply to magazine doors that are adequately secured on the inside by means of a bolt, lock, or bar that cannot be actuated from the outside.

(10) Ventilation. Ventilation is to be provided to prevent dampness and heating of stored explosive materials. Ventilation openings must be screened to prevent the entrance of sparks. Ventilation openings in side walls and foundations must be offset or shielded for bullet-resistant purposes. Magazines having foundation and roof ventilators with the air circulating between the side walls and the floors and between the side walls and the ceiling must have a wooden lattice lining or equivalent to prevent the packages of explosive materials from being stacked against the side walls and blocking the air circulation.

(11) Exposed metal. No sparking material is to be exposed to contact with the stored explosive materials. All ferrous metal nails in the floor and side walls, which might be exposed to contact with explosive materials, must be blind nailed, countersunk, or covered with a nonsparking lattice work or other nonsparking material.

(b) Igloos, "Army-type structures", tunnels, and dugouts. Igloo, "Army-type structure", tunnel, and dugout magazines are to be constructed of reinforced concrete, masonry, metal, or a combination of these materials. They must have an earthmound covering of not less than 24 inches on the top, sides and rear unless the magazine meets the requirements of paragraph (a)(7) of this section. Interior walls and floors must be constructed of, or covered with, a nonsparking material. Magazines of this type are also to be constructed in conformity with the requirements of paragraph (a)(4) and paragraphs (a)(8) through (11) of this section.

§ 555.208 Construction of type 2 magazines.

A type 2 magazine is a box, trailer, semitrailer, or other mobile facility.

(a) Outdoor magazines—

(1) General. Outdoor magazines are to be bullet-resistant, fire-resistant, weather-resistant, theft-resistant, and ventilated. They are to be supported to prevent direct contact with the ground and, if less than one cubic yard in size, must be securely fastened to a fixed object. The ground around outdoor magazines must slope away for drainage or other adequate drainage provided. When unattended, vehicular magazines must have wheels removed or otherwise effectively immobilized by kingpin locking devices or other methods approved by the Director.

(2) Exterior construction. The exterior and doors are to be constructed of not less than ¼-inch steel and lined with at least two inches of hardwood. Magazines with top openings will have lids with water-resistant seals or which overlap the sides by at least one inch when in a closed position.

(3) Hinges and hasps. Hinges and hasps are to be attached to doors by welding, riveting, or bolting (nuts on inside of door). Hinges and hasps must be installed so that they cannot be removed when the doors are closed and locked.

(4) Locks. Each door is to be equipped with (i) two mortise locks; (ii) two padlocks fastened in separate hasps and staples; (iii) a combination of a mortise lock and a padlock; (iv) a mortise lock that requires two keys to open; or (v) a three-point lock. Padlocks must have at least five tumblers and a case-hardened shackle of at least ⅜-inch diameter. Padlocks must be protected with not less than ¼-inch steel hoods constructed so as to prevent sawing or lever action on the locks, hasps, and staples. These requirements do not apply to magazine doors that are adequately secured on the inside by means of a bolt, lock, or bar that cannot be actuated from the outside.

(b) Indoor magazines—

(1) General. Indoor magazines are to be fire-resistant and theft-resistant. They need not be bullet-resistant and weather-resistant if the buildings in which they are stored provide protection from the weather and from bullet penetration. No indoor magazine is to be located in a residence or dwelling. The indoor storage of high explosives must not exceed a quantity of 50 pounds. More than one indoor magazine may be located in the same building if the total quantity of explosive materials stored does not exceed 50 pounds. Detonators must be stored in a separate magazine (except as provided in §555.213) and the total quantity of detonators must not exceed 5,000.

(2) Exterior construction. Indoor magazines are to be constructed of wood or metal according to one of the following specifications:

(i) Wood indoor magazines are to have sides, bottoms and doors constructed of at least two inches of hardwood and are to be well braced at the corners. They are to be covered with sheet metal of not less than number 26-gauge (.0179 inches). Nails exposed to the interior of magazines must be countersunk.

(ii) Metal indoor magazines are to have sides, bottoms and doors constructed of not less than number 12-gauge (.1046 inches) metal and be lined inside with a nonsparking material. Edges of metal covers must overlap sides at least one inch.

(3) Hinges and hasps. Hinges and hasps are to be attached to doors by welding, riveting, or bolting (nuts on inside of door). Hinges and hasps must be installed so that they cannot be removed when the doors are closed and locked.

(4) Locks. Each door is to be equipped with (i) two mortise locks; (ii) two padlocks fastened in separate hasps and staples; (iii) a combination of a mortise lock and a padlock; (iv) a mortise lock that requires two keys to open; or (v) a three-point lock. Padlocks must have at least five tumblers and a case-hardened shackle of at least ⅜-inch diameter. Padlocks must be protected with not less than ¼-inch steel hoods constructed so as to prevent sawing or lever action on the locks, hasps, and staples. Indoor magazines located in secure rooms that are locked as provided in this subparagraph may have each door locked with one steel padlock (which need not be protected by a steel hood) having at least five tumblers and a case-hardened shackle of at least ⅜-inch diameter, if the door hinges and lock hasp are securely fastened to the magazine.

These requirements do not apply to magazine doors that are adequately secured on the inside by means of a bolt, lock, or bar that cannot be actuated from the outside.

(c) Detonator boxes. Magazines for detonators in quantities of 100 or less are to have sides, bottoms and doors constructed of not less than number 12-gauge (.1046 inches) metal and lined with a nonsparking material. Hinges and hasps must be attached so they cannot be removed from the outside. One steel padlock (which need not be protected by a steel hood) having at least five tumblers and a case-hardened shackle of at least ⅜-inch diameter is sufficient for locking purposes.

§ 555.209 Construction of type 3 magazines.

A type 3 magazine is a "day-box" or other portable magazine. It must be fire-resistant, weather-resistant, and theft-resistant. A type 3 magazine is to be constructed of not less than number 12-gauge (.1046 inches) steel, lined with at least either ½-inch plywood or ½-inch Masonite-type hardboard.

Doors must overlap sides by at least one inch. Hinges and hasps are to be attached by welding, riveting or bolting (nuts on inside).

One steel padlock (which need not be protected by a steel hood) having at least five tumblers and a case-hardened shackle of at least ⅜-inch diameter is sufficient for locking purposes. Explosive materials are not to be left unattended in type 3 magazines and must be removed to type 1 or 2 magazines for unattended storage.

§ 555.210 Construction of type 4 magazines.

A type 4 magazine is a building, igloo or "Army-type structure", tunnel, dugout, box, trailer, or a semitrailer or other mobile magazine.

(a) Outdoor magazines—

(1) General. Outdoor magazines are to be fire-resistant, weather-resistant, and theft-resistant. The ground around outdoor magazines must slope away for drainage or other adequate drainage be provided. When unattended, vehicular magazines must have wheels removed or otherwise be effectively immobilized by kingpin locking devices or other methods approved by the Director.

(2) Construction. Outdoor magazines are to be constructed of masonry, metal-covered wood, fabricated metal, or a combination of these materials. Foundations are to be constructed of brick, concrete, cement block, stone, or metal or wood posts. If piers or posts are used, in lieu of a continuous foundation, the space under the building is to be enclosed with fire-resistant material. The walls and floors are to be constructed of, or covered with, a nonsparking material or lattice work. The doors must be metal or solid wood covered with metal.

(3) Hinges and hasps. Hinges and hasps are to be attached to doors by welding, riveting, or bolting (nuts on inside of door). Hinges and hasps must be installed so that they cannot be removed when the doors are closed and locked.

(4) Locks. Each door is to be equipped with (i) two mortise locks; (ii) two padlocks fastened in separate hasps and staples; (iii) a combination of a mortise lock and a padlock; (iv) a mortise lock that requires two keys to open; or (v) a three-point lock. Padlocks must have at least five tumblers and case-hardened shackle of at least ⅜ inch diameter. Padlocks must be protected with not less than ¼ inch steel hoods constructed so as to prevent sawing or lever action on the locks, hasps, and staples.

These requirements do not apply to magazine doors that are adequately secured on the inside by means of a bolt, lock, or bar that cannot be actuated from the outside.

(b) Indoor magazine—

(1) General. Indoor magazines are to be fire-resistant and theft-resistant. They need not be weather-resistant if the buildings in which they are stored provide protection from the weather. No indoor magazine is to be located in a residence or dwelling. The indoor storage of low explosives must not exceed a quantity of 50 pounds. More than one indoor magazine may be located in the same building if the total quantity of explosive materials stored does not exceed 50 pounds. Detonators that will not mass detonate must be stored in a separate magazine and the total number of electric detonators must not exceed 5,000.

(2) Construction. Indoor magazines are to be constructed of masonry, metal-covered wood, fabricated metal, or a combination of these materials. The walls and floors are to be constructed of, or covered with, a nonsparking material. The doors must be metal or solid wood covered with metal.

(3) Hinges and hasps. Hinges and hasps are to be attached to doors by welding, riveting, or bolting (nuts on inside of door). Hinges and hasps must be installed so that they cannot be removed when the doors are closed and locked.

(4) Locks. Each door is to be equipped with (i) two mortise locks; (ii) two padlocks fastened in separate hasps and staples; (iii) a combination of a mortise lock and padlock; (iv) a mortise lock that requires two keys to open; or (v) a three-point lock.

Padlocks must have at least five tumblers and a case-hardened shackle of at least ⅜ inch diameter. Padlocks must be protected with not less than ¼ inch steel hoods constructed so as to prevent sawing or lever action on the locks, hasps, and staples. Indoor magazines located in secure rooms that are locked as provided in this subparagraph may have each door locked with one steel padlock (which need not be protected by a steel hood) having at least five tumblers and a case-hardened shackle of at least ⅜ inch diameter, if the door hinges and lock hasp are securely fastened to the magazine.

These requirements do not apply to magazine doors that are adequately secured on the inside by means of a bolt, lock, or bar that cannot be actuated from the outside.

§ 555.211 Construction of type 5 magazines.

A type 5 magazine is a building, igloo or "Army-type structure", tunnel, dugout, bin, box, trailer, or a semitrailer or other mobile facility.

(a) Outdoor magazines—

(1) General. Outdoor magazines are to be weather-resistant and theft-resistant. The ground around magazines must slope away for drainage or other adequate drainage be provided. When unattended, vehicular magazines must have wheels removed or otherwise be effectively immobilized by kingpin locking devices or other methods approved by the Director.

(2) **Construction.** The doors are to be constructed of solid wood or metal.

(3) **Hinges and hasps.** Hinges and hasps are to be attached to doors by welding, riveting, or bolting (nuts on inside of door). Hinges and hasps must be installed so that they cannot be removed when the doors are closed and locked.

(4) **Locks.** Each door is to be equipped with (i) two mortise locks; (ii) two padlocks fastened in separate hasps and staples; (iii) a combination of a mortise lock and a padlock; (iv) a mortise lock that requires two keys to open; or (v) a three-point lock. Padlocks must have at least five tumblers and a case-hardened shackle of at least ⅜ inch diameter. Padlocks must be protected with not less than ¼ inch steel hoods constructed so as to prevent sawing or lever action on the locks, hasps, and staples. Trailers, semitrailers, and similar vehicular magazines may, for each door, be locked with one steel padlock (which need not be protected by a steel hood) having at least five tumblers and a case-hardened shackle of at least ⅜ inch diameter, if the door hinges and lock hasp are securely fastened to the magazine and to the door frame.

These requirements do not apply to magazine doors that are adequately secured on the inside by means of a bolt, lock, or bar that cannot be actuated from the outside.

(5) **Placards.** The placards required by Department of Transportation regulations at 49 CFR part 172, subpart F, for the transportation of blasting agents shall be displayed on all magazines.

(b) Indoor magazines—

(1) **General.** Indoor magazines are to be theft-resistant. They need not be weather-resistant if the buildings in which they are stored provide protection from the weather. No indoor magazine is to be located in a residence or dwelling. Indoor magazines containing quantities of blasting agents in excess of 50 pounds are subject to the requirements of §555.206 of this subpart.

(2) **Construction.** The doors are to be constructed of wood or metal.

(3) **Hinges and hasps.** Hinges and hasps are to be attached to doors by welding, riveting, or bolting (nuts on inside). Hinges and hasps must be installed so that they cannot be removed when the doors are closed and locked.

(4) **Locks.** Each door is to be equipped with (i) two mortise locks; (ii) two padlocks fastened in separate hasps and staples; (iii) a combination of a mortise lock and a padlock; (iv) a mortise lock that requires two keys to open; or (v) a three-point lock.

Padlocks must have at least five tumblers and a case-hardened shackle of at least ⅜ inch diameter. Padlocks must be protected with not less than ¼ inch steel hoods constructed so as to prevent sawing or lever action on the locks, hasps, and staples. Indoor magazines located in secure rooms that are locked as provided in this subparagraph may have each door locked with one steel padlock (which need not be protected by a steel hood) having at least five tumblers and a case-hardened shackle of at least ⅜ inch diameter, if the door hinges and lock hasps are securely fastened to the magazine and to the door frame.

These requirements do not apply to magazine doors that are adequately secured on the inside by means of a bolt, lock, or bar that cannot be actuated from the outside.

[T.D. ATF–87, 46 FR 40384, Aug. 7, 1981, as amended by T.D. ATF–298, 55 FR 21863, May 30, 1990]

§ 555.212 Smoking and open flames.

Smoking, matches, open flames, and spark producing devices are not permitted:

(a) In any magazine;

(b) Within 50 feet of any outdoor magazine; or

(c) Within any room containing an indoor magazine.

§ 555.213 Quantity and storage restrictions.

(a) Explosive materials in excess of 300,000 pounds or detonators in excess of 20 million are not to be stored in one magazine unless approved by the Director.

(b) Detonators are not to be stored in the same magazine with other explosive materials, except under the following circumstances:

(1) In a type 4 magazine, detonators that will not mass detonate may be stored with electric squibs, safety fuse, igniters, and igniter cord.

(2) In a type 1 or type 2 magazine, detonators may be stored with delay devices and any of the items listed in paragraph (b)(1) of this section.

§ 555.214 Storage within types 1, 2, 3, and 4 magazines.

(a) Explosive materials within a magazine are not to be placed directly against interior walls and must be stored so as not to interfere with ventilation. To prevent contact of stored explosive materials with walls, a nonsparking lattice work or other nonsparking material may be used.

(b) Containers of explosive materials are to be stored so that marks are visible. Stocks of explosive materials are to be stored so they can be easily counted and checked upon inspection.

(c) Except with respect to fiberboard or other nonmetal containers, containers of explosive materials are not to be unpacked or repacked inside a magazine or within 50 feet of a magazine, and must not be unpacked or repacked close to other explosive materials. Containers of explosive materials must be closed while being stored.

(d) Tools used for opening or closing containers of explosive materials are to be of nonsparking materials, except that metal slitters may be used for opening fiberboard containers. A wood wedge and a fiber, rubber, or wooden mallet are to be used for opening or closing wood containers of explosive materials. Metal tools other than nonsparking transfer conveyors are not to be stored in any magazine containing high explosives.

§ 555.215 Housekeeping.

Magazines are to be kept clean, dry, and free of grit, paper, empty packages and containers, and rubbish. Floors are to be regularly swept. Brooms and other utensils used in the cleaning and maintenance of magazines must have no spark-producing metal parts, and may be kept in magazines. Floors stained by leakage from explosive materials are to be cleaned according to instructions of the explosives manufacturer. When any explosive material has deteriorated it is to be destroyed in accordance with the advice or instructions of the manufacturer. The area surrounding magazines is to be kept clear of rubbish, brush, dry grass, or trees (except live trees more than 10 feet tall), for not less than 25 feet in all directions. Volatile materials are to be kept a distance of not less than 50 feet from outdoor magazines. Living foliage which is used to stabilize the earthen covering of a magazine need not be removed.

§ 555.216 Repair of magazines.

Before repairing the interior of magazines, all explosive materials are to be removed and the interior cleaned. Before repairing the exterior of magazines, all explosive materials must be removed if there exists any possibility that repairs may produce sparks or flame. Explosive materials removed from magazines under repair must be

(a) placed in other magazines appropriate for the storage of those explosive materials under this subpart, or

(b) placed a safe distance from the magazines under repair where they are to be properly guarded and protected until the repairs have been completed.

§ 555.217 Lighting.

(a) Battery-activated safety lights or battery-activated safety lanterns may be used in explosives storage magazines.

(b) Electric lighting used in any explosives storage magazine must meet the standards prescribed by the "National Electrical Code," (National Fire Protection Association, NFPA 70–81), for the conditions present in the magazine at any time. All electrical switches are to be located outside of the magazine and also meet the standards prescribed by the National Electrical Code.

(c) Copies of invoices, work orders or similar documents which indicate the lighting complies with the National Electrical Code must be available for inspection by ATF officers.

§ 555.218 Table of distances for storage of explosive materials.

Quantity of Explosives		Distances in feet							
Pounds over	Pounds not over	Inhabited buildings		Public highways with traffic volume 3000 or fewer vehicles/day		Passenger railways-public highways with traffic volume more than 3,000 vehicles/ day		Separation of magazines	
		Barricaded	Unbarricaded	Barricaded	Unbarricaded	Barricaded	Unbarricaded	Barricaded	Unbarricaded
0	5	70	140	30	60	51	102	6	12
5	10	90	180	35	70	64	128	8	16
10	20	110	220	45	90	81	162	10	20
20	30	125	250	50	100	93	186	11	22
30	40	140	280	55	110	103	206	12	24
40	50	150	300	60	120	110	220	14	28
50	75	170	340	70	140	127	254	15	30
75	100	190	380	75	150	139	278	16	32
100	125	200	400	80	160	150	300	18	36
125	150	215	430	85	170	159	318	19	38
150	200	235	470	95	190	175	350	21	42
200	250	255	510	105	210	189	378	23	46
250	300	270	540	110	220	201	402	24	48
300	400	295	590	120	240	221	442	27	54
400	500	320	640	130	260	238	476	29	58
500	600	340	680	135	270	253	506	31	62
600	700	355	710	145	290	266	532	32	64
700	800	375	750	150	300	278	556	33	66
800	900	390	780	155	310	289	578	35	70
900	1,000	400	800	160	320	300	600	36	72
1,000	1,200	425	850	165	330	318	636	39	78
1,200	1,400	450	900	170	340	336	672	41	82
1,400	1,600	470	940	175	350	351	702	43	86
1,600	1,800	490	980	180	360	366	732	44	88
1,800	2,000	505	1,010	185	370	378	756	45	90
2,000	2,500	545	1,090	190	380	408	816	49	98
2,500	3,000	580	1,160	195	390	432	864	52	104
3,000	4,000	635	1,270	210	420	474	948	58	116
4,000	5,000	685	1,370	225	450	513	1,026	61	122
5,000	6,000	730	1,460	235	470	546	1,092	65	130
6,000	7,000	770	1,540	245	490	573	1,146	68	136
7,000	8,000	800	1,600	250	500	600	1,200	72	144
8,000	9,000	835	1,670	255	510	624	1,248	75	150
9,000	10,000	865	1,730	260	520	645	1,290	78	156
10,000	12,000	875	1,750	270	540	687	1,374	82	164
12,000	14,000	885	1,770	275	550	723	1,446	87	174
14,000	16,000	900	1,800	280	560	756	1,512	90	180
16,000	18,000	940	1,880	285	570	786	1,572	94	188
18,00	20,000	975	1,950	290	580	813	1,626	98	196
20,000	25,000	1,055	2,000	315	630	876	1,752	105	210
25,000	30,000	1,130	2,000	340	680	933	1,866	112	224
30,000	35,000	1,205	2,000	360	720	981	1,962	119	238
35,000	40,000	1,275	2,000	380	760	1,026	2,000	124	248
40,000	45,000	1,340	2,000	400	800	1,068	2,000	129	258
45,000	50,000	1,400	2,000	420	840	1,104	2,000	135	270
50,000	55,000	1,460	2,000	440	880	1,140	2,000	140	280
55,000	60,000	1,515	2,000	455	910	1,173	2,000	145	290
60,000	65,000	1,565	2,000	470	940	1,206	2,000	150	300
65,000	70,000	1,610	2,000	485	970	1,236	2,000	155	310
70,000	75,000	1,655	2,000	500	1,000	1,263	2,000	160	320
75,000	80,000	1,695	2,000	510	1,020	1,293	2,000	165	330
80,000	85,000	1,730	2,000	520	1,040	1,317	2,000	170	340
85,000	90,000	1,760	2,000	530	1,060	1,344	2,000	175	350
90,000	95,000	1,790	2,000	540	1,080	1,368	2,000	180	360
95,000	100,000	1,815	2,000	545	1,090	1,392	2,000	185	370
100,000	110,000	1,835	2,000	550	1,100	1,437	2,000	195	390
110,000	120,000	1,855	2,000	555	1,110	1,479	2,000	205	410
120,000	130,000	1,875	2,000	560	1,120	1,521	2,000	215	430
130,000	140,000	1,890	2,000	565	1,130	1,557	2,000	225	450
140,000	150,000	1,900	2,000	570	1,140	1,593	2,000	235	470
150,000	160,000	1,935	2,000	580	1,160	1,629	2,000	245	490
160,000	170,000	1,965	2,000	590	1,180	1,662	2,000	255	510
170,000	180,000	1,990	2,000	600	1,200	1,695	2,000	265	530
180,000	190,000	2,010	2,010	605	1,210	1,725	2,000	275	550
190,000	200,000	2,030	2,030	610	1,220	1,755	2,000	285	570
200,000	210,000	2,055	2,055	620	1,240	1,782	2,000	295	590
210,000	230,000	2,100	2,100	635	1,270	1,836	2,000	315	630
230,000	250,000	2,155	2,155	650	1,300	1,890	2,000	335	670
250,000	275,000	2,215	2,215	670	1,340	1,950	2,000	360	720
275,000	300,000	2,275	2,275	690	1,380	2,000	2,000	385	770

Table: American Table of Distances for Storage of Explosives (December 1910), as Revised and Approved by the Institute of Makers of Explosives—July, 1991.

Notes to the Table of Distances for Storage of Explosives

(1) Terms found in the table of distances for storage of explosive materials are defined in §555.11.

(2) When two or more storage magazines are located on the same property, each magazine must comply with the minimum distances specified from inhabited buildings, railways, and highways, and, in addition, they should be separated from each other by not less than the distances shown for "Separation of Magazines," except that the quantity of explosives contained in cap magazines shall govern in regard to the spacing of said cap magazines from magazines containing other explosives. If any two or more magazines are separated from each other by less than the specified "Separation of Magazines" distances, then such two or more magazines, as a group, must be considered as one magazine, and the total quantity of explosives stored in such group must be treated as if stored in a single magazine located on the site of any magazine of the group, and must comply with the minimum of distances specified from other magazines, inhabited buildings, railways, and highways.

(3) All types of blasting caps in strengths through No. 8 cap should be rated at 1½ lbs. (1.5 lbs.) of explosives per 1,000 caps. For strengths higher than No. 8 cap, consult the manufacturer.

(4) For quantity and distance purposes, detonating cord of 50 or 60 grains per foot should be calculated as equivalent to 9 lbs. of high explosives per 1,000 feet. Heavier or lighter core loads should be rated proportionately.

[T.D. ATF–87, 46 FR 40384, Aug. 7, 1981, as amended by T.D. ATF–400, 63 FR 45003, Aug. 24, 1998; T.D. ATF–446, 66 FR 16602, Mar. 27, 2001; T.D. ATF–446a, 66 FR 19089, Apr. 13, 2001]

§ 555.219 Table of distances for storage of low explosives.

Pounds		From Inhabited building distance (feet)	From public railroad and highway distance (feet)	From above ground magazine (feet)
Over	Not over			
0	1,000	75	75	50
1,000	5,000	115	115	75
5,000	10,000	150	150	100
10,000	20,000	190	190	125
20,000	30,000	215	215	145
30,000	40,000	235	235	155
40,000	50,000	250	250	165
50,000	60,000	260	260	175
60,000	70,000	270	270	185
70,000	80,000	280	280	190
80,000	90,000	295	295	195
90,000	100,000	300	300	200
100,000	200,000	375	375	250
200,000	300,000	450	450	300

Table: Department of Defense Ammunition and Explosives Standards, Table 5–4.1 Extract; 4145.27 M, March 1969

§ 555.220 Table of separation distances of ammonium nitrate and blasting agents from explosives or blasting agents.

Donor weight (pounds)		Minimum separation distance of acceptor from donor when barricaded (feet)		Minimum thickness of artificial barricades (inches)
Over	Not over	Ammonium nitrate	Blasting agent	
0	100	3	11	12
100	300	4	14	12
300	600	5	18	12
600	1,000	6	22	12
1,000	1,600	7	25	12
1,600	2,000	8	29	12
2,000	3,000	9	32	15
3,000	4,000	10	36	15
4,000	6,000	11	40	15
6,000	8,000	12	43	20
8,000	10,000	13	47	20
10,000	12,000	14	50	20
12,000	16,000	15	54	25
16,000	20,000	16	58	25
20,000	25,000	18	65	25
25,000	30,000	19	68	30
30,000	35,000	20	72	30
35,000	40,000	21	76	30
40,000	45,000	22	79	35
45,000	50,000	23	83	35
50,000	55,000	24	86	35
55,000	60,000	25	90	35
60,000	70,000	26	94	40
70,000	80,000	28	101	40
80,000	90,000	30	108	40
90,000	100,000	32	115	40
100,000	120,000	34	122	50
120,000	140,000	37	133	50
140,000	160,000	40	144	50
160,000	180,000	44	158	50
180,000	200,000	48	173	50
200,000	220,000	52	187	60
220,000	250,000	56	202	60
250,000	275,000	60	216	60
275,000	300,000	64	230	60

Table: National Fire Protection Association (NFPA) Official Standard No. 492, 1968

Notes of Table of Separation Distances of Ammonium Nitrate and Blasting Agents from Explosives or Blasting Agents

(1) This table specifies separation distances to prevent explosion of ammonium nitrate and ammonium nitrate-based blasting agents by propagation from nearby stores of high explosives or blasting agents referred to in the table as the "donor." Ammonium nitrate, by itself, is not considered to be a donor when applying this table. Ammonium nitrate, ammonium nitrate-fuel oil or combinations thereof are acceptors. If stores of ammonium nitrate are located within the sympathetic detonation distance of explosives or blasting agents, one-half the mass of the ammonium nitrate is to be included in the mass of the donor.

(2) When the ammonium nitrate and/or blasting agent is not barricaded, the distances shown in the table must be multiplied by six. These distances allow for the possibility of high velocity metal fragments from mixers, hoppers, truck bodies, sheet metal structures, metal containers, and the like which may enclose the "donor." Where explosives storage is in bullet-resistant magazines or where the storage is protected by a bullet-resistant wall, distances and barricade thicknesses in excess of those prescribed in the table in §555.218 are not required.

(3) These distances apply to ammonium nitrate that passes the insensitivity test prescribed in the definition of ammonium nitrate fertilizer issued by the Fertilizer Institute. 1 Ammonium nitrate failing to pass the test must be stored at separation distances in accordance with the table in §555.218.

[1] Definition and Test Procedures for Ammonium Nitrate Fertilizer, Fertilizer Institute 1015–18th St. N.W. Washington, DC 20036.

(4) These distances apply to blasting agents which pass the insensitivity test prescribed in regulations of the U.S. Department of Transportation (49 CFR part 173).

(5) Earth or sand dikes, or enclosures filled with the prescribed minimum thickness of earth or sand are acceptable artificial barricades. Natural barricades, such as hills or timber of sufficient density that the surrounding exposures which require protection cannot be seen from the "donor" when the trees are bare of leaves, are also acceptable.

(6) For determining the distances to be maintained from inhabited buildings, passenger railways, and public highways, use the table in §555.218.

§ 555.221 Requirements for display fireworks, pyrotechnic compositions, and explosive materials used in assembling fireworks or articles pyrotechnic.

(a) Display fireworks, pyrotechnic compositions, and explosive materials used to assemble fireworks and articles pyrotechnic shall be stored at all times as required by this Subpart unless they are in the process of manufacture, assembly, packaging, or are being transported.

(b) No more than 500 pounds (227 kg) of pyrotechnic compositions or explosive materials are permitted at one time in any fireworks mixing building, any building or area in which the pyrotechnic compositions or explosive materials are pressed or otherwise prepared for finishing or assembly, or any finishing or assembly building. All pyrotechnic compositions or explosive materials not in immediate use will be stored in covered, non-ferrous containers.

(c) The maximum quantity of flash powder permitted in any fireworks process building is 10 pounds (4.5 kg).

(d) All dry explosive powders and mixtures, partially assembled display fireworks, and finished display fireworks shall be removed from fireworks process buildings at the conclusion of a day's operations and placed in approved magazines.
[T.D. ATF–293, 55 FR 3722, Feb. 5, 1990, as amended by T.D. ATF–400, 63 FR 45004, Aug. 24, 1998]

§ 555.222 Table of distances between fireworks process buildings and between fireworks process and fireworks nonprocess buildings.

Net weight of fireworks1 (pounds)	Display fireworks2 (feet)	Consumer fireworks3 (feet)
0-100	57	37
101-200	69	37
201-300	77	37
301-400	85	37
401-500	91	37
Above 500	Not permitted4 5	Not permitted4 5

[1] Net weight is the weight of all pyrotechnic compositions, and explosive materials and fuse only.

[2] The distances in this column apply only with natural or artificial barricades. If such barricades are not used, the distances must be doubled.

[3] While consumer fireworks or articles pyrotechnic in a finished state are not subject to regulation, explosive materials used to manufacture or assemble such fireworks or articles are subject to regulation. Thus, fireworks process buildings where consumer fireworks or articles pyrotechnic are being processed shall meet these requirements.

[4] A maximum of 500 pounds of in-process pyrotechnic compositions, either loose or in partially-assembled fireworks, is permitted in any fireworks process building. Finished display fireworks may not be stored in a fireworks process building.

[5] A maximum of 10 pounds of flash powder, either in loose form or in assembled units, is permitted in any fireworks process building. Quantities in excess of 10 pounds must be kept in an approved magazine.
[T.D. ATF–293, 55 FR 3723, Feb. 5, 1990, as amended by T.D. ATF–400, 63 FR 45004, Aug. 24, 1998]

§ 555.223 Table of distances between fireworks process buildings and other specified areas.

Net weight of fireworks1 (pounds)	Display fireworks1 (feet)	Consumer fireworks2 (feet)
0-100	200	25
101-200	200	50
201-300	200	50
301-400	200	50
401-500	200	50
Above 500	Not permitted	Not permitted

Distance from Passenger Railways, Public Highways, Fireworks Plant Buildings used to Store Consumer Fireworks and Articles Pyrotechnic, Magazines and Fireworks Shipping Buildings, and Inhabited Buildings. [3 4 5]

[1] Net weight is the weight of all pyrotechnic compositions, and explosive materials and fuse only.

[2] While consumer fireworks or articles pyrotechnic in a finished state are not subject to regulation, explosive materials used to manufacture or assemble such fireworks or articles are subject to regulation. Thus, fireworks process buildings where consumer fireworks or articles pyrotechnic are being processed shall meet these requirements.

[3] This table does not apply to the separation distances between fireworks process buildings (see § 555.222) and between magazines (see §§ 555.218 and 555.224).

[4] The distances in this table apply with or without artificial or natural barricades or screen barricades. However, the use of barricades is highly recommended.

[5] No work of any kind, except to place or move items other than explosive materials from storage, shall be conducted in any building designated as a warehouse. A fireworks plant warehouse is not subject to § 555.222 or this section, tables of distances.
[T.D. ATF–293, 55 FR 3723, Feb. 5, 1990, as amended by T.D. ATF–400, 63 FR 45004, Aug. 24, 1998]

§ 555.224 Table of distances for the storage of display fireworks (except bulk salutes).

Net weight of fireworks1 (pounds)	Distance between magazine and inhabited building, passenger railway, or public highway3 4 (feet)	Distance between magazines2 3 (feet)
0-1000	150	100
1,001-5,000	230	150
5,001-10,000	300	200
Above 10,000	Use Table § 55.218	

[1] Net weight is the weight of all pyrotechnic compositions, and explosive materials and fuse only.

[2] For the purposes of applying this table, the term "magazine" also includes fireworks shipping buildings for display fireworks.

[3] For fireworks storage magazines in use prior to (30 days from the date of publication of the final rule in the Federal Register), the distances in this table may be halved if properly barricaded between the magazine and potential receptor sites.

[4] This table does not apply to the storage of bulk salutes. Use table at § 555.218.

[T.D. ATF–293, 55 FR 3723, Feb. 5, 1990, as amended by T.D. ATF–400, 63 FR 45004, Aug. 24, 1998]

Questions and Answers (Revised 1/07)

18 U.S.C. Chapter 40 and 27 CFR Part 555

Introduction

The following list of Questions and Answers are intended to aid you in gaining a better understanding of:

18 U.S.C. Chapter 40 – Importation, Manufacture, Distribution and Storage of Explosive Materials and the implementing regulations issued within:

27 CFR Part 555 – Commerce in Explosives

This listing is not all-inclusive. However it contains some of the most frequently asked questions that ATF receives. These questions and answers are intended only as a general overview. To determine how the law and regulations apply to your specific circumstances, you must refer directly to the applicable law and regulation or contact your local ATF Office of Industry Operations. Also, please be aware that both the law and regulations are subject to change. Please contact your local ATF office for the most up-to-date information. You can also find the latest Federal explosives regulations on the ATF website: www.atf.gov.

Unless otherwise stated, these Questions and Answers apply only to Federal law and regulations. States and local jurisdictions have, in many cases, enacted their own requirements relating to explosives. Check with appropriate State or local authorities for information on those requirements. Compliance with Federal law and regulations does not exempt any person from compliance with any State or local requirements.

A Table of Contents and a Subject Index has been included for your convenience. The Index is located at the end of the Questions and Answers section.

Table of Contents

General Questions

1. Who is affected by the Federal explosives law?

The law affects all persons who import, manufacture, deal in, purchase, use, store, or possess explosive materials. It also affects those who ship, transport or cause to be transported, or receive explosive materials. Also, see 18 U.S.C. 845 and 27 CFR 555.141 for exemptions.

2. What changes were brought about by the Safe Explosives Act?

Among other things, the Act mandated that all persons who wish to receive or transport explosive materials must first obtain a Federal explosives license or permit. In addition, the act imposed new restrictions on who may lawfully receive and possess explosive materials (See question 4). All Federal explosive licensees and permittees and their responsible persons and employees authorized to possess explosives are affected by the new requirements and background checks mandated by the Act.

3. Can I obtain general information from ATF on the Internet?

Yes. ATF maintains a website on the Internet at <u>*www.atf.gov*</u>.

4. Does the law make some classes of persons ineligible to receive a Federal license to import, manufacture, or deal in explosive materials or to receive a Federal explosives permit?

Yes. A license or permit will not be issued to any person who:

(a) Is under indictment for, or who has been convicted in any court, of a crime punishable by imprisonment for a term exceeding one year;

(b) Is a fugitive from justice;

(c) Is an unlawful user of or addicted to any controlled substance (as defined in section 102 of the Controlled Substances Act (21 U.S.C. 802));

(d) Has been adjudicated a mental defective or who has been committed to a mental institution;

(e) Is an alien (with certain exceptions);

(f) Has been discharged from the armed forces under dishonorable conditions; or,

(g) Having been a citizen of the United States, has renounced his citizenship. [18 U.S.C. 842(i), 843(b)(1); 27 CFR 555.49(b)(2)(i)]

5. Are there classes of persons to whom the distribution of explosive materials by licensees is prohibited?

Yes. No person shall knowingly distribute explosive materials to any individual listed in Question 4, or to an individual who is under 21 years of age. [18 U.S.C. 842(d); 27 CFR 555.26(d)]

6. What other distributions of explosive materials by licensees and permittees are prohibited?

A licensee or permittee shall not knowingly distribute any explosive materials to any person who:

(a) Is not a licensee [18 U.S.C. 842(b), 27 CFR 555.105, 555.106];

(b) Is not a holder of a user permit [18 U.S.C. 842(b); 27 CFR 555.105, 555.106];

(c) Is not a holder of a limited permit who resides in the same State where distribution is made and in which premises of the transferor are located. [18 U.S.C. 842(b); 27 CFR 555.105, 555.106];

(d) The licensee has reason to believe intends to transport such explosive materials into a State where the purchase, possession, or use of explosive materials is prohibited or which does not permit its residents to transport or ship explosive materials into the State or to receive explosive materials in the State. [18 U.S.C. 842(c); 27 CFR 555.106]

(e) Is in any State where the purchase, possession, or use by such person of such explosive materials would be in violation of any State law or any published ordinance applicable at the place of distribution. [18 U.S.C. 842(e); 27 CFR 555.106(b)(2)]

7. Does Federal law prohibit certain persons from receiving or possessing explosive materials?

Yes. The law prohibits the receipt or possession of explosive materials by any person listed in question 4. [18 U.S.C. 842(i); 27 CFR 555.26, 555.49(b)]

8. May a licensed manufacturer, importer, or dealer distribute explosive materials to nonlicensees and nonpermittees?

No. Every person who receives explosive materials must first obtain a Federal explosives license or permit. Distribution of explosive materials to persons who do not hold a license or permit is unlawful. Also, see 18 U.S.C. 845 and 27 CFR 555.141 for exemptions. [18 U.S.C. 842(a), 842(b); 27 CFR 555.26(a), 555.106]

9. Does Federal law provide penalties for purchasers who give false information at the time of purchasing explosive materials?

Yes. The penalty for knowingly providing false information or misrepresented identification is a maximum 10 years' imprisonment and/or a fine not exceeding $250,000. [18 U.S.C. 842(a)(2), 844(a)]

10. Is the theft of explosive materials, as well as the possession of stolen explosive materials, a Federal crime?

Yes. It is a Federal crime for any person to steal any explosive materials. It is also a Federal crime for any person to receive, possess, transport, ship, conceal, store, barter, sell, dispose of, or pledge or accept as security for a loan any stolen explosive materials. [18 U.S.C. 842(h), 844(k), (l)]

11. Are thefts and losses of explosive materials required to be reported to ATF?

Yes. Any licensee or permittee who has knowledge of the theft or loss of any explosive materials from his or her stock shall, within 24 hours of discovery, report the theft or loss by telephoning 800-461-8841 (Monday-Friday 8:00a.m.– 5:00 p.m. Eastern Time) or 888-283-2662 (after hours and weekends) (nationwide toll free number) and on ATF Form 5400.5, "Report of Theft or Loss – Explosive Materials", in accordance with the instructions on the form. The theft or loss shall also be reported to appropriate local authorities. The same requirements are imposed upon persons other than licensees and permittees, except that nonlicensees and nonpermittees, other than carriers, need not report a theft or loss on Form 5400.5, but must report the theft or loss by telephone, using the same numbers: 800-461-8841 (Monday-Friday 8:00a.m.–5:00 p.m. Eastern Time) or 888-283-2662 (after hours and weekends) and in writing to the nearest ATF office. The theft or loss shall also be reported to appropriate local authorities. Carriers of explosive materials must report a theft or loss by telephone but need not make the report on the ATF form or in writing. See 27 CFR 555.30 for the specific information required to be reported in connection with a theft or loss. [18 U.S.C. 842(k), 18 U.S.C. 844(p); 27 CFR 555.30]

12. My company holds a Federal explosives license and after conducting an inventory of our explosives on hand, we noticed one case of dynamite missing. After double-checking all Daily Summaries of Magazine Transactions, invoices, and delivery sheets, we still cannot reconcile the discrepancy. What are we required to do?

This should be considered a theft or loss of explosive materials. As stated in the answer to Question 11, you must report the theft or loss of explosive materials, within 24 hours of discovery, to ATF by telephone (toll free: 800-461-8841 (Monday-Friday 8:00 a.m.– 5:00 p.m. Eastern Time) or 888-283-2662 (after hours and weekends). ATF Form 5400.5, "Report of Theft or Loss – Explosive Materials", must then be completed and forwarded in accordance with the instructions on the form. [18 U.S.C. 842(k); 27 CFR 555.30]

13. May ATF conduct warrantless inspections of licensees' and permittees' records of explosives materials, stocks of such materials, and magazines?

Except for limited permit holders, any ATF officer may, without a warrant, enter during business hours the premises, including places of storage, of any licensee or permittee for the purpose of inspecting or examining any records or documents required to be kept by the law and regulations and any explosive materials kept or stored at the premises. For inspection purposes, "business hours" includes hours during which business is actually conducted, not just those hours stated on license applications. Any licensee or permittee who refuses to permit the inspection or examination is subject to having his or her license or permit revoked, as well as to denial of an application to renew the license or permit. For limited permit holders, an ATF officer may inspect the places of storage for explosive materials of either an applicant for a limited permit or at the time of renewal of such permit, but in no event shall such inspection occur more than once every three years. [18 U.S.C. 843(b)(4), 18 U.S.C. 843(f)]

14. Will ATF investigate accidents involving explosive materials?

ATF is authorized to inspect the site of any accident or fire where there is reason to believe that explosive materials were involved. Other Federal agencies, or State or local agencies, may also investigate such incidents, depending on the circumstances. [18 U.S.C. 846(a); 27 CFR 555.31]

15. Is black powder subject to regulation under Federal explosives laws?

Black powder is an explosive material for purposes of Federal explosives laws and regulations. However, the law exempts from regulation commercially manufactured black powder in quantities not exceeding 50 pounds (as well as percussion caps, safety and pyrotechnic fuses, quills, quick and slow matches, and friction primers) intended to be used solely for sporting, recreational, or cultural purposes in antique firearms as defined in 18 U.S.C. 921(a)(16) or in antique devices exempted from the term "destructive device" in 18 U.S.C. 921(a)(4). However, persons engaged in the business of importing, manufacturing, or dealing in black powder in any quantity must have a Federal explosives license. [18 U.S.C. 841(c), 841(d), 845(a)(5); 27 CFR 555.11: definitions of "explosives" and "explosive materials", 555.141(b)]

16. Is small arms ammunition subject to regulation under Federal explosives laws?

No. The law specifically exempts small arms ammunition and components thereof. (See also Question 81.) [18 U.S.C. 845(a)(4)]

17. Are binary explosives subject to regulation under Federal explosives laws?

Until the compounds are mixed, they are not classified as explosives and, therefore, are not subject to control. However, once mixed, binary explosives are "explosive materials" and are subject to all applicable Federal requirements. A person who mixes or combines compounds of binary explosives for the purpose of sale or distribution or for the person's own business use is a "manufacturer" of explosive materials and must be licensed as a manufacturer under the law. [18 U.S.C. 841(h); 27 CFR 555.11: definition of "manufacturer"]

18. Does ATF have any regulations governing the actual transportation of explosive materials?

Federal explosives laws and regulations generally prohibit any person from transporting explosive materials interstate or intrastate unless the person has a Federal explosives license or permit. Also, the transportation of stolen explosives materials is a Federal crime (see also Question 10). However, the law exempts from regulation under 18 U.S.C. Chapter 40 and 27 CFR Part 555 aspects of the transportation of explosive materials via railroad, water, highway, or air which are regulated by the United States Department of Transportation, and the Department of Homeland Security, and agencies thereof and which pertain to safety and security. [18 U.S.C. 842(a)(3), 842(h), 845(a)(1); 27 CFR 555.26, 555.28, 555.141(a)(1)]

19. Are common or contract carriers required to obtain a Federal explosives license or permit to transport explosive materials?

No. The actual transportation of explosive materials by carriers is subject to Department of Transportation or Department of Homeland Security regulations. [18 U.S.C. 845(a)(1); 27 CFR 555.141(a)(1)]

20. What is the "Explosives List"?

The Explosives List is a comprehensive (but not all-inclusive) listing of explosive materials which have been determined to be within the coverage of Chapter 40. The list is published annually by ATF (the most recent list can be found under the Explosives Laws and Regulations link on the ATF website). [18 U.S.C. 841(d); 27 CFR 555.23]

21. May a person under the age of 21 be lawfully employed by an explosives business and lawfully receive, possess, and use explosive materials on behalf of the business?

Yes. Federal explosives law prohibits any person from distributing explosive materials to persons under 21 years of age. However, it does not prohibit the delivery to or possession of explosive materials by persons under the age of 21 who are receiving or using the materials on behalf of their employers to whom the materials were lawfully sold. [18 U.S.C. 842(d), (i); 27 CFR 555.11 (definition of "distribute"), 555.26, 555.106(b)(1)]

22. ATF regulations require explosive materials to be stored at certain minimum distances from a "public highway". What is a "public highway" for purposes of the regulations?

The term "highway" is defined in 27 CFR 555.11 as "any public street, public alley, or public road, including a privately financed, constructed, or maintained road that is regularly and openly traveled by the general public." Privately financed, constructed, or maintained roads that are marked and barricaded in a manner that prevents access by the general public do not fall within the meaning of the term and would, therefore, be exempt from table of distance requirements. [27 CFR 555.11: definition of "highway"](See also ATF Ruling 2005-2)

23. Is an airport runway or taxiway considered a public highway for purposes of the Table of Distances for storage of explosive materials?

No. However, airport terminals are considered inhabited buildings for Table of Distance requirements.

24. How is shock tube regulated by ATF?

Shock tube contains highly explosive material. However, it may be stored as a low explosive when not attached to a detonator. [27 CFR 555.202(b), 555.213]

25. What is an EX number?

An EX number is a number, preceded by the prefix "EX-", which is issued and used by the Department of Transportation (DOT) to identify an explosive which has been tested and classified by DOT. See U.S. Department of Transportation regulations at 49 CFR 171.8 and 49 CFR 173.56.

26. What is a UN number?

A UN (United Nations) number is used by DOT as a method of identification and classification of products for shipping purposes. UN numbers are different from the hazard class or division designations used by DOT (for example, 1.1, 1.2, 1.3, 1.4, and 1.5). ATF regulations in 27 CFR Part 555 also use UN numbers to help identify certain explosives. [27 CFR 555.11]

27. Can Federal explosives disabilities resulting from a conviction of a crime punishable by imprisonment for a term exceeding one year be removed if the conviction is expunged or set aside or the convicted person has received a pardon for the offense or has had his or her civil rights restored?

A person convicted of, or under indictment for, a "crime punishable by imprisonment for a term exceeding one year" may not lawfully receive or possess explosive materials or be issued a Federal explosives license or permit. The term "crime punishable by imprisonment for a term exceeding one year" does not include offenses pertaining to antitrust violations, unfair trade practices, restraints of trade, or any State offense (other than one involving a firearm or explosive) classified as a misdemeanor and punishable by imprisonment for 2 years or less. There are only 3 means by which Federal explosives disabilities resulting from a conviction of, or indictment for, a "crime punishable by imprisonment for a term exceeding one year" can be removed:

(a) A decision of a court invalidating a conviction on the basis that the conviction was unconstitutional;

(b) In the case of a Federal conviction, a presidential pardon; and

(c) The granting of relief from Federal explosives disabilities by ATF pursuant to the filing of a relief application with the Director. Information on how to apply for relief and contact information for the Relief of Disabilities Section is available on the ATF website. [18 U.S.C. 841(l), 842(d),(i), 845(b); 27 CFR 555.11: definition of "crime punishable by imprisonment for a term exceeding one year", 555.26(c), 555.142]

Licenses and Permits

28. Who needs a Federal explosives license or permit?

All persons who wish to transport, ship, cause to be transported, or receive explosive materials must first obtain a Federal explosives license or permit. Certain exemptions apply. [18 U.S.C. 842(b); 18 U.S.C. 845; 27 CFR 555.26(a), 27 CFR 555.141]

29. Who is eligible for a Federal explosives license or permit?

The Chief, Federal Explosives Licensing Center, will approve a properly completed application for a license or permit on ATF Form 5400.13/5400.16 if the applicant:

(a) Is not a person prohibited from possessing or receiving explosive materials under 18 U.S.C. 842(i) and none of the applicant's "responsible persons" are prohibited under section 842(1); (see also Question 4 in General Q&A);

(b) Has not willfully violated any provision of Chapter 40 or the regulations in 27 CFR Part 555;

(c) Has not knowingly withheld information or has not made any false or fictitious statement intended or likely to deceive, in connection with the application,

(d) Has premises in a State from which he intends to conduct business or operations;

(e) Has storage for the class (as described in 27 CFR 555.202) of explosive materials described on the application;

(f) Has certified in writing that he is familiar with and understands all published State laws and local ordinances relating to explosive materials for the location in which he intends to do business;

(g) Has submitted the certificate required by section 21 of the Federal Water Pollution Control Act, as amended (33 U.S.C. 1341) [18 U.S.C. 843(b); 27 CFR 555.49(b)];

(h) None of the applicant's employees authorized to possess explosives are prohibited persons under 18 U.S.C. 842(i); and

(i) In the case of an applicant for a limited permit, the applicant has certified that the applicant will not receive explosive materials on more than 6 occasions during the 12-month period for which the limited permit is valid.

30. What activities are covered by licenses and permits?

Licenses allow persons to engage in the business of importing, manufacturing, or dealing in explosive materials. Any individual or business entity intending to engage in any of these activities must first obtain a license. A user permit allows the receipt and transportation of explosive materials. A limited permit allows the receipt of explosive materials from a licensee or permittee within the permittee's state of residence only, and on no more than six occasions in the 12-month period during which the limited permit is valid. A limited permit does not authorize the receipt or transportation of explosive materials in interstate or foreign commerce. [27 CFR 555.11: definitions of "importer", "manufacturer", "dealer", "limited permit", and "user permit", 555.41]

A separate license is needed for each business premises where an explosives business or activity is conducted. Only a single user permit is needed by a permittee who uses explosives in more than one location. [27 CFR 555.41]

31. What is a Limited Permit?

A limited permit is for persons who wish to transport, ship, cause to be transported, or receive explosive materials in intrastate commerce only. This permit is designed for the infrequent receipt of explosive materials by intrastate users. The limited permit will allow a purchaser to receive explosive materials on no more than six separate occasions from in-state licensees or permittees during the 12-month period of the permit. The limited permit does not allow the holder to transport, ship, cause to be transported, or receive explosive materials in interstate commerce.

32. What is the duration of a license or permit?

(a) A user license or permit is valid for a period of 3 years.

(b) The user-limited permit is valid only for a single purchase transaction.

(c) Limited permits are valid for no more than six separate receipts of explosive materials during a 12-month period. [27 CFR 555.51]

33. What are the fees for licenses and permits?

Each license applicant must pay a fee of $200 for obtaining a 3-year license, a separate license and fee being required for each business premises. The fee for renewal of a license is $100 for a 3-year license. [27 CFR 555.42]

Each applicant for a user permit must pay a fee of $100 for a 3-year permit, and each applicant for a user-limited permit (nonrenewable) must pay a fee of $75. The fee for renewal of a user permit is $50 for a 3-year permit. [27 CFR 555.43]

Each applicant for a limited permit must pay a fee of $25 for a 1-year limited permit. The fee for renewal of a limited permit is $12 for a 1-year limited permit. [27 CFR 555.43]

34. Will the Government investigate an application for a license or permit?

ATF will investigate any applicant before issuing a license or permit. Additionally, ATF must inspect places of storage and conduct background checks on responsible persons and employee possessors authorized to possess explosives [18 U.S.C. 843(b); 18 U.S.C. 843(h); 27 CFR 555.33, 27 CFR 555.49(b)]

35. What may a licensed explosives dealer do?

A licensed dealer may engage in the business of distributing explosive materials at wholesale or retail [27 CFR 555.11: definition of "dealer"]

36. What may a licensed explosives importer do?

A licensed importer may engage in the business of importing or bringing explosive materials into the United States for purposes of sale or distribution. It is not necessary for a licensed importer to also obtain a dealer's license to engage in business on his or her licensed premises as a dealer in explosive materials (see also Question 52 and 53) [27 CFR 555.11: definition of "importer", 27 CFR 555.41(b)(2)]

37. When is a manufacturer's license required?

A manufacturer's license is required by persons engaged in the business of manufacturing explosive materials for sale, distribution, or for their own business use. For example, persons engaged in the business of providing a blasting service using explosives of their own manufacture would be required to have a manufacturer's license. Persons who manufacture explosives for their personal, non-business use are not required to have a manufacturer's license. However, no person may ship, transport, cause to be transported, or receive explosive materials unless such person holds a license or permit. [27 CFR 555.11: definition of "manufacturer", 555.41(b)] A separate manufacturer's license is not required by a licensed manufacturer for the purpose of on-site manufacture, for example, mixing binary explosives or making blasting agents at a quarry or other job site. It is not necessary for a licensed manufacturer to also obtain a dealer's license to engage in business on his or her licensed premises as a dealer in explosive materials (see also Question 52 and 53) [27 CFR 555.11: definition of "manufacturer", 555.41(b)(2)]

38. How do I apply for a Federal explosives license or permit?

You can request an application for a Federal explosives license or permit from the Federal Explosives Licensing Center at 877-283-3352 or from the ATF Distribution Center at 703-455-7801. As part of the application process, you must complete and submit an ATF Form 5400.13/5400.16, Application for Explosives License or Permit. You must also submit the names, identifying information, fingerprints, and photographs of all responsible persons. In addition, you must submit the names and identifying information of all employees who are authorized to possess explosive materials in the course of their employment on ATF Form 5400.28, Employee Possessor Questionnaire. [27 CFR 555.45(c)]

39. Who is a "responsible person"?

Federal explosives laws define a "responsible person" as an individual who has the power to direct the management and policies of the applicant pertaining to explosive materials. Responsible persons generally include sole proprietors and explosives facility site managers. In the case of a corporation, association, or similar organization, responsible persons generally include only those corporate directors/officers, and stockholders, who have the power to direct management and policies as they pertain to explosive materials.

For example, a corporate vice president whose duties include acquiring and approving contracts with explosives distributors would be considered a responsible person. Other corporate officials whose duties do not include the power to direct the management and policies of the applicant pertaining to explosive materials, for example, a vice president responsible solely for human resources, would not typically be considered a responsible person. Each applicant for a license or permit must assess the corporate and other management responsibilities for all key personnel and determine whether or not these duties place the individual in the position of being a responsible person. [18 U.S.C. 841(s), 27 CFR 555.11: definition of "responsible person"]

40. Who is a "possessor of explosives"?

A possessor of explosives is any employee of a license or permit holder or any employee of an applicant for a license or permit who has or will have actual physical possession of explosive materials or who has or will have constructive possession of explosive materials. For example, persons who physically handle explosive materials would be considered to be actual possessors of explosive materials. This would include employees who directly handle explosive materials as part of the production process; employees who handle explosive materials in order to ship, transport, or sell them; and employees, such as blasters and their helpers who actually use explosive materials. A constructive possessor is any person who has access to explosive materials, without physically handling them. For example, a supervisor at a construction site who keeps keys for storage magazines in which explosives are stored or who directs the use of explosive materials by other employees has constructive possession of explosives.

41. Why is it necessary to provide new and additional information on responsible persons and employee possessors of explosives?

The law requires this information for ATF to conduct background checks on all responsible persons and employee possessors to restrict the availability of explosives to authorized persons only and to reduce the risk of prohibited persons acquiring explosive materials. [18 U.S.C. 843(h); 27 CFR 555.33, 555.45(c)]

42. When will I need to submit the identifying information for my responsible persons and employee possessors of explosives?

(a) All license and permit applicants and any renewal applicants must submit identifying information for responsible persons and employee possessors (and fingerprints and photographs for responsible persons) upon submission of an original or renewal application.

(b) Any new responsible person added after a license or permit has been issued by ATF must be reported to ATF within 30 days. However, the submission of fingerprints and photographs by the new responsible person is required only at the time of any subsequent renewal.

(c) For all licenses and permits (new and renewal), any new employee possessors must be reported to ATF within 30 days of hire on the Employee Possessor Questionnaire form (ATF F 5400.28). [27 CFR 555.45(c), 27 CFR 555.57(b)]

43. How do I get my fingerprints taken?

Fingerprints must be submitted on Fingerprint Identification Cards, FD-258 that have been issued by ATF. The fingerprint cards must contain the following ORI information: WVATF0900; ATF-NATL EXPL LIC, MARTINSBURG WV. These fingerprint cards may be obtained by contacting the Federal Explosives Licensing Center at 877-283-3352 or the ATF Distribution Center at 703-455-7801. The fingerprint cards must be completed by your local law enforcement authority.

44. Will ATF notify me whether or not my responsible persons and employee possessors have passed their background checks?

Yes. A "Notification of Clearance" will be issued directly to all license or permit holders advising whether their responsible persons and employee possessors have been cleared to possess explosive materials, or are or may be prohibited from possessing explosives. These notices must be retained as part of the license or permit holders permanent records. In addition, letters of clearance or denial will be issued directly to responsible persons and employee possessors. [27 CFR 555.33]

45. What notification will I receive if one of my responsible persons or employee possessors does not pass their ATF background check?

If an individual does not pass the background check, a letter will be sent to the licensee or permittee who submitted the individual's name indicating that the individual was denied. A letter will also be sent to that individual explaining the prohibition and outlining appeal and relief procedures, as may be applicable. Unless and until an appeal overturns the denial or relief from disabilities is granted, that individual may not lawfully possess explosives. [27 CFR 555.33]

46. Who will conduct the background checks on applicants, responsible persons, and possessors?

ATF will perform the background checks. If employers wish to require their own background checks as a condition of employment, they may do so. However, such a background check will not be accepted in place of the ATF background check. [27 CFR 555.33]

47. May I sell black powder without a license?

No. Anyone who engages in the business of selling black powder, regardless of quantity, must be licensed as an explosives dealer. [27 CFR 555.41(b)]

48. Is a manufacturer's license required to acquire and mix binary explosives?

If the individual purchasing the binary explosives is engaged in the business of manufacturing explosives, i.e., mixes and uses them in the operation of a commercial business (for example, operating a quarry, or providing the service of removing stumps or boulders from a farm field), then a manufacturer's license is required.

An individual farmer who merely wishes to mix the binary explosives to remove obstacles from his field and provides no other outside service would not need a manufacturer's license.

Please note, however: A Federal explosives license or permit would be required to obtain any explosive device, such as detonators, used to initiate the mixed binary explosives. In addition, transportation of any explosive material, including mixed binary explosives, without a Federal license or permit is prohibited. [27 CFR 555.11: Definition of "manufacturer"; 27 CFR 555.26, 555.41(b)]

49. What is theatrical flash powder and is there a license for its manufacture?

Theatrical flash powder is flash powder commercially manufactured in premeasured kits not exceeding 1 ounce in weight, and mixed immediately prior to use and intended for use in events such as theatrical shows, stage plays, band concerts, magic acts, thrill shows, and clown acts in circuses. A manufacturer's license allows on-site manufacturers to operate nationally on one license issued to their principal place of business. [27 CFR 555.11: definitions of "flash powder" and "theatrical flash powder", 555.41(b)]

50. Is a separate license required for each location where business is conducted?

Yes. A separate license is required for each location where business is conducted. However, a separate license is not required for:

(a) Facilities used only for the storage of explosive materials;

(b) Locations used solely for the storage of records relating to the business; and

(c) Licensed manufacturers' on-site manufacturing. [27 CFR 555.41(b)]

51. Must a person who engages in the business of both manufacturing and importing at the same location have both licenses?

Yes. The licenses for manufacturing and importing allow a person to engage in separate and distinct activities and a separate license is required for each activity. However, a manufacturer or an importer does not need a separate dealer's license to also distribute explosive materials from the licensed premises. [27 CFR 555.41(b)]

52. Does a licensed manufacturer, importer, or dealer need a permit to use explosive materials?

No. No licensee will be required to obtain a user permit to lawfully transport, ship, or receive explosive materials in interstate or foreign commerce. [27 CFR 555.41(b)(2)]

53. Does a Federal license or permit exempt the holder from State or local requirements?

No. A license or permit confers no right or privilege to conduct business or operations, including storage, contrary to State or other law. All legal requirements must be followed, whether Federal, State, or local. [18 U.S.C. 848; 27 CFR 555.62]

54. Who is authorized to import explosive materials?

Any licensed importer is authorized to engage in the business of importing explosive materials for sale, distribution, or their own use. Any licensed manufacturer, dealer, or holder of a user permit may import explosive materials for their own use only. Licensees and user permittees importing explosive materials must provide to the U.S. Customs and Border Protection (CBP) a copy of the license or permit. Note, however, that in the case of certain military explosives or propellant powder or other components of small arms ammunition, Federal firearms regulations require the importer to provide an approved ATF Form 6 to the CBP. [27 CFR 555.41(b)(2), 555.41(b)(3), 447.21, 555.108(a), 555.183, 478.113]

55. How may an employee of an explosives licensee or permittee qualify to accept delivery of explosive materials for the employer?

The employee must be on the current list of representatives or agents authorized to accept delivery of explosive materials on behalf of the employer and be an authorized employee possessor of explosives. [27 CFR 555.103(b), 555.105(b)]

56. When an explosives licensee or permittee sends one of their truck drivers to the distributor's premises to pick up explosive materials that have been purchased by the licensee or permittee, will the driver be required to sign any forms?

No, however the driver is required to furnish the seller with an identification document as defined in 27 CFR Part 555.11. [27 CFR 555.103(b), 555.105(b)]

57. Will a licensee or permittee be notified in advance when the license or user permit needs to be renewed?

Generally, prior to expiration of the license or permit, a licensee or permittee will be notified. The application form must be completed and filed with ATF before expiration of the current license or permit for the renewal to be considered timely. However, if a licensee or permittee does not receive a renewal notification, it is still that licensee's or permittee's responsibility to ensure that an application is filed prior to expiration of the current license or permit. [27 CFR 555.46]

58. I have timely filed my application for renewal of my license (or user's permit) but I have not received my new license (or permit). May I continue in business even though the expiration date shown on my license or permit has passed? If so, how long?

Yes. You may continue to operate the business pursuant to your current license or permit until the application for renewal is acted upon. [5 U.S.C. 558]

59. Can a license or permit be revoked?

Yes. The Director, Industry Operations for the ATF Field Division in which a licensee or permittee is located may revoke a license or permit if the holder has violated any provision of 18 U.S.C. Chapter 40 or its implementing regulations or has become ineligible to receive explosive materials under 18 U.S.C. 842(i). [18 U.S.C. 843(d); 27 CFR 555.71, 555.74]

60. If a Federal explosives licensee or permittee is indicted for or convicted of a "crime punishable by imprisonment for a term exceeding one year", may he or she continue operations under the license or permit?

As stated in the answer to Question 4 in General Q&A, a person under indictment for, or convicted of, a crime punishable by imprisonment for a term exceeding one year is not eligible to be issued a license or permit. However, a licensee or permittee who is indicted for, or convicted of, such a crime during the term of his or her existing license or permit is not barred from licensed or permit operations for 30 days after the date of the indictment or the date the conviction becomes final. If the licensee or permittee files an application for relief from disabilities within such 30-day period, he or she may continue licensed or permit operations while the application is pending. If a relief application is not filed during that period, the licensee or permittee may not continue operations beyond such 30-day period. The right of a licensee to continue licensed or permitted operations beyond such 30-day period is also conditioned on the licensee or permittee timely filing a license or permit renewal application disclosing that the applicant has been indicted for, or convicted of, the crime. A licensee or permittee may not continue operations beyond 30 days following the date the Director issues notification that the relief application has been denied. [18 U.S.C. 845(b); 27 CFR 555.142]

61. May a licensed dealer make a sale to a holder of a limited permit in an adjoining State?

No. Sales may not be made to limited permittees who are out-of-State residents. [18 U.S.C. 842(a); 27 CFR 555.11: definition of "limited permit", 555.41(b)(3)]

Recordkeeping

62. Does a licensee or permittee have to keep records of the acquisition, distribution, and storage of explosive materials?

Yes. Licensees and permittees must keep records of acquisitions, dispositions, and storage of explosive materials. [18 U.S.C. 842(f), 847; 27 CFR 555.107, 555.122-.125, and 555.127, Subpart G]

63. How do licensees and permittees account for explosive quantities in their records?

If acquisitions are recorded by weight, then distribution must also be recorded by weight. If acquisitions are recorded by physical count (e.g., by units), then distribution must also be recorded by physical count. [27 CFR 555.122-.125]

64. Must a licensee or permittee maintain a daily summary of magazine transactions?

Yes. After the initial inventory required by regulations has been taken, the inventory shall be entered in a record of daily transactions. Not later than the close of the next business day, each licensee and permittee shall record by manufacturer's name or brand name the total quantity received in and removed from each magazine during the day and the total remaining on hand at the end of the day. [27 CFR 555.127]

65. Where must a licensee or permittee keep the daily summary of magazine transactions?

The records must either be kept at each magazine or at one central location on the business premises, provided a separate record of daily transactions is maintained for each magazine. [27 CFR 555.127]

66. How can I obtain additional copies of ATF Forms?

Forms are available on-line at www.atf.gov/forms/5000.htm#explosives. Requests for forms should be mailed to the ATF Distribution Center, 7943 Angus Court, Springfield, Virginia 22153. You may also have forms mailed to you by submitting an on-line request at www.atf.gov/dcof/index.htm, or by telephoning your request to 703-455-7801. [27 CFR 555.21(b)]

67. Does a purchaser of black powder have to sign any forms at the time of purchase?

If 50 pounds or less of commercially manufactured black powder is being purchased, and the powder is intended to be used solely for sporting, recreational, or cultural purposes in antique firearms as defined in 18 U.S.C. 921(a)(16) or in antique devices exempt from the term "destructive device" in 18 U.S.C. 921(a)(4), no form is required. However, if the black powder is being purchased for any other purpose (regardless of quantity), the purchaser or other transferee must possess a Federal explosives license or permit. [18 U.S.C. 845(a)(5); 18 U.S.C. 926(c); 27 CFR 555.141(b), 555.26(a)]

68. Is there a requirement for licensees and permittees to make an annual inventory of explosive materials on hand?

Yes. An inventory is required to be taken at least once a year. [27 CFR 555.122-.125]

69. When must ATF Form 5400.4, "Limited Permittee Transaction Report (LPTR)", be executed?

Before distribution of explosive materials to a limited permittee, the licensee or permittee must obtain an executed ATF F 5400.4 from the limited permittee with an original unaltered and unexpired Intrastate Purchase of Explosives Coupon (IPEC) attached. Except when delivery of explosive materials is made by a common or contract carrier who is an agent of the limited permittee, the licensee or permittee must verify the identity of the holder of the limited permit by examining an identification document (as defined in 555.11) and noting on the ATF F 5400.4 the type of document presented. The licensee or permittee must complete the appropriate section on ATF F 5400.4 to indicate the type and quantity of explosive materials distributed, the license or permit number of the seller, and the date of the transaction. The licensee or permittee must sign and date the form. [27 CFR 555.126(b)]

70. Do the ATF Forms 5400.4 have to be maintained by the licensee or permittee making the sale?

Yes. One copy of ATF F 5400.4 must be retained by the seller as part of his permanent records in chronological order by date of disposition, or in alphabetical order by name of limited permittee. They must be maintained for a period of five years. [27 CFR 555.126]

71. May I keep computerized records?

Yes. See ATF Ruling 2007-1.

Storage

72. Who must comply with the storage requirements?

Except for those items and activities given exempt status under 18 U.S.C. 845 (also see 27 CFR 555.141), or exempted under 27 CFR 555.32, Special Explosives Devices, all persons who store explosive materials must store them in conformity with the provisions of Subpart K of the regulations, unless the person or the materials are exempt from regulation. [18 U.S.C. 842(j); 27 CFR 555.29, 555.141, 555.201(a)]

73. What are the classes of explosive materials for storage purposes?

There are 3 classes of explosive materials:

(a) High explosives (for example, dynamite, flash powders, and bulk salutes);

(b) Low explosives (for example, black powder, safety fuses, igniters, igniter cords, fuse lighters, and "display fireworks", except for bulk salutes); and

(c) Blasting agents (for example, ammonium nitrate-fuel oil and certain water gels). [27 CFR 555.202]

74. May a person store explosive materials in a residence or dwelling?

No. Storage of explosive materials in a residence or dwelling is prohibited. [27 CFR 555.208(b), 555.210(b), 555.211(b)]

75. What is the "Table of Distances"?

This table lists the minimum acceptable distances separating explosives magazines from inhabited buildings, passenger railroads, public highways, and other explosives magazines. The table is contained in 27 CFR 555.218.

76. When low and high explosives are stored together, how is the distance determined to meet the table of distance requirements?

The table of distances at 27 CFR 555.218 would be applied using the total weight of explosive materials in the magazine. [27 CFR 555.218]

77. Is it necessary to inspect my explosives magazines on a regular basis?

Yes. Any person storing explosives must inspect the magazines at least once every 7 days to determine whether there has been unauthorized entry or attempted entry into the magazines or unauthorized removal of the contents of the magazines. [27 CFR 555.204]

78. What are the requirements for making changes or additions to an approved storage facility?

Making changes in construction to an approved explosives magazine or adding a magazine requires that ATF be notified. However, mobile or portable type 5 magazines and magazines used for the temporary (under 24 hours) storage of explosive materials are exempt from this requirement. See 27 CFR 555.63 for details.

79. Is any type of black powder fuse exempt from storage requirements?

Yes, 3/32-inch and other external burning pyrotechnic hobby fuses are exempt from the requirements of Federal explosives laws and regulations. [18 U.S.C. 845(a)(4-5); 27 CFR 555.11: definition of "ammunition", 555.141(a)(4), 555.141(b)]

80. With the exception of 3/32-inch pyrotechnic safety fuse for use in small arms, must black powder fuses generally be stored in approved explosives magazines?

Yes. Generally igniter fuses, time fuses, blasting fuses, safety fuses, or other black powder fuses by whatever name known, must be stored in approved magazines.

81. Is smokeless powder designed for use in small arms ammunition subject to the explosives storage requirements?

Smokeless propellants designed for use in small arms ammunition are exempt from regulation under 18 U.S.C. Chapter 40 and the regulations in 27 CFR Part 555. However, it should be noted that persons engaged in the business of importing or manufacturing smokeless propellants must have a Federal explosives license. Additionally, smokeless propellant designed for use other than small arms ammunition is not exempt. Therefore, explosives products such as squibs, fireworks, theatrical special effects, or other articles that may be utilizing smokeless propellants are regulated and must be stored accordingly.

82. My office building, in which several company employees work during the day in connection with my explosives business, is located in the general area of my explosives magazine. Do the regulations and the Table of Distances apply to this building as an "inhabited building"?

No. A building such as an office building or repair shop which is part of the premises of an explosives business and is used by the business in connection with the manufacture, transportation, storage, or use of explosive materials is not considered to be an "inhabited building". [27 CFR 555.11: definition of "inhabited building", 555.218]

83. Am I required to notify my State or local authorities about my explosives storage magazines?

Yes. All persons who store explosive materials must notify the fire department having jurisdiction over the site where explosive materials are manufactured or stored. Notification must be made orally by the end of the day on which storage begins and in writing within 48 hours from the time storage began. The notification must include the type of explosive materials, magazine capacity, and the location of each storage site. [27 CFR 555.11: Definition of "authority having jurisdiction for fire safety", 27 CFR 555.201(f)]

84. What is the definition of a "case hardened shackle?"

Case hardening involves putting carbon (or a combination of carbon and nitrogen) into the surface of the steel to make it a high-carbon steel, which can be hardened by heat treatment. Only the outer skin gets hard in this manner. The center is still tough and malleable. This makes for a strong lock with a tough surface.

85. Can detonators be stored with detonating cord?

No. However, products which are manufactured with a detonator attached to the detonating cord as an integral part need not be disassembled and stored separately. [27 CFR 555.213]

86. Are there storage requirements for oxidizers, such as ammonium nitrate?

In general, no. However, when a magazine or bin containing ammonium nitrate is located within the sympathetic detonation distance of other explosives or blasting agents, it must be stored in accordance with the table of distances in 27 CFR 555.220.

87. Are State and local government agencies required to store their explosive materials in conformity with Federal storage regulations?

Yes. There is no exemption in the law or regulations for the storage of explosive materials by any State or political subdivision thereof. [18 U.S.C. 842(j), 845(a)(6); 27 CFR 555.141(a)(3), (a)(5)]

Fireworks

Fireworks are defined in the Federal explosives regulations as any composition or device designed to produce a visible or an audible effect by combustion, deflagration, or detonation. Fireworks are further divided into two broad classifications, consumer fireworks or display fireworks as defined at 27 CFR Part 555.11.

88. Are "consumer fireworks" subject to regulation under the Federal explosives laws?

No. The importation, distribution, and storage of fireworks defined as consumer fireworks are exempted from the provisions of the Federal explosives laws. However, because they contain pyrotechnic compositions classed by ATF as explosive materials, the manufacture of consumer fireworks requires a manufacturer's license. In addition, pyrotechnic compositions used in the manufacture of consumer fireworks must be stored in accordance with regulations in 27 CFR Subpart K. Consumer fireworks are defined as "any small firework device designed to produce visible effects by combustion and which must comply with the construction, chemical composition, and labeling regulations of the U.S. Consumer Product Safety Commission, as set forth in title 16, Code of Federal Regulations, parts 1500 and 1507. Some small devices designed to produce audible effects are included, such as whistling devices, ground devices containing 50 mg or less of explosive materials, and aerial devices containing 130 mg or less of explosive materials. Consumer fireworks are classified as fireworks UN0336 and UN0337 by the U.S. Department of Transportation at 49 CFR 172.101. This term does not include fused set pieces containing components which together exceed 50 mg of salute powder." [27 CFR 555.11: definition of "consumer fireworks"; definition of "licensed manufacturer", 555.141(a)(7)]

89. Are "display fireworks" considered to be explosive materials subject to regulation under Federal explosives laws and regulations?

Yes. Display fireworks include, but are not limited to, salutes containing more than 2 grains (130 mg) of explosive materials, aerial shells containing more than 40 grams of pyrotechnic compositions, and other display pieces which exceed the limits of explosive materials for classification as "consumer fireworks". These fireworks are classified as fireworks UN0333, UN0334, or UN0335 by regulations of the U.S. Department of Transportation at 49 CFR 172.101. Display fireworks also include fused set pieces containing components which together exceed 50 mg of salute powder. [27 CFR 555.11: definition of "display fireworks"]

90. How must display fireworks be stored?

Display fireworks, with the exception of bulk salutes, are considered low explosives and, at a minimum, must be stored in type 4 storage magazines. They may also be stored in type 1 or type 2 magazines. Bulk salutes, which are defined as either salute components prior to final assembly into aerial shells, (or) finished salute shells held separately prior to being packed with other types of display fireworks, are classified as high explosives. As such, bulk salutes may only be stored in type 1 or type 2 magazines specifically constructed for the storage of high explosives. [27 CFR 555.11, 555.202(b), 555.203(d), 555.207, 555.208, 555.210]

91. Are "Articles Pyrotechnic" subject to the requirements of the Federal explosives regulations?

The importation, distribution, and storage of fireworks defined as "Articles Pyrotechnic", are exempt from the Federal explosives laws and regulations. However, because they contain pyrotechnic compositions classed by ATF as explosive materials, the manufacture of items defined as "articles pyrotechnic" requires an ATF manufacturer's license. In addition, pyrotechnic compositions used in the manufacture of articles pyrotechnic must be stored in accordance with regulations in 27 CFR Subpart K. [27 CFR 555.11: definitions of "articles pyrotechnic" and "consumer fireworks", 555.141(a)(7)]

92. Must partially assembled display fireworks be removed from a drying building for overnight storage?

Yes. At the end of a day's manufacturing operations, all dry explosive powders and mixtures and partially assembled and finished display fireworks must be removed from fireworks process buildings and stored in a magazine meeting the storage requirements in 27 CFR Part 555, Subpart K. [27 CFR 555.205, 555.221]

93. What areas of a fireworks manufacturing plant are considered to be "fireworks process buildings?"

Fireworks process buildings include any buildings in which pyrotechnic compositions or explosives materials are mixed, pressed, finished, or assembled. Fireworks process buildings do not include plant warehouses, office buildings, or other buildings and areas in which no fireworks, pyrotechnic compositions, or explosive materials are processed or stored. [27 CFR 555.11: definition of "fireworks process building"]

94. Under what conditions may I temporarily store display fireworks (including low explosives for choreographed shows) on trucks?

See ATF Ruling 2007-2.

95. What types of fireworks require an ATF license or permit in order to be lawfully transported or received?

Any fireworks defined as "display fireworks" in 27 CFR 555.11 may be lawfully received or transported only by persons who hold a valid license or permit. No ATF license or permit is required to receive or transport "consumer fireworks" or "articles pyrotechnic". [18 U.S.C. 842(a)(3); 27 CFR 555.26, 555.141(a)(7)]

Plastic Explosives

96. What is a plastic explosive?

A plastic explosive is defined as "an explosive material in flexible or elastic sheet form formulated with one or more high explosives which in their pure form has a vapor pressure less than 10^{-4} Pa at a temperature of 250 C., is formulated with a binder material, and is as a mixture malleable or flexible at normal room temperature." [18 U.S.C. 841(q); 27 CFR 555.180(d)(4)]

97. What plastic explosives are required to contain detection agents?

All plastic explosives manufactured or imported on or after April 24, 1996, must contain a detection agent. Federal law enforcement agencies and the military may possess unmarked plastic explosives if they meet the requirements of the use-up period described in Question 103. [18 U.S.C. 841(q), 842(n); 27 CFR 555.180]

98. What are the permissible detection agents for marking plastic explosives?

These agents are listed in the law and regulations at 18 U.S.C. 841(p) and 27 CFR 555.180(d)(3).

99. Is it lawful to manufacture plastic explosives that do not contain a detection agent?

No. [18 U.S.C. 842(l); 27 CFR 555.180(a)]

100. Is it lawful to import into the United States plastic explosives that do not contain a detection agent?

No. The importation of plastic explosives into the United States requires that the importer file ATF Form 6 certifying that the imported plastic explosives contain the required detection agent, or is exempted from the marking requirements as provided in the regulations. [18 U.S.C. 842(m); 27 CFR 555.180(b), 555.182, 555.183]

101. Is it lawful to ship, transport, transfer, receive, or possess any plastic explosive that does not contain a detection agent?

No. However, a 15-year use-up period is provided for Federal law enforcement agencies and the military for unmarked plastic explosives imported into or manufactured in the U.S. prior to April 24, 1996. [18 U.S.C. 842(n); 27 CFR 555.180(c)]

102. If a person acquired plastic explosives not containing a detection agent before April 24, 1996, may he or she continue to lawfully possess the explosives?

No. With the exception of the use-up period provided by law for Federal law enforcement agencies or the military, the time period for lawful possession of unmarked plastic explosives terminated on April 24, 1999. [18 U.S.C. 842(n); 27 CFR 555.180(c)]

103. Are police departments exempt from the prohibition against possessing unmarked plastic explosives after April 24, 1999?

No. Police departments and other State or local law enforcement agencies could lawfully possess unmarked plastic explosives acquired on or before April 24, 1996, until April 24, 1999. Such agencies still possessing unmarked plastic explosives should destroy them or abandon them to ATF. Contact the nearest ATF field office for information. [18 U.S.C. 842(n); 27 CFR 555.180(c)(1)]

U.S. Military Explosives

104. Would an ATF license or permit be needed to demilitarize (demil) U.S. military explosives?

As long as the demil operator has a valid Department of Defense contract to perform such operations, the operations would be exempt from 27 CFR Part 555 and no license or permit would be required. However, if title to the explosive materials has passed from the military to the demil operator and the operator intends to resell the explosives on the commercial market, then such operations may be regulated by ATF (e.g., storage, sales, manufacturing) and an ATF license or permit may be needed. Contact the nearest ATF field office for further information. [18 U.S.C. 845(a)(3), (a)(6) and 27 CFR 555.141(a)(3), (a)(5)]

105. Would a civilian contractor who is manufacturing explosive materials pursuant to a government contract for or on behalf of the United States military be entitled to the exemptions from the explosives laws and regulations?

Yes, provided that all the explosive materials in question are manufactured under a government contract. Any explosive materials manufactured in anticipation of receiving a government contract would not qualify for this exemption.

If the contractor manufactures any explosive materials not pursuant to a U.S. military contract, the manufacture and the explosive materials are subject to all requirements of the law and regulations. [18 U.S.C. 845(a)(3), (a)(6); 27 CFR 555.26, 555.41, 555.141(b)]

106. Is an ATF licensee or permittee, whose licensed premises are located on a U.S. military installation, subject to the regulations in 27 CFR Part 555?

All activities conducted outside the scope of a U.S. Government contract are subject to the requirements of Part 555, even if the activities are conducted on property owned by the military. [18 U.S.C. 845(a)(3), (a)(6); 27 CFR 555.26, 555.29, 555.41, 555.141(a)(3), (a)(5)]

Index to Questions and Answers

ATF Explosives Rulings (Revised 1/07)

Table of Contents

1. 27 CFR 55.11: Meaning of Terms (Also § 55.206)

An office or repair shop used in connection with the manufacture, etc. of explosive materials is not an "inhabited building."

ATF Ruling 75-20

ATF has held that a building, such as an office or repair shop, which is a part of the premises of an explosives manufacturer and is used in connection with the manufacture, transportation, storage, or use of explosive materials, is not an "inhabited building." Section 55.11 of 27 CFR defines inhabited building as "any building regularly occupied in whole or in part as a habitation for human beings, or any church, schoolhouse, railroad station, store, or other structure where people are accustomed to assemble, except any building occupied in connection with the manufacture, transportation, storage, or use of explosive materials."

Regulations in 27 CFR §§ 55.206 and 55.218 set forth provisions concerning the location of storage facilities and the minimum distances such storage facilities may be located from, among other things, "inhabited buildings." These provisions are intended to provide protection to persons who inhabit buildings located near premises where explosives are manufactured, stored, etc. However, it is the intent of § 55.11 to exempt buildings used by the explosives industry in connection with the manufacture, transportation, storage, or use of explosive materials from the table of distance requirements on "inhabited buildings."
[75 CB 64]

2. 27 CFR 55.207: Construction of Type 1 Magazines (Also § 55.210)

Certain explosives storage facilities meeting standards of construction prescribed by the Department of Defense Explosives Safety Board for such storage are approved by the Bureau.

ATF Ruling 75-21

ATF has held that explosives storage facilities with smooth-finished concrete floors that were constructed under contract for the use of the Department of Defense (DOD) and that are presently being leased to licensees and permittees for the storage of commercial explosives are considered to be in compliance with the requirements for nonsparking floors, as set forth in 27 CFR §§ 55.207(a)(4), 55.207(b), and 55.210, for the storage of all types of fully packaged explosives, pyrotechnics and propellants, with the exception of black powder. Any other such magazines which have smooth finished concrete floors and which meet or exceed DOD construction specifications will also be considered to be in compliance with the requirements of Part 555 with respect to nonsparking floors. It is the responsibility of the licensee or permittee to provide verification that such facilities were manufactured under DOD specifications or that the facilities meet or exceed such specification standards. If the Division Director determines that the concrete floors of type 1 or type 4 magazines do not meet the preceding requirements, he will require such floors to be covered with a nonsparking material, such as epoxy paint or mastic. [75 CB 67]

3. 27 CFR 55.41: Licenses and Permits-General

Certain companies that manufacture explosive materials for use in their own operations are required to obtain licenses as manufacturers of explosive materials.

ATF Ruling 75-31

ATF has held that companies, such as public utility companies engaged in line and facility construction, which manufacture explosives on a regular or continual basis are considered to be engaged in the business of manufacturing explosive materials and must be appropriately licensed as required by 18 U.S.C. 842. The term "manufacturer" is defined in 18 U.S.C. 841(h) as "any person engaged in the business of manufacturing explosive materials for purposes of sale or distribution or for his own use." Although the term "engaged in the business" is not susceptible to a rigid definition within 18 U.S.C. §§ 841-848, it is interpreted to imply an element of continuity or habitual practice; an element clearly present in the operations of companies described herein. Therefore, these companies are considered to be "engaged in the business" and must be licensed as explosives manufacturers. [75 CB 65]

4. 27 CFR 555.109: Identification of Explosive Materials

Methods of marking containers of explosive materials are prescribed.

ATF Ruling 75-35

Editors note: ATF Ruling 75-35 was rendered obsolete pursuant to ATF 5F, 70 Federal Register 30626 (May 27, 2005), and effective July 26, 2005.

5. 27 CFR 55.11: Meaning of Terms- State of Residence

"State of residence" of business entities who use explosive materials; distribution of explosive materials by licensees to out-of-State business entities other than licensees and permittees; and distribution to nonresident employees of such entities are discussed.

Editor's Note: Provisions of ATF Ruling 76-4 were modified, in part, by the Safe Explosives Act.

Effective May 24, 2003, it is unlawful for any person to receive explosive materials unless such person holds an ATF license of permit. It is also unlawful for any licensee or permittee to knowingly distribute explosive materials to any person who does not hold a license or permit. The only relevance remaining in the term "State of residence" is for distribution of explosive materials to, and receipt by, limited permit holders. Pursuant to 18 U.S.C. 842(a)(3) and (a)(4), limited permit holders may, on not more than 3 separate occasions, lawfully receive explosive materials from a licensee or permittee whose premises are located within the state of residence of the limited permit holder. ATF Rule 76-4 continues to apply in determining whether a limited permit holder has acquired a "State of residence" for purposes of receipt of explosives under 18 U.S.C. 842(a)(4)(B).

6. 27 CFR 55.126: Explosives Transaction Record

Under certain conditions, a single Form 5400.4 may be used to cover a series of deliveries.

Editor's Note: The provisions of ATF Rule 76-10 were rendered obsolete by ATF No. 1, 68 FR 13791, Mar. 20, 2003.

7. 27 CFR 55.207: Construction of Type 1 Storage Facilities (Also § 55.208)

Alternate construction standards for storage facilities for explosive materials are prescribed.

ATF Ruling 76-18

Section 842(j) of 18 U.S.C. states: "It shall be unlawful for any person to store any explosive material in a manner not in conformity with regulations promulgated by the Secretary. In promulgating such regulations, the Secretary shall take into consideration the class, type, and quantity of explosive materials to be stored, as well as the standards of safety and security recognized in the explosives industry." The regulations in 27 CFR §§ 55.207 and 55.208 prescribe types of storage facilities for explosive materials and provide (among other things) that such storage facilities shall be bullet resistant.

Section 55.201(b) provides that alternate storage facilities may be authorized for the storage of explosive materials when it is shown that such alternate facilities are or will be constructed in a manner substantially equivalent to the standards of construction contained in the applicable regulations.

The term "bullet-resistant" means resistant to penetration of a bullet of 150 grain M2 ball ammunition having a nominal muzzle velocity of 2700 feet per second fired from a .30 caliber rifle from a distance of 100 feet perpendicular to the wall or door. It has been determined that a wide range of construction criteria meet the bullet-resistant requirements of regulations for construction of storage facilities for explosive materials. In order to promote standards of safety and security in the storage of explosive materials while allowing the industry a wide latitude in the selection of construction materials, it is held that storage facilities (magazines) that are constructed according to the following minimum specifications are bullet-resistant and meet the requirements of the regulations as set forth in 27 CFR Part 55 (All steel and wood dimensions are actual thicknesses. To meet the concrete block and brick dimensions indicated, the manufacturers' represented thicknesses may be used).

(a) Exterior of ⅝ inch steel, lined with an interior of any type of nonsparking material.

(b) Exterior of ½ inch steel, lined with an interior of not less than ⅜ inch plywood.

(c) Exterior of ⅜ inch steel, lined with an interior of two inches of hardwood.

(d) Exterior of ⅜ inch steel, lined with an interior of three inches of softwood or 2¼ inches of plywood.

(e) Exterior of ¼ inch steel, lined with an interior of three inches of hardwood.

(f) Exterior of ¼ inch steel, lined with an interior of five inches of softwood or 5 ¼ inches of plywood.

(g) Exterior of ¼ inch steel, lined with an intermediate layer of two inches of hardwood and an interior lining of 1½ inches of plywood.

(h) Exterior of 3⁄16 inch steel, lined with an interior of four inches of hardwood.

(i) Exterior of 3⁄16 inch steel, lined with an interior of seven inches of softwood or 6¾ inches of plywood.

(j) Exterior of 3⁄16 inch steel, lined with an intermediate layer of three inches of hardwood and an interior lining of ¾ inch of plywood.

(k) Exterior of ⅛ inch steel, lined with an interior of five inches of hardwood.

(l) Exterior of ⅛ inch steel, lined with an interior of nine inches of softwood.

(m) Exterior of ⅛ inch steel, lined with an intermediate layer of four inches of hardwood and an interior lining of ¾ inch plywood.

(n) Exterior of any type of fire-resistant material which is structurally sound, lined with an intermediate layer of four inches of solid concrete block, OR four inches of solid brick OR four inches of solid concrete; AND, an interior lining of ½ inch plywood placed securely against the masonry lining.

(o) Standard eight inch concrete block with voids filled with well-tamped sand/cement mixture.

(p) Standard eight inch solid brick.

(q) Exterior of any type of fire-resistant material which is structurally sound, lined with an intermediate six inch space filled with well-tamped dry sand or well-tamped sand/cement mixture.

(r) Exterior of 1/8 inch steel, lined with a first intermediate layer of ¾ inch plywood, a second intermediate layer of 3⅝ inches of well-tamped dry sand or sand/cement mixture and an interior lining of ¾ inch plywood.

(s) Exterior of any type of fire-resistant material, lined with a first intermediate layer of ¾ inch plywood, a second intermediate layer of 3⅝ inches well-tamped dry sand or sand/cement mixture, a third intermediate layer of ¾ inch plywood, and a fourth intermediate layer of two inches of hardwood OR 14 gauge steel AND an interior lining of ¾ inch plywood.

(t) Eight inch thick solid concrete. [76 CB 106]

8. 27 CFR 55.213: Quantity and Storage Restrictions(Also § 55.208)

Alternate magazine construction standards for storage of electric blasting caps with other explosive materials are prescribed.

ATF Ruling 77-24

Section 842(j) of 18 U.S.C. states: "It shall be unlawful for any person to store any explosive material in a manner not in conformity with regulations promulgated by the Secretary. In promulgating such regulations, the Secretary shall take into consideration the class, type, and quantity of explosive materials to be stored, as well as the standards of safety and security recognized in the explosives industry." The regulations in 27 CFR § 55.213 restrict the storage of blasting caps with other explosive materials. Section 55.201(b) provides that alternate storage magazines may be authorized for the storage of explosive materials when it is shown that such alternate magazines are or will be constructed in a manner substantially equivalent to the standards of construction contained in the applicable regulations.

ATF recognizes that the transportation and storage of explosive materials in the same

vehicle along with electric blasting caps is often desired. The Institute of Makers of Explosives established a recommended standard for such transport in their Safety Library Publication No. 22, dated November 5, 1971 [revised January 1985]. This standard prescribes the minimum construction criteria for:

(a) A container securely attached—

(1) Above the cab of the vehicle (see Figure 1, Appendix A), and

(2) To the vehicle frame under the cargo space (see Figure 2, Appendix A), or

(b) A built-in compartment in the cargo space of the vehicle (see Appendix B).

In addition to motorized vehicles, consideration was also given for the use of similar criteria on portable wheeled trailers being used as magazines under § 55.208(a) of the regulations (see

Appendix E). In order to insure standards of safety and security in the storage of explosive materials while allowing the industry a proper latitude in the construction of magazines, it is held that vehicles used for transporting and for storing explosive materials that are constructed in conformity with the standards listed below, and in compliance with all other safety and security provisions contained in Part 55 (e.g., effectively immobilized when unattended) will meet the requirements of ATF regulations. Even though constructed on the same vehicle, each compartment will be considered as a separate magazine. The two magazines on the vehicle will, however, be considered as one magazine when applying the American Table of Distances [see Table at § 55.218].

Construction Standards For Storage of Electric Blasting Caps (Non Mass-Detonating)

a. The container or compartment must provide for total enclosure of the electric blasting caps.

b. The partition between the explosives storage compartment and the electric blasting cap compartment must be of laminate construction consisting of A/C grade or better exterior plywood, gypsum board [sheetrock] and low carbon steel plates. In order of arrangement, the laminate must conform to the following, with minimum thickness of each lamination as indicated:
½ inch plywood
½ inch gypsum board [sheetrock],
⅛ inch low carbon steel, and,
¼ inch plywood, with the ¼ inch plywood facing the explosives storage compartment. See Appendix C for details of laminate construction. The door to the electric blasting cap compartment must be of metal construction or solid wood covered with metal; the outside walls and top must be of the same construction as the rest of the vehicle or trailer. If high explosives or bullet sensitive explosive materials are stored in the vehicle, then the storage compartment of the vehicle must be constructed so as to be bullet-resistant.

c. As an alternative to the construction requirements shown in paragraph b, a container for use only as illustrated in Appendix A may be used when constructed as follows:

1. The top, lid or door, and the sides and bottom of each container must be of laminate construction consisting of A/C grade or better exterior plywood, solid hardwood, gypsum board [sheetrock], and sheet metal. In order of arrangement, the laminate must conform to the following, with minimum thickness of each lamination as indicated:
¼ inch plywood,
1 inch solid hardwood,
½ inch plywood,
½ inch gypsum board [sheetrock] (OR ¼ inch particle board), and 22 gauge sheet metal, constructed inside to outside in that order. See Appendix D for details of laminate construction.

2. The hardwood must be fastened together with wood screws, the ½ inch plywood must be fastened to the hardwood with wood screws, the inner ¼ inch plywood must be fastened to the hardwood with adhesive, and the 22 gauge sheet metal must be attached to the exterior of the container with screws.

d. The laminate composite material must be securely bound together by waterproof adhesive or other equally effective means.

e. The steel plates at the joints of laminations must be secured by continuous fillet welds.

f. All interior surfaces of the container or compartment must be constructed so as to prevent contact of contents with any sparking metal.

g. There must be direct access to the container or into a compartment from outside the vehicle.

h. Each container or compartment must have a snug fitting continuous piano-type hinged lid or door equipped with a locking device (or devices).

i. Without permitting direct access to contents under normal conditions, the locking or hinging mechanisms must permit at least one edge of the lid or door to rise or move outward at least ½ inch when subjected to internal pressure.

j. The exterior of the container or compartment must be weather-resistant. [77 CB 191]

9. 18 U.S.C. 842(j): Storage of Explosives

27 CFR 55.208(b)(1), 55.210(b)(1), and 55.211(b)(1): Indoor Storage of Explosives in a Residence or Dwelling

ATF will approve variances to store explosives in a residence or dwelling only upon certain conditions including, but not limited to, receipt of a certification of compliance with State and local law, and documentation that local fire safety officials have received a copy of the certification.

ATF Ruling 2002-3

Section 842(j) of 18 U.S.C. states: "It shall be unlawful for any person to store any explosive material in a manner not in conformity with regulations promulgated by the Secretary."

The regulations in 27 CFR 55.208(b)(1), 55.210(b)(1), and 55.211(b)(1) specify that no indoor magazine is to be located in a residence or dwelling. Section 55.22 specifies that the Director may allow alternate methods or procedures in lieu of a method or procedure specifically prescribed in the regulations. Specifically, section 55.22(a)(3) provides that such "variances" are permissible only in certain circumstances, including where "[t]he alternate method or procedure will not be contrary to any provision of law and will not . . . hinder the effective administration of this part."

ATF has been advised that certain variances previously approved for storage of explosives in residences or dwellings are in violation of State or local zoning law. ATF believes it is important to ensure that approval of variances is in compliance with all State and local provisions.

To obtain a variance for indoor storage of explosives in a residence or dwelling, ATF has determined that a person must submit to ATF a certification signed under penalty of perjury along with the request for the variance. The certification must:

1. State that the proposed alternative storage method will comply with all applicable State and local law;

2. Provide the name, title, address, and phone number of the authority having jurisdiction for fire safety of the locality in which the explosive materials are being stored; and,

3. Demonstrate that the person has mailed or delivered the certification to the authority identified in (2).

When required by the Director, such persons must furnish other documentation as may be necessary to determine whether a variance should be approved.

Held, ATF will approve variances to store explosives in a residence or dwelling only upon certain conditions including, but not limited to, receipt of a certification of compliance with State and local law, and documentation that local fire safety officials have received a copy of the certification.

Date signed: August 23, 2002

10. 18 U.S.C. 842(j): Storage of Explosives 27 CFR 55.208(b)(1), 55.210(b)(1), and 55.211(b)(1): Indoor Storage of Explosives in Business Premises Directly Adjacent to a Residence or Dwelling

ATF Ruling 2002-4

ATF requires approval of variances for indoor storage of explosives in business premises directly adjacent to a residence or dwelling.

Section 842(j) of 18 U.S.C. states: "It shall be unlawful for any person to store any explosive material in a manner not in conformity with regulations promulgated by the Secretary."

The regulations in 27 CFR 55.208(b)(1), 55.210(b)(1), and 55.211(b)(1) specify that no indoor magazine is to be located in a residence or dwelling. Section 55.22 specifies that the Director may allow alternate methods or procedures in lieu of a method or procedure specifically prescribed in the regulations. Specifically, section 55.22(a)(3) provides that such "variances" are permissible only in certain circumstances, including where "[t]he alternate method or procedure will not be contrary to any provision of law and will not … hinder the effective administration of this part."

ATF has been asked whether businesses that are directly adjacent to living quarters may lawfully store explosive materials in the business premises. The issue presented is whether the premises amount to a "residence or dwelling" within the meaning of the regulations cited above.

Even where the business premises are segregable from the living quarters by the existence of a door or a common wall, the business premises retain their character as a residence or dwelling. Accordingly, indoor storage of explosives in such premises is generally prohibited and can be allowed only pursuant to an approved variance.

Held, ATF requires approval of variances for indoor storage of explosives in business premises directly adjacent to a residence or dwelling. ATF may approve such variances upon receipt of all appropriate certification and other documentation as may be requested.

Date signed: August 23, 2002

11. 18 U.S.C. 842(f): Unlawful Acts, 27 CFR 555.105(b)(6)(iii): Distribution of Explosives to Limited Permittees

Distributors distributing explosive materials to holders of limited permits via common or contract carrier may verify receipt of the explosive materials by telephone, facsimile, e-mail, or other means within three business days of shipment in lieu of requiring the common or contract carrier to verify the identity of the person accepting delivery of the explosives. The distributor shall make a notation on ATF Form 5400.4 indicating whether the shipment was received and the date and time of the contact with the distributee.

ATF Ruling 2003-5

The Bureau of Alcohol, Tobacco, Firearms and Explosives (ATF) has received questions from the explosives industry regarding the requirement under 27 CFR 555.105(b)(6)(iii) that, effective May 24, 2003, a common or contract carrier hired by a Federal explosives licensee or permittee verify the identity of the distributee, note the type and number of the identification document, and provide this information to the distributor. The distributor is required to record this information on ATF Form 5400.4, Limited Permittee Transaction Report (LPTR). Industry members have informed ATF that this requirement places an undue burden on common and contract carriers. Drivers are concerned that verifying the identity of persons accepting delivery of explosive materials by examining an identification document and providing the identification information to the distributor will be overly time consuming. Drivers are also concerned that they could be held personally liable for delivering explosives to persons not authorized to receive them.

ATF imposed the verification requirements of section 555.105(b)(6)(iii) to ensure that when explosive materials are sold by a distributor to a holder of a limited permit and transported by a common or contract carrier hired by the distributor, the explosive materials are delivered only to a person authorized to receive them. ATF continues to believe it is important that sellers of explosive materials verify that such materials are delivered to persons authorized to receive them.

However, it was not ATF's intention to impose an undue burden on common or contract carriers.

Section 555.22, Title 27, CFR, provides that the Director may approve an alternate method or procedure in lieu of a method or procedure specifically prescribed in the regulations when he finds that:

(1) Good cause is shown for the use of the alternate method or procedure;

(2) The alternate method or procedure is within the purpose of, and consistent with the effect intended by, the specifically prescribed method or procedure and that the alternate method or procedure is substantially equivalent to that specifically prescribed method or procedure; and

(3) The alternate method or procedure will not be contrary to any provision of law and will not result in an increase in cost to the Government or hinder the effective administration of Part 555.

ATF finds that there is good cause to authorize a variance to the provisions of section 555.105(b)(6)(iii) due to the undue burden placed on common or contract carriers by the verification requirement. Accordingly, ATF authorizes the following alternate method or procedure to the identification verification requirements of section 555.105(b)(6)(iii):

The distributor shall, no later than three business days after shipment of the explosive materials, contact the distributee by telephone, facsimile, e-mail or any other means to ensure that the shipment has been received. The distributor shall make a notation on ATF Form 5400.4 indicating whether the shipment was received and the date and time of the contact with the distributee.

ATF finds that the above alternate method is consistent with the verification provisions of section 555.105(b)(6)(iii), because it will ensure that delivery has taken place and document the information in the distributor's records.

The alternate method is not contrary to any provision of law, will not increase the costs to ATF, and will not hinder the effective administration of the regulations in 27 CFR Part 555.

Held, pursuant to 27 CFR 555.22, ATF authorizes a variance from the requirements of 27 CFR 555.105(b)(6)(iii) for Federal explosives licensees and permittees making distributions of explosive materials to holders of limited permits via common or contract carrier. As an alternate method or procedure, distributors distributing explosive materials to holders of limited permits via common or contract carrier may verify receipt of the explosive materials by telephone, facsimile, e-mail or other means within three business days of shipment in lieu of requiring the common or contract carrier to verify the identity of the person accepting delivery of the explosives. The distributor shall make a notation on ATF Form 5400.4 indicating whether the shipment was received and the date and time of the contact with the distributee.

Date signed: May 23, 2003.

12. 27 CFR 555.11: Meaning of Terms

ATF provides guidance on three different private roads and whether they are "highways" as defined in 27 CFR 555.11.

ATF Ruling 2005-2

The Federal explosives laws, 18 U.S.C. Chapter 40, require all persons to store explosive materials in a manner in conformity with regulations issued by the Attorney General. 18 U.S.C. 842(j). The Attorney General has delegated his authority to administer and enforce the Federal explosives laws to the Director, ATF. 28 CFR 0.130. Regulations in 27 CFR Part 555 implement the provisions of the Federal explosives laws. Storage regulations in 27 CFR Part 555, Subpart K, provide that outdoor magazines in which high explosives are stored must be located no closer to inhabited buildings, passenger railways, public highways, or other magazines in which high explosives are stored than the minimum distances specified in the table of distances for storage of explosive materials in section 555.218 of the regulations. 27 CFR 555.206.

Section 555.11 of the regulations defines the term "highway" as "any public street, public alley, or public road, including a privately financed, constructed, or maintained road that is regularly and openly traveled by the general public."

In Scenario A, a private road owned by a corporation is used by the general public as an access road to a parking lot owned by the corporation. The road is near an explosives magazine. The road does not have a gate, sign, or other means of restricting access to the road. The road is also used by the general public on a daily basis to gain access to other public streets.

In Scenario B, a company that manufactures display fireworks, a logging company, and an individual who owns buildings utilized to store his collection of automobiles all occupy property to which the only access is a privately owned road. A separate party that leases to these three entities owns the property. The road is located on private property, and a locked gate at the entrance to the road prevents access by the general public. The display fireworks company, the logging company, and the individual storing automobiles all have keys to unlock the gate and travel on the road when needed. The gate is locked at all times, and there is no evidence that the road is open to anyone other than the two businesses and one individual who require access to their property.

In Scenario C, an explosives company maintains explosives magazines in a quarry area that has a roadway traversing through the quarry. The quarry owns the property, and the road is maintained by the quarry. The road has a gate and there are signs advising no trespassing. However, when ATF officials visited the location on several occasions, the gate was left open and members of the public regularly utilized the roadway as a shortcut between two major highways. There were no indications the owner of the property took any steps to prevent members of the public from utilizing the roadway.

Applying the regulatory definition of "highway" to the three scenarios, the road in Scenario A is clearly a highway that is subject to the tables of distance in Part 555. Although it is privately owned, it is regularly and openly traveled by members of the general public without restriction.

The roadway in Scenario B is not a "highway" as defined. Access is restricted at all times and there is no evidence the general public regularly travels on the roadway. Access to the road is provided to only a limited number of persons who have a legal right to travel the road. Accordingly, this road is not regularly and openly traveled by members of the general public.

ATF concludes that the roadway described in Scenario C is a "highway" as defined in 27 CFR 555.11. Although access to the roadway is restricted by a gate and "No trespassing" signs are posted, the gate is not closed at all times. Furthermore, ATF observation indicates that the roadway is regularly and openly traveled by members of the general public. Based on these facts, the roadway is a highway which is subject to the tables of distance in 27 CFR Part 555.

Held, a private road with no gate, signs, or other means of restricting access that is used by the general public as an access road to a parking lot and as access to other public streets is a "highway" as defined in 27 CFR 555.11.

Held further, a private road with a locked gate at the entrance that is locked at all times and used by a limited number of persons leasing or owning property accessed by the road is not a "highway" as defined in 27 CFR 555.11.

Held further, a private roadway traversing a quarry with a gate restricting access and a "no trespassing" sign is a "highway," as defined in 27 CFR 555.11, because the gate is not locked at all times and the general public regularly utilizes the roadway as a shortcut between two public highways.

Date approved: September 8, 2005

13. 27 CFR 555.11: Meaning of Terms

ATF provides guidance on two situations involving structures and whether they are "inhabited buildings" as defined in 27 CFR Part 555.

ATF Ruling 2005-3

The Federal explosives laws, 18 U.S.C. Chapter 40, require all persons to store explosive materials in a manner in conformity with regulations issued by the Attorney General. 18 U.S.C. 842(j). The Attorney General has delegated his authority to administer and enforce the Federal explosives laws to the Director, ATF. 28 CFR 0.130. Regulations in 27 CFR Part 555 implement the provisions of the Federal explosives laws. Storage regulations in 27 CFR Part 555, Subpart K, provide that outdoor magazines in which high explosives are stored must be located no closer to inhabited buildings, passenger railways, public highways, or other magazines in which high explosives are stored than the minimum distances specified in the table of distances for storage of explosive materials in section 555.218 of the regulations. 27 CFR 555.206.

The regulation at 27 CFR 555.11 defines the term "inhabited building" as "any building regularly occupied in whole or in part as a habitation for human beings, or any church, schoolhouse, railroad station, store, or other structure where people are accustomed to assemble, except any building occupied in connection with the manufacture, transportation, storage, or use of explosive materials."

In Scenario A, an explosives licensee leases explosives magazines to an individual who uses the magazines for storage of goods other than explosive materials. The magazines are located adjacent to magazines used by the explosives licensee for the storage of explosive materials. The magazines used by the lessee are not separated by the minimum distances required for the separation of magazines from "inhabited buildings" as required by the regulations in 27 CFR Part 555.

The magazines are visited regularly by the individual who stores property in the magazines, but no additional persons accompany the individual when he is present at the magazine. However, the individual hires contractors to repair equipment stored in the facilities from time to time, and 1-3 employees of the contractor may occasionally be present for short periods of time at the storage site. However, such visits occur no more than 3-5 times per year. In addition, an employee from the water company visits the storage site once a month to read the water meter, an employee from the power company reads the power meter once a month, and other vendors may be present at the site for short periods of time for other legitimate purposes.

In Scenario B, a licensed manufacturer of explosives X leases a unit in an industrial park that shares a common wall with a unit leased by licensed manufacturer Y. Both licensees store explosives in magazines located inside and outside the units. The magazines of Manufacturer X and the building used by Manufacturer Y are not separated by the minimum distance prescribed in 27 CFR Part 555 for the separation of magazines and inhabited buildings. Likewise, the magazines of Manufacturer Y and the building used by Manufacturer X are not separated by the minimum distance prescribed in 27 CFR Part 555 for the separation of magazines and inhabited buildings. In both structures, employees and contractors are regularly present during work hours for purposes of carrying on the manufacturing and distributing businesses of the two licensees. This includes personnel who work in the manufacturing plant, those who work on the loading dock to load and ship explosives products to customers, and those who work in the office taking orders, sending out invoices, and handling other clerical work for the businesses.

Applying the law and regulations to the facts of Scenario A, ATF concludes that the leased structures used by the individual to store items other than explosives are not being used as a habitation for human beings and are not buildings occupied in connection with the manufacture, transportation, storage, or use of explosive materials. Accordingly, the sole issue remaining is whether the structure is one where people are accustomed to assemble.

Noteworthy, the regulation uses the term "people," which is the plural version of "person." Thus, ATF believes that more than one person must "assemble" at the structure for it to be an "inhabited building." In addition, the word "assemble" is defined, in part, as "To bring or gather together in a group or whole." The word "assembly" is defined, in part, as "A group of persons gathered for a common purpose." The American Heritage Dictionary, Second College Edition, Houghton Mifflin Co., 1982. It is clear that the presence of one person at a structure or location cannot be an assembly of any sort. Accordingly, in situations where one person is present at a particular structure, whether on a regular or infrequent basis, such a structure is not an "inhabited building" as defined in 27 CFR 555.11.

Likewise, occasional visits to the storage facility by mail delivery persons or employees of public utility companies for brief periods of time would not be an "assembly" that would make the facility an inhabited building. However, if 2 or more repair persons are present at the facility to make repairs to equipment stored there, such persons would be there for a common purpose, and would have "assembled" at the structure. However, the structure would be an "inhabited building" only if it is a structure where people are accustomed to assemble. The word "accustom" is defined as "To familiarize, as by constant practice, use, or habit: accustomed himself to working long hours." The word "accustomed" is defined as "Usual, characteristic, or normal: worked with her accustomed thoroughness." The American Heritage Dictionary, Second College Edition, Houghton Mifflin Co., 1982. These definitions indicate that a structure will be one where people are accustomed to assemble only if there is some degree of continuity, regularity, or frequency to such assembly.

Infrequent, occasional visits to the storage site by 2 or more repair persons would not make the storage facility an "inhabited building," because such intermittent visits would not be "customary." Only where 2 or more persons are present at the site for a common purpose and on a regular basis would the building fit within the definition of "inhabited building."

To address Scenario B, it is apparent that the building leased by Manufacturer X is exempt from the definition of "inhabited building" as to the magazines of Manufacturer X, and the building leased by Manufacturer Y is likewise exempt as to the magazines of Manufacturer Y. This is because both buildings are occupied in connection with the manufacture, transportation, and storage of explosive materials. Clearly, the employees of both licensees are aware that explosive materials are present on the premises and they assume the risk of any such operation. A more difficult question is presented by the buildings of Manufacturer X and the magazines of Manufacturer Y and vice versa. ATF cannot assume that all employees are cognizant of the activities of their neighbors in the industrial park. Thus, it cannot be assumed that the employees are knowingly assuming the risk of explosive materials stored in magazines owned by the other licensee.

Given the plain language of the regulation, however, ATF does not believe it is appropriate to deny the coverage of the regulatory exemption to adjoining licensees on the basis of an assumption of the risk analysis. The current regulatory definition excludes from the definition of "inhabited building" any building occupied in connection with the manufacture, transportation, storage, or use of explosive materials, regardless of the knowledge of the building's occupants. Accordingly, ATF concludes that the industrial units occupied by Manufacturer X and Manufacturer Y are both exempted from the definition of "inhabited building" as to the magazines of each other as well as to their own magazines.

Held, a structure used to store items other than explosive materials that is visited on a regular basis by one individual is not an "inhabited building" as defined in 27 CFR 555.11, because it is not a structure where people are accustomed to assemble. Where 2 or more persons are present at the structure to repair equipment stored therein and such visits to the site are occasional and infrequent, the structure is not an "inhabited building" because the visits are not "customary." However, where 2 or more persons make regular visits to the structure for a common purpose, the structure is an "inhabited building," and explosives magazines may not be stored closer to the structure than the minimum distances specified in the regulations in 27 CFR Part 555.

Held further, buildings occupied by licensed explosives manufacturers in connection with the manufacture, transportation, storage, or use of explosive materials are not included within the definition of "inhabited building" as to magazines located on their own premises. In addition, buildings occupied by licensed explosives manufacturers in connection with the manufacture, transportation, storage, or use of explosives are not included within the definition of "inhabited building" as to magazines located on property owned by another licensee.

Date approved: November 25, 2005

14. 18 U.S.C. 842(f): Records Required for Explosives Licensees and Permittees

27 CFR 555, Subpart G: Records and Reports

27 CFR 555.22: Alternate Methods or Procedures; Emergency Variations from Requirements

Under specified conditions, approval is granted to utilize computerized records as required records under 27 CFR 555, Subpart G.

ATF Ruling 2007-1

Section 842(f), Title 18, United States Code, makes it unlawful for any licensee or permittee to willfully manufacture, import, purchase, distribute, or receive explosive materials without making such records as the Attorney General may by regulation require, including, but not limited to, a statement of intended use, the name, date, place of birth, social security number or taxpayer identification number, and place of residence of any natural person to whom explosive materials are distributed.

Regulations implementing section 842(f) are in 27 CFR Part 555, Subpart G. The regulations in this subpart specify the records required to be created and maintained by licensed importers (section 555.122), licensed manufacturers (section 555.123), licensed dealers (section 555.124), and permittees (section 555.125). The regulation in section 555.121 provides that licensees and permittees must keep records pertaining to explosive materials in permanent form (i.e., commercial invoices, record books) and in the manner required in Subpart G. In addition, sections 555.122-555.125 specifically allow licensees and permittees to use an alternate record to record the distribution of explosive materials when it is shown that the alternate records would accurately and readily disclose the information required by the regulations. These regulations require licensees and permittees who propose to use alternate records to submit a letter application to ATF describing the proposed alternate records and the need for them. Alternate records are not to be employed until approval from ATF is received.

Regulations at 27 CFR 555.22 allow for the approval and use of an alternate method or procedure in lieu of a method or procedure specifically prescribed in Part 555. ATF may approve an alternate method or procedure when it is found that—

(1) Good cause is shown for the use of the alternate method or procedure;

(2) The alternate method or procedure is within the purpose of, and consistent with the effect intended by, the specifically prescribed method or procedure and that the alternate method or procedure is substantially equivalent to that specifically prescribed method or procedure; and

(3) The alternate method or procedure will not be contrary to any provision of law and will not result in an increase in cost to the Government or hinder the effective administration of 27 CFR Part 555.

With advances in technology and the dramatic decrease in the cost of computers, many businesses rely upon computers to maintain records of their inventory, sales, customer lists, and other business information. Even the smallest home-based business utilizes computers to record and maintain business information. Creating and maintaining records in a computer database, rather than paper form, makes it easier to ensure accuracy of records and makes it less likely that records will be lost or misplaced. In addition, maintaining records via computer generally saves time and money in bookkeeping and auditing expenses. This utilization of computers has allowed companies to automate inventories, utilizing technology such as bar codes or RFID (radio frequency identification) chips. This facilitates better accountability of product overall, reducing the potential of everyday accounting errors. Over the years ATF has seen a significant increase in the number of requests from explosives licensees and permittees for authorization to utilize computerized records rather than paper records of acquisition and distribution and other required records, such as magazine transaction records. ATF routinely approves requests to utilize computerized records, with certain conditions, finding that the use of such records is substantially equivalent to methods of record keeping set forth in the regulations in 27 CFR Part 555, Subpart G.

Several explosives industry members have asked whether computerized records may be maintained without obtaining written approval from ATF if they contain all the required information specified in the regulations and are maintained in a permanent form. Additionally, industry members have questioned whether computer records in combination with paper records may be maintained if they are permanent and contain all the information required by the regulations.

ATF has determined that records of acquisition and disposition, magazine summary records, and the other records required by 27 CFR Part 555, Subpart G, satisfy the standard of permanency and are substantially equivalent to paper records if they meet the following criteria:

1. All data entered into the computer system must be recorded into the database and cannot be capable of being edited or modified at a later date. The software system must retain any correction of errors as an entirely new entry, without deleting or modifying the original entry. The system may allow for entries in a notes column to explain any correction.

2. The system must have a reliable daily memory backup capability to protect the data from accidental deletion or other system failure.

It is also acceptable for licensees/permittees to maintain required records using a combination of a computer program, commercial invoices, and other documents, provided that all of the required information is maintained in the records in permanent form. Any use of a computer for any portion of the required records must comply with the standards outlined above. However, each particular transaction must be self-contained with all the required information in the same recordkeeping medium. As one example, dispositions of explosives by a dealer cannot be separated by keeping the dates of disposition and the manufacturer's name or brand name in the computer, and all the other required information for that disposition on separate written documents.

ATF finds that good cause exists for authorizing the use of a computer to create and maintain the records required by 27 CFR Part 555, Subpart G, as the use of computers is accepted throughout the business community as a reliable, cost-efficient means of maintaining business records. ATF also finds that the use of a computer to maintain required records, contingent upon the requirements outlined above, is consistent with the effect intended by the requirements of Subpart G, as it will result in a permanent, reliable record that will accurately indicate acquisitions and dispositions of explosive materials.

Finally, ATF finds that the use of computer records properly containing all the required information should not hinder the effective administration of the Federal explosives laws or regulations – use of such records generally makes it easier for ATF to conduct inventories of product on hand and to audit required records. Accordingly, ATF concludes that the requirements for approval of an alternate method or procedure in accordance with 27 CFR Part 555, sections 555.22 and 555.122-555.125, are met.

Held, persons holding licenses and permits issued under 18 U.S.C., Chapter 40, may use computers to create and maintain all or any portion of the records required by 18 U.S.C. 842(f) and 27 CFR Part 555, Subpart G, if the following conditions are satisfied:

1. All data entered into the computer system must be recorded into the database and cannot be capable of being edited or modified at a later date. The software system must retain any correction of errors as an entirely new entry, without deleting or modifying the original entry. The system may allow for entries in a notes column to explain any correction.

2. The system must have a reliable daily memory backup capability to protect the data from accidental deletion or other system failure.

Held further, licensees and permittees who wish to use computers to create and maintain all or a portion of their required records in accordance with the requirements set forth in this ruling are not required to obtain advance approval in accordance with 27 CFR 555.22 or 555.121-125.

Held further, licensees and permittees utilizing a combination of a computer program, commercial invoices, and other paper documents as required records must ensure that the required information for a particular transaction is fully contained in the same recordkeeping medium.

Held further, this ATF approved alternate method or procedure for computerized records shall not be withdrawn unless the holder of said variance is so advised by ATF in writing or no longer holds a Federal explosives license or permit.

Date approved: January 18, 2007

15. 18 U.S.C. 842(j): Storage of Explosives

27 CFR 555.210: Construction of Type 4 Magazines

27 CFR 555.215: Housekeeping

27 CFR 555.201: Notification of Local Fire Officials

27 CFR 555.63: Explosives Magazine Changes

27 CFR 555.22: Alternative Methods or Procedures; Emergency Variations from Requirements

Under specified conditions, display fireworks may be temporarily stored in locked and attended motor vehicles at the explosives magazine site(s) and at fireworks display site(s) without meeting the locking requirements of 27 CFR 555.210 provided certain additional security measures are in place. Additionally, allowance per 27 CFR 555.215 is made for the fuel tanks containing volatile materials that may be on the temporary storage vehicles. Finally, slight variation is provided for notification requirements to ATF and local fire officials.

ATF Rule. 2007-2

The Federal explosives laws, 18 U.S.C. Chapter 40, require all persons to store explosive materials in a manner in conformity with regulations issued by the Attorney General. 18 U.S.C. 842(j). The Attorney General has delegated the authority to administer and enforce the Federal explosives laws to the Director, ATF. 28 CFR 0.130. Regulations in 27 CFR Part 555, implement the provisions of the Federal explosives laws.

The regulation at 27 CFR 555.210(a) states, in part, "[o]utdoor magazines are to be fire-resistant, weather-resistant, and theft-resistant." Partly to satisfy the theft-resistant requirement, this section requires that each door be equipped with two mortise locks; two padlocks fastened in separate hasps and staples; a combination of a mortise lock and a padlock; a mortise lock that requires two keys to open; or a three-point lock. Padlocks must have at least five tumblers and a case-hardened shackle of at least

⅜-inch diameter. Padlocks must be protected with not less than ¼-inch steel hoods constructed so as to prevent sawing or lever action on the locks, hasps, and staples.

The regulation at 27 CFR 555.215 states, in part, "[v]olatile materials are to be kept a distance of not less than 50 feet from outdoor magazines."

The regulation at 27 CFR 555.201 requires, in part, that any person storing explosive materials notify local fire authorities orally before the end of the day on which storage of the explosive materials began and in writing within 48 hours from the time such storage began.

The regulation at 27 CFR 555.63 requires that any licensee or permittee who acquires (adds) a storage magazine must notify ATF at least five business days in advance of using any added explosives storage magazine.

Regulations at 27 CFR 555.22 allow for the approval and use of an alternate method or procedure in lieu of a method or procedure specifically prescribed in Part 555. ATF may approve an alternate method or procedure when it is found that—

(1) Good cause is shown for the use of the alternate method or procedure;

(2) The alternate method or procedure is within the purpose of, and consistent with the effect intended by, the specifically prescribed method or procedure and that the alternate method or procedure is substantially equivalent to that specifically prescribed method or procedure; and

(3) The alternate method or procedure will not be contrary to any provision of law and will not result in an increase in cost to the Government or hinder the effective administration of 27 CFR Part 555.

ATF has approved a significant number of variances for temporary storage for a specified amount of time before a display fireworks event, as well as during and after the event until the remaining explosive materials can be placed back into the appropriate storage magazine.

Preparation of display fireworks shows and the transportation of explosive materials to numerous show sites often take place over a period of several days. Preparing and temporarily storing the fireworks for these shows ordinarily take place on delivery trucks and trailers in one storage location where the proprietor already maintains storage of explosives materials with a high degree of security and safety by complying with the provisions of 27 CFR Part 555.

Many display fireworks shows also take several days to prepare at the show site. During preparation and after the show is completed, explosive materials frequently must be temporarily stored. This is often either extra product that was brought to the show or misfires that have been maintained and must be returned to permanent storage.

Allowing flexibility through alternate methods or procedures for specific regulations increases both safety and security at these show sites. These procedures are needed to increase public safety, as well as facilitate smooth operations for the display fireworks industry. The highest risk of incidents involving the accidental ignition of display fireworks is during handling, with the next highest risk being transportation. Providing no flexibility to allow storage in the delivery vehicles would require the industry to dangerously load and unload from storage magazines to vehicles and back into a storage magazine. Additionally, ATF believes that providing this guidance allows for preplanning by the proprietor and consistency of regulatory application nationwide.

One of the major dangers around explosives is fire. Therefore, the regulations require that volatile materials be maintained a distance of not less than 50 feet from outdoor explosives storage magazines. ATF believes that requiring attended storage for display fireworks temporarily stored in vehicles will ensure public safety, in lieu of the 50 foot separation requirement. The attendee should be able to alert the proper authorities if needed to ensure that a fire does not compromise this storage, or may relocate these temporary storage magazines to a safe location away from an identified fire.

Held, ATF will approve alternate methods or procedures for the temporary storage of display fireworks in locked and attended vehicles at explosives magazine site(s), as well as at the fireworks display site(s), under the following conditions:

1. The doors to each storage compartment containing explosive materials must be locked with at least one steel padlock having at least five tumblers and a casehardened shackle of at least ⅜-inch diameter. The padlock does not need to be protected by a steel hood. However, each temporary storage magazine must be attended at all times for security purposes. The vehicle is considered "attended" when an authorized individual is within 100 feet of all temporary storage and has an unobstructed view of the vehicle(s) containing the explosive materials. The individual must remain awake and observant of activities around the vehicle(s).

2. The person who temporarily stores the explosive materials must notify in writing the authority having jurisdiction for fire safety in the locality in which the explosive materials are stored no less than 3 Federal office business days prior to utilizing the additional temporary storage magazine(s).

3. The person who temporarily stores the explosive materials must notify ATF in writing of the location of this storage no less than 3 Federal office business days prior to utilizing the additional temporary storage magazine(s).

All other provisions of 27 CFR Part 555 must be complied with as prescribed.

Held further, this ATF-approved alternate method or procedure for the temporary storage of display fireworks in locked and attended vehicles shall not expire unless the holder of said variance is so advised by ATF or no longer holds a Federal explosives license or permit.

Date approved: January 18, 2007

16. 18 U.S.C. 842(j): Storage of Explosives

27 CFR 555.211(a): Immobilization of Outdoor Type 5 Mobile Storage Magazines

27 CFR 555.215: Housekeeping

27 CFR 555.22: Alternative Methods or Procedures; Emergency Variations from Requirements

Under specified conditions, blasting agents may be stored in mobile type 5 magazines (bulk delivery trucks) without meeting the prescribed immobilization requirements of 27 CFR 555.211.

ATF Rule 2007-3

The Federal explosives laws, 18 U.S.C. Chapter 40, require all persons to store explosive materials in a manner in conformity with regulations issued by the Attorney General. 18 U.S.C. 842(j). The Attorney General has delegated his authority to administer and enforce the Federal explosives laws to the Director, ATF. 28 CFR 0.130. Regulations in 27 CFR

Part 555 implement the provisions of the Federal explosives laws.

The regulation at 27 CFR 555.211(a)(1) states, in part, "[o]utdoor magazines are to be weather-resistant and theft-resistant." This section further states, "[w]hen unattended, vehicular magazines must have wheels removed or otherwise be effectively immobilized by kingpin locking devices or other methods approved by the Director."

The regulation at 27 CFR 555.215 states, in part, "[v]olatile materials are to be kept a distance of not less than 50 feet from outdoor magazines."

The regulation at 27 CFR 555.22 allows for the approval and use of an alternate method or procedure in lieu of a method or procedure specifically prescribed in Part 555. ATF may approve an alternate method or procedure when it is found that—

(1) Good cause is shown for the use of the alternate method or procedure;

(2) The alternate method or procedure is within the purpose of, and consistent with the effect intended by, the specifically prescribed method or procedure and that the alternate method or procedure is substantially equivalent to that specifically prescribed method or procedure; and

(3) The alternate method or procedure will not be contrary to any provision of law and will not result in an increase in cost to the Government or hinder the effective administration of 27 CFR Part 555.

ATF has approved a significant number of variances for an alternate means of immobilizing preloaded bulk delivery vehicles when some additional security measures were put in place.

Bulk delivery vehicles are routinely utilized for on-site delivery of blasting services. Often these trucks contain a blasting agent as defined under 27 CFR 555.11. Utilization of these bulk products delivered on-site and used immediately has increased safety and security by reducing the number of remotely located storage trailers containing packaged blasting agent products. These delivery vehicles generally leave the explosives storage locations at unusual times of day or night, and for safety reasons they are loaded the day before for next day's delivery. These trucks are incapable of being disabled by a kingpin locking device, and the requirement to remove the wheels for immobilization is obviously not feasible. Additionally, there are times when explosive material remains on the vehicle when it returns from use. Most of these products degrade during handling, therefore, removing them from the bulk vehicle would not be an option.

One of the major dangers surrounding explosives is fire. Therefore, the regulations at 27

CFR 555.215 require that volatile materials be maintained a distance of not less than 50 feet from outdoor explosive storage magazines. Because these preloaded storage vehicles contain a fuel tank filled with volatile materials, they must remain in an area protected from fire such as gravel, paved, or closely mowed designated parking area.

ATF believes that the following alternate method of operation is substantially equivalent to the prescribed methods. Increased safety for employees and the public provides good cause for this alternate method. It is not contrary to law and will not result in any increased cost to the Government. Overall, ATF believes allowing for this flexibility assists with the effective administration of 27 CFR Part 555.

Held, ATF will approve alternate methods or procedures for the preloading and temporary storage of bulk blasting agents in delivery vehicles at explosive magazine site(s), when the security and immobilization meets the following criteria:

1. All doors on the vehicle are locked, the ignition key is removed, and the key is secured away from the truck.

2. When the site is not in operation, outer perimeter security is established. This may be by a variety of means such as a locked gate, security guards, fence, natural features, or a combination of these.

3. Each potential access point to explosive materials on a storage vehicle will be secured with a minimum of one padlock that has at least five tumblers and casehardened shackle of at least ⅜-inch diameter.

4. Each vehicle shall be immobilized through the use of a steering wheel locking device, lockable battery disconnect switch, or both.

5. All vehicles preloaded with blasting agents shall be parked in a company designated area not susceptible to fire propagation such as bare dirt, gravel, rock, paving, or closely mowed parking lot.

All other provisions of 27 CFR Part 555 must be complied with as prescribed.

Held further, licensees and permittees who wish to use the alternate method or procedure set forth in this ruling are not required to obtain advance approval in accordance with 27 CFR 555.22.

Held further, this ATF-approved alternate method or procedure for the temporary storage of bulk blasting agent products in locked and properly secured vehicles shall not be withdrawn unless the holder of said variance is so advised by ATF in writing or no longer holds a Federal explosives license or permit.

Date approved: January 26, 2007

APPENDIX A

PERMANENTLY MOUNTED CONTAINERS

Figure 1

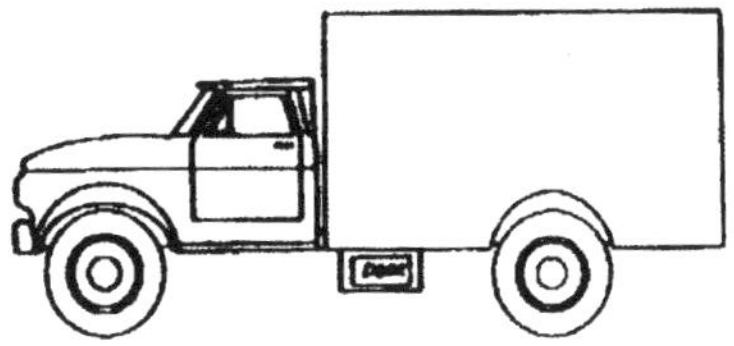

Figure 2

NOTE: The configurations shown in Figures 1 and 2 are equally applicable to multi-axle and "cab-over" vehicles.

[Diagrams: Courtesy of IME]

APPENDIX B

COMPARTMENTS

Figure 1

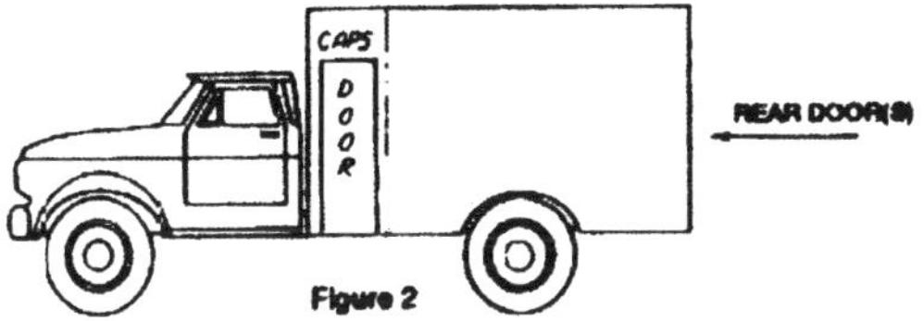

Figure 2

Figure 3

NOTE: The configurations shown in Figures 1 and 2 are equally applicable to multi-axle and "cab-over" vehicles.

[Diagrams: Courtesy of IME]

ATF Rul. 77–24, Appendix A and B

APPENDIX C

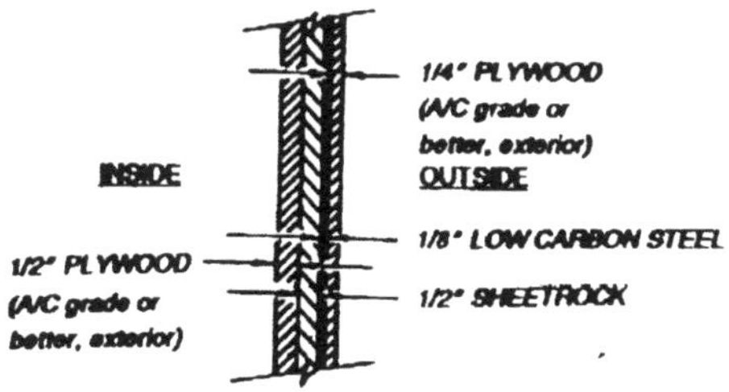

Sketch of laminate construction for container or compartment for electric blasting caps use, as illustrated in Appendix A, B, and E.

APPENDIX D

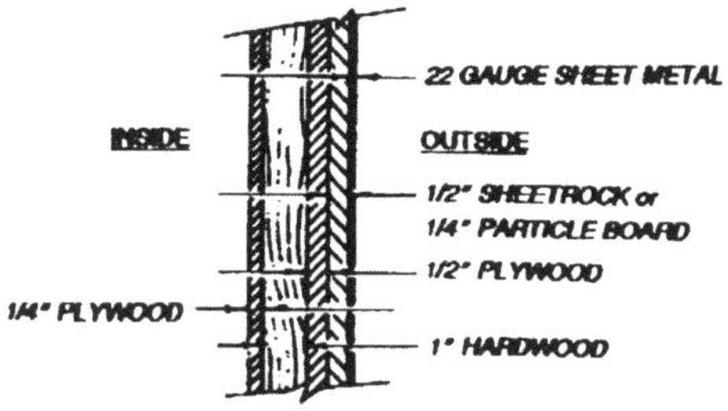

Sketch of laminate construction for container or compartment for electric blasting caps; restricted to use as illustrated in Appendix A.

APPENDIX E

PORTABLE WHEELED TRAILERS

Figure 1

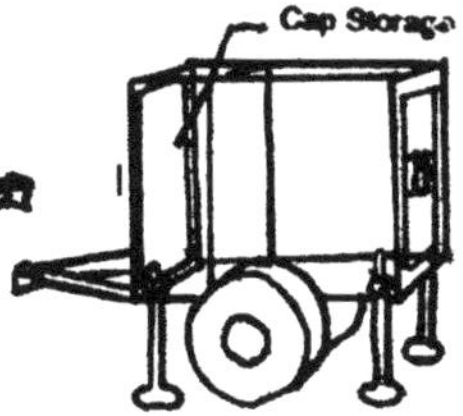

Figure 3

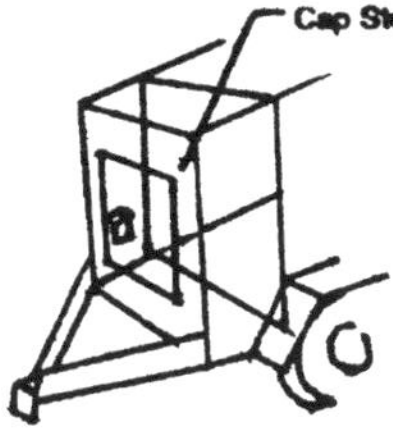

Figure 2

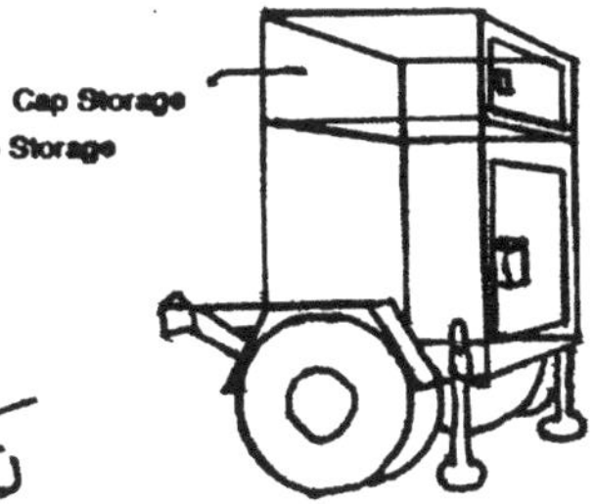

Figure 4

[Diagrams: Courtesy of IME]

General Information (Revised 8/06)

Effect of 18 U.S.C. Chapter 40 On the Fireworks Industry

[Caution! This item discusses Federal requirements only. Please contact your State or local authorities for any additional requirements.]

Title XI of the Organized Crime Control Act of 1970 (18 U.S.C. Chapter 40) establishes controls over explosive materials, including black powder and other pyrotechnic compositions commonly used in fireworks. Part 555 of Title 27, Code of Federal Regulations (CFR), contains the regulations which implement Title XI. Section 555.141(a)(7) exempts "the importation, distribution, and storage of fireworks classified as UN0336, UN0337, UN0431, or UN0432 explosives by the U.S. Department of Transportation at 49 CFR 172.101 and generally known as 'consumer fireworks' or 'articles pyrotechnic.'" Section 555.141(a)(7) does not exempt "display fireworks," as defined in 555.11.

With Respect to Fireworks: Who needs a license?

1. Manufacturers of black powder;

2. Manufacturers of any other explosive material used in manufacturing consumer fireworks or display fireworks; and

3. Importers of, or dealers in, display fireworks.

With Respect to Fireworks: Who needs a permit?

1. All persons transporting, shipping, causing to be transported, or receiving display fireworks, regardless of whether for their own use or for commercial display purposes (Certain exemptions apply, e.g. agencies of the United States or of any State or political subdivisions thereof are exempt from permit requirements); and

2. A person, other than a licensee, transporting, shipping, causing to be transported, or receiving explosive materials for use in manufacturing display fireworks or consumer fireworks.

With Respect to Fireworks: Who may not need a license or permit?

Frequently, persons contracting for display fireworks (e.g., for Fourth of July observances) from a Federal explosives licensee or permittee receive a total service, including the services of a pyrotechnician who transports display fireworks in interstate or intrastate commerce to the site of the display and conducts and supervises the display. In these instances, the customers purchase and receive the contractor's services and not the explosive materials themselves (i.e. the cost of the services includes the contractor's expense in providing the fireworks utilized), and the cost of the services includes the dealer's expense in providing the fireworks utilized. When business is transacted in this manner, the customers purchasing and receiving the services need not obtain Federal explosives licenses or permits under Part 555 as long as they are not transporting, shipping, causing to be transported, or receiving explosive materials. Note: the transportation of explosive materials to the display sites would be authorized by the Federal explosives license or permit of the licensee or permittee providing the services.

With Respect to Fireworks: Types of permits

1. User permit: Allows the permit holder to transport, ship, cause to be transported, and receive display fireworks in interstate or foreign commerce for his or her own use and not for resale. This permit is issued at a cost of $100 for a 3-year period and is renewable at a cost of $50 for a 3-year period.

2. User-limited permit: Identical to the user permit but issued for a single purchase transaction, only. The fee is $75; the permit is nonrenewable.

With Respect to Fireworks: Storage

The law prohibits any person from storing any explosive materials in a manner not in conformity with the regulations promulgated by the Attorney General (18 U.S.C. 842(j)). Pursuant to this section, the Attorney General has prescribed storage regulations in 27 CFR Part 555, Subpart K. Display fireworks must be stored in conformity with the regulations. Display fireworks generally contain perchlorate mixture explosives, potassium chlorate base explosive mixtures, and black powder, which are entered on the List of Explosive Materials with numerous others. (The List, which is not all-inclusive, is annually compiled and readily available without charge from the address set out in 27 CFR 555.23 or online at www.atf.gov.) Display fireworks must be stored as low explosives in magazines meeting, at a minimum, the requirements for type 4 storage magazines prescribed by 27 CFR 555.210 unless they contain other classes of explosives. Bulk salutes must be stored as high explosives in type 1 or type 2 magazines. The net weight of the explosive materials contained in the display fireworks may be used in determining compliance with table of distance requirements. To determine the actual weight of the materials, it may be necessary to contact their manufacturers. The manufacturer of exempt or nonexempt fireworks having stocks of explosive materials on hand to be used in the manufacture of fireworks must store the stocks in conformity with applicable storage requirements.

Explosives Dealer's and User's Guide to Federal Explosives Regulation

Explosives May Not be Distributed by Licensees (Or by Any Person) to Any Person Who:

1. Is under indictment for, or who has been convicted of a crime punishable by imprisonment for a term exceeding one year.

2. Is an unlawful user of, or addicted to, marijuana or any depressant or stimulant drug or narcotic drug (as these terms are defined in section 102 of the Controlled Substances Act).

3. Has been adjudicated as a mental defective or has been committed to a mental institution.

4. Is a fugitive from justice.

5. Is an alien (with certain exceptions)

6. Has been discharged from the armed forces under dishonorable conditions; or

7. Having been a citizen of the United States, has renounced citizenship.

8. Is less than 21 years of age.

Dealers in Explosives Must:

Have a current and valid Federal explosives license.

Have proper storage facilities.

Keep accurate and complete records.

Verify that each buyer has a Federal explosives license or permit.

Verify buyers' identities.

Users of Explosives

Federal permits are required of those who transport, ship, cause to be transported, or receive explosive materials. Among other things, the permittee must keep complete and accurate records of the acquisitions and dispositions of explosives materials. Unless otherwise exempted by law, no person may receive or transport any explosive materials without a permit.

No person shall store any explosive material in a matter not in conformity with applicable regulations.

All persons must report to ATF and local authorities any loss or theft of their explosive materials within 24 hours of discovery.

A Federal license or permit does not confer any right or privilege to violate any state law or local ordinance.

The above summary is general and does not purport to fully convey the Federal explosives law and regulations pertaining to dealers and users.

Black Powder Transactions

Public Law 93-639 (1975) allows nonlicensees/nonpermittees to purchase commercially manufactured black powder, in quantities of 50 pounds or less, solely for sporting, recreational or cultural purposes for use in antique firearms or antique devices. A nonlicensee or nonpermittee purchasing black powder under the exemption need not be a resident of the State in which the dealer is located. Also, the categories of persons to whom the distribution of explosive materials is prohibited do not apply to black powder transactions made under the exemption. Acquisitions of black powder not qualifying under this exemption are subject to the same regulatory requirements that govern any other low explosive.

All persons who distribute black powder, regardless of quantity, must be licensed as explosives dealers and, among other things, must provide adequate storage.

Explosives Security

Through prompt reporting of losses and thefts of explosives and increased emphasis on physical security, explosives licensees and permittees can contribute greatly to efforts by Federal, State and local authorities to reduce the incidence of bombings and other criminal misuse of explosives in the United States. The following actions are of prime importance and in some instances required:

Report . . . any thefts or losses of explosives within 24 hours of discovery, by telephone, to ATF (toll free: 1-800-800-3855) and to appropriate local authorities. Because the States and many municipalities have designated specific agencies to investigate the theft or loss of explosives, licensees and permittees are urged to be familiar with State and local reporting procedures and appropriate contact points.

Follow . . . telephone notification with a written report on ATF Form 5400.5, "Report of Theft or Loss—Explosive Materials," to the nearest ATF Division Office, and in accordance with the instructions on the form.

Observe . . . activity around magazines, within business premises, and on job sites, particularly if strangers appear to be loitering in the area in which explosives are being kept. On-site users should take special care to assure that explosives removed from storage for use on the job are either detonated or accounted for and unused items returned to storage.

Review . . . recordkeeping practices to assure that no discrepancies exist and that no figures in reported inventories have been manipulated, and correct any clerical errors promptly. Should any questions arise concerning explosives security procedures or any aspect of explosives regulation coming under the jurisdiction of ATF, do not hesitate to contact ATF.

Note: For Q&A's on regulatory requirements governing recordkeeping and storage, see "Questions and Answers" numbers 62-87.

Additional Information

The flow of useful information is an essential ingredient in the effective administration of regulatory programs. The Bureau of Alcohol, Tobacco, Firearms and Explosives is the Federal agency charged with the responsibility of administering laws impacting the explosives industry. We call your attention to the following publication distributed by ATF that merits your attention:

The Explosives Newsletter

During 1989 ATF developed the Explosives Newsletter, an information service for Federal explosives licensees and permittees which is intended to help explosives industry members better understand the Federal laws under which they must operate. It also includes other items of particular interest to the explosives industry. There is no charge for the Explosives Newsletter; licensees and permittees automatically receive copies when new editions are published.

Explosives industry members having questions on the Federal explosives laws and regulations may address their inquiries to:

Bureau of Alcohol, Tobacco, Firearms and Explosives
Explosives Industry Programs Branch
99 New York Avenue, N.E.
Mailstop 6E405
Washington, DC 20226

Direct e-mail inquiries on general questions or variance requests may be sent to the branch at EIPB@atf.gov.

List of Explosive Materials

Pursuant to the provisions of section 841(d) of title 18, U.S.C., and 27 CFR 555.23, the Director, Bureau of Alcohol, Tobacco, Firearms and Explosives, must revise and publish in the Federal Register at least annually a list of explosives determined to be within the coverage of 18 U.S.C. Chapter 40, Importation, Manufacture, Distribution and Storage of Explosive Materials. This chapter covers not only explosives, but also blasting agents and detonators, all of which are defined as explosive materials in section 841(c) of title 18, U.S.C. Accordingly, the following is the current List of Explosive Materials subject to regulation under 18 U.S.C. Chapter 40. Materials constituting blasting agents are marked by an asterisk. While the list is comprehensive, it is not all-inclusive. The fact that an explosive material may not be on the list does not mean that it is not within the coverage definitions in section 841 of title 18, U.S.C. Explosive materials are listed alphabetically by their common names, followed by chemical names and synonyms in brackets. This revised list is effective as of September 18, 2006.

List of Explosive Materials

A

Acetylides of heavy metals.
Aluminum containing polymeric propellant.
Aluminum ophorite explosive.
Amatex.
Amatol.
Ammonal.
Ammonium nitrate explosive mixtures (cap sensitive).
*Ammonium nitrate explosive mixtures (non-cap sensitive).
Ammonium perchlorate having particle size less than 15 microns.
Ammonium perchlorate composite propellant.
Ammonium perchlorate explosive mixtures.
Ammonium picrate [picrate of ammonia, Explosive D].
Ammonium salt lattice with isomorphously substituted inorganic salts.
*ANFO [ammonium nitrate-fuel oil].
Aromatic nitro-compound explosive mixtures.
Azide explosives.

B

Baranol.
Baratol.
BEAF [1,2-bis (2, 2-difluoro-2-nitroacetoxyethane)].
Black powder.
Black powder based explosive mixtures.
*Blasting agents, nitro-carbo-nitrates, including non-cap sensitive slurry and water gel explosives.
Blasting caps.
Blasting gelatin.
Blasting powder.
BTNEC [bis (trinitroethyl) carbonate].
BTNEN [bis (trinitroethyl) nitramine].
BTTN [1,2,4 butanetriol trinitrate].
Bulk salutes.
Butyl tetryl.

C

Calcium nitrate explosive mixture.
Cellulose hexanitrate explosive mixture.
Chlorate explosive mixtures.
Composition A and variations.
Composition B and variations.
Composition C and variations.
Copper acetylide.
Cyanuric triazide.
Cyclonite [RDX].
Cyclotetramethylenetetranitramine [HMX].
Cyclotol.
Cyclotrimethylenetrinitramine [RDX].

D

DATB [diaminotrinitrobenzene].
DDNP [diazodinitrophenol].
DEGDN [diethyleneglycol dinitrate].
Detonating cord.
Detonators.
Dimethylol dimethyl methane dinitrate composition.
Dinitroethyleneurea.
Dinitroglycerine [glycerol dinitrate].
Dinitrophenol.
Dinitrophenolates.
Dinitrophenyl hydrazine.
Dinitroresorcinol.
Dinitrotoluene-sodium nitrate explosive mixtures.
DIPAM [dipicramide; diaminohexanitrobiphenyl].
Dipicryl sulfone.
Dipicrylamine.
Display fireworks.
DNPA [2,2-dinitropropyl acrylate].
DNPD [dinitropentano nitrile].
Dynamite.

E

EDDN [ethylene diamine dinitrate].
EDNA [ethylenedinitramine].
Ednatol.
EDNP [ethyl 4,4-dinitropentanoate].
EGDN [ethylene glycol dinitrate].
Erythritol tetranitrate explosives.
Esters of nitro-substituted alcohols.
Ethyl-tetryl.
Explosive conitrates.
Explosive gelatins.
Explosive liquids.
Explosive mixtures containing oxygen-releasing inorganic salts and hydrocarbons.
Explosive mixtures containing oxygen-releasing inorganic salts and nitro bodies.
Explosive mixtures containing oxygen-releasing inorganic salts and water insoluble fuels.
Explosive mixtures containing oxygen-releasing inorganic salts and water soluble fuels.
Explosive mixtures containing sensitized nitromethane.

Explosive mixtures containing tetranitromethane (nitroform).
Explosive nitro compounds of aromatic hydrocarbons.
Explosive organic nitrate mixtures.
Explosive powders.

F

Flash powder.
Fulminate of mercury.
Fulminate of silver.
Fulminating gold.
Fulminating mercury.
Fulminating platinum.
Fulminating silver.

G

Gelatinized nitrocellulose.
Gem-dinitro aliphatic explosive mixtures.
Guanyl nitrosamino guanyl tetrazene.
Guanyl nitrosamino guanylidene hydrazine.
Guncotton.

H

Heavy metal azides.
Hexanite.
Hexanitrodiphenylamine.
Hexanitrostilbene.
Hexogen [RDX].
Hexogene or octogene and a nitrated Nmethylaniline.
Hexolites.
HMTD
[hexamethylenetriperoxidediamine].
HMX [cyclo-1,3,5,7-tetramethylene 2,4,6,8-tetranitramine; Octogen].
Hydrazinium nitrate/hydrazine/aluminum explosive system.
Hydrazoic acid.

I

Igniter cord.
Igniters.
Initiating tube systems.

K

KDNBF [potassium dinitrobenzofuroxane].

L

Lead azide.
Lead mannite.
Lead mononitroresorcinate.
Lead picrate.
Lead salts, explosive.
Lead styphnate [styphnate of lead, lead trinitroresorcinate].
Liquid nitrated polyol and trimethylolethane.
Liquid oxygen explosives.

M

Magnesium ophorite explosives.
Mannitol hexanitrate.
MDNP [methyl 4,4-dinitropentanoate].
MEAN [monoethanolamine nitrate].
Mercuric fulminate.
Mercury oxalate.
Mercury tartrate.
Metriol trinitrate.
Minol-2 [40% TNT, 40% ammonium nitrate, 20% aluminum].
MMAN [monomethylamine nitrate]; methylamine nitrate.
Mononitrotoluene-nitroglycerin mixture.
Monopropellants.

N

NIBTN [nitroisobutametriol trinitrate].
Nitrate explosive mixtures.
Nitrate sensitized with gelled nitroparaffin.
Nitrated carbohydrate explosive.
Nitrated glucoside explosive.
Nitrated polyhydric alcohol explosives.
Nitric acid and a nitro aromatic compound explosive.
Nitric acid and carboxylic fuel explosive.
Nitric acid explosive mixtures.
Nitro aromatic explosive mixtures.
Nitro compounds of furane explosive mixtures.
Nitrocellulose explosive.
Nitroderivative of urea explosive mixture.
Nitrogelatin explosive.
Nitrogen trichloride.
Nitrogen tri-iodide.
Nitroglycerine [NG, RNG, nitro, glyceryl trinitrate, trinitroglycerine].
Nitroglycide.
Nitroglycol [ethylene glycol dinitrate, EGDN].
Nitroguanidine explosives.
Nitronium perchlorate propellant mixtures.
Nitroparaffins Explosive Grade and ammonium nitrate mixtures.
Nitrostarch.
Nitro-substituted carboxylic acids.
Nitrourea.

O

Octogen [HMX].
Octol [75 percent HMX, 25 percent TNT].
Organic amine nitrates.
Organic nitramines.

P

PBX [plastic bonded explosives].
Pellet powder.
Penthrinite composition.
Pentolite.
Perchlorate explosive mixtures.
Peroxide based explosive mixtures.
PETN [nitropentaerythrite, pentaerythrite tetranitrate, pentaerythritol tetranitrate].
Picramic acid and its salts.
Picramide.
Picrate explosives.
Picrate of potassium explosive mixtures.
Picratol.
Picric acid (manufactured as an explosive).
Picryl chloride.

Picryl fluoride.
PLX [95% nitromethane, 5% ethylenediamine].
Polynitro aliphatic compounds.
Polyolpolynitrate-nitrocellulose explosive gels.
Potassium chlorate and lead sulfocyanate explosive.
Potassium nitrate explosive mixtures.
Potassium nitroaminotetrazole.
Pyrotechnic compositions.
PYX [2,6-bis(picrylamino)] 3,5-dinitropyridine.

R

RDX [cyclonite, hexogen, T4, cyclo-1,3,5,-trimethylene-2,4,6,-trinitramine; hexahydro-1,3,5-trinitro-S-triazine].

S

Safety fuse.
Salts of organic amino sulfonic acid explosive mixture.
Salutes (bulk).
Silver acetylide.
Silver azide.
Silver fulminate.
Silver oxalate explosive mixtures.
Silver styphnate.
Silver tartrate explosive mixtures.
Silver tetrazene.
Slurried explosive mixtures of water, inorganic oxidizing salt, gelling agent, fuel, and sensitizer (cap sensitive).
Smokeless powder.
Sodatol.
Sodium amatol.
Sodium azide explosive mixture.
Sodium dinitro-ortho-cresolate.
Sodium nitrate explosive mixtures.
Sodium nitrate-potassium nitrate explosive mixture.
Sodium picramate.
Special fireworks.
Squibs.
Styphnic acid explosives.

T

Tacot [tetranitro-2,3,5,6-dibenzo-1,3a,4,6a tetrazapentalene].
TATB [triaminotrinitrobenzene].
TATP [triacetonetriperoxide].
TEGDN [triethylene glycol dinitrate].
Tetranitrocarbazole.
Tetrazene [tetracene, tetrazine, 1(5-tetrazolyl)-4-guanyl tetrazene hydrate].
Tetrazole explosives.
Tetryl [2,4,6 tetranitro-N-methylaniline].
Tetrytol.
Thickened inorganic oxidizer salt slurried explosive mixture.
TMETN [trimethylolethane trinitrate].
TNEF [trinitroethyl formal].
TNEOC [trinitroethylorthocarbonate].
TNEOF [trinitroethylorthoformate].
TNT [trinitrotoluene, trotyl, trilite, triton].
Torpex.
Tridite.
Trimethylol ethyl methane trinitrate composition.
Trimethylolthane trinitratenitrocellulose.
Trimonite.
Trinitroanisole.
Trinitrobenzene.
Trinitrobenzoic acid.
Trinitrocresol.
Trinitro-meta-cresol.
Trinitronaphthalene.
Trinitrophenetol.
Trinitrophloroglucinol.
Trinitroresorcinol.
Tritonal.

U

Urea nitrate.

W

Water-bearing explosives having salts of oxidizing acids and nitrogen bases, sulfates, or sulfamates (cap sensitive).
Water-in-oil emulsion explosive compositions.

X

Xanthamonas hydrophilic colloid explosive mixture.

Approved: September 18, 2006.
Michael J. Sullivan, Acting Director.

[FR Doc. E6–15850 Filed 9–26–06; 8:45 am]

ATF Federal Firearms Regulations Reference Guide

SPECIAL MESSAGE from the Director

Bureau of Alcohol, Tobacco, Firearms and Explosives
WASHINGTON, DC 20226

Dear Federal Firearms Licensee:

The Bureau of Alcohol, Tobacco, Firearms and Explosives (ATF) is the primary agency responsible for enforcing the Federal firearms laws. Our mission is to prevent terrorism, reduce violent crime, and protect the public. With respect to firearms, ATF works to take armed, violent offenders off the streets and to ensure criminals and other prohibited persons do not possess firearms.

Licensees play a critical role in protecting America from violent firearms crime through responsible and lawful distribution of firearms and accurate recordkeeping which assists ATF in tracing guns used to commit crimes. By complying with Federal law, licensees prevented the sale of more than 870,000 firearms to prohibited persons from November 1998 through September 2005. An untold number of lives have been saved and countless crimes have been prevented by stopping such transactions from taking place.

Dedicated licensees also make significant contributions to crime prevention efforts through active support of such collaborative efforts as the National Shooting Sports Foundation (NSSF) campaign “Don’t Lie for the Other Guy.” We are grateful to the scores of licensees who cooperate with ATF in preventing illegal straw purchases, which helps to disrupt firearms trafficking organizations. We encourage your continued support and we thank you for reporting suspicious criminal activity and firearms thefts to ATF. Please note, to report a theft, you should call 1-888-930-9275, and to report other criminal activity, you should call 1-800-ATF-GUNS.

The 2005 edition of the Federal Firearms Regulations Reference Guide provides information designed to help you comply with all of the laws and regulations governing the manufacture, importation, and distribution of firearms and ammunition. Accordingly, it contains the relevant Federal firearms laws and regulations, some of which have changed since the 2000 edition. It also contains rulings, general information, and questions and answers to give you further guidance on the Federal firearms laws. The laws and regulations are in a different order from previous versions so that this publication will be easier to use. This new edition also contains additional points of contact to help you get in touch with ATF more easily.

Since firearms laws can change over time, please be aware that the information in this book may change after the date of publication. The ATF Web site at www.atf.gov is an excellent source for up-to-date information. As always, you are welcome to contact your local ATF field office for information or assistance.

Sincerely yours,

Carl J. Truscott
Director

TABLE OF CONTENTS

EDITOR'S NOTE

The cross references, bracketed notes, and Editor's notes seen in the laws and regulations are for guidance and assistance only and do not appear in the official United States Code and Code of Federal Regulations.

THE GUN CONTROL ACT OF 1968

TITLE 18, UNITED STATE CODE, CHAPTER 44

TITLE I : STATE FIREARMS CONTROL ASSISTANCE

PURPOSE

Sec. 101. The Congress hereby declares that the purpose of this title is to provide support to Federal, State, and local law enforcement officials in their fight against crime and violence, and it is not the purpose of this title to place any undue or unnecessary Federal restrictions or burdens on law-abiding citizens with respect to the acquisition, possession, or use of firearms appropriate to the purpose of hunting, trapshooting, target shooting, personal protection, or any other lawful activity, and that this title is not intended to discourage or eliminate the private ownership or use of firearms by law-abiding citizens for lawful purposes, or provide for the imposition by Federal regulations of any procedures or requirements other than those reasonably necessary to implement and effectuate the provisions of this title.

Chapter 44 – Firearms

Editor's Note:

On September 13, 1994, Congress passed the Violent Crime Control and Law Enforcement Act of 1994, Public Law 103-322. Title IX, Subtitle A, Section 110105 of this Act generally made it unlawful to manufacture, transfer and possess semiautomatic assault weapons (SAWs) and to transfer and possess large capacity ammunition feeding devices (LCAFDs). The law also required importers and manufacturers to place certain markings on SAWs and LCAFDs, designating they were for export or law enforcement/government use. Significantly, the law provided that it would expire 10 years from the date of enactment. Accordingly, effective 12:01 a.m. on September 13, 2004, the provisions of the law ceased to apply. These provisions are marked "repealed" in this publication.

§ 921 Definitions.

(a) As used in this chapter—

(1) The term **"person"** and the term **"whoever"** include any individual, corporation, company, association, firm, partnership, society, or joint stock company.

(2) The term **"interstate or foreign commerce"** includes commerce between any place in a State and any place outside of that State, or within any possession of the United States (not including the Canal Zone) or the District of Columbia, but such term does not include commerce between places within the same State but through any place outside of that State. The term **"State"** includes the District of Columbia, the Commonwealth of Puerto Rico, and the possessions of the United States (not including the Canal Zone).

(3) The term **"firearm"** means (A) any weapon (including a starter gun) which will or is designed to or may readily be converted to expel a projectile by the action of an explosive; (B) the frame or receiver of any such weapon; (C) any firearm muffler or firearm silencer; or (D) any destructive device. Such term does not include an antique firearm.

(4) The term **"destructive device"** means—

(A) any explosive, incendiary, or poison gas—

(i) bomb,

(ii) grenade,

(iii) rocket having a propellant charge of more than four ounces,

(iv) missile having an explosive or incendiary charge of more than one-quarter ounce,

(v) mine, or

(vi) device similar to any of the devices described in the preceding clauses;

(B) any type of weapon (other than a shotgun or a shotgun shell which the Attorney General finds is generally recognized as particularly suitable for sporting purposes) by whatever name known which will, or which may be readily converted to, expel a projectile by the action of an explosive or other propellant, and which has any barrel with a bore of more than one-half inch in diameter; and

(C) any combination of parts either designed or intended for use in converting any device into any destructive device described in subparagraph (A) or (B) and from which a destructive device may be readily assembled.

The term **"destructive device"** shall not include any device which is neither designed nor redesigned for use as a weapon; any device, although originally designed for use as a weapon, which is redesigned for use as a signaling, pyrotechnic, line throwing, safety, or similar device; surplus ordinance sold, loaned, or given by the Secretary of the Army pursuant to the provisions of section 4684(2), 4685, or 4686 of title 10; or any other device which the Attorney General finds is not likely to be used as a weapon, is an antique, or is a rifle which the owner intends to use solely for sporting, recreational or cultural purposes.

(5) The term **"shotgun"** means a weapon designed or redesigned, made or remade, and intended to be fired from the shoulder and designed or redesigned and made or remade to use the energy of an explosive to fire through a smooth bore either a number of ball shot or a single projectile for each single pull of the trigger.

(6) The term **"short-barreled shotgun"** means a shotgun having one or more barrels less than eighteen inches in length and any weapon made from a shotgun (whether by alteration, modification, or otherwise) if such weapon as modified has an overall length of less than twenty-six inches.

(7) The term **"rifle"** means a weapon designed or redesigned, made or remade, and intended to be fired from the shoulder and designed or redesigned and made or remade to use the energy of an explosive to fire only a single projectile through a rifled bore for each single pull of the trigger.

(8) The term **"short-barreled rifle "** means a rifle having one or more barrels less than sixteen inches in length and any weapon made from a rifle (whether by alteration, modification, or otherwise) if such weapon, as modified, has an overall length of less than twenty-six inches.

(9) The term **"importer"** means any person engaged in the business of importing or bringing firearms or ammunition into the United States for purposes of sale or distribution; and the term **"licensed imp orter"** means any such person licensed under the provisions of this chapter

(10) The term **"manufacturer"** means any person engaged in the business of manufacturing firearms or ammunition for purposes of sale or distribution; and the term **"licensed manufacturer"** means any such person licensed under the provisions of this chapter.

(11) The term **"dealer"** means (A) any person engaged in the business of selling firearms at wholesale or retail, (B) any person engaged in the business of repairing firearms or of making or fitting special barrels, stocks, or trigger mechanisms to firearms, or (C) any person who is a pawnbroker. The term **"licensed d ealer"** means any dealer who is licensed under the provisions of this chapter.

(12) The term **"pawnbroker"** means any person whose business or occupation includes the taking or receiving, by way of pledge or pawn, of any firearm as security for the payment or repayment of money.

(13) The term **"collector"** means any person who acquires, holds, or disposes of firearms as curios or relics, as the Attorney General shall by regulation define, and the term **"licensed collector"** means any such person licensed under the provisions of this chapter.

(14) The term **"indictment"** includes an indictment or information in any court under which a crime punishable by imprisonment for a term exceeding one year may be prosecuted.

(15) The term **"fugitive from justice"** means any person who has fled from any State to avoid prosecution for a crime or to avoid giving testimony in any criminal proceeding.

(16) The term **"antique fire arm"** means—

(A) any firearm (including any firearm with a matchlock, flintlock, percussion cap, or similar type of ignition system) manufactured in or before 1898; or

(B) any replica of any firearm described in subparagraph (A) if such replica—

(i) is not designed or redesigned for using rimfire or conventional centerfire fixed ammunition, or

(ii) uses rimfire or conventional centerfire fixed ammunition which is no longer manufactured in the United States and which is not readily available in the ordinary channels of commercial trade; or

(C) any muzzle loading rifle, muzzle loading shotgun, or muzzle loading pistol, which is designed to use black powder, or a black powder substitute, and which cannot use fixed ammunition. For purposes of this subparagraph, the term "antique firearm" shall not include any weapon which incorporates a firearm frame or receiver, any firearm which is converted into a muzzle loading weapon, or any muzzle loading weapon which can be readily converted to fire fixed ammunition by replacing the barrel, bolt, breechblock, or any combination thereof.

(17) (A) The term **"ammunition"** means ammunition or cartridge cases, primers, bullets, or propellant powder designed for use in any firearm.

(B) The term **"armor piercing ammunition"** means—

(i) a projectile or projectile core which may be used in a handgun and which is constructed entirely (excluding the presence of traces of other substances) from one or a combination of tungsten alloys, steel, iron, brass, bronze, beryllium copper, or depleted uranium; or

(ii) a full jacketed projectile larger than .22 caliber designed and intended for use in a handgun and whose jacket has a weight of more than 25 percent of the total weight of the projectile.

(C) The term **"armor piercing ammunition"** does not include shotgun shot required by Federal or State environmental or game regulations for hunting purposes, a frangible projectile designed for target shooting, a projectile which the Attorney General finds is primarily intended to be used for sporting purposes, or any other projectile or projectile core which the Attorney General finds is intended to be used for industrial purposes, including a charge used in an oil and gas well perforating device.

(18) The term **"Attorney General"** means the Attorney General of the United States.

(19) The term **"published ordinance"** means a published law of any political subdivision of a State which the Attorney General determines to be relevant to the enforcement of this chapter and which is contained on a list compiled by the Attorney General, which list shall be published in the Federal Register, revised annually, and furnished to each licensee under this chapter.

(20) The term **"crime punishable by imprisonment for a term exceedin g one year"** does not include—

(A) any Federal or State offenses pertaining to antitrust violations, unfair trade practices, restraints of trade, or other similar offenses relating to the regulation of business practices, or

(B) any State offense classified by the laws of the State as a misdemeanor and punishable by a term of imprisonment of two years or less.

What constitutes a conviction of such a crime shall be determined in accordance with the law of the jurisdiction in which the proceedings were held. Any conviction which has been expunged, or set aside, or for which a person has been pardoned or has had civil rights restored, shall not be considered a conviction for purposes of this chapter, unless such pardon, expungement, or restoration of civil rights expressly provides that the person may not ship, transport, possess, or receive firearms.

(21) The term **"engaged in the business"** means—

(A) as applied to a manufacturer of firearms, a person who devotes time, attention, and labor to manufacturing firearms as a regular course of trade or business with the principal objective of livelihood and profit through the sale or distribution of the firearms manufactured;

(B) as applied to a manufacturer of ammunition, a person who devotes time, attention, and labor to manufacturing ammunition as a regular course of trade or business with the principal objective of livelihood and profit through the sale or distribution of the ammunition manufactured;

(C) as applied to a dealer in firearms, as defined in section 921(a)(11)(A), a person who devotes time, attention, and labor to dealing in firearms as a regular course of trade or business with the principal objective of livelihood and profit through the repetitive purchase and resale of firearms, but such term shall not include a person who makes occasional sales, exchanges, or purchases of firearms for the enhancement of a personal collection or for a hobby, or who sells all or part of his personal collection of firearms;

(D) as applied to a dealer in firearms, as defined in section 921(a)(11)(B), a person who devotes time, attention, and labor to engaging in such activity as a regular course of trade or business with the principal objective of livelihood and profit, but such term shall not include a person who makes occasional repairs of firearms, or who occasionally fits special barrels, stocks, or trigger mechanisms to firearms;

(E) as applied to an importer of firearms, a person who devotes time, attention, and labor to importing firearms as a regular course of trade or business with the principal objective of livelihood and profit through the sale or distribution of the firearms imported; and

(F) as applied to an importer of ammunition, a person who devotes time, attention, and labor to importing ammunition as a regular course of trade or business with the principal objective of livelihood and profit through the sale or distribution of the ammunition imported.

(22) The term **"with t he principal objective of liv elihood a nd profit"** means that the intent underlying the sale or disposition of firearms is predominantly one of obtaining livelihood and pecuniary gain, as opposed to other intents, such as improving or liquidating a personal firearms collection: **Provided**, That proof of profit shall not be required as to a person who engages in the regular and repetitive purchase and disposition of firearms for criminal purposes or terrorism. For purposes of this paragraph, the term **"terrorism"** means activity, directed against United States persons, which—

(A) is committed by an individual who is not a national or permanent resident alien of the United States;

(B) involves violent acts or acts dangerous to human life which would be a criminal violation if committed within the jurisdiction of the United States; and

(C) is intended—

(i) to intimidate or coerce a civilian population;

(ii) to influence the policy of a government by intimidation or coercion; or

(iii) to affect the conduct of a government by assassination or kidnapping.

(23) The term **"machinegun"** has the meaning given such term in section 5845(b) of the National Firearms Act (26 U.S.C. 5845(b)).

(24) The terms **"firearm silencer"** and **"firearm muffler"** mean any device for silencing, muffling, or diminishing the report of a portable firearm, including any combination of parts, designed or redesigned, and intended for use in assembling or fabricating a firearm silencer or firearm muffler, and any part intended only for use in such assembly or fabrication.

(25) The term **"school zone"** means—

(A) in, or on the grounds of, a public, parochial or private school; or

(B) within a distance of 1,000 feet from the grounds of a public, parochial or private school.

(26) The term **"school"** means a school which provides elementary or secondary education, as determined under State law.

(27) The term **"motor v ehicle"** has the meaning given such term in section 13102 of title 49, United States Code.

(28) The term **"semiautomatic rifle"** means any repeating rifle which utilizes a portion of the energy of a firing cartridge to extract the fired cartridge case and chamber the next round, and which requires a separate pull of the trigger to fire each cartridge.

(29) The term **"handgun"** means—

(A) a firearm which has a short stock and is designed to be held and fired by the use of a single hand; and

(B) any combination of parts from which a firearm described in subparagraph (A) can be assembled.

(30) Repealed.

(31) Repealed.

(32) The term **"intimate pa rtner"** means, with respect to a person, the spouse of the person, a former spouse of the person, an individual who is a parent of a child of the person, and an individual who cohabitates or has cohabited with the person.

(33) (A) Except as provided in subparagraph (C), the term **"misdemeanor crime of d omestic violence"** means an offense that—

(i) is a misdemeanor under Federal or State law; and

(ii) has, as an element, the use or attempted use of physical force, or the threatened use of a deadly weapon, committed by a current or former spouse, parent, or guardian of the victim, by a person with whom the victim shares a child in common, by a person who is cohabiting with or has cohabited with the victim as a spouse, parent, or guardian, or by a person similarly situated to a spouse, parent, or guardian of the victim.

(B) **(i)** A person shall not be considered to have been convicted of such an offense for purposes of this chapter, unless—

(I) the person was represented by counsel in the case, or knowingly and intelligently waived the right to counsel in the case; and

(II) in the case of a prosecution for an offense described in this paragraph for which a person was entitled to a jury trial in the jurisdiction in which the case was tried, either

(aa) the case was tried by a jury, or

(bb) the person knowingly and intelligently waived the right to have the case tried by a jury, by guilty plea or otherwise.

(ii) A person shall not be considered to have been convicted of such an offense for purposes of this chapter if the conviction has been expunged or set aside, or is an offense for which the person has been pardoned or has had civil rights restored (if the law of the applicable jurisdiction provides for the loss of civil rights under such an offense) unless the pardon, expungement, or restoration of civil rights expressly provides that the person may not ship, transport, possess, or receive firearms.

Editor's Note:

Subparagraph (C) referenced in 921(a)(33)(A) never was enacted. We presume the reference should have been to subparagraph (B).

(34) The term **"secure gun storage or safety device"** means—

(A) a device that, when installed on a firearm, is designed to prevent the firearm from being operated without first deactivating the device;

(B) a device incorporated into the design of the firearm that is designed to prevent the operation of the firearm by anyone not having access to the device; or

(C) a safe, gun safe, gun case, lock box, or other device that is designed to be or can be used to store a firearm and that is designed to be unlocked only by means of a key, a combination, or other similar means.

(35) The term **"body armor"** means any product sold or offered for sale, in interstate or foreign commerce, as personal protective body covering intended to protect against gunfire, regardless of whether the product is to be worn alone or is sold as a complement to another product or garment.

(b) For the purposes of this chapter, a member of the Armed Forces on active duty is a resident of the State in which his permanent duty station is located.

§ 922 Unlawful acts.

(a) It shall be unlawful—

(1) for any person—

(A) except a licensed importer, licensed manufacturer, or licensed dealer, to engage in the business of importing, manufacturing, or dealing in firearms, or in the course of such business to ship, transport, or receive any firearm in interstate or foreign commerce; or

(B) except a licensed importer or licensed manufacturer, to engage in the business of importing or manufacturing ammunition, or in the course of such business, to ship, transport, or receive any ammunition in interstate or foreign commerce;

(2) for any importer, manufacturer, dealer, or collector licensed under the provisions of this chapter to ship or transport in interstate or foreign commerce any firearm to any person other than a licensed importer, licensed manufacturer, licensed dealer, or licensed collector, except that—

(A) this paragraph and subsection (b)(3) shall not be held to preclude a licensed importer, licensed manufacturer, licensed dealer, or licensed collector from returning a firearm or replacement firearm of the same kind and type to a person from whom it was received; and this paragraph shall not be held to preclude an individual from mailing a firearm owned in compliance with Federal, State, and local law to a licensed importer, licensed manufacturer, licensed dealer, or licensed collector;

(B) this paragraph shall not be held to preclude a licensed importer, licensed manufacturer, or licensed dealer from depositing a firearm for conveyance in the mails to any officer, employee, agent, or watchman who, pursuant to the provisions of section 1715 of this title, is eligible to receive through the mails pistols, revolvers, and other firearms capable of being concealed on the person, for use in connection with his official duty; and

(C) nothing in this paragraph shall be construed as applying in any manner in the District of Columbia, the Commonwealth of Puerto Rico, or any possession of the United States differently than it would apply if the District of Columbia, the Commonwealth of Puerto Rico, or the possession were in fact a State of the United States;

(3) for any person other than a licensed importer, licensed manufacturer, licensed dealer, or licensed collector to transport into or receive in the State where he resides (or if the person is a corporation or other business entity, the State where it maintains a place of business) any firearm purchased or otherwise obtained by such person outside that State, except that this paragraph (A) shall not preclude any person who lawfully acquires a firearm by bequest or intestate succession in a State other than his State of residence from transporting the firearm into or receiving it in that State, if it is lawful for such person to purchase or possess such firearm in that State, (B) shall not apply to the transportation or receipt of a firearm obtained in conformity with subsection (b)(3) of this section, and (C) shall not apply to the transportation of any firearm acquired in any State prior to the effective date of this chapter;

(4) for any person, other than a licensed importer, licensed manufacturer, licensed dealer, or licensed collector, to transport in interstate or foreign commerce any destructive device, machinegun (as defined in section 5845 of the Internal Revenue Code of 1954), short-barreled shotgun, or short-barreled rifle, except as specifically authorized by the Attorney General consistent with public safety and necessity;

(5) for any person (other than a licensed importer, licensed manufacturer, licensed dealer, or licensed collector) to transfer, sell, trade, give, transport, or deliver any firearm to any person (other than a licensed importer, licensed manufacturer, licensed dealer, or licensed collector) who the transferor knows or has reasonable cause to believe does not reside in (or if the person is a corporation or other business entity, does not maintain a place of business in) the State in which the transferor resides; except that this paragraph shall not apply to (A) the transfer, transportation, or delivery of a firearm made to carry out a bequest of a firearm to, or an acquisition by intestate succession of a firearm by, a person who is permitted to acquire or possess a firearm under the laws of the State of his residence, and (B) the loan or rental of a firearm to any person for temporary use for lawful sporting purposes;

(6) for any person in connection with the acquisition or attempted acquisition of any firearm or ammunition from a licensed importer, licensed manufacturer, licensed dealer, or licensed collector, knowingly to make any false or fictitious oral or written statement or to furnish or exhibit any false, fictitious, or misrepresented identification, intended or likely to deceive such importer, manufacturer, dealer, or collector with respect to any fact material to the lawfulness of the sale or other disposition of such firearm or ammunition under the provisions of this chapter;

(7) for any person to manufacture or import armor piercing ammunition, unless--

(A) the manufacture of such ammunition is for the use of the United States, any department or agency of the United States, any State, or any department, agency, or political subdivision of a State;

(B) the manufacture of such ammunition is for the purpose of exportation; or

(C) the manufacture or importation of such ammunition is for the purpose of testing or experimentation and has been authorized by the Attorney General;

(8) for any manufacturer or importer to sell or deliver armor piercing ammunition, unless such sale or delivery--

(A) is for the use of the United States, any department or agency of the United States, any State, or any department, agency, or political sudivision of a State;

(B) is for the purpose of exportation; or

(C) is for the purpose of testing or experimentation and has been authorized by the Attorney General;

(9) for any person, other than a licensed importer, licensed manufacturer, licensed dealer, or licensed collector, who does not reside in any State to receive any firearms unless such receipt is for lawful sporting purposes.

(b) It shall be unlawful for any licensed importer, licensed manufacturer, licensed dealer, or licensed collector to sell or deliver—

(1) any firearm or ammunition to any individual who the licensee knows or has reasonable cause to believe is less than eighteen years of age, and, if the firearm or ammunition is other than a shotgun or rifle, or ammunition for a shotgun or rifle, to any individual who the licensee knows or has reasonable cause to believe is less than twenty-one years of age;

(2) any firearm to any person in any State where the purchase or possession by such person of such firearm would be in violation of any State law or any published ordinance applicable at the place of sale, delivery or other disposition, unless the licensee knows or has reasonable cause to believe that the purchase or possession would not be in violation of such State law or such published ordinance;

(3) any firearm to any person who the licensee knows or has reasonable cause to believe does not reside in (or if the person is a corporation or other business entity, does not maintain a place of business in) the State in which the licensee's place of business is located, except that this paragraph (A) shall not apply to the sale or delivery of any rifle or shotgun to a resident of a State other than a State in which the licensee's place of business is located if the transferee meets in person with the transferor to accomplish the transfer, and the sale, delivery, and receipt fully comply with the legal conditions of sale in both such States (and any licensed manufacturer, importer or dealer shall be presumed, for purposes of this subparagraph, in the absence of evidence to the contrary, to have had actual knowledge of the State laws and published ordinances of both States), and (B) shall not apply to the loan or rental of a firearm to any person for temporary use for lawful sporting purposes;

(4) to any person any destructive device, machinegun (as defined in section 5845 of the Internal Revenue Code of 1954), short-barreled shotgun, or short-barreled rifle, except as specifically authorized by the Attorney General consistent with public safety and necessity; and

(5) any firearm or armor-piercing ammunition to any person unless the licensee notes in his records, required to be kept pursuant to section 923 of this chapter, the name, age, and place of residence of such person if the person is an individual, or the identity and principal and local places of business of such person if the person is a corporation or other business entity.

Paragraphs (1), (2), (3), and (4) of this subsection shall not apply to transactions between licensed importers, licensed manufacturers, licensed dealers, and licensed collectors. Paragraph (4) of this subsection shall not apply to a sale or delivery to any research organization designated by the Attorney General.

(c) In any case not otherwise prohibited by this chapter, a licensed importer, licensed manufacturer, or licensed dealer may sell a firearm to a person who does not appear in person at the licensee's business premises (other than another licensed importer, manufacturer, or dealer) only if—

(1) the transferee submits to the transferor a sworn statement in the following form:

"Subject to penalties provided by law, I swear that, in the case of any firearm other than a shotgun or a rifle, I am twenty-one years or more of age, or that, in the case of a shotgun or a rifle, I am eighteen years or more of age; that I am not prohibited by the provisions of chapter 44 of title 18, United States Code, from receiving a firearm in interstate or foreign commerce; and that my receipt of this firearm will not be in violation of any statute of the State and published ordinance applicable to the locality in which I reside. Further, the true title, name, and address of the principal law enforcement officer of the locality to which the firearm will be delivered are ____________.

____________________________.
Signature

Date ______________ ."

and containing blank spaces for the attachment of a true copy of any permit or other information required pursuant to such statute or published ordinance;

(2) the transferor has, prior to the shipment or delivery of the firearm, forwarded by registered or certified mail (return receipt requested) a copy of the sworn statement, together with a description of the firearm, in a form prescribed by the Attorney General, to the chief law enforcement officer of the transferee's place of residence, and has received a return receipt evidencing delivery of the statement or has had the statement returned due to the refusal of the named addressee to accept such letter in accordance with United States Post Office Department regulations; and

(3) the transferor has delayed shipment or delivery for a period of at least seven days following receipt of the notification of the acceptance or refusal of delivery of the statement.

A copy of the sworn statement and a copy of the notification to the local law enforcement officer, together with evidence of receipt or rejection of that notification, shall be retained by the licensee as a part of the records required to be kept under section 923(g).

(d) It shall be unlawful for any person to sell or otherwise dispose of any firearm or ammunition to any person knowing or having reasonable cause to believe that such person—

(1) is under indictment for, or has been convicted in any court of, a crime punishable by imprisonment for a term exceeding one year;

(2) is a fugitive from justice;

(3) is an unlawful user of or addicted to any controlled substance (as defined in section 102 of the Controlled Substances Act (21 U.S.C. 802));

(4) has been adjudicated as a mental defective or has been committed to any mental institution;

(5) who, being an alien—

(A) is illegally or unlawfully in the United States; or

(B) except as provided in subsection (y)(2), has been admitted to the United States under a nonimmigrant visa (as that term is defined in section 101(a)(26) of the Immigration and Nationality Act (8 U.S.C. 1101(a)(26));

(6) who has been discharged from the Armed Forces under dishonorable conditions;

(7) who, having been a citizen of the United States, has renounced his citizenship;

(8) is subject to a court order that restrains such person from harassing, stalking, or threatening an intimate partner of such person or child of such intimate partner or person, or engaging in other conduct that would place an intimate partner in reasonable fear of bodily injury to the partner or child, except that this paragraph shall only apply to a court order that—

(A) was issued after a hearing of which such person received actual notice, and at which such person had the opportunity to participate; and

(B) (i) includes a finding that such person represents a credible threat to the physical safety of such intimate partner or child; or

(ii) by its terms explicitly prohibits the use, attempted use, or threatened use of physical force against such intimate partner or child that would reasonably be expected to cause bodily injury; or

(9) has been convicted in any court of a misdemeanor crime of domestic violence.

This subsection shall not apply with respect to the sale or disposition of a firearm or ammunition to a licensed importer, licensed manufacturer, licensed dealer, or licensed collector who pursuant to subsection (b) of section 925 of this chapter is not precluded from dealing in firearms or ammunition, or to a person who has been granted relief from disabilities pursuant to subsection (c) of section 925 of this chapter.

(e) It shall be unlawful for any person knowingly to deliver or cause to be delivered to any common or contract carrier for transportation or shipment in interstate or foreign commerce, to persons other than licensed importers, licensed manufacturers, licensed dealers, or licensed collectors, any package or other container in which there is any firearm or ammunition without written notice to the carrier that such firearm or ammunition is being transported or shipped; except that any passenger who owns or legally possesses a firearm or ammunition being transported aboard any common or contract carrier for movement with the passenger in interstate or foreign commerce may deliver said firearm or ammunition into the custody of the pilot, captain, conductor or operator of such common or contract carrier for the duration of the trip without violating any of the provisions of this chapter. No common or contract carrier shall require or cause any label, tag, or other written notice to be placed on the outside of any package, luggage, or other container that such package, luggage, or other container contains a firearm.

(f) (1) It shall be unlawful for any common or contract carrier to transport or deliver in interstate or foreign commerce any firearm or ammunition with knowledge or reasonable cause to believe that the shipment, transportation, or receipt thereof would be in violation of the provisions of this chapter.

(2) It shall be unlawful for any common or contract carrier to deliver in interstate or foreign commerce any firearm without obtaining written acknowledgement of receipt from the recipient of the package or other container in which there is a firearm.

(g) It shall be unlawful for any person—

(1) who has been convicted in any court of, a crime punishable by imprisonment for a term exceeding one year;

(2) who is a fugitive from justice;

(3) who is an unlawful user of or addicted to any controlled substance (as defined in section 102 of the Controlled Substances Act (21 U.S.C. 802));

(4) who has been adjudicated as a mental defective or who has been committed to a mental institution;

(5) who, being an alien—

(A) is illegally or unlawfully in the United States; or

(B) except as provided in subsection (y)(2), has been admitted to the United States under a nonimmigrant visa (as that term is defined in section 101(a)(26) of the Immigration and Nationality Act (8 U.S.C. 1101(a)(26)));

(6) who has been discharged from the Armed Forces under dishonorable conditions;

(7) who, having been a citizen of the United States, has renounced his citizenship;

(8) who is subject to a court order that—

(A) was issued after a hearing of which such person received actual notice, and at which such person had an opportunity to participate;

(B) restrains such person from harassing, stalking, or threatening an intimate partner of such person or child of such intimate partner or person, or engaging in other conduct that would place an intimate partner in reasonable fear of bodily injury to the partner or child; and

(C) (i) includes a finding that such person represents a credible threat to the physical safety of such intimate partner or child; or

(ii) by its terms explicitly prohibits the use, attempted use, or threatened use of physical force against such intimate partner or child that would reasonably be expected to cause bodily injury; or

(9) who has been convicted in any court of a misdemeanor crime of domestic violence, to ship or transport in interstate or foreign commerce, or possess in or affecting commerce, any firearm or ammunition; or to receive any firearm or ammunition which has been shipped or transported in interstate or foreign commerce.

(h) It shall be unlawful for any individual, who to that individual's knowledge and while being employed for any person described in any paragraph of subsection (g) of this section, in the course of such employment—

(1) to receive, possess, or transport any firearm or ammunition in or affecting interstate or foreign commerce; or

(2) to receive any firearm or ammunition which has been shipped or transported in interstate or foreign commerce.

(i) It shall be unlawful for any person to transport or ship in interstate or foreign commerce, any stolen firearm or stolen ammunition, knowing or having reasonable cause to believe that the firearm or ammunition was stolen.

(j) It shall be unlawful for any person to receive, possess, conceal, store, barter, sell, or dispose of any stolen firearm or stolen ammunition, or pledge or accept as security for a loan any stolen firearm or stolen ammunition, which is moving as, which is a part of, which constitutes, or which has been shipped or transported in, interstate or foreign commerce, either before or after it was stolen, knowing or having reasonable cause to believe that the firearm or ammunition was stolen.

(k) It shall be unlawful for any person knowingly to transport, ship, or receive, in interstate or foreign commerce, any firearm which has had the importer's or manufacturer's serial number removed, obliterated, or altered, or to possess or receive any firearm which has had the importer's or manufacturer's serial number removed, obliterated, or altered and has, at any time, been shipped or transported in interstate or foreign commerce.

(l) Except as provided in section 925(d) of this chapter, it shall be unlawful for any person knowingly to import or bring into the United States or any possession thereof any firearm or ammunition; and it shall be unlawful for any person knowingly to receive any firearm or ammunition which has been imported or brought into the United States or any possession thereof in violation of the provisions of this chapter.

(m) It shall be unlawful for any licensed importer, licensed manufacturer, licensed dealer, or licensed collector knowingly to make any false entry in, to fail to make appropriate entry in, or to fail to properly maintain, any record which he is required to keep pursuant to section 923 of this chapter or regulations promulgated thereunder.

(n) It shall be unlawful for any person who is under indictment for a crime punishable by imprisonment for a term exceeding one year to ship or transport in interstate or foreign commerce any firearm or ammunition or receive any firearm or ammunition which has been shipped or transported in interstate or foreign commerce.

(o) (1) Except as provided in paragraph (2), it shall be unlawful for any person to transfer or possess a machinegun.

(2) This subsection does not apply with respect to—

(A) a transfer to or by, or possession by or under the authority of, the United States or any department or agency thereof or a State, or a department, agency, or political subdivision thereof; or

(B) any lawful transfer or lawful possession of a machinegun that was lawfully possessed before the date this subsection takes effect.

(p) (1) It shall be unlawful for any person to manufacture, import, sell, ship, deliver, possess, transfer, or receive any firearm—

(A) that, after removal of grips, stocks, and magazines, is not as detectable as the Security Exemplar, by walk-through metal detectors calibrated and operated to detect the Security Exemplar; or

(B) any major component of which, when subjected to inspection by the types of x-ray machines commonly used at airports, does not generate an image that accurately depicts the shape of the component. Barium sulfate or other compounds may be used in the fabrication of the component.

(2) For purposes of this subsection—

(A) the term **"firearm"** does not include the frame or receiver of any such weapon;

(B) the term **"major component"** means, with respect to a firearm, the barrel, the slide or cylinder, or the frame or receiver of the firearm; and

(C) the term **"Security Exemplar"** means an object, to be fabricated at the direction of the Attorney General, that is—

(i) constructed of, during the 12-month period beginning on the date of the enactment of this subsection, 3.7 ounces of material type 17-4 PH stainless steel in a shape resembling a handgun; and

(ii) suitable for testing and calibrating metal detectors:

Provided, however, That at the close of such 12-month period, and at appropriate times thereafter the Attorney General shall promulgate regulations to permit the manufacture, importation, sale, shipment, delivery, possession, transfer, or receipt of firearms previously prohibited under this subparagraph that are as detectable as a "Security Exemplar" which contains 3.7 ounces of material type 17-4 PH stainless steel, in a shape resembling a handgun, or such lesser amount as is detectable in view of advances in state-of-the-art developments in weapons detection technology.

(3) Under such rules and regulations as the Attorney General shall prescribe, this subsection shall not apply to the manufacture, possession, transfer, receipt, shipment, or delivery of a firearm by a licensed manufacturer or any person acting pursuant to a contract with a licensed manufacturer, for the purpose of examining and testing such firearm to determine whether paragraph (1) applies to such firearm. The Attorney General shall ensure that rules and regulations adopted pursuant to this paragraph do not impair the manufacture of prototype firearms or the development of new technology.

(4) The Attorney General shall permit the conditional importation of a firearm by a licensed importer or licensed manufacturer, for examination and testing to determine whether or not the unconditional importation of such firearm would violate this subsection.

(5) This subsection shall not apply to any firearm which—

(A) has been certified by the Secretary of Defense or the Director of Central Intelligence, after consultation with the Attorney General and the Administrator of the Federal Aviation Administration, as necessary for military or intelligence applications; and

(B) is manufactured for and sold exclusively to military or intelligence agencies of the United States.

(6) This subsection shall not apply with respect to any firearm manufactured in, imported into, or possessed in the United States before the date of the enactment of the Undetectable Firearms Act of 1988.

(q) (1) The Congress finds and declares that—

(A) crime, particularly crime involving drugs and guns, is a pervasive, nationwide problem;

(B) crime at the local level is exacerbated by the interstate movement of drugs, guns, and criminal gangs;

(C) firearms and ammunition move easily in interstate commerce and have been found in increasing numbers in and around schools, as documented in numerous hearings in both the Committee on the Judiciary of the House of Representatives and the Committee on the Judiciary of the Senate;

(D) in fact, even before the sale of a firearm, the gun, its component parts, ammunition, and the raw materials from which they are made have considerably moved in interstate commerce;

(E) while criminals freely move from State to State, ordinary citizens and foreign visitors may fear to travel to or through certain parts of the country due to concern about violent crime and gun violence, and parents may decline to send their children to school for the same reason;

(F) the occurrence of violent crime in school zones has resulted in a decline in the quality of education in our country;

(G) this decline in the quality of education has an adverse impact on interstate commerce and the foreign commerce of the United States;

(H) States, localities, and school systems find it almost impossible to handle gun-related crime by themselves--even States, localities, and school systems that have made strong efforts to prevent, detect, and punish gun-related crime find their efforts unavailing due in part to the failure or inability of other States or localities to take strong measures; and

(I) the Congress has the power, under the interstate commerce clause and other provisions of the Constitution, to enact measures to ensure the integrity and safety of the Nation's schools by enactment of this subsection.

(2) (A) It shall be unlawful for any individual knowingly to possess a firearm that has moved in or that otherwise affects interstate or foreign commerce at a place that the individual knows, or has reasonable cause to believe, is a school zone.

(B) Subparagraph (A) does not apply to the possession of a firearm—

(i) on private property not part of school grounds;

(ii) if the individual possessing the firearm is licensed to do so by the State in which the school zone is located or a political subdivision of the State, and the law of the State or political subdivision requires that, before an individual obtains such a license, the law enforcement authorities of the State or political subdivision verify that the individual is qualified under law to receive the license;

(iii) that is—

(I) not loaded; and

(II) in a locked container, or a locked firearms rack that is on a motor vehicle;

(iv) by an individual for use in a program approved by a school in the school zone;

(v) by an individual in accordance with a contract entered into between a school in the school zone and the individual or an employer of the individual;

(vi) by a law enforcement officer acting in his or her official capacity; or

(vii) that is unloaded and is possessed by an individual while traversing school premises for the purpose of gaining access to public or private lands open to hunting, if the entry on school premises is authorized by school authorities.

(3) (A) Except as provided in subparagraph (B), it shall be unlawful for any person, knowingly or with reckless disregard for the safety of another, to discharge or attempt to discharge a firearm that has moved in or that otherwise affects interstate or foreign commerce at a place that the person knows is a school zone.

(B) Subparagraph (A) does not apply to the discharge of a firearm—

(i) on private property not part of school grounds;

(ii) as part of a program approved by a school in the school zone, by an individual who is participating in the program;

(iii) by an individual in accordance with a contract entered into between a school in a school zone and the individual or an employer of the individual; or

(iv) by a law enforcement officer acting in his or her official capacity.

(4) Nothing in this subsection shall be construed as preempting or preventing a State or local government from enacting a statute establishing gun free school zones as provided in this subsection.

(r) It shall be unlawful for any person to assemble from imported parts any semiautomatic rifle or any shotgun which is identical to any rifle or shotgun prohibited from importation under section 925(d)(3) of this chapter as not being particularly suitable for or readily adaptable to sporting purposes except that this subsection shall not apply to—

(1) the assembly of any such rifle or shotgun for sale or distribution by a licensed manufacturer to the United States or any department or agency thereof or to any State or any department, agency, or political subdivision thereof; or

(2) the assembly of any such rifle or shotgun for the purposes of testing or experimentation authorized by the Attorney General.

(s) (1) Beginning on the date that is 90 days after the date of enactment of this subsection and ending on the day before the date that is 60 months after such date of enactment, it shall be unlawful for any licensed importer, licensed manufacturer, or licensed dealer to sell, deliver, or transfer a handgun (other than the return of a handgun to the person from whom it was received) to an individual who is not licensed under section 923, unless—

(A) after the most recent proposal of such transfer by the transferee—

(i) the transferor has—

(I) received from the transferee a statement of the transferee containing the information described in paragraph (3);

(II) verified the identity of the transferee by examining the identification document presented;

(III) within 1 day after the transferee furnishes the statement, provided notice of the contents of the statement to the chief law enforcement officer of the place of residence of the transferee; and

(IV) within 1 day after the transferee furnishes the statement, transmitted a copy of the statement to the chief law enforcement officer of the place of residence of the transferee; and

(ii) (I) 5 business days (meaning days on which State offices are open) have elapsed from the date the transferor furnished notice of the contents of the statement to the chief law enforcement officer, during which period the transferor has not received information from the chief law enforcement officer that receipt or possession of the handgun by the transferee would be in violation of Federal, State, or local law; or

(II) the transferor has received notice from the chief law enforcement officer that the officer has no information indicating that receipt or possession of the handgun by the transferee would violate Federal, State, or local law;

(B) the transferee has presented to the transferor a written statement, issued by the chief law enforcement officer of the place of residence of the transferee during the 10-day period ending on the date of the most recent proposal of such transfer by the transferee, stating that the transferee requires access to a handgun because of a threat to the life of the transferee or of any member of the household of the transferee;

(C) (i) the transferee has presented to the transferor a permit that—

(I) allows the transferee to possess or acquire a handgun; and

(II) was issued not more than 5 years earlier by the State in which the transfer is to take place; and

(ii) the law of the State provides that such a permit is to be issued only after an authorized government official has verified that the information available to such official does not indicate that possession of a handgun by the transferee would be in violation of the law;

(D) the law of the State requires that, before any licensed importer, licensed manufacturer, or licensed dealer completes the transfer of a handgun to an individual who is not licensed under section 923, an authorized government official verify that the information available to such official does not indicate that possession of a handgun by the transferee would be in violation of law;

(E) the Attorney General has approved the transfer under section 5812 of the Internal Revenue Code of 1954; or

(F) on application of the transferor, the Attorney General has certified that compliance with subparagraph (A)(i)(III) is impracticable because—

(i) the ratio of the number of law enforcement officers of the State in which the transfer is to occur to the number of square miles of land area of the State does not exceed 0.0025;

(ii) the business premises of the transferor at which the transfer is to occur are extremely remote in relation to the chief law enforcement officer; and

(iii) there is an absence of telecommunications facilities in the geographical area in which the business premises are located.

(2) A chief law enforcement officer to whom a transferor has provided notice pursuant to paragraph (1)(A)(i)(III) shall make a reasonable effort to ascertain within 5 business days whether receipt or possession would be in violation of the law, including research in whatever State and local recordkeeping systems are available and in a national system designated by the Attorney General.

(3) The statement referred to in paragraph (1)(A)(i)(I) shall contain only—

(A) the name, address, and date of birth appearing on a valid identification document (as defined in section 1028(d)(1)) of the transferee containing a photograph of the transferee and a description of the identification used;

(B) a statement that the transferee—

(i) is not under indictment for, and has not been convicted in any court of, a crime punishable by imprisonment for a term exceeding 1 year, and has not been convicted in any court of a misdemeanor crime of domestic violence;

(ii) is not a fugitive from justice;

(iii) is not an unlawful user of or addicted to any controlled substance (as defined in section 102 of the Controlled Substances Act);

(iv) has not been adjudicated as a mental defective or been committed to a mental institution;

(v) is not an alien who—

(I) is illegally or unlawfully in the United States; or

(II) subject to subsection (y)(2), has been admitted to the United States under a nonimmigrant visa (as that term is defined in section 101(a)(26) of the Immigration and Nationality Act (8 U.S.C. 1101(a)(26)));

(vi) has not been discharged from the Armed Forces under dishonorable conditions; and

(vii) is not a person who, having been a citizen of the United States, has renounced such citizenship;

(C) the date the statement is made; and

(D) notice that the transferee intends to obtain a handgun from the transferor.

(4) Any transferor of a handgun who, after such transfer, receives a report from a chief law enforcement officer containing information that receipt or possession of the handgun by the transferee violates Federal, State, or local law shall, within 1 business day after receipt of such request, communicate any information related to the transfer that the transferor has about the transfer and the transferee to—

(A) the chief law enforcement officer of the place of business of the transferor; and

(B) the chief law enforcement officer of the place of residence of the transferee.

(5) Any transferor who receives information, not otherwise available to the public, in a report under this subsection shall not disclose such information except to the transferee, to law enforcement authorities, or pursuant to the direction of a court of law.

(6) **(A)** Any transferor who sells, delivers, or otherwise transfers a handgun to a transferee shall retain the copy of the statement of the transferee with respect to the handgun transaction, and shall retain evidence that the transferor has complied with subclauses (III) and (IV) of paragraph (1)(A)(i) with respect to the statement.

(B) Unless the chief law enforcement officer to whom a statement is transmitted under paragraph (1)(A)(i)(IV) determines that a transaction would violate Federal, State, or local law—

(i) the officer shall, within 20 business days after the date the transferee made the statement on the basis of which the notice was provided, destroy the statement, any record containing information derived from the statement, and any record created as a result of the notice required by paragraph (1)(A)(i)(III);

(ii) the information contained in the statement shall not be conveyed to any person except a person who has a need to know in order to carry out this subsection; and

(iii) the information contained in the statement shall not be used for any purpose other than to carry out this subsection.

(C) If a chief law enforcement officer determines that an individual is ineligible to receive a handgun and the individual requests the officer to provide the reason for such determination, the officer shall provide such reasons to the individual in writing within 20 business days after receipt of the request.

(7) A chief law enforcement officer or other person responsible for providing criminal history background information pursuant to this subsection shall not be liable in an action at law for damages—

(A) for failure to prevent the sale or transfer of a handgun to a person whose receipt or possession of the handgun is unlawful under this section; or

(B) for preventing such a sale or transfer to a person who may lawfully receive or possess a handgun.

(8) For purposes of this subsection, the term **"chief law enforcement officer"** means the chief of police, the sheriff, or an equivalent officer or the designee of any such individual.

(9) The Attorney General shall take necessary actions to ensure that the provisions of this subsection are published and disseminated to licensed dealers, law enforcement officials, and the public.

(t) **(1)** Beginning on the date that is 30 days after the Attorney General notifies licensees under section 103(d) of the Brady Handgun Violence Prevention Act that the national instant criminal background check system is established, a licensed importer, licensed manufacturer, or licensed dealer shall not transfer a firearm to any other person who is not licensed under this chapter, unless—

(A) before the completion of the transfer, the licensee contacts the national instant criminal background check system established under section 103 of that Act;

(B) **(i)** the system provides the licensee with a unique identification number; or

(ii) 3 business days (meaning a day on which State offices are open) have elapsed since the licensee contacted the system, and the system has not notified the licensee that the receipt of a firearm by such other person would violate subsection (g) or (n) of this section; and

(C) the transferor has verified the identity of the transferee by examining a valid identification document (as defined in section 1028(d) of this title) of the transferee containing a photograph of the transferee.

(2) If receipt of a firearm would not violate section 922 (g) or (n) or State law, the system shall—

(A) assign a unique identification number to the transfer;

(B) provide the licensee with the number; and

(C) destroy all records of the system with respect to the call (other than the identifying number and the date the number was assigned) and all records of the system relating to the person or the transfer.

(3) Paragraph (1) shall not apply to a firearm transfer between a licensee and another person if—

(A) **(i)** such other person has presented to the licensee a permit that—

(I) allows such other person to possess or acquire a firearm; and

(II) was issued not more than 5 years earlier by the State in which the transfer is to take place; and

(ii) the law of the State provides that such a permit is to be issued only after an authorized government official has verified that the information available to such official does not indicate that possession of a firearm by such other person would be in violation of law;

(B) the Attorney General has approved the transfer under section 5812 of the Internal Revenue Code of 1954; or

(C) on application of the transferor, the Attorney General has certified that compliance with paragraph (1)(A) is impracticable because—

(i) the ratio of the number of law enforcement officers of the State in which the transfer is to occur to the number of square miles of land area of the State does not exceed 0.0025;

(ii) the business premises of the licensee at which the transfer is to occur are extremely remote in relation to the chief law enforcement officer (as defined in subsection (s)(8)); and

(iii) there is an absence of telecommunications facilities in the geographical area in which the business premises are located.

(4) If the national instant criminal background check system notifies the licensee that the information available to the system does not demonstrate that the receipt of a firearm by such other person would violate subsection (g) or (n) or State law, and the licensee transfers a firearm to such other person, the licensee shall include in the record of the transfer the unique identification number provided by the system with respect to the transfer.

(5) If the licensee knowingly transfers a firearm to such other person and knowingly fails to comply with paragraph (1) of this subsection with respect to the transfer and, at the time such other person most recently proposed the transfer, the national instant criminal background check system was operating and information was available to the system demonstrating that receipt of a firearm by such other person would violate subsection (g) or (n) of this section or State law, the Attorney General may, after notice and opportunity for a hearing, suspend for not more than 6 months or revoke any license issued to the licensee under section 923, and may impose on the licensee a civil fine of not more than $5,000.

(6) Neither a local government nor an employee of the Federal Government or of any State or local government, responsible for providing information to the national instant criminal background check system shall be liable in an action at law for damages –

(A) for failure to prevent the sale or transfer of a firearm to a person whose receipt or possession of the firearm is unlawful under this section; or

(B) for preventing such a sale or transfer to a person who may lawfully receive or possess a firearm.

(u) It shall be unlawful for a person to steal or unlawfully take or carry away from the person or the premises of a person who is licensed to engage in the business of importing, manufacturing, or dealing in firearms, any firearm in the licensee's business inventory that has been shipped or transported in interstate or foreign commerce.

(v) Repealed.

(w) Repealed.

(x) **(1)** It shall be unlawful for a person to sell, deliver, or otherwise transfer to a person who the transferor knows or has reasonable cause to believe is a juvenile—

(A) a handgun; or

(B) ammunition that is suitable for use only in a handgun.

(2) It shall be unlawful for any person who is a juvenile to knowingly possess—

(A) a handgun; or

(B) ammunition that is suitable for use only in a handgun.

(3) This subsection does not apply to—

(A) a temporary transfer of a handgun or ammunition to a juvenile or to the possession or use of a handgun or ammunition by a juvenile if the handgun and ammunition are possessed and used by the juvenile—

(i) in the course of employment, in the course of ranching or farming related to activities at the residence of the juvenile (or on property used for ranching or farming at which the juvenile, with the permission of the property owner or lessee, is performing activities related to the operation of the farm or ranch), target practice, hunting, or a course of instruction in the safe and lawful use of a handgun;

(ii) with the prior written consent of the juvenile's parent or guardian who is not prohibited by Federal, State, or local law from possessing a firearm, except—

(I) during transportation by the juvenile of an unloaded handgun in a locked container directly from the place of transfer to a place at which an activity described in clause (i) is to take place and transportation by the juvenile of that handgun, unloaded and in a locked container, directly from the place at which such an activity took place to the transferor; or

(II) with respect to ranching or farming activities as described in clause (i), a juvenile may possess and use a handgun or ammunition with the prior written approval of the juvenile's parent or legal guardian and at the direction of an adult who is not prohibited by Federal, State or local law from possessing a firearm;

(iii) the juvenile has the prior written consent in the juvenile's possession at all times when a handgun is in the possession of the juvenile; and

(iv) in accordance with State and local law;

(B) a juvenile who is a member of the Armed Forces of the United States or the National Guard who possesses or is armed with a handgun in the line of duty;

(C) a transfer by inheritance of title (but not possession) of a handgun or ammunition to a juvenile; or

(D) the possession of a handgun or ammunition by a juvenile taken in defense of the juvenile or other persons against an intruder into the residence of the juvenile or a residence in which the juvenile is an invited guest.

(4) A handgun or ammunition, the

possession of which is transferred to a juvenile in circumstances in which the transferor is not in violation of this subsection shall not be subject to permanent confiscation by the Government if its possession by the juvenile subsequently becomes unlawful because of the conduct of the juvenile, but shall be returned to the lawful owner when such handgun or ammunition is no longer required by the Government for the purposes of investigation or prosecution.

(5) For purposes of this subsection, the term **"juvenile"** means a person who is less than 18 years of age.

(6) (A) In a prosecution of a violation of this subsection, the court shall require the presence of a juvenile defendant's parent or legal guardian at all proceedings.

(B) The court may use the contempt power to enforce subparagraph (A).

(C) The court may excuse attendance of a parent or legal guardian of a juvenile defendant at a proceeding in a prosecution of a violation of this subsection for good cause shown.

(y) Provisions relating to aliens admitted under nonimmigrant visas.

(1) Definitions. In this subsection—

(A) the term **"alien"** has the same meaning as in section 101(a)(3) of the Immigration and Nationality Act (8 U.S.C. 1101(a)(3)); and

(B) the term **"nonimmigrant visa"** has the same meaning as in section 101(a)(26) of the Immigration and Nationality Act (8 U.S.C. 1101(a)(26)).

(2) Exceptions. Subsections (d)(5)(B), (g)(5)(B), and (s)(3)(B)(v)(II) do not apply to any alien who has been lawfully admitted to the United States under a nonimmigrant visa, if that alien is—

(A) admitted to the United States for lawful hunting or sporting purposes or is in possession of a hunting license or permit lawfully issued in the United States;

(B) an official representative of a foreign government who is—

(i) accredited to the United States Government or the Government's mission to an international organization having its headquarters in the United States; or

(ii) en route to or from another country to which that alien is accredited;

(C) an official of a foreign government or a distinguished foreign visitor who has been so designated by the Department of State; or

(D) a foreign law enforcement officer of a friendly foreign government entering the United States on official law enforcement business.

(3) Waiver.

(A) Conditions for waiver. Any individual who has been admitted to the United States under a nonimmigrant visa may receive a waiver from the requirements of subsection (g)(5), if—

(i) the individual submits to the Attorney General a petition that meets the requirements of subparagraph (C); and

(ii) the Attorney General approves the petition.

(B) Petition. Each petition under subparagraph (B) shall—

(i) demonstrate that the petitioner has resided in the United States for a continuous period of not less than 180 days before the date on which the petition is submitted under this paragraph; and

(ii) include a written statement from the embassy or consulate of the petitioner, authorizing the petitioner to acquire a firearm or ammunition and certifying that the alien would not, absent the application of subsection (g)(5)(B), otherwise be prohibited from such acquisition under subsection (g).

(C) Approval of petition. The Attorney General shall approve a petition submitted in accordance with this paragraph, if the Attorney General determines that waiving the requirements of subsection (g)(5)(B) with respect to the petitioner—

(i) would be in the interests of justice; and

(ii) would not jeopardize the public safety.

Editor's Note:

18 U.S.C. § 922(z) was added to the Gun Control Act as part of the "Protection of Lawful Commerce in Arms Act" enacted on October 26, 2005. It is effective April 24, 2006.

(z) SECURE GUN STORAGE OR SAFTEY DEVICE—

(1) IN GENERAL- Except as Provided under paragraph (2), it shall be unlawful for any licensed importer, licensed manufacturer, or licensed dealer to sell, deliver, or transfer any handgun to any person other than any person licensed under this chapter, unless the transferee is provided with a secure gun storage or safety device (as defined in section 921(a)(34)) for than handgun.

(2) EXCEPTIONS – Paragraph shall not apply to--

(A)(i) the manufacture for, transfer to, or possession by, the United States, a department or agency of the United States, a State, or a department, agency, or political subdivision of a State, of a handgun; or

(ii) the transfer to, or possession by, a law enforcement officer employed by an entity referred to in clause (i) of a handgun for law enforcement purposes (whether on or off duty); or

(B) the transfer to, or possession by a rail police officer employed by a rail carrier and certified or commissioned as a police officer under the laws of a State of a handgun for purposes of law enforcement (whether on or off duty);

(C) the transfer to any person of a handgun listed as a curio or relic by the Secretary pursuant to section 921(a)(13); or

(D) the transfer to any person of a handgun for which a secure gun storage or safety device is temporarily unavailable for the reasons described in the exceptions stated in section 923(e), if the licensed manufacturer, licensed importer, or licensed dealer delivers to the transferee within 10 calendar days from the date of the delivery of the handgun to the transferee a secure gun storage or safety device for the handgun.

(3) LIABILITY FOR USE –

(A) IN GENERAL– Nothwithstanding any other provision of law, a person who has lawful possession and control of a handgun, and who uses a secure gun storage or safety device with the handgun, shall be entitled to immunity from a qualified civil liability action.

(B) PROSPECTIVE ACTIONS- A qualified civil liability action may not be brought in any Federal or State court.

(C) DEFINED TERM – As used in this paragraph, the term qualified civil liability action--

(i) means a civil action brought by any person against a person described in subparagraph (A) for damages resulting from the criminal or unlawful misuse of the handgun by a third party, if—

(I) the handgun was accessed by another person who did not have the permission or authorization of the person having lawful possession and control of the handgun to have access to it; and

(II) at the time access was gained by the person not so authorized, the handgun had been made in operable by use of a secure gun storage or safety device; and

(ii) shall not include an action brought against the person having lawful possession and control of the handgun for negligent entrustment or negligence per se.

APPENDIX A – Repealed.

§ 923 Licensing.

(a) No person shall engage in the business of importing, manufacturing, or dealing in firearms, or importing or manufacturing ammunition, until he has filed an application with and received a license to do so from the Attorney General. The application shall be in such form and contain only that information necessary to determine eligibility for licensing as the Attorney General shall by regulation prescribe and shall include a photograph and finger-prints of the applicant. Each applicant shall pay a fee for obtaining such a license, a separate fee being required for each place in which the applicant is to do business, as follows:

(1) If the applicant is a manufacturer—

(A) of destructive devices, ammunition for destructive devices or armor piercing ammunition, a fee of $1,000 per year;

(B) of firearms other than destructive devices, a fee of $50 per year; or

(C) of ammunition for firearms, other than ammunition for destructive devices or armor piercing ammunition, a fee of $10 per year.

(2) If the applicant is an importer—

(A) of destructive devices, ammunition for destructive devices or armor piercing ammunition, a fee of $1,000 per year; or

(B) of firearms other than destructive devices or ammunition for firearms other than destructive devices, or ammunition other than armor piercing ammunition, a fee of $50 per year.

(3) If the applicant is a dealer—

(A) in destructive devices or ammunition for destructive devices, a fee of $1,000 per year; or

(B) who is not a dealer in destructive devices, a fee of $200 for 3 years, except that the fee for renewal of a valid license shall be $90 for 3 years.

(C) Repealed by Pub. L. 103-159, Title III, 303(4), Nov. 30, 1993, 107 Stat. 1546.

(b) Any person desiring to be licensed as a collector shall file an application for such license with the Attorney General. The application shall be in such form and contain only that information necessary to determine eligibility as the Attorney General shall by regulation prescribe. The fee for such license shall be $10 per year. Any license granted under this subsection shall only apply to transactions in curios and relics.

(c) Upon the filing of a proper application and payment of the prescribed fee, the Attorney General shall issue to a qualified applicant the appropriate license which, subject to the provisions of this chapter and other applicable provisions of law, shall entitle the licensee to transport, ship, and receive firearms and ammunition covered by such license in interstate or foreign commerce during the period stated in the license. Nothing in this chapter shall be construed to prohibit a licensed manufacturer, importer, or dealer from maintaining and disposing of a personal collection of firearms, subject only to such restrictions as apply in this chapter to dispositions by a person other than a licensed manufacturer, importer, or dealer. If any firearm is so disposed of by a licensee within one year after its transfer from his business inventory into such licensee's personal collection or if such disposition or any other acquisition is made for the purpose of willfully evading the restrictions placed upon licensees by this chapter, then such firearm shall be deemed part of such licensee's business inventory, except that any licensed manufacturer, importer, or dealer who has maintained a firearm as part of a personal collection for one year and who sells or otherwise disposes of such firearm shall record the description of the firearm in a bound volume, containing the name and place of residence and date of birth of the transferee if the transferee is an individual, or the identity and principal and local places of business of the transferee if the transferee is a corporation or other business entity: **Provided**, That no other recordkeeping shall be required.

(d) (1) Any application submitted under subsection (a) or (b) of this section shall be approved if—

(A) the applicant is twenty-one years of age or over;

(B) the applicant (including, in the case of a corporation, partnership, or association, any individual possessing, directly or indirectly, the power to direct or cause the direction of the management and policies of the corporation, partnership, or association) is not prohibited from transporting, shipping, or receiving firearms or ammunition in interstate or foreign commerce under section 922(g) and (n) of this chapter;

(C) the applicant has not willfully violated any of the provisions of this chapter or regulations issued thereunder;

(D) the applicant has not willfully failed to disclose any material information required, or has not made any false statement as to any material fact, in connection with his application;

(E) the applicant has in a State (i) premises from which he conducts business subject to license under this chapter or from which he intends to conduct such business within a reasonable period of time, or (ii) in the case of a collector, premises from which he conducts his collecting subject to license under this chapter or from which he intends to conduct such collecting within a reasonable period of time;

(F) the applicant certifies that—

(i) the business to be conducted under the license is not prohibited by State or local law in the place where the licensed premise is located;

(ii)(I) within 30 days after the application is approved the business will comply with the requirements of State and local law applicable to the conduct of the business; and

(II) the business will not be conducted under the license until the requirements of State and local law applicable to the business have been met; and

(iii) that the applicant has sent or delivered a form to be prescribed by the Attorney General, to the chief law enforcement officer of the locality in which the premises are located, which indicates that the applicant intends to apply for a Federal firearms license; and

(G) in the case of an application to be licensed as a dealer, the applicant certifies that secure gun storage or safety devices will be available at any place in which firearms are sold under the license to persons who are not licensees (subject to the exception that in any case in which a secure gun storage or safety device is temporarily unavailable because of theft, casualty loss, consumer sales, backorders from a manufacturer, or any other similar reason beyond the control of the licensee, the dealer shall not be considered to be in violation of the requirement under this subparagraph to make available such a device).

(2) The Attorney General must approve or deny an application for a license within the 60-day period beginning on the date it is received. If the Attorney General fails to act within such period, the applicant may file an action under section 1361 of title 28 to compel the Attorney General to act. If the Attorney General approves an applicant's application, such applicant shall be issued a license upon the payment of the prescribed fee.

(e) The Attorney General may, after notice and opportunity for hearing, revoke any license issued under this section if the holder of such license has willfully violated any provision of this chapter or any rule or regulation prescribed by the Attorney General under this chapter or fails to have secure gun storage or safety devices available at any place in which firearms are sold under the license to persons who are not licensees (except that in any case in which a secure gun storage or safety device is temporarily unavailable because of theft, casualty loss, consumer sales, backorders from a manufacturer, or any other similar reason beyond the control of the licensee, the dealer shall not be considered to be in violation of the requirement to make available such a device). The Attorney General may, after notice and opportunity for hearing, revoke the license of a dealer who willfully transfers armor piercing ammunition. The Attorney General's action under this subsection may be reviewed only as provided in subsection (f) of this section.

(f) (1) Any person whose application for a license is denied and any holder of a license which is revoked shall receive a written notice from the Attorney General stating specifically the grounds upon which the application was denied or upon which the license was revoked. Any notice of a revocation of a license shall be given to the holder of such license before the effective date of the revocation.

(2) If the Attorney General denies an application for, or revokes, a license, he shall, upon request by the aggrieved party, promptly hold a hearing to review his denial or revocation. In the case of a revocation of a license, the Attorney General shall upon the request of the holder of the license stay the effective date of the revocation. A hearing held under this paragraph shall be held at a location convenient to the aggrieved party.

(3) If after a hearing held under paragraph (2) the Attorney General decides not to reverse his decision to deny an application or revoke a license, the Attorney General shall give notice of his decision to the aggrieved party. The aggrieved party may at any time within sixty days after the date notice was given under this paragraph file a petition with the United States district court for the district in which he resides or has his principal place of business for a de novo judicial review of such denial or revocation. In a proceeding conducted under this subsection, the court may consider any evidence submitted by the parties to the proceeding whether or not such evidence was considered at the hearing held under paragraph (2). If the court decides that the Attorney General was not authorized to deny the application or to revoke the license, the court shall order the Attorney General to take such action as may be necessary to comply with the judgment of the court.

(4) If criminal proceedings are instituted against a licensee alleging any violation of this chapter or of rules or regulations prescribed under this chapter, and the licensee is acquitted of such charges, or such proceedings are terminated, other than upon motion of the Government before trial upon such charges, the Attorney General shall be absolutely barred from denying or revoking any license granted under this chapter where such denial or revocation is based in whole or in part on the facts which form the basis of such criminal charges. No proceedings for the revocation of a license shall be instituted by the Attorney General more than one year after the filing of the indictment or information.

(g) (1) (A) Each licensed importer, licensed manufacturer, and licensed dealer shall maintain such records of importation, production, shipment, receipt, sale, or other disposition of firearms at his place of business for such period, and in such form, as the Attorney General may by regulations prescribe. Such importers, manufacturers, and dealers shall not be required to submit to the Attorney General reports and information with respect to such records and the contents thereof, except as expressly required by this section. The Attorney General, when he has reasonable cause to believe a violation of this chapter has occurred and that evidence thereof may be found on such premises, may, upon demonstrating such cause before a Federal magistrate and securing from such magistrate a warrant authorizing entry, enter during business hours the premises (including places of storage) of any licensed firearms importer, licensed manufacturer, licensed dealer, licensed collector, or any licensed importer or manufacturer of ammunition, for the purpose of inspecting or examining—

(i) any records or documents required to be kept by such licensed importer, licensed manufacturer, licensed dealer, or licensed collector under this chapter or rules or regulations under this chapter, and

(ii) any firearms or ammunition kept or stored by such licensed importer, licensed manufacturer, licensed dealer, or licensed collector, at such premises.

(B) The Attorney General may inspect or examine the inventory and records of a licensed importer, licensed manufacturer, or licensed dealer without such reasonable cause or warrant—

(i) in the course of a reasonable inquiry during the course of a criminal investigation of a person or persons other than the licensee;

(ii) for ensuring compliance with the record keeping requirements of this chapter—

(I) not more than once during any 12-month period; or

(II) at any time with respect to records relating to a firearm involved in a criminal investigation that is traced to the licensee; or

(iii) when such inspection or examination may be required for determining the disposition of one or more particular firearms in the course of a bona fide criminal investigation.

(C) The Attorney General may inspect the inventory and records of a licensed collector without such reasonable cause or warrant—

(i) for ensuring compliance with the record keeping requirements of this chapter not more than once during any twelve-month period; or

(ii) when such inspection or examination may be required for determining the disposition of one or more particular firearms in the course of a bona fide criminal investigation.

(D) At the election of a licensed collector, the annual inspection of records and inventory permitted under this paragraph shall be performed at the office of the Attorney General designed for such inspections which is located in closest proximity to the premises where the inventory and records of such licensed collector are maintained. The inspection and examination authorized by this paragraph shall not be construed as authorizing the Attorney General to seize any records or other documents other than those records or documents constituting material evidence of a violation of law. If the Attorney General seizes such records or documents, copies shall be provided the licensee within a reasonable time. The Attorney General may make available to any Federal, State, or local law enforcement agency any information which he may obtain by reason of this chapter with respect to the identification of persons prohibited from purchasing or receiving firearms or ammunition who have purchased or received firearms or ammunition, together with a description of such firearms or ammunition, and he may provide information to the extent such information may be contained in the records required to be maintained by this chapter, when so requested by any Federal, State, or local law enforcement agency.

(2) Each licensed collector shall maintain in a bound volume the nature of which the Attorney General may by regulations prescribe, records of the receipt, sale, or other disposition of firearms. Such records shall include the name and address of any person to whom the collector sells or otherwise disposes of a firearm. Such collector shall not be required to submit to the Attorney General reports and information with respect to such records and the contents thereof, except as expressly required by this section.

(3) (A) Each licensee shall prepare a report of multiple sales or other dispositions whenever the licensee sells or otherwise disposes of, at one time or during any five consecutive business days, two or more pistols, or revolvers, or any combination of pistols and revolvers totaling two or more, to an unlicensed person. The report shall be prepared on a form specified by the Attorney General and forwarded to the office specified thereon and to the department of State police or State law enforcement agency of the State or local law enforcement agency of the local jurisdiction in which the sale or other disposition took place, not later than the close of business on the day that the multiple sale or other disposition occurs.

(B) Except in the case of forms and contents thereof regarding a purchaser who is prohibited by subsection (g) or (n) of section 922 of this title from receipt of a firearm, the department of State police or State law enforcement agency or local law enforcement agency of the local jurisdiction shall not disclose any such form or the contents thereof to any person or entity, and shall destroy each such form and any record of the contents thereof no more than 20 days from the date such form is received. No later than the date that is 6 months after the effective date of this subparagraph, and at the end of each 6-month period thereafter, the department of State police or State law enforcement agency or local law enforcement agency of the local jurisdiction shall certify to the Attorney General of the United States that no disclosure contrary to this subparagraph has been made and that all forms and any record of the contents thereof have been destroyed as provided in this subparagraph.

(4) Where a firearms or ammunition business is discontinued and succeeded by a new licensee, the records required to be kept by this chapter shall appropriately reflect such facts and shall be delivered to the successor. Where discontinuance of the business is absolute, such records shall be delivered within thirty days after the business discontinuance to the Attorney General. However, where State law or local ordinance requires the delivery of records to other responsible authority, the Attorney General may arrange for the delivery of such records to such other responsible authority.

(5) (A) Each licensee shall, when required by letter issued by the Attorney General, and until notified to the contrary in writing by the Attorney General, submit on a form specified by the Attorney General, for periods and at the times specified in such letter, all record information required to be kept by this chapter or such lesser record information as the Attorney General in such letter may specify.

(B) The Attorney General may authorize such record information to be submitted in a manner other than that prescribed in subparagraph (A) of this paragraph when it is shown by a licensee that an alternate method of reporting is reasonably necessary and will not unduly hinder the effective administration of this chapter. A licensee may use an alternate method of reporting if the licensee describes the proposed alternate method of reporting and the need therefor in a letter application submitted to the Attorney General, and the Attorney General approves such alternate method of reporting.

(6) Each licensee shall report the theft or loss of a firearm from the licensee's inventory or collection, within 48 hours after the theft or loss is discovered, to the Attorney General and to the appropriate local authorities.

(7) Each licensee shall respond immediately to, and in no event later than 24 hours after the receipt of, a request by the Attorney General for information contained in the records required to be kept by this chapter as may be required for determining the disposition of 1 or more firearms in the course of a bona fide criminal investigation. The requested information shall be provided orally or in writing, as the Attorney General may require. The Attorney

General shall implement a system whereby the licensee can positively identify and establish that an individual requesting information via telephone is employed by and authorized by the agency to request such information.

(h) Licenses issued under the provisions of subsection (c) of this section shall be kept posted and kept available for inspection on the premises covered by the license.

(i) Licensed importers and licensed manufacturers shall identify, by means of a serial number engraved or cast on the receiver or frame of the weapon, in such manner as the Attorney General shall by regulations prescribe, each firearm imported or manufactured by such importer or manufacturer.

(j) A licensed importer, licensed manufacturer, or licensed dealer may, under rules or regulations prescribed by the Attorney General, conduct business temporarily at a location other than the location specified on the license if such temporary location is the location for a gun show or event sponsored by any national, State, or local organization, or any affiliate of any such organization devoted to the collection, competitive use, or other sporting use of firearms in the community, and such location is in the State which is specified on the license. Records of receipt and disposition of firearms transactions conducted at such temporary location shall include the location of the sale or other disposition and shall be entered in the permanent records of the licensee and retained on the location specified on the license. Nothing in this subsection shall authorize any licensee to conduct business in or from any motorized or towed vehicle. Notwithstanding the provisions of subsection (a) of this section, a separate fee shall not be required of a licensee with respect to business conducted under this subsection. Any inspection or examination of inventory or records under this chapter by the Attorney General at such temporary location shall be limited to inventory consisting of, or records relating to, firearms held or disposed at such temporary location. Nothing in this subsection shall be construed to authorize the Attorney General to inspect or examine the inventory or records of a licensed importer, licensed manufacturer, or licensed dealer at any location other than the location specified on the license. Nothing in this subsection shall be construed to diminish in any manner any right to display, sell, or otherwise dispose of firearms or ammunition, which is in effect before the date of the enactment of the Firearms Owners' Protection Act, including the right of a licensee to conduct "curios or relics" firearms transfers and business away from their business premises with another licensee without regard as to whether the location of where the business is conducted is located in the State specified on the license of either licensee.

(k) Licensed importers and licensed manufacturers shall mark all armor piercing projectiles and packages containing such projectiles for distribution in the manner prescribed by the Attorney General by regulation. The Attorney General shall furnish information to each dealer licensed under this chapter defining which projectiles are considered armor piercing ammunition as defined in section 921(a)(17)(B).

(l) The Attorney General shall notify the chief law enforcement officer in the appropriate State and local jurisdictions of the names and addresses of all persons in the State to whom a firearms license is issued.

§ 924 Penalties.

(a) (1) Except as otherwise provided in this subsection, subsection (b), (c), (f), or (p) of this section, or in section 929, whoever—

(A) knowingly makes any false statement or representation with respect to the information required by this chapter to be kept in the records of a person licensed under this chapter or in applying for any license or exemption or relief from disability under the provisions of this chapter;

(B) knowingly violates subsection (a)(4), (f), (k), or (q) of section 922;

(C) knowingly imports or brings into the United States or any possession thereof any firearm or ammunition in violation of section 922(l); or

(D) willfully violates any other provision of this chapter, shall be fined under this title, imprisoned not more than five years, or both.

Editor's Note:

The reference to subsection (p) in the introductory paragraph of 18 U.S.C. section 924(a)(1) was added to the Gun Control Act as part of the "Protection of Lawful Commerce in Arms Act" enacted on October 26, 2005. It is effective April 24, 2006.

(2) Whoever knowingly violates subsection (a)(6), (d), (g), (h), (i), (j), or (o) of section 922 shall be fined as provided in this title, imprisoned not more than 10 years, or both.

(3) Any licensed dealer, licensed importer, licensed manufacturer, or licensed collector who knowingly—

(A) makes any false statement or representation with respect to the information required by the provisions of this chapter to be kept in the records of a person licensed under this chapter, or

(B) violates subsection (m) of section 922, shall be fined under this title, imprisoned not more than one year, or both.

(4) Whoever violates section 922(q) shall be fined under this title, imprisoned for not more than 5 years, or both. Notwithstanding any other provision of law, the term of imprisonment imposed under this paragraph shall not run concurrently with any other term of imprisonment imposed under any other provision of law. Except for the authorization of a term of imprisonment of not more than 5 years made in this paragraph, for the purpose of any other law a violation of section 922(q) shall be deemed to be a misdemeanor.

(5) Whoever knowingly violates subsection (s) or (t) of section 922 shall be fined not more than $1,000 under this title, imprisoned for not more than 1 year, or both.

(6) (A) (i) A juvenile who violates section 922(x) shall be fined under this title, imprisoned not more than 1 year, or both, except that a juvenile described in clause (ii) shall be sentenced to probation on appropriate conditions and shall not be incarcerated unless the juvenile fails to comply with a condition of probation.

(ii) A juvenile is described in this clause if—

(I) the offense of which the juvenile is charged is possession of a handgun or ammunition in violation of section 922(x)(2); and

(II) the juvenile has not been convicted in any court of an offense (including an offense under section 922(x) or a similar State law, but not including any other offense consisting of conduct that if engaged in by an adult would not constitute an offense) or adjudicated as a juvenile delinquent for conduct that if engaged in by an adult would constitute an offense.

(B) A person other than a juvenile who knowingly violates section 922(x)—

(i) shall be fined under this title, imprisoned not more than 1 year, or both; and

(ii) if the person sold, delivered, or otherwise transferred a handgun or ammunition to a juvenile knowing or having reasonable cause to know that the juvenile intended to carry or otherwise possess or discharge or otherwise use the handgun or ammunition in the commission of a crime of violence, shall be fined under this title, imprisoned not more than 10 years, or both.

(7) Whoever knowingly violates section 931 shall be fined under this title, imprisoned not more than 3 years, or both.

(b) Whoever, with intent to commit therewith an offense punishable by imprisonment for a term exceeding one year, or with knowledge or reasonable cause to believe that an offense punishable by imprisonment for a term exceeding one year is to be committed therewith, ships, transports, or receives a firearm or any ammunition in interstate or foreign commerce shall be fined under this title, or imprisoned not more than ten years, or both.

(c) (1) (A) Except to the extent that a greater minimum sentence is otherwise provided by this subsection or by any other provision of law, any person who, during and in relation to any crime of violence or drug trafficking crime (including a crime of violence or drug trafficking crime that provides for an enhanced punishment if committed by the use of a deadly or dangerous weapon or device) for which the person may be prosecuted in a court of the United States, uses or carries a firearm, or who, in furtherance of any such crime, possesses a firearm, shall, in addition to the punishment provided for such crime of violence or drug trafficking crime—

(i) be sentenced to a term of imprisonment of not less than 5 years;

(ii) if the firearm is brandished, be sentenced to a term of imprisonment of not less than 7 years; and

(iii) if the firearm is discharged, be sentenced to a term of imprisonment of not less than 10 years.

(B) If the firearm possessed by a person convicted of a violation of this subsection—

(i) is a short-barreled rifle, short-barreled shotgun, or semiautomatic assault weapon, the person shall be sentenced to a term of imprisonment of not less than 10 years; or

Editor's Note:

The reference to semiautomatic assault weapons in § 924 (c)(1)(B)(i) was repealed when the semiautomatic assault weapon provision ceased to apply on September 13, 2004.

(ii) is a machinegun or a destructive device, or is equipped with a firearm silencer or firearm muffler, the person shall be sentenced to a term of imprisonment of not less than 30 years.

(C) In the case of a second or subsequent conviction under this subsection, the person shall—

(i) be sentenced to a term of imprisonment of not less than 25 years; and

(ii) if the firearm involved is a machinegun or a destructive device, or is equipped with a firearm silencer or firearm muffler, be sentenced to imprisonment for life.

(D) Notwithstanding any other provision of law—

(i) a court shall not place on probation any person convicted of a violation of this subsection; and

(ii) no term of imprisonment imposed on a person under this subsection shall run concurrently with any other term of imprisonment imposed on the person, including any term of imprisonment imposed for the crime of violence or drug trafficking crime during which the firearm was used, carried, or possessed.

(2) For purposes of this subsection, the term **"drug trafficking crime"** means any felony punishable under the Controlled Substances Act (21 U.S.C. 801 et seq.), the Controlled Substances Import and Export Act (21 U.S.C. 951 et seq.), or the Maritime Drug Law Enforcement Act (46 U.S.C. App. 1901 et seq.).

(3) For purposes of this subsection the term **"crime of violence"** means an offense that is a felony and—

(A) has as an element the use, attempted use, or threatened use of physical force against the person or property of another, or

(B) that by its nature, involves a substantial risk that physical force against the person or property of another may be used in the course of committing the offense.

(4) For purposes of this subsection, the term **"brandish"** means, with respect to a firearm, to display all or part of the firearm, or otherwise make the presence of the firearm known to another person, in order to intimidate that person, regardless of whether the firearm is directly visible to that person.

(5) Except to the extent that a greater minimum sentence is otherwise provided under this subsection, or by any other provision of law, any person who, during and in relation to any crime of violence or drug trafficking crime (including a crime of violence or drug trafficking crime that provided for an enhanced punishment if committed by the use of a deadly or dangerous weapon or device) for which the person may be prosecuted in a court of the United States, uses or carries armor piercing ammunition, or who, in furtherance of any such crime, possesses armor piercing ammunition, shall, in addition to the punishment provided for such crime of violence or drug trafficking crime or conviction under this section –

(A) be sentenced to a term of imprisonment of not less than 15 years; and

(B) if death results from the use of such ammunition—

(i) if the killing is murder (as defined in section 1111), be punished by death or sentenced to a term of imprisonment for any term of years or for life; and

(ii) if the killing is manslaughter (as defined in section 1112), be punished as provided in section 1112.

(d) (1) Any firearm or ammunition involved in or used in any knowing violation of subsection (a)(4), (a)(6), (f), (g), (h), (i), (j), or (k) of section 922, or knowing importation or bringing into the United States or any possession thereof any firearm or ammunition in violation of section 922(l), or knowing violation of section 924, or willful violation of any other provision of this chapter or any rule or regulation promulgated thereunder, or any violation of any other criminal law of the United States, or any firearm or ammunition intended to be used in any offense referred to in paragraph (3) of this subsection, where such intent is demonstrated by clear and convincing evidence, shall be subject to seizure and forfeiture, and all provisions of the Internal Revenue Code of 1954 relating to the seizure, forfeiture, and disposition of firearms, as defined in section 5845(a) of that Code, shall, so far as applicable, extend to seizures and forfeitures under the provisions of this chapter: **Provided**, That upon acquittal of the owner or possessor, or dismissal of the charges against him other than upon motion of the Government prior to trial, or lapse of or court termination of the restraining order to which he is subject, the seized or relinquished firearms or ammunition shall be returned forthwith to the owner or possessor or to a person delegated by the owner or possessor unless the return of the firearms or ammunition would place the owner or possessor or his delegate in violation of law. Any action or proceeding for the forfeiture of firearms or ammunition shall be commenced within one hundred and twenty days of such seizure.

(2) (A) In any action or proceeding for the return of firearms or ammunition seized under the provisions of this chapter, the court shall allow the prevailing party, other than the United States, a reasonable attorney's fee, and the United States shall be liable therefor.

(B) In any other action or proceeding under the provisions of this chapter, the court, when it finds that such action was without foundation, or was initiated vexatiously, frivolously, or in bad faith, shall allow the prevailing party, other than the United States, a reasonable attorney's fee, and the United States shall be liable therefor.

(C) Only those firearms or quantities of ammunition particularly named and individually identified as involved in or used in any violation of the provisions of this chapter or any rule or regulation issued thereunder, or any other criminal law of the United States or as intended to be used in any offense referred to in paragraph (3) of this subsection, where such intent is demonstrated by clear and convincing evidence, shall be subject to seizure, forfeiture, and disposition.

(D) The United States shall be liable for attorneys' fees under this paragraph only to the extent provided in advance by appropriation Acts.

(3) The offenses referred to in paragraphs (1) and (2)(C) of this subsection are—

(A) any crime of violence, as that term is defined in section 924(c)(3) of this title;

(B) any offense punishable under the Controlled Substances Act (21 U.S.C. 801 et seq.) or the Controlled Substances Import and Export Act (21 U.S.C. 951 et seq.);

(C) any offense described in section 922(a)(1), 922(a)(3), 922(a)(5), or 922(b)(3) of this title where the firearm or ammunition intended to be used in any such offense is involved in a pattern of activities which includes a violation of any offense described in section 922(a)(1), 922(a)(3), 922(a)(5), or 922(b)(3) of this title;

(D) any offense described in section 922(d) of this title where the firearm or ammunition is intended to be used in such offense by the transferor of such firearm or ammunition;

(E) any offense described in section 922(i), 922(j), 922(l), 922(n), or 924(b) of this title; and

(F) any offense which may be prosecuted in a court of the United States which involves the exportation of firearms or ammunition.

(e) (1) In the case of a person who violates section 922(g) of this title and has three previous convictions by any court referred to in section 922(g)(1) of this title for a violent felony or a serious drug offense, or both, committed on occasions different from one another, such person shall be fined under this title and imprisoned not less than fifteen years, and, notwithstanding any other provision of law, the court shall not suspend the sentence of, or grant a probationary sentence to, such person with respect to the conviction under section 922(g).

(2) As used in this subsection—

(A) the term **"serious dru g offense"** means—

(i) an offense under the Controlled Substances Act (21 U.S.C. 801 et seq.), the Controlled Substances Import and Export Act (21 U.S.C. 951 et seq.), or the Maritime Drug Law Enforcement Act (46 U.S.C. App. 1901 et seq.), for which a maximum term of imprisonment of ten years or more is prescribed by law; or

(ii) an offense under State law, involving manufacturing, distributing, or possessing with intent to manufacture or distribute, a controlled substance (as defined in section 102 of the Controlled Substances Act (21 U.S.C. 802)), for which a maximum term of imprisonment of ten years or more is prescribed by law;

(B) the term **"violent felony "** means any crime punishable by imprisonment for a term exceeding one year, or any act of juvenile delinquency involving the use or carrying of a firearm, knife, or destructive device that would be punishable by imprisonment for such term if committed by an adult, that—

(i) has as an element the use, attempted use, or threatened use of physical force against the person of another; or

(ii) is burglary, arson, or extortion, involves use of explosives, or otherwise involves conduct that presents a serious potential risk of physical injury to another; and

(C) the term **"conviction"** includes a finding that a person has committed an act of juvenile delinquency involving a violent felony.

(f) In the case of a person who knowingly violates section 922(p), such person shall be fined under this title, or imprisoned not more than 5 years, or both.

(g) Whoever, with the intent to engage in conduct which—

(1) constitutes an offense listed in section 1961(1),

(2) is punishable under the Controlled Substances Act (21 U.S.C. 802 et seq.), the Controlled Substances Import and Export Act (21 U.S.C. 951 et seq.), or the Maritime Drug Law Enforcement Act (46 U.S.C. App. 1901 et seq.),

(3) violates any State law relating to any controlled substance (as defined in section 102(6) of the Controlled Substances Act (21 U.S.C. 802(6))), or

(4) constitutes a crime of violence (as defined in subsection (c)(3)),

travels from any State or foreign country into any other State and acquires, transfers, or attempts to acquire or transfer, a firearm in such other State in furtherance of such purpose, shall be imprisoned not more than 10 years, fined in accordance with this title, or both.

(h) Whoever knowingly transfers a firearm, knowing that such firearm will be used to commit a crime of violence (as defined in subsection (c)(3)) or drug trafficking crime (as defined in subsection (c)(2)) shall be imprisoned not more than 10 years, fined in accordance with this title, or both.

(i) (1) A person who knowingly violates section 922(u) shall be fined under this title, imprisoned not more than 10 years, or both.

(2) Nothing contained in this subsection shall be construed as indicating an intent on the part of Congress to occupy the field in which provisions of this subsection operate to the exclusion of State laws on the same subject matter, nor shall any provision of this subsection be construed as invalidating any provision of State law unless such provision is inconsistent with any of the purposes of this subsection.

(j) A person who, in the course of a violation of subsection (c), causes the death of a person through the use of a firearm, shall—

(1) if the killing is a murder (as defined in section 1111), be punished by death or by imprisonment for any term of years or for life; and

(2) if the killing is manslaughter (as defined in section 1112), be punished as provided in that section.

(k) A person who, with intent to engage in or to promote conduct that—

(1) is punishable under the Controlled Substances Act (21 U.S.C. 801 et seq.), the Controlled Substances Import and Export Act (21 U.S.C. 951 et seq.), or the Maritime Drug Law Enforcement Act (46 U.S.C. App. 1901 et seq.);

(2) violates any law of a State relating to any controlled substance (as defined in section 102 of the Controlled Substances Act, 21 U.S.C. 802); or

(3) constitutes a crime of violence (as defined in subsection (c) (3)), smuggles or knowingly brings into the United States a firearm, or attempts to do so, shall be imprisoned not more than 10 years, fined under this title, or both.

(l) A person who steals any firearm which is moving as, or is a part of, or which has moved in, interstate or foreign commerce shall be imprisoned for not more than 10 years, fined under this title, or both.

(m) A person who steals any firearm from a licensed importer, licensed manufacturer, licensed dealer, or licensed collector shall be fined under this title, imprisoned not more than 10 years, or both.

(n) A person who, with the intent to engage in conduct that constitutes a violation of section 922(a)(1)(A), travels from any State or foreign country into any other State and acquires, or attempts to acquire, a firearm in such other State in furtherance of such purpose shall be imprisoned for not more than 10 years.

(o) A person who conspires to commit an offense under subsection (c) shall be imprisoned for not more than 20 years, fined under this title, or both; and if the firearm is a machinegun or destructive device, or is equipped with a firearm silencer or muffler, shall be imprisoned for any term of years or life.

Editor's Note

18 U.S.C. § 924(p) was added to the Gun Control Act as part of the "Protection of Lawful Commerce in Arms Act" enacted on October 26, 2005. It is effective April 24, 2006.

(p) PENALTIES RELATING TO SECURE GUN STORAGE OR SAFETY DEVICE—

(1) IN GENERAL—

(A) SUSPENSION OR REVOCATION OF LICENSE; CIVIL PENALTIES – With respect to each violation of section 922(z)(1) by a li censed manufacturer, licensed importer, or licensed dealer, the Secretary may, after notice and opportunity for hearing--

(i) suspend for not more than 6 months, or revoke, the license issued to the licensee under this chapter that was used to conduct the firearm transfer; or

(ii) subject the licensee to a civil penalty in an amount equal to not more than $2, 500.

(B) REVIEW – An action of the Secretary under this paragraph may be reviewed only as provided under section 923(f).

(2) ADMINISTRATIVE REMEDIES The suspension or revocation of a license or the imposition of a civil penalty under paragraph (1) shall not preclude any administrative remedy that is otherwise available to the Secretary..

§ 925 Exceptions: Relief from disabilities.

(a) (1) The provisions of this chapter, except for sections 922(d)(9) and 922(g)(9) and provisions relating to firearms subject to the prohibitions of section 922(p), shall not apply with respect to the transportation, shipment, receipt, possession, or importation of any firearm or ammunition imported for, sold or shipped to, or issued for the use of, the United States or any department or agency thereof or any State or any department, agency, or political subdivision thereof.

(2) The provisions of this chapter, except for provisions relating to firearms subject to the prohibitions of section 922(p), shall not apply with respect to (A) the shipment or receipt of firearms or ammunition when sold or issued by the Secretary of the Army pursuant to section 4308 of title 10 before the repeal of such section by section 1624(a) of the Corporation for the Promotion of Rifle Practice and Firearms Safety Act, and (B) the transportation of any such firearm or ammunition carried out to enable a person, who lawfully received such firearm or ammunition from the Secretary of the Army, to engage in military training or in competitions.

(3) Unless otherwise prohibited by this chapter, except for provisions relating to firearms subject to the prohibitions of section 922(p), or any other Federal law, a licensed importer, licensed manufacturer, or licensed dealer may ship to a member of the United States Armed Forces on active duty outside the United States or to clubs, recognized by the Department of Defense, whose entire membership is composed of such members, and such members or clubs may receive a firearm or ammunition determined by the Attorney General to be generally recognized as particularly suitable for sporting purposes and intended for the personal use of such member or club.

(4) When established to the satisfaction of the Attorney General to be consistent with the provisions of this chapter, except for provisions relating to firearms subject to the prohibitions of section 922(p), and other applicable Federal and State laws and published ordinances, the Attorney General may authorize the transportation, shipment, receipt, or importation into the United States to the place of residence of any member of the United States Armed Forces who is on active duty outside the United States (or who has been on active duty outside the United States within the sixty day period immediately preceding the transportation, shipment, receipt, or importation), of any firearm or ammunition which is (A) determined by the Attorney General to be generally recognized as particularly suitable for sporting purposes, or determined by the Department of Defense to be a type of firearm normally classified as a war souvenir, and (B) intended for the personal use of such member.

(5) For the purpose of paragraph (3) of this subsection, the term **"United States"** means each of the several States and the District of Columbia.

(b) A licensed importer, licensed manufacturer, licensed dealer, or licensed collector who is indicted for a crime punishable by imprisonment for a term exceeding one year, may, notwithstanding any other provision of this chapter, continue operation pursuant to his existing license (if prior to the expiration of the term of the existing license timely application is made for a new license) during the term of such indictment and until any conviction pursuant to the indictment becomes final.

(c) A person who is prohibited from possessing, shipping, transporting, or receiving firearms or ammunition may make application to the Attorney General for relief from the disabilities imposed by Federal laws with respect to the acquisition, receipt, transfer, shipment, transportation, or possession of firearms, and the Attorney General may grant such relief if it is established to his satisfaction that the circumstances regarding the disability, and the applicant's record and reputation, are such that the applicant will not be likely to act in a manner dangerous to public safety and that the granting of the relief would not be contrary to the public interest. Any person whose application for relief from disabilities is denied by the Attorney General may file a petition with the United States district court for the district in which he resides for a judicial review of such denial. The court may in its discretion admit additional evidence where failure to do so would result in a miscarriage of justice. A licensed importer, licensed manufacturer, licensed dealer, or licensed collector conducting operations under this chapter, who makes application for relief from the disabilities incurred under this chapter, shall not be barred by such disability from further operations under his license pending final action on an application for relief filed pursuant to this section. Whenever the Attorney General grants relief to any person pursuant to this section he shall promptly publish in the Federal Register notice of such action, together with the reasons therefore .

Editor's Note:

As of the date of this publication, ATF is prohibited from acting upon applications for relief from individuals because of an appropriations restriction. Please contact ATF to determine if the restriction is still in effect.

(d) The Attorney General shall authorize a firearm or ammunition to be imported or brought into the United States or any possession thereof if the firearm or ammunition—

(1) is being imported or brought in for scientific or research purposes, or is for use in connection with competition or training pursuant to chapter 401 of title 10;

(2) is an unserviceable firearm, other than a machinegun as defined in section 5845(b) of the Internal Revenue Code of 1954 (not readily restorable to firing condition), imported or brought in as a curio or museum piece;

(3) is of a type that does not fall within the definition of a firearm as defined in section 5845(a) of the Internal Revenue Code of 1954 and is generally recognized as particularly suitable for or readily adaptable to sporting purposes, excluding surplus military firearms, except in any case where the Attorney General has not authorized the importation of the firearm pursuant to this paragraph, it shall be unlawful to import any frame, receiver, or barrel of such firearm which would be prohibited if assembled; or

(4) was previously taken out of the United States or a possession by the person who is bringing in the firearm or ammunition.

The Attorney General shall permit the conditional importation or bringing in of a firearm or ammunition for examination and testing in connection with the making of a determination as to whether the importation or bringing in of such firearm or ammunition will be allowed under this subsection.

(e) Notwithstanding any other provision of this title, the Attorney General shall authorize the importation of any licensed importer, by the following:

(1) All rifles and shotguns listed as curios or relics by the Attorney General pursuant to section 921(a)(13), and

(2) All handguns listed as curios or relics by the Attorney General pursuant to section 921(a)(13), provided that such handguns are generally recognized as particularly suitable for or readily adaptable to sporting purposes.

(f) The Attorney General shall not authorize, under subsection (d), the importation of any firearm the importation of which is prohibited by section 922(p).

§ 925A Remedy for erroneous denial of firearm.

Any person denied a firearm pursuant to subsection (s) or (t) of section 922—

(1) due to the provision of erroneous information relating to the person by any State or political subdivision thereof, or by the national instant criminal background check system established under section 103 of the Brady Handgun Violence Prevention Act; or

(2) who was not prohibited from receipt of a firearm pursuant to subsection (g) or (n) of section 922, may bring an action against the State or political subdivision responsible for providing the erroneous information, or responsible for denying the transfer, or against the United States, as the case may be, for an order directing that the erroneous information be corrected or that the transfer be approved, as the case may be. In any action under this section, the court, in its discretion, may allow the prevailing party a reasonable attorney's fee as part of the costs.

§ 926 Rules and regulations.

(a) The Attorney General may prescribe only such rules and regulations as are necessary to carry out the provisions of this chapter, including—

(1) regulations providing that a person licensed under this chapter, when dealing with another person so licensed, shall provide such other licensed person a certified copy of this license;

(2) regulations providing for the issuance, at a reasonable cost, to a person licensed under this chapter, of certified copies of his license for use as provided under regulations issued under paragraph (1) of this subsection; and

(3) regulations providing for effective receipt and secure storage of firearms relinquished by or seized from persons described in subsection (d)(8) or (g)(8) of section 922.

No such rule or regulation prescribed after the date of the enactment of the Firearms Owners' Protection Act may require that records required to be maintained under this chapter or any portion of the contents of such records, be recorded at or transferred to a facility owned, managed, or controlled by the United States or any State or any political subdivision thereof, nor that any system of registration of firearms, firearms owners, or firearms transactions or dispositions be established. Nothing in this section expands or restricts the Attorney General's authority to inquire into the disposition of any firearm in the course of a criminal investigation.

(b) The Attorney General shall give not less than ninety days public notice, and shall afford interested parties opportunity for hearing, before prescribing such rules and regulations.

(c) The Attorney General shall not prescribe rules or regulations that require purchasers of black powder under the exemption provided in section 845(a)(5) of this title to complete affidavits or forms attesting to that exemption.

§ 926A Interstate transportation of firearms.

Notwithstanding any other provisions of any law or any rule or regulation of a State or any political subdivision therof, any person who is not otherwise prohibited by this chapter from transporting, shipping, or receiving a firearm shall be entitled to transport a firearm for any lawful purpose from any place where he may lawfully possess and carry such firearm to any other place where he may lawfully possess and carry such firearm if, during such transportation the firearm is unloaded, and neither the firearm nor any ammunition being transported is readily accessible or is directly accessible from the passenger compartment of such transporting vehicle: **Provided**, That in the case of a vehicle without a compartment separate from the driver's compartment, the firearm or ammunition shall be contained in a locked container other than the glove compartment or console.

§ 926B Carrying of concealed firearms by qualified law enforcement officers.

(a) Nothwithstanding any other provision of the law of any State or any political subdivision thereof, an individual who is a qualified law enforcement officer and who is carrying the indentification required by subsection (d) may carry a concealed firearm that has been shipped or transported in interstate or foreign commerce, subject to subsection (b).

(b) This section shall not be construed to supersede or limit the laws of any State that —

(1) permit private persons or entities to prohibit or restrict the possession of concealed firearms on their property; or

(2) prohibit or restrict the possession of firearms on any State or local government property, installation, building, base, or park.

(c) As used in this section, the term **"qualified la w enforceme nt officer"** means an employee of a governmental agency who —

(1) is authorized by law to engage in or supervise the prevention, detection, investigation, or prosecution of, or the incarceration of any person for, any violation of law, and has statutory powers of arrest;

(2) is authorized by the agency to carry a firearm;

(3) is not the subject of any disciplinary action by the agency;

(4) meets standards, if any, established by the agency which require the employee to regularly qualify in the use of a firearm;

(5) is not under the influence of alcohol or another intoxicating or hallucinatory drug or substance; and

(6) is not prohibited by Federal law from receiving a firearm.

(d) The identification required by this subsection is the photographic identification issued by the governmental agency for which the individual is employed as a law enforcement officer.

(e) As used in this section, the term **"firearm"** does not include—

(1) any machinegun (as defined in section 5845 of the National Firearms Act);

(2) any firearm silencer (as defined in section 921 of this title); and

(3) any destructive device (as defined in section 921 of this title).

§ 926C Carrying of concealed firearms by qualified retired law enforcement officers.

(a) Notwithstanding any other provisions of the law of any State or any political subdivision thereof, an individual who is a qualified retired law enforcement officer and who is carrying the identification required by subsection (d) may carry a concealed firearm that has been shipped or transported in interstate or foreign commerce, subject to subsection (b).

(b) This section shall not be construed to supersede or limit the laws of any State that—

(1) permit private persons or entities to prohibit or restrict the possession of concealed firearms on their property; or

(2) prohibit or restrict the possession of firearms on any State or local government property, installation, building, base, or park.

(c) As used in this section, the term **"qualified retired la w enforcement officer"** means an individual who—

(1) retired in good standing from service with a public agency as a law enforcement officer, other than for reasons of mental instability;

(2) before such retirement, was authorized by law to engage in or supervise the prevention, detection, investigation, or prosecution of, or the incarceration of any person for, any violation of law, and had statutory powers of arrest;

(3) **(A)** before such retirement, was regularly employed as a law enforcement officer for an aggregate of 15 years or more; or

(B) retired from service with such agency, after completing any applicable probationary period of such service, due to a service-connected disability, as determined by such agency;

(4) has a nonforfeitable right to benefits under the retirement plan of the agency;

(5) during the most recent 12-month period, has met, at the expense of the individual, the State's standards for training and qualification for active law enforcement officers to carry firearms;

(6) is not under the influence of alcohol or another intoxicating or hallucinatory drug or substance; and

(7) is not prohibited by Federal law from receiving a firearm.

(d) The identification required by this subsection is—

(1) a photographic identification issued by the agency from which the individual retired from service as a law enforcement officer that indicates that the individual has, not less recently than one year before the date the individual is carrying the concealed firearm, been tested or otherwise found by the agency to meet the standards established by the agency for training and qualification for active law enforcement officers to carry a firearm of the same type as the concealed firearm; or

(2) (A) a photographic identification issued by the agency from which the individual retired from service as a law enforcement officer; and

(B) a certification issued by the State in which the individual resides that indicates that the individual has, not less recently than one year before the date the individual is carrying the concealed firearm, been tested or otherwise found by the State to meet the standards established by the State for training and qualification for active law enforcement officers to carry a firearm of the same type as the concealed firearm.

(e) As used in this section, the term **"firearm"** does not include—

(1) any machineguns (as defined in section 5845 of the National Firearms Act);

(2) any firearm silencer (as defined in section 921 of this title); and

(3) a destructive device (as defined in section 921 of this title).

§ 927 Effect on State law.

No provision of this chapter shall be construed as indicating an intent on the part of the Congress to occupy the field in which such provision operates to the exclusion of the law of any State on the same subject matter, unless there is a direct and positive conflict between such provision and the law of the State so that the two cannot be reconciled or consistently stand together.

§ 928 Separability.

If any provision of this chapter or the application thereof to any person or circumstance is held invalid, the remainder of the chapter and the application of such provision to other persons not similarly situated or to other circumstances shall not be affected thereby.

§ 929 Use of restricted ammunition.

(a) (1) Whoever, during and in relation to the commission of a crime of violence or drug trafficking crime (including a crime of violence or drug trafficking crime which provides for an enhanced punishment if committed by the use of a deadly or dangerous weapon or device) for which he may be prosecuted in a court of the United States, uses or carries a firearm and is in possession of armor piercing ammunition capable of being fired in that firearm, shall, in addition to the punishment provided for the commission of such crime of violence or drug trafficking crime be sentenced to a term of imprisonment for not less than five years.

(2) For purposes of this subsection, the term **"drug trafficking crime"** means any felony punishable under the Controlled Substances Act (21 U.S.C. 801 et seq.), the Controlled Substances Import and Export Act (21 U.S.C. 951 et seq.), or the Maritime Drug Law Enforcement Act (46 U.S.C. App. 1901 et seq.).

(b) Notwithstanding any other provision of law, the court shall not suspend the sentence of any person convicted of a violation of this section, nor place the person on probation, nor shall the terms of imprisonment run concurrently with any other terms of imprisonment, including that imposed for the crime in which the armor piercing ammunition was used or possessed.

§ 930 Possession of firearms and dangerous weapons in Federal facilities.

(a) Except as provided in subsection (d), whoever knowingly possesses or causes to be present a firearm or other dangerous weapon in a Federal facility (other than a Federal court facility), or attempts to do so, shall be fined under this title or imprisoned not more than 1 year, or both.

(b) Whoever, with intent that a firearm or other dangerous weapon be used in the commission of a crime, knowingly possesses or causes to be present such firearm or dangerous weapon in a Federal facility, or attempts to do so, shall be fined under this title or imprisoned not more than 5 years, or both.

(c) A person who kills any person in the course of a violation of subsection (a) or (b), or in the course of an attack on a Federal facility involving the use of a firearm or other dangerous weapon, or attempts or conspires to do such an act, shall be punished as provided in sections 1111, 1112, 1113, and 1117.

(d) Subsection (a) shall not apply to—

(1) the lawful performance of official duties by an officer, agent, or employee of the United States, a State, or a political subdivision thereof, who is authorized by law to engage in or supervise the prevention, detection, investigation, or prosecution of any violation of law;

(2) the possession of a firearm or other dangerous weapon by a Federal official or a member of the Armed Forces if such possession is authorized by law; or

(3) the lawful carrying of firearms or other dangerous weapons in a Federal facility incident to hunting or other lawful purposes.

(e) (1) Except as provided in paragraph (2), whoever knowingly possesses or causes to be present a firearm in a Federal court facility, or attempts to do so, shall be fined under this title, imprisoned not more than 2 years, or both.

(2) Paragraph (1) shall not apply to conduct which is described in paragraph (1) or (2) of subsection (d).

(f) Nothing in this section limits the power of a court of the United States to punish for contempt or to promulgate rules or orders regulating, restricting, or prohibiting the possession of weapons within any building housing such court or any of its proceedings, or upon any grounds appurtenant to such building.

(g) As used in this section:

(1) The term **"Federal facilit y"** means a building or part thereof owned or leased by the Federal Government, where Federal employees are regularly present for the purpose of performing their official duties.

(2) The term **"dangerous w eapon"** means a weapon, device, instrument, material, or substance, animate or inanimate, that is used for, or is readily capable of, causing death or serious bodily injury, except that such term does not include a pocket knife with a blade of less than 2 1/2 inches in length.

(3) The term **"Federal court facility"** means the courtroom, judges' chambers, witness rooms, jury deliberation rooms, attorney conference rooms, prisoner holding cells, offices of the court clerks, the United States attorney, and the United States marshal, probation and parole offices, and adjoining corridors of any court of the United States.

(h) Notice of the provisions of subsections (a) and (b) shall be posted conspicuously at each public entrance to each Federal facility, and notice of subsection (e) shall be posted conspicuously at each public entrance to each Federal court facility, and no person shall be convicted of an offense under subsection (a) or (e) with respect to a Federal facility if such notice is not so posted at such facility, unless such person had actual notice of subsection (a) or (e), as the case may be.

§ 931 Prohibition on purchase, ownership, or possession of body armor by violent felons.

(a) In general. Except as provided in subsection (b), it shall be unlawful for a person to purchase, own, or possess body armor, if that person has been convicted of a felony that is—

(1) a crime of violence (as defined in section 16); or

(2) an offense under State law that would constitute a crime of violence under paragraph (1) if it occurred within the special maritime and territorial jurisdiction of the United States.

(b) Affirmative defense.

(1) In general. It shall be an affirmative defense under this section that—

(A) the defendant obtained prior written certification from his or her employer that the defendant's purchase, use, or possession of body armor was necessary for the safe performance of lawful business activity; and

(B) the use and possession by the defendant were limited to the course of such performance.

(2) Employer. In this subsection, the term **"employer"** means any other individual employed by the defendant's business that supervises defendant's activity. If that defendant has no supervisor, prior written certification is acceptable from any other employee of the business.

Editor's Note:

The following provisions of the GCA were repealed when the semiautomatic assault weapon and large capacity ammunition feeding device bans sunset on September 13, 2004.

18 U.S.C. § 921 (30): The term "**semiautomatic assault weapon**" means -

(A) any of the firearms, or copies or duplicates of the firearms in any caliber, known as –

(i) Norinco, Mitchell, and Poly Technologies Avtomat Kalashnikovs (all models);

(ii) Action Arms Israeli Military Industries UZI and Galil;

(iii) Beretta Ar70 (SC-70);

(iv) Colt AR-15;

(v) Fabrique National FN/FAL, FN/LAR, and FNC;

(vi) SWD M-10, M-11, M-11/9, and M-12;

(vii) Steyr AUG;

(viii) INTRATEC TEC-9, TEC-DC9 and TEC-22; and

(ix) revolving cylinder shotguns, such as (or similar to) the Street Sweeper and Striker 12;

(B) a **semiautomatic rifle** that has an ability to accept a detachable magazine and has at least 2 of -

(i) a folding or telescoping stock;

(ii) a pistol grip that protrudes conspicuously beneath the action of the weapon;

(iii) a bayonet mount;

(iv) a flash suppressor or threaded barrel designed to accommodate a flash suppressor; and

(v) a grenade launcher;

(C) a **semiautomatic pistol** that has an ability to accept a detachable magazine and has at least 2 of -

(i) an ammunition magazine that attaches to the pistol outside of the pistol grip;

(ii) a threaded barrel capable of accepting a barrel extender, flash suppressor, forward handgrip, or silencer;

(iii) a shroud that is attached to, or partially or completely encircles, the barrel and that permits the shooter to hold the firearm with the nontrigger hand without being burned;

(iv) a manufactured weight of 50 ounces or more when the pistol is unloaded; and

(v) a semiautomatic version of an automatic firearm; and

(D) a **semiautomatic shotgun** that has at least 2 of –

(i) a folding or telescoping stock;

(ii) a pistol grip that protrudes conspicuously beneath the action of the weapon;

(iii) a fixed magazine capacity in excess of 5 rounds; and

(iv) an ability to accept a detachable magazine.

18 U.S.C. § 921 (31) The term "**large capacity ammunition feeding device**" -

(A) means a magazine, belt, drum, feed strip, or similar device manufactured after the date of enactment of the Violent Crime Control and Law Enforcement Act of 1994 that has a capacity of, or that can be readily restored or converted to accept, more than 10 rounds of ammunition; but

(B) does not include an attached tubular device designed to accept, and capable of operating only with, .22 caliber rimfire ammunition.

18 U. S.C. § 9 22 (v) (1) It shall be unlawful for a person to manufacture, transfer, or possess a semiautomatic assault weapon.

(2) Paragraph (1) shall not apply to the possession or transfer of any semiautomatic assault weapon otherwise lawfully possessed under Federal law on the date of the enactment of this subsection.

(3) Paragraph (1) shall not apply to—

(A) any of the firearms, or replicas or duplicates of the firearms, specified in Appendix A to this section, as such firearms were manufactured on October 1, 1993;

(B) any firearm that—

(i) is manually operated by bolt, pump, lever, or slide action;

(ii) has been rendered permanently inoperable; or

(iii) is an antique firearm;

(C) any semiautomatic rifle that cannot accept a detachable magazine that holds more than 5 rounds of ammunition; or

(D) any semiautomatic shotgun that cannot hold more than 5 rounds of ammunition in a fixed or detachable magazine.

The fact that a firearm is not listed in Appendix A shall not be construed to mean that paragraph (1) applies to such firearm. No firearm exempted by this subsection may be deleted from Appendix A so long as this subsection is in effect.

(4) Paragraph (1) shall not apply to—

(A) the manufacture for, transfer to, or possession by the United States or a department or agency of the United States or a State or a department, agency, or political subdivision of a State, or a transfer to or possession by a law enforcement officer employed by such an entity for purposes of law enforcement (whether on or off duty);

(B) the transfer to a licensee under title I of the Atomic Energy Act of 1954 for purposes of establishing and maintaining an on-site physical protection system and security organization required by Federal law, or possession by an employee or contractor of such licensee on-site for such purposes or off-site for purposes of licensee-authorized training or transportation of nuclear materials;

(C) the possession, by an individual who is retired from service with a law enforcement agency and is not otherwise prohibited from receiving a firearm, of a semiautomatic assault weapon transferred to the individual by the agency upon such retirement; or

(D) the manufacture, transfer, or possession of a semiautomatic assault weapon by a licensed manufacturer or licensed importer for the purposes of testing or experimentation authorized by the Attorney General.

(w) (1) Except as provided in paragraph (2), it shall be unlawful for a person to transfer or possess a large capacity ammunition feeding device.

(2) Paragraph (1) shall not apply to the possession or transfer of any large capacity ammunition feeding device otherwise lawfully possessed on or before the date of the enactment of this subsection.

(3) This subsection shall not apply to—

(A) the manufacture for, transfer to, or possession by the United States or a department or agency of the United States or a State or a department, agency, or political subdivision of a State, or a transfer to or possession by a law enforcement officer employed by such an entity for purposes of law enforcement (whether on or off duty);

(B) the transfer to a licensee under title I of the Atomic Energy Act of 1954 for purposes of establishing and maintaining an on-site physical protection system and security organization required by Federal law, or possession by an employee or contractor of such licensee on-site for such purposes or off-site for purposes of licensee-authorized training or transportation of nuclear materials;

(C) the possession, by an individual who is retired from service with a law enforcement agency and is not otherwise prohibited from receiving ammunition, of a large capacity ammunition feeding device transferred to the individual by the agency upon such retirement; or

(D) the manufacture, transfer, or possession of any large capacity ammunition feeding device by a licensed manufacturer or licensed importer for the purposes of testing or experimentation authorized by the Attorney General.

(4) If a person charged with violating paragraph (1) asserts that paragraph (1) does not apply to such person because of paragraph (2) or (3), the Government shall have the burden of proof to show that such paragraph (1) applies to such person. The lack of a serial number as described in section 923 (i) of this title shall be a presumption that the large capacity ammunition feeding device is not subject to the prohibition of possession in paragraph (1).

APPENDIX

Centerfire Rifles — Autoloaders

Browning BAR Mark II Safari Semi-Auto Rifle
Browning BAR Mark II Safari Magnum Rifle
Browning High-Power Rifle
Heckler & Koch Model 300 Rifle
Iver Johnson M-1 Carbine
Iver Johnson 50th Anniversary M-1 Carbine
Marlin Model 9 Camp Carbine
Marlin Model 45 Carbine
Remington Nylon 66 Auto-Loading Rifle
Remington Model 7400 Auto Rifle
Remington Model 7400 Rifle
Remington Model 7400 Special Purpose Auto Rifle
Ruger Mini-14 Autoloading Rifle (w/o folding stock)
Ruger Mini Thirty Rifle

Centerfire Rifles — Lever & Slide

Browning Model 81 BLR Lever-Action Rifle
Browning Model 81 Long Action BLR
Browning Model 1886 Lever-Action Carbine
Browning Model 1886 High Grade Carbine
Cimarron 1860 Henry Replica
Cimarron 1866 Winchester Replicas
Cimarron 1873 Short Rifle
Cimarron 1873 Sporting Rifle
Cimarron 1873 30" Express Rifle
Dixie Engraved 1873 Rifle
E.M.F. 1866 Yellowboy Lever Actions
E.M.F. 1860 Henry Rifle
E.M.F. Model 73 Lever-Action Rifle
Marlin Model 336CS Lever-Action Carbine
Marlin Model 30AS Lever-Action Carbine
Marlin Model 444SS Lever-Action Sporter
Marlin Model 1894S Lever-Action Carbine
Marlin Model 1894CS Carbine
Marlin Model 1894CL Classic
Marlin Model 1895SS Lever-Action Rifle
Mitchell 1858 Henry Replica
Mitchell 1866 Winchester Replica
Mitchell 1873 Winchester Replica
Navy Arms Military Henry Rifle
Navy Arms Henry Trapper
Navy Arms Iron Frame Henry
Navy Arms Henry Carbine
Navy Arms 1866 Yellowboy Rifle
Navy Arms 1873 Winchester-Style Rifle
Navy Arms 1873 Sporting Rifle
Remington 7600 Slide Action

Remington Model 7600 Special Purpose Slide Action

Rossi M92 SRC Saddle-Ring Carbine
Rossi M92 SRS Short Carbine
Savage 99C Lever-Action Rifle
Uberti Henry Rifle
Uberti 1866 Sporting Rifle
Uberti 1873 Sporting Rifle
Winchester Model 94 Side Eject Lever-Action Rifle
Winchester Model 94 Trapper Side Eject
Winchester Model 94 Big Bore Side Eject
Winchester Model 94 Ranger Side Eject Lever-Action Rifle
Winchester Model 94 Wrangler Side Eject

Centerfire Rifles —Bolt Action

Alpine Bolt-Action Rifle
A-Square Caesar Bolt-Action Rifle
A-Square Hannibal Bolt-Action Rifle
Anschutz 1700D Classic Rifles
Anschutz 1700D Custom Rifles
Anschutz 1700D Bavarian Bolt-Action Rifle
Anschutz 1733D Mannlicher Rifle
Barret Model 90 Bolt-Action Rifle
Beeman/HW 60J Bolt-Action Rifle
Blaser R84 Bolt-Action Rifle
BRNO 537 Sporter Bolt-Action Rifle
BRNO ZKB 527 Fox Bolt-Action Rifle
BRNO ZKK 600, 601, 602 Bolt-Action Rifles
Browning A-Bolt Rifle
Browning A-Bolt Stainless Stalker
Browning A-Bolt Left Hand
Browning A-Bolt Short Action
Browning Euro-Bolt Rifle
Browning A-Bolt Gold Medallion
Browning A-Bolt Micro Medallion
Century Centurion 14 Sporter
Century Enfield Sporter #4
Century Swedish Sporter #38
Century Mauser 98 Sporter
Cooper Model 38 Centerfire Sporter
Dakota 22 Sporter Bolt-Action Rifle
Dakota 76 Classic Bolt-Action Rifle
Dakota 76 Short Action Rifles
Dakota 76 Safari Bolt-Action Rifle
Dakota 416 Rigby African
E.A.A./Sabatti Rover 870 Bolt-Action Rifle
Auguste Francotte Bolt-Action Rifles
Carl Gustaf 2000 Bolt-Action Rifle
Heym Magnum Express Series Rifle
Howa Lightning Bolt-Action Rifle
Howa Realtree Camo Rifle
Interarms Mark X Viscount Bolt-Action Rifle
Interarms Mini-Mark X Rifle
Interarms Mark X Whitworth Bolt-Action Rifle
Interarms Whitworth Express Rifle
Iver Johnson Model 5100A1 Long-Range Rifle
KDF K15 American Bolt-Action Rifle
Krico Model 600 Bolt-Action Rifle
Krico Model 700 Bolt-Action Rifles
Mauser Model 66 Bolt-Action Rifle
Mauser Model 99 Bolt-Action Rifle
McMillan Signature Classic Sporter
McMillan Signature Super Varminter
McMillan Signature Alaskan
McMillan Signature Titanium Mountain Rifle
McMillan Classic Stainless Sporter
McMillan Talon Safari Rifle
McMillan Talon Sporter Rifle
Midland 1500S Survivor Rifle
Navy Arms TU-33/40 Carbine
Parker-Hale Model 81 Classic Rifle
Parker-Hale Model 81 Classic African Rifle
Parker-Hale Model 1000 Rifle
Parker-Hale Model 1100M African Magnum
Parker-Hale Model 1100 Lightweight Rifle
Parker-Hale Model 1200 Super Rifle
Parker-Hale Model 1200 Super Clip Rifle
Parker-Hale Model 1300C Scout Rifle
Parker-Hale Model 2100 Midland Rifle
Parker-Hale Model 2700 Lightweight Rifle
Parker-Hale Model 2800 Midland Rifle
Remington Model Seven Bolt-Action Rifle
Remington Model Seven Youth Rifle
Remington Model Seven Custom KS
Remington Model Seven Custom MS Rifle
Remington 700 ADL Bolt-Action Rifle
Remington 700 BDL Bolt-Action Rifle
Remington 700 BDL Varmint Special
Remington 700 BDL European Bolt-Action Rifle
Remington 700 Varmint Synthetic Rifle
Remington 700 BDL SS Rifle
Remington 700 Stainless Synthetic Rifle
Remington 700 MTRSS Rifle
Remington 700 BDL Left Hand
Remington 700 Camo Synthetic Rifle
Remington 700 Safari
Remington 700 Mountain Rifle
Remington 700 Custom KS Mountain Rifle
Remington 700 Classic Rifle
Ruger M77 Mark II Rifle
Ruger M77 Mark II Magnum Rifle
Ruger M77RL Ultra Light
Ruger M77 Mark II All-Weather Stainless Rifle
Ruger M77 RSI International Carbine
Ruger M77 Mark II Express Rifle
Ruger M77VT Target Rifle
Sako Hunter Rifle
Sako Fiberclass Sporter
Sako Safari Grade Bolt Action
Sako Hunter Left-Hand Rifle
Sako Classic Bolt Action
Sako Hunter LS Rifle
Sako Deluxe Lightweight
Sako Super Deluxe Sporter
Sako Mannlicher-Style Carbine
Sako Varmint Heavy Barrel
Sako TRG-S Bolt-Action Rifle
Sauer 90 Bolt-Action Rifle
Savage 110G Bolt-Action Rifle
Savage 110CY Youth/Ladies Rifle
Savage 110WLE One of One Thousand Limited Edition Rifle
Savage 110GXP3 Bolt-Action Rifle
Savage 110F Bolt-Action Rifle
Savage 110FXP3 Bolt-Action Rifle
Savage 110GV Varmint Rifle
Savage 112FV Varmint Rifle
Savage Model 112FVS Varmint Rifle
Savage Model 112BV Heavy Barrel Varmint Rifle
Savage 116FSS Bolt-Action Rifle
Savage Model 116FSK Kodiak Rifle
Savage 110FP Police Rifle
Steyr-Mannlicher Sporter Models SL, L, M, S, S/T
Steyr-Mannlicher Luxus Model L, M, S
Steyr-Mannlicher Model M Professional Rifle
Tikka Bolt-Action Rifle
Tikka Premium Grade Rifles
Tikka Varmint/Continental Rifle
Tikka Whitetail/Battue Rifle
Ultra Light Arms Model 20 Rifle
Ultra Light Arms Model 28, Model 40 Rifles
Voere VEC 91 Lightning Bolt-Action Rifle
Voere Model 2165 Bolt-Action Rifle
Voere Model 2155, 2150 Bolt-Action Rifles
Weatherby Mark V Deluxe Bolt-Action Rifle
Weatherby Lasermark V Rifle
Weatherby Mark V Crown Custom Rifles
Weatherby Mark V Sporter Rifle
Weatherby Mark V Safari Grade Custom Rifles
Weatherby Weathermark Rifle
Weatherby Weathermark Alaskan Rifle
Weatherby Classicmark No. 1 Rifle
Weatherby Weatherguard Alaskan Rifle
Weatherby Vanguard VGX Deluxe Rifle
Weatherby Vanguard Classic Rifle
Weatherby Vanguard Classic No. 1 Rifle
Weatherby Vanguard Weatherguard Rifle
Wichita Classic Rifle
Wichita Varmint Rifle
Winchester Model 70 Sporter
Winchester Model 70 Sporter WinTuff
Winchester Model 70 SM Sporter
Winchester Model 70 Stainless Rifle
Winchester Model 70 Varmint
Winchester Model 70 Synthetic Heavy Varmint Rifle
Winchester Model 70 DBM Rifle
Winchester Model 70 DBM-S Rifle
Winchester Model 70 Featherweight
Winchester Model 70 Featherweight WinTuff
Winchester Model 70 Featherweight Classic
Winchester Model 70 Lightweight Rifle
Winchester Ranger Rifle
Winchester Model 70 Super Express

Magnum
Winchester Model 70 Super Grade
Winchester Model 70 Custom Sharpshooter
Winchester Model 70 Custom Sporting Sharpshooter Rifle

Centerfire Rifles — Single Shot

Armsport 1866 Sharps Rifle, Carbine
Brown Model One Single Shot Rifle
Browning Model 1885 Single Shot Rifle
Dakota Single Shot Rifle
Desert Industries G-90 Single Shot Rifle
Harrington & Richardson Ultra Varmint Rifle
Model 1885 High Wall Rifle
Navy Arms Rolling Block Buffalo Rifle
Navy Arms #2 Creedmoor Rifle
Navy Arms Sharps Cavalry Carbine
Navy Arms Sharps Plains Rifle
New England Firearms Handi-Rifle
Red Willow Armory Ballard No. 5 Pacific
Red Willow Armory Ballard No. 1.5 Hunting Rifle
Red Willow Armory Ballard No. 8 Union Hill Rifle
Red Willow Armory Ballard No. 4.5 Target Rifle
Remington-Style Rolling Block Carbine
Ruger No. 1B Single Shot
Ruger No. 1A Light Sporter
Ruger No. 1H Tropical Rifle
Ruger No. 1S Medium Sporter
Ruger No. 1 RSI International
Ruger No. 1V Special Varminter
C.Sharps Arms New Model 1874 Old Reliable
C.Sharps Arms New Model 1875 Rifle
C.Sharps Arms 1875 Classic Sharps
C.Sharps Arms New Model 1875 Target & Long Range
Shiloh Sharps 1874 Long Range Express
Shiloh Sharps 1874 Montana Roughrider
Shiloh Sharps 1874 Military Carbine
Shiloh Sharps 1874 Business Rifle
Shiloh Sharps 1874 Military Rifle
Sharps 1874 Old Reliable
Thompson/Center Contender Carbine
Thompson/Center Stainless Contender Carbine
Thompson/Center Contender Carbine Survival System
Thompson/Center Contender Carbine Youth Model
Thompson/Center TCR '87 Single Shot Rifle
Uberti Rolling Block Baby Carbine

Drillings, Combination Guns, Double Rifles

Beretta Express SSO O/U Double Rifles
Beretta Model 455 SxS Express Rifle
Chapuis RGExpress Double Rifle
Auguste Francotte Sidelock Double Rifles
Auguste Francotte Boxlock Double Rifle
Heym Model 55B O/U Double Rifle
Heym Model 55FW O/U Combo Gun
Heym Model 88b Side-by-Side Double Rifle
Kodiak Mk. IV Double Rifle
Kreighoff Teck O/U Combination Gun
Kreighoff Trumpf Drilling
Merkel Over/Under Combination Guns
Merkel Drillings
Merkel Model 160 Side-by-Side Double Rifles
Merkel Over/Under Double Rifles
Savage 24F O/U Combination Gun
Savage 24F-12T Turkey Gun
Springfield Inc. M6 Scout Rifle/Shotgun
Tikka Model 412s Combination Gun
Tikka Model 412s Double Fire
A. Zoli Rifle-Shotgun O/U Combo

Rimfire Rifles — Autoloaders

AMT Lightning 25/22 Rifle
AMT Lightning Small-Game Hunting Rifle II
AMT Magnum Hunter Auto Rifle
Anschutz 525 Deluxe Auto
Armscor Model 20P Auto Rifle
Browning Auto-22 Rifle
Browning Auto-22 Grade VI
Krico Model 260 Auto Rifle
Lakefield Arms Model 64B Auto Rifle
Marlin Model 60 Self-Loading Rifle
Marlin Model 60ss Self-Loading Rifle
Marlin Model 70 HC Auto
Marlin Model 990l Self-Loading Rifle
Marlin Model 70P Papoose
Marlin Model 922 Magnum Self-Loading Rifle
Marlin Model 995 Self-Loading Rifle
Norinco Model 22 ATD Rifle
Remington Model 522 Viper Autoloading Rifle
Remington 552BDL Speedmaster Rifle
Ruger 10/22 Autoloading Carbine (w/o folding stock)
Survival Arms AR-7 Explorer Rifle
Texas Remington Revolving Carbine
Voere Model 2115 Auto Rifle

Rimfire Rifles — Lever & Slide Action

Browning BL-22 Lever-Action Rifle
Marlin 39TDS Carbine
Marlin Model 39AS Golden Lever-Action Rifle
Remington 572BDL Fieldmaster Pump Rifle
Norinco EM-321 Pump Rifle
Rossi Model 62 SA Pump Rifle
Rossi Model 62 SAC Carbine
Winchester Model 9422 Lever-Action Rifle
Winchester Model 9422 Magnum Lever-Action Rifle

Rimfire Rifles — Bolt Actions & Single Shots

Anschutz Achiever Bolt-Action Rifle
Anschutz 1416D/1516D Classic Rifles
Anschutz 1418D/1518D Mannlicher Rifles
Anschutz 1700D Classic Rifles
Anschutz 1700D Custom Rifles
Anschutz 1700 FWT Bolt-Action Rifle
Anschutz 1700D Graphite Custom Rifle
Anschutz 1700D Bavarian Bolt-Action Rifle
Armscor Model 14P Bolt-Action Rifle
Armscor Model 1500 Rifle
BRNO ZKM-452 Deluxe Bolt-Action Rifle
BRNO ZKM 452 Deluxe
Beeman/HW 60-J-ST Bolt-Action Rifle
Browning A-Bolt 22 Bolt-Action Rifle
Browning A-Bolt Gold Medallion
Cabanas Phaser Rifle
Cabanas Master Bolt-Action Rifle
Cabanas Espronceda IV Bolt-Action Rifle
Cabanas Leyre Bolt-Action Rifle
Chipmunk Single Shot Rifle
Cooper Arms Model 36S Sporter Rifle
Dakota 22 Sporter Bolt-Action Rifle
Krico Model 300 Bolt-Action Rifles
Lakefield Arms Mark II Bolt-Action Rifle
Lakefield Arms Mark I Bolt-Action Rifle
Magtech Model MT-22C Bolt-Action Rifle
Marlin Model 880 Bolt-Action Rifle
Marlin Model 881 Bolt-Action Rifle
Marlin Model 882 Bolt-Action Rifle
Marlin Model 883 Bolt-Action Rifle
Marlin Model 883SS Bolt-Action Rifle
Marlin Model 25MN Bolt-Action Rifle
Marlin Model 25N Bolt-Action Repeater
Marlin Model 15YN "Little Buckaroo"
Mauser Model 107 Bolt-Action Rifle
Mauser Model 201 Bolt-Action Rifle
Navy Arms TU-KKW Training Rifle
Navy Arms TU-33/40 Carbine
Navy Arms TU-KKW Sniper Trainer
Norinco JW-27 Bolt-Action Rifle
Norinco JW-15 Bolt-Action Rifle
Remington 541-T
Remington 40-XR Rimfire Custom Sporter
Remington 541-T HB Bolt-Action Rifle
Remington 581-S Sportsman Rifle
Ruger 77/22 Rimfire Bolt-Action Rifle
Ruger K77/22 Varmint Rifle
Ultra Light Arms Model 20 RF Bolt-Action Rifle
Winchester Model 52B Sporting Rifle

Competition Rifles — Centerfire & Rimfire

Anschutz 64-MS Left Silhouette
Anschutz 1808D RT Super Match 54 Target
Anschutz 1827B Biathlon Rifle
Anschutz 1903D Match Rifle
Anschutz 1803D Intermediate Match
Anschutz 1911 Match Rifle

Anschutz 54.18MS REP Deluxe Silhouette Rifle
Anschutz 1913 Super Match Rifle
Anschutz 1907 Match Rifle
Anschutz 1910 Super Match II
Anschutz 54.18MS Silhouette Rifle
Anschutz Super Match 54 Target Model 2013
Anschutz Super Match 54 Target Model 2007
Beeman/Feinwerkbau 2600 Target Rifle
Cooper Arms Model TRP-1 ISU Standard Rifle
E.A.A./Weihrauch HW 60 Target Rifle
E.A.A./HW 660 Match Rifle
Finnish Lion Standard Target Rifle
Krico Model 360 S2 Biathlon Rifle
Krico Model 400 Match Rifle
Krico Model 360S Biathlon Rifle
Krico Model 500 Kricotronic Match Rifle
Krico Model 600 Sniper Rifle
Krico Model 600 Match Rifle
Lakefield Arms Model 90B Target Rifle
Lakefield Arms Model 91T Target Rifle
Lakefield Arms Model 92S Silhouette Rifle
Marlin Model 2000 Target Rifle
Mauser Model 86-SR Specialty Rifle
McMillan M-86 Sniper Rifle
McMillan Combo M-87/M-88 50-Caliber Rifle
McMillan 300 Phoenix Long Range Rifle
McMillan M-89 Sniper Rifle
McMillan National Match Rifle
McMillan Long Range Rifle
Parker-Hale M-87 Target Rifle
Parker-Hale M-85 Sniper Rifle
Remington 40-XB Rangemaster Target Centerfire
Remington 40-XR KS Rimfire Position Rifle
Remington 40-XBBR KS
Remington 40-XC KS National Match Course Rifle
Sako TRG-21 Bolt-Action Rifle
Steyr-Mannlicher Match SPG-UIT Rifle
Steyr-Mannlicher SSG P-I Rifle
Steyr-Mannlicher SSG P-III Rifle
Steyr-Mannlicher SSG P-IV Rifle
Tanner Standard UIT Rifle
Tanner 50 Meter Free Rifle
Tanner 300 Meter Free Rifle
Wichita Silhouette Rifle

Shotguns — Autoloaders

American Arms/Franchi Black Magic 48/AL
Benelli Super Black Eagle Shotgun
Benelli Super Black Eagle Slug Gun
Benelli M1 Super 90 Field Auto Shotgun
Benelli Montefeltro Super 90 20-Gauge Shotgun
Benelli Montefeltro Super 90 Shotgun
Benelli M1 Sporting Special Auto Shotgun
Benelli Black Eagle Competition Auto Shotgun
Beretta A-303 Auto Shotgun
Beretta 390 Field Auto Shotgun
Beretta 390 Super Trap, Super Skeet Shotguns
Beretta Vittoria Auto Shotgun
Beretta Model 1201F Auto Shotgun
Browning BSA 10 Auto Shotgun
Browning BSA 10 Stalker Auto Shotgun
Browning A-500R Auto Shotgun
Browning A-500G Auto Shotgun
Browning A-500G Sporting Clays
Browning Auto-5 Light 12 and 20
Browning Auto-5 Stalker
Browning Auto-5 Magnum 20
Browning Auto-5 Magnum 12
Churchill Turkey Automatic Shotgun
Cosmi Automatic Shotgun
Maverick Model 60 Auto Shotgun
Mossberg Model 5500 Shotgun
Mossberg Model 9200 Regal Semi-Auto Shotgun
Mossberg Model 9200 USST Auto Shotgun
Mossberg Model 9200 Camo Shotgun
Mossberg Model 6000 Auto Shotgun
Remington Model 1100 Shotgun
Remington 11-87 Premier Shotgun
Remington 11-87 Sporting Clays
Remington 11-87 Premier Skeet
Remington 11-87 Premier Trap
Remington 11-87 Special Purpose Magnum
Remington 11-87 SPS-T Camo Auto Shotgun
Remington 11-87 Special Purpose Deer Gun
Remington 11-87 SPS-BG-Camo Deer/Turkey Shotgun
Remington 11-87 SPS-Deer Shotgun
Remington 11-87 Special Purpose Synthetic Camo
Remington SP-10 Magnum-Camo Auto Shotgun
Remington SP-10 Magnum Auto Shotgun
Remington SP-10 Magnum Turkey Combo
Remington 1100 LT-20 Auto
Remington 1100 Special Field
Remington 1100 20-Gauge Deer Gun
Remington 1100 LT-20 Tournament Skeet
Winchester Model 1400 Semi-Auto Shotgun

Shotguns — Slide Actions

Browning Model 42 Pump Shotgun
Browning BPS Pump Shotgun
Browning BPS Stalker Pump Shotgun
Browning BPS Pigeon Grade Pump Shotgun
Browning BPS Pump Shotgun (Ladies and Youth Model)
Browning BPS Game Gun Turkey Special
Browning BPS Game Gun Deer Special
Ithaca Model 87 Supreme Pump Shotgun
Ithaca Model 87 Deerslayer Shotgun
Ithaca Deerslayer II Rifled Shotgun
Ithaca Model 87 Turkey Gun
Ithaca Model 87 Deluxe Pump Shotgun
Magtech Model 586-VR Pump Shotgun
Maverick Models 88, 91 Pump Shotguns
Mossberg Model 500 Sporting Pump
Mossberg Model 500 Camo Pump
Mossberg Model 500 Muzzleloader Combo
Mossberg Model 500 Trophy Slugster
Mossberg Turkey Model 500 Pump
Mossberg Model 500 Bantam Pump
Mossberg Field Grade Model 835 Pump Shotgun
Mossberg Model 835 Regal Ulti-Mag Pump
Remington 870 Wingmaster
Remington 870 Special Purpose Deer Gun
Remington 870 SPS-BG-Camo Deer/Turkey Shotgun
Remington 870 SPS-Deer Shotgun
Remington 870 Marine Magnum
Remington 870 TC Trap
Remington 870 Special Purpose Synthetic Camo
Remington 870 Wingmaster Small Gauges
Remington 870 Express Rifle Sighted Deer Gun
Remington 879 SPS Special Purpose Magnum
Remington 870 SPS-T Camo Pump Shotgun
Remington 870 Special Field
Remington 870 Express Turkey
Remington 870 High Grades
Remington 870 Express
Remington Model 870 Express Youth Gun
Winchester Model 12 Pump Shotgun
Winchester Model 42 High Grade Shotgun
Winchester Model 1300 Walnut Pump
Winchester Model 1300 Slug Hunter Deer Gun
Winchester Model 1300 Ranger Pump Gun Combo & Deer Gun
Winchester Model 1300 Turkey Gun
Winchester Model 1300 Ranger Pump Gun

Shotguns — Over/Unders

American Arms/Franchi Falconet 2000 O/U
American Arms Silver I O/U
American Arms Silver II Shotgun
American Arms Silver Skeet O/U
American Arms/Franchi Sporting 2000 O/U
American Arms Silver Sporting O/U
American Arms Silver Trap O/U
American Arms WS/OU 12, TS/OU 12 Shotguns
American Arms WT/OU 10 Shotgun
Armsport 2700 O/U Goose Gun
Armsport 2700 Series O/U
Armsport 2900 Tri-Barrel Shotgun
Baby Bretton Over/Under Shotgun
Beretta Model 686 Ultralight O/U
Beretta ASE 90 Competition O/U Shotgun
Beretta Over/Under Field Shotguns

Beretta Onyx Hunter Sport O/U Shotgun
Beretta Model SO5, SO6, SO9 Shotguns
Beretta Sporting Clay Shotguns
Beretta 687EL Sporting O/U
Beretta 682 Super Sporting O/U
Beretta Series 682 Competition Over/Unders
Browning Citori O/U Shotgun
Browning Superlight Citori Over/Under
Browning Lightning Sporting Clays
Browning Micro Citori Lightning
Browning Citori Plus Trap Combo
Browning Citori Plus Trap Gun
Browning Citori O/U Skeet Models
Browning Citori O/U Trap Models
Browning Special Sporting Clays
Browning Citori GTI Sporting Clays
Browning 325 Sporting Clays
Centurion Over/Under Shotgun
Chapuis Over/Under Shotgun
Connecticut Valley Classics Classic Sporter O/U
Connecticut Valley Classics Classic Field Waterfowler
Charles Daly Field Grade O/U
Charles Daly Lux Over/Under
E.A.A./Sabatti Sporting Clays Pro-Gold O/U
E.A.A/Sabatti Falcon-Mon Over/Under
Kassnar Grade I O/U Shotgun
Krieghoff K-80 Sporting Clays O/U
Krieghoff K-80 Skeet Shotgun
Krieghoff K-80 International Skeet
Krieghoff K-80 Four-Barrel Skeet Set
Krieghoff K-80/RT Shotguns
Krieghoff K-80 O/U Trap Shotgun
Laurona Silhouette 300 Sporting Clays
Laurona Silhouette 300 Trap
Laurona Super Model Over/Unders
Ljutic LM-6 Deluxe O/U Shotgun
Marocchi Conquista Over/Under Shotgun
Marocchi Avanza O/U Shotgun
Merkel Model 200E O/U Shotgun
Merkel Model 200E Skeet, Trap Over/Unders
Merkel Model 203E, 303E Over/Under Shotguns
Perazzi Mirage Special Sporting O/U
Perazzi Mirage Special Four-Gauge Skeet
Perazzi Sporting Classic O/U
Perazzi MX7 Over/Under Shotguns
Perazzi Mirage Special Skeet Over/Under
Perazzi MX8/MX8 Special Trap, Skeet
Perazzi MX 8/20 Over/Under Shotgun
Perazzi MX9 Single Over/Under Shotguns
Perazzi MX12 Hunting Over/Under
Perazzi MX28, MX410 Game O/U Shotguns
Perazzi MX20 Hunting Over/Under
Piotti Boss Over/Under Shotgun
Remington Peerless Over/Under Shotgun
Ruger Red Label O/U Shotgun
Ruger Sporting Clays O/U Shotgun
San Marco 12-Ga. Wildflower Shotgun
San Marco Field Special O/U Shotgun
San Marco 10-Ga. O/U Shotgun
SKB Model 505 Deluxe Over/Under Shotgun
SKB Model 685 Over/Under Shotgun
SKB Model 885 Over/Under Trap, Skeet, Sporting Clays
Stoeger/IGA Condor I O/U Shotgun
Stoeger/IGA ERA 2000 Over/Under Shotgun
Techni-Mec Model 610 Over/Under
Tikka Model 412S Field Grade Over/Under
Weatherby Athena Grade IV O/U Shotguns
Weatherby Athena Grade V Classic Field O/U
Weatherby Orion O/U Shotguns
Weatherby II, III Classic Field O/Us
Weatherby Orion II Classic Sporting Clays O/U
Weatherby Orion II Sporting Clays O/U
Winchester Model 1001 O/U Shotgun
Winchester Model 1001 Sporting Clays O/U
Pietro Zanoletti Model 2000 Field O/U

Shotguns — Side by Sides

American Arms Brittany Shotgun
American Arms Gentry Double Shotgun
American Arms Derby Side-by-Side
American Arms Grulla #2 Double Shotgun
American Arms WS/SS 10
American Arms TS/SS 10 Double Shotgun
American Arms TS/SS 12 Side-by-Side
Arrieta Sidelock Double Shotguns
Armsport 1050 Series Double Shotguns
Arizaga Model 31 Double Shotgun
AYA Boxlock Shotguns
AYA Sidelock Double Shotguns
Beretta Model 452 Sidelock Shotgun
Beretta Side-by-Side Field Shotguns
Crucelegui Hermanos Model 150 Double
Chapuis Side-by-Side Shotgun
E.A.A./Sabatti Saba-Mon Double Shotgun
Charles Daly Model Dss Double
Ferlib Model F VII Double Shotgun
Auguste Francotte Boxlock Shotgun
Auguste Francotte Sidelock Shotgun
Garbi Model 100 Double
Garbi Model 101 Side-by-Side
Garbi Model 103A, B Side-by-Side
Garbi Model 200 Side-by-Side
Bill Hanus Birdgun Doubles
Hatfield Uplander Shotgun
Merkel Model 8, 47E Side-by-Side Shotguns
Merkel Model 47LSC Sporting Clays Double
Merkel Model 47S, 147S Side-by-Sides
Parker Reproductions Side-by-Side
Piotti King No. 1 Side-by-Side
Piotti Lunik Side-by-Side
Piotti King Extra Side-by-Side
Piotti Piuma Side-by-Side
Precision Sports Model 600 Series Doubles
Rizzini Boxlock Side-by-Side
Rizzini Sidelock Side-by-Side

Stoeger/IGA Uplander Side-by-Side Shotgun
Ugartechea 10-Ga. Magnum Shotgun

Shotguns — Bolt Actions & Single Shots

Armsport Single Barrel Shotgun
Browning BT-99 Competition Trap Special
Browning BT-99 Plus Trap Gun
Browning BT-99 Plus Micro
Browning Recoilless Trap Shotgun
Browning Micro Recoilless Trap Shotgun
Desert Industries Big Twenty Shotgun
Harrington & Richardson Topper Model 098
Harrington & Richardson Topper Classic Youth Shotgun
Harrington & Richardson N.W.T.F. Turkey Mag
Harrington & Richardson Topper Deluxe Model 098
Krieghoff KS-5 Trap Gun
Krieghoff KS-5 Special
Krieghoff K-80 Single Barrel Trap Gun
Ljutic Mono Gun Single Barrel
Ljutic LTX Super Deluxe Mono Gun
Ljutic Recoilless Space Gun Shotgun
Marlin Model 55 Goose Gun Bolt Action
New England Firearms Turkey and Goose Gun
New England Firearms N.W.T.F. Shotgun
New England Firearms Tracker Slug Gun
New England Firearms Standard Pardner
New England Firearms Survival Gun
Perazzi TM1 Special Single Trap
Remington 90-T Super Single Shotgun
Snake Charmer II Shotgun
Stoeger/IGA Reuna Single Barrel Shotgun
Thompson/Center TCR '87 Hunter Shotgun

TITLE 27 CFR CHAPTER II

PART 478—COMMERCE IN FIREARMS AND AMMUNITION

(This Part was formerly designated as Part 178)

Editor's Note:

Effective January 24, 2003, the Homeland Security Act transferred the Bureau of Alcohol, Tobacco and Firearms from the Department of the Treasury to the Department of Justice. In addition, the agency's name was changed to the Bureau of Alcohol, Tobacco, Firearms and Explosives. The regulations, as printed in this publication, do not yet reflect this change. The regulations will be amended to change the references from the "Bureau of Alcohol, Tobacco and Firearms," the "Department of the Treasury," and the "Secretary of the Treasury" to the "Bureau of Alcohol, Tobacco, Firearms and Explosives," the "Department of Justice" and the "Attorney General," respectively.

NOTICE: ASSAULT WEAPONS BAN

On September 13, 1994, Congress passed the Violent Crime Control and Law Enforcement Act of 1994, Public Law 103-322. Title IX, Subtitle A, Section 110105 of this Act generally made it unlawful to manufacture, transfer, and possess semiautomatic assault weapons (SAWs) and to transfer and possess large capacity ammunition feeding devices (LCAFDs). The law also required importers and manufacturers to place certain markings on SAWs and LCAFDs, designating they were for export or law enforcement/government use. Significantly, the law provided that it would expire 10 years from the date of enactment. Accordingly, effective 12:00 am on September 13, 2004, these provisions of the law ceased to apply and the following provisions of the regulations in Part 478 no longer apply:

Section 478.11 –	Definitions of the terms "semiautomatic assault weapon" and "large capacity ammunition feeding device;"
Section 478.40 –	Entire section;
Section 478.40a –	Entire section;
Section 478.57 –	Paragraphs (b) and (c);
Section 478.92 –	Paragraphs (a)(3) and (c);
Section 478.119 –	Entire section (Note: an import permit is still needed pursuant to the Arms Export Control Act – see 27 CFR 447.41(a));
Section 478.132 –	Entire section; and
Section 478.153 –	Entire section.

References to "ammunition feeding device" in section 478.116 are not applicable on or after September 13, 2004. References to "semiautomatic assault weapons" in section 478.171 are not applicable on or after September 13, 2004.

The regulations will be amended to reflect these changes.

Subpart A—Introduction

§ 478.1 Scope of regulations.

(a) General. The regulations contained in this part relate to commerce in firearms and ammunition and are promulgated to implement Title I, State Firearms Control Assistance.

(b) Proce dural an d substantive requirements. This part contains the procedural and substantive requirements relative to:

(1) The interstate or foreign commerce in firearms and ammunition;

(2) The licensing of manufacturers and importers of firearms and ammunition, collectors of firearms, and dealers in firearms;

(3) The conduct of business or activity by licensees;

(4) The importation of firearms and ammunition;

(5) The records and reports required of licensees;

(6) Relief from disabilities under this part;

(7) Exempt interstate and foreign commerce in firearms and ammunition; and

(8) Restrictions on armor piercing ammunition.

§ 478.2 Relation to other provisions of law.

The provisions in this part are in addition to, and are not in lieu of, any other provision of law, or regulations, respecting commerce in firearms or ammunition. For regulations applicable to traffic in machine guns, destructive devices, and certain other firearms, see Part 479 of this chapter. For statutes applicable to the registration and licensing of persons engaged in the business of manufacturing, importing or exporting arms, ammunition, or implements of war, see section 38 of the Arms Export Control Act (22 U.S.C. 2778) and regulations thereunder and Part 447 of this chapter. For statutes applicable to nonmailable firearms, see 18 U.S.C. 1715 and regulations thereunder.

Subpart B – Definitions

§ 478.11 Meaning of terms.

When used in this part and in forms prescribed under this part, where not otherwise distinctly expressed or manifestly incompatible with the intent thereof, terms shall have the meanings ascribed in this section. Words in the plural form shall include the singular, and vice versa, and words importing the masculine gender shall include the feminine. The terms **"includes"** and **"including"** do not exclude other things not enumerated which are in the same general class or are otherwise within the scope thereof.

Act. 18 U.S.C. Chapter 44.

Adjudicated as a mental defective.

(a) A determination by a court, board, commission, or other lawful authority that a person, as a result of marked subnormal intelligence, or mental illness, incompetency, condition, or disease:

(1) Is a danger to himself or to others; or

(2) Lacks the mental capacity to contract or manage his own affairs.

(b) The term shall include—

(1) A finding of insanity by a court in a criminal case; and

(2) Those persons found incompetent to stand trial or found not guilty by reason of lack of mental responsibility pursuant to articles 50a and 72b of the Uniform Code of Military Justice, 10 U.S.C. 850a, 876b.

Admitted to the United States f or lawful hunting or sporting purposes. (a) Is entering the United States to participate in a competitive target shooting event sponsored by a national, State, or local organization, devoted to the competitive use or other sporting use of firearms; or **(b)** Is entering the United States to display firearms at a sports or hunting trade show sponsored by a national, State, or local firearms trade organization, devoted to the competitive use or other sporting use of firearms.

Alien. Any person not a citizen or national of the United States.

Alien illegally or unlawfully in the United States. Aliens who are unlawfully in the United States are not in valid immigrant, nonimmigrant or parole status. The term includes any alien—

(a) Who unlawfully entered the United States without inspection and authorization by an immigration officer and who has not been paroled into the United States under section 212(d)(5) of the Immigration and Nationality Act (INA);

(b) Who is a nonimmigrant and whose authorized period of stay has expired or who has violated the terms of the nonimmigrant category in which he or she was admitted;

(c) Paroled under INA section 212(d)(5) whose authorized period of parole has expired or whose parole status has been terminated; or

(d) Under an order of deportation, exclusion, or removal, or under an order to depart the United States voluntarily, whether or not he or she has left the United States.

Ammunition. Ammunition or cartridge cases, primers, bullets, or propellent powder designed for use in any firearm other than an antique firearm. The term shall not include **(a)** any shotgun shot or pellet not designed for use as the single, complete projectile load for one shotgun hull or casing, nor **(b)** any unloaded, non-metallic shotgun hull or casing not having a primer.

Antique firearm. **(a)** Any firearm (including any firearm with a matchlock, flintlock, percussion cap, or similar type of ignition system) manufactured in or before 1898; and **(b)** any replica of any firearm described in paragraph (a) of this definition if such replica **(1)** is not designed or redesigned for using rimfire or conventional centerfire fixed ammunition, or **(2)** uses rimfire or conventional centerfire fixed ammunition which is no longer manufactured in the United States and which is not readily available in the ordinary channels of commercial trade.

Armor piercing ammunition. Projectiles or projectile cores which may be used in a handgun and which are constructed entirely (excluding the presence of traces of other substances) from one or a combination of tungsten alloys, steel, iron, brass, bronze, beryllium copper, or depleted uranium; or full jacketed projectiles larger than .22 caliber designed and intended for use in a handgun and whose jacket has a weight of more than 25 percent of the total weight of the projectile. The term does not include shotgun shot required by Federal or State environmental or game regulations for hunting purposes, frangible projectiles designed for target shooting, projectiles which the Director finds are primarily intended to be used for sporting purposes, or any other projectiles or projectile cores which the Director finds are intended to be used for industrial purposes, including charges used in oil and gas well perforating devices.

ATF officer. An officer or employee of the Bureau of Alcohol, Tobacco and Firearms (ATF) authorized to perform any function relating to the administration or enforcement of this part.

Business premises. The property on which the manufacturing or importing of firearms or ammunition or the dealing in firearms is or will be conducted. A private dwelling, no part of which is open to the public, shall not be recognized as coming within the meaning of the term.

Chief, National Licensing Center. The ATF official responsible for the issuance and renewal of licenses under this part.

Collector. Any person who acquires, holds, or disposes of firearms as curios or relics.

Collection premises. The premises described on the license of a collector as the location at which he maintains his collection of curios and relics.

Commerce. Travel, trade, traffic, commerce, transportation, or communication among the several States, or between the District of Columbia and any State, or between any foreign country or any territory or possession and any State or the District of Columbia, or between points in the same State but through any other State or the District of Columbia or a foreign country.

Committed to a mental institution. A formal commitment of a person to a mental institution by a court, board, commission, or other lawful authority. The term includes a commitment to a mental institution involuntarily. The term includes commitment for mental defectiveness or mental illness. It also includes commitments for other reasons, such as for drug use. The term does not include a person in a mental institution for observation or a voluntary admission to a mental institution.

Controlled substance. A drug or other substance, or immediate precursor, as defined in section 102 of the Controlled Substances Act, 21 U.S.C. 802. The term includes, but is not limited to, marijuana, depressants, stimulants, and narcotic drugs. The term does not include distilled spirits, wine, malt beverages, or tobacco, as those terms are defined or used in Subtitle E of the Internal Revenue Code of 1954, as amended.

Crime punishable by imprisonment for a term exceeding 1 year. Any Federal, State or foreign offense for which the maximum penalty, whether or not imposed, is capital punishment or imprisonment in excess of 1 year. The term shall not include **(a)** any Federal or State offenses pertaining to antitrust violations, unfair trade practices, restraints of trade, or other similar offenses relating to the regulation of business practices or **(b)** any State offense classified by the laws of the State as a misdemeanor and punishable by a term of imprisonment of 2 years or less. What constitutes a conviction of such a crime shall be determined in accordance with the law of the jurisdiction in which the proceedings were held. Any conviction which has been expunged or set aside or for which a person has been pardoned or has had civil rights restored shall not be considered a conviction for the purposes of the Act or this part, unless such pardon, expunction, or restoration of civil rights expressly provides that the person may not ship, transport, possess, or receive firearms, or unless the person is prohibited by the law of the jurisdiction in which the proceedings were held from receiving or possessing any firearms.

Editor's Note:

Foreign offenses no longer qualify as crimes punishable by imprisonment for a term exceeding 1 year. The regulation will be changed to reflect this.

Curios or relics. Firearms which are of special interest to collectors by reason of some quality other than is associated with firearms intended for sporting use or as offensive or defensive weapons. To be recognized as curios or relics, firearms must fall within one of the following categories:

(a) Firearms which were manufactured at least 50 years prior to the current date, but not including replicas thereof;

(b) Firearms which are certified by the curator of a municipal, State, or Federal museum which exhibits firearms to be curios or relics of museum interest; and

(c) Any other firearms which derive a substantial part of their monetary value from the fact that they are novel, rare, bizarre, or because of their association with some historical figure, period, or event. Proof of qualification of a particular firearm under this category may be established by evidence of present value and evidence that like firearms are not available except as collector's items, or that the value of like firearms available in ordinary commercial channels is substantially less.

Editor's Note:

ATF Publication 5300.11, *Firearms Curios and Relics List*, consists of lists of those firearms determined to be curios or relics from 1972 to the present.

Customs officer. Any officer of the Customs Service or any commissioned, warrant, or petty officer of the Coast Guard, or any agent or other person authorized by law or designated by the Secretary of the Treasury to perform any duties of an officer of the Customs Service.

Dealer. Any person engaged in the business of selling firearms at wholesale or retail; any person engaged in the business of repairing firearms or of making or fitting special barrels, stocks, or trigger mechanisms to firearms; or any person who is a pawnbroker. The term shall include any person who engages in such business or occupation on a part-time basis.

Destructive device. (a) Any explosive, incendiary, or poison gas (1) bomb, (2) grenade, (3) rocket having a propellant charge of more than 4 ounces, (4) missile having an explosive or incendiary charge of more than one-quarter ounce, (5) mine, or (6) device similar to any of the devices described in the preceding paragraphs of this definition; **(b)** any type of weapon (other than a shotgun or a shotgun shell which the Director finds is generally recognized as particularly suitable for sporting purposes) by whatever name known which will, or which may be readily converted to, expel a projectile by the action of an explosive or other propellant, and which has any barrel with a bore of more than one-half inch in diameter; and **(c)** any combination of parts either designed or intended for use in converting any device into any destructive device described in paragraph (a) or (b) of this section and from which a destructive device may be readily assembled. The term shall not include any device which is neither designed nor redesigned for use as a weapon; any device, although originally designed for use as a weapon, which is redesigned for use as a signaling, pyrotechnic, line throwing, safety, or similar device; surplus ordnance sold, loaned, or given by the Secretary of the Army pursuant to the provisions of section 4684(2), 4685, or 4686 of title 10, United States Code; or any other device which the Director finds is not likely to be used as a weapon, is an antique, or is a rifle which the owner intends to use solely for sporting, recreational, or cultural purposes.

Director. The Director, Bureau of Alcohol, Tobacco and Firearms, the Department of the Treasury, Washington, D.C.

Director of Industry Operations. The principal ATF official in a Field Operations division responsible for administering regulations in this part.

Discharged under dishonorable conditions. Separation from the U.S. Armed Forces resulting from a dishonorable discharge or dismissal adjudged by a general court-martial. The term does not include any separation from the Armed Forces resulting from any other discharge, e.g., a bad conduct discharge.

Division. A Bureau of Alcohol, Tobacco and Firearms Division.

Engaged in the business—(a) Manufacturer of firearms. A person who devotes time, attention, and labor to manufacturing firearms as a regular course of trade or business with the principal objective of livelihood and profit through the sale or distribution of the firearms manufactured;

(b) Manufacturer of ammunition. A person who devotes time, attention, and labor to manufacturing ammunition as a regular course of trade or business with the principal objective of livelihood and profit through the sale or distribution of the ammunition manufactured;

(c) Dealer in firearms other than a gunsmith or a pawnbroker. A person who devotes time, attention, and labor to dealing in firearms as a regular course of trade or business with the principal objective of livelihood and profit through the repetitive purchase and resale of firearms, but such a term shall not include a person who makes occasional sales, exchanges, or purchases of firearms for the enhancement of a personal collection or for a hobby, or who sells all or part of his personal collection of firearms;

(d) Gunsmith. A person who devotes time, attention, and labor to engaging in such activity as a regular course of trade or business with the principal objective of livelihood and profit, but such a term shall not include a person who makes occasional repairs of firearms or who occasionally fits special barrels, stocks, or trigger mechanisms to firearms;

(e) Importer of firearms. A person who devotes time, attention, and labor to importing firearms as a regular course of trade or business with the principal objective of livelihood and profit through the sale or distribution of the firearms imported; and,

(f) Importer of ammunition. A person who devotes time, attention, and labor to importing ammunition as a regular course of trade or business with the principal objective of livelihood and profit through the sale or distribution of the ammunition imported.

Executed under penalties of perjury. Signed with the prescribed declaration under the penalties of perjury as provided on or with respect to the return form, or other document or, where no form of declaration is prescribed, with the declaration:

> **"I declare under the penalties of perjury that this—(insert type of document, such as, statement, application, request, certificate), including the documents submitted in support thereof, has been examined by me and, to the best of my knowledge and belief, is true, correct, and complete."**

Federal Firearms Act. 15 U.S.C. Chapter 18.

Firearm. Any weapon, including a starter gun, which will or is designed to or may readily be converted to expel a projectile by the action of an explosive; the frame or receiver of any such weapon; any firearm muffler or firearm silencer; or any destructive device; but the term shall not include an antique firearm. In the case of a licensed collector, the term shall mean only curios and relics.

Firearm frame or receiver. That part of a firearm which provides housing for the hammer, bolt or breechblock, and firing mechanism, and which is usually threaded at its forward portion to receive the barrel.

Firearm muffler or firearm silencer. Any device for silencing, muffling, or diminishing the report of a portable firearm, including any combination of parts, designed or redesigned, and intended for use in assembling or fabricating a firearm silencer or firearm muffler, and any part intended only for use in such assembly or fabrication.

Friendly foreign government. Any government with whom the United States has diplomatic relations and whom the United States has not identified as a State sponsor of terrorism.

Fugitive from justice. Any person who

has fled from any State to avoid prosecution for a felony or a misdemeanor; or any person who leaves the State to avoid giving testimony in any criminal proceeding. The term also includes any person who knows that misdemeanor or felony charges are pending against such person and who leaves the State of prosecution.

Handgun. (a) Any firearm which has a short stock and is designed to be held and fired by the use of a single hand; and

(b) Any combination of parts from which a firearm described in paragraph (a) can be assembled.

Hunting license or permit lawfully issued in the United States. A license or permit issued by a State for hunting which is valid and unexpired.

Identification document. A document containing the name, residence address, date of birth, and photograph of the holder and which was made or issued by or under the authority of the United States Government, a State, political subdivision of a State, a foreign government, political subdivision of a foreign government, an international governmental or an international quasi-governmental organization which, when completed with information concerning a particular individual, is of a type intended or commonly accepted for the purpose of identification of individuals.

Importation. The bringing of a firearm or ammunition into the United States; except that the bringing of a firearm or ammunition from outside the United States into a foreign trade zone for storage pending shipment to a foreign country or subsequent importation into this country, pursuant to this part, shall not be deemed importation.

Importer. Any person engaged in the business of importing or bringing firearms or ammunition into the United States. The term shall include any person who engages in such business on a part-time basis.

Indictment. Includes an indictment or information in any court, under which a crime punishable by imprisonment for a term exceeding 1 year (as defined in this section) may be prosecuted, or in military cases to any offense punishable by imprisonment for a term exceeding 1 year which has been referred to a general court-martial. An information is a formal accusation of a crime, differing from an indictment in that it is made by a prosecuting attorney and not a grand jury.

Interstate or foreign commerce. Includes commerce between any place in a State and any place outside of that State, or within any possession of the United States (not including the Canal Zone) or the District of Columbia. The term shall not include commerce between places within the same State but through any place outside of that State.

Intimate partner. With respect to a person, the spouse of the person, a former spouse of the person, an individual who is a parent of a child of the person, and an individual who cohabitates or has cohabitated with the person.

Large capacity ammunition feeding device. A magazine, belt, drum, feed strip, or similar device for a firearm manufactured after September 13, 1994, that has a capacity of, or that can be readily restored or converted to accept, more than 10 rounds of ammunition. The term does not include an attached tubular device designed to accept, and capable of operating only with, .22 caliber rimfire ammunition, or a fixed device for a manually operated firearm, or a fixed device for a firearm listed in 18 U.S.C. 922, Appendix A.

Licensed collector. A collector of curios and relics only and licensed under the provisions of this part.

Licensed dealer. A dealer licensed under the provisions of this part.

Licensed importer. An importer licensed under the provisions of this part.

Licensed manufacturer. A manufacturer licensed under the provisions of this part.

Machinegun. Any weapon which shoots, is designed to shoot, or can be readily restored to shoot, automatically more than one shot, without manual reloading, by a single function of the trigger. The term shall also include the frame or receiver of any such weapon, any part designed and intended solely and exclusively, or combination of parts designed and intended, for use in converting a weapon into a machine gun, and any combination of parts from which a machine gun can be assembled if such parts are in the possession or under the control of a person.

Manufacturer. Any person engaged in the business of manufacturing firearms or ammunition. The term shall include any person who engages in such business on a part-time basis.

Mental institution. Includes mental health facilities, mental hospitals, sanitariums, psychiatric facilities, and other facilities that provide diagnoses by licensed professionals of mental retardation or mental illness, including a psychiatric ward in a general hospital.

Misdemeanor crime of domestic violence. (a) Is a Federal, State or local offense that:

(1) Is a misdemeanor under Federal or State law or, in States which do not classify offenses as misdemeanors, is an offense punishable by imprisonment for a term of one year or less, and includes offenses that are punishable only by a fine. (This is true whether or not the State statute specifically defines the offense as a "misdemeanor" or as a "misdemeanor crime of domestic violence." The term includes all such misdemeanor convictions in Indian Courts established pursuant to 25 CFR Part 11.);

(2) Has, as an element, the use or attempted use of physical force (e.g., assault and battery), or the threatened use of a deadly weapon; and

(3) Was committed by a current or former spouse, parent, or guardian of the victim, by a person with whom the victim shares a child in common, by a person who is cohabiting with or has cohabited with the victim as a spouse, parent, or guardian, (e.g., the equivalent of a "common law" marriage even if such relationship is not recognized under the law), or a person similarly situated to a spouse, parent, or guardian of the victim (e.g., two persons who are residing at the same location in an intimate relationship with the intent to make that place their home would be similarly situated to a spouse).

(b) A person shall not be considered to have been convicted of such an offense for purposes of this part unless:

(1) The person is considered to have been convicted by the jurisdiction in which the proceedings were held.

(2) The person was represented by counsel in the case, or knowingly and intelligently waived the right to counsel in the case; and

(3) In the case of a prosecution for which a person was entitled to a jury trial in the jurisdiction in which the case was tried, either

(i) The case was tried by a jury, or

(ii) The person knowingly and intelligently waived the right to have the case tried by a jury, by guilty plea or otherwise.

(c) A person shall not be considered to have been convicted of such an offense for purposes of this part if the conviction has been expunged or set aside, or is an offense for which the person has been pardoned or has had civil rights restored (if the law of the jurisdiction in which the proceedings were held provides for the loss of civil rights upon conviction for such an offense) unless the pardon, expunction, or restoration of civil rights expressly provides that the person may not ship, transport, possess, or receive firearms, and the person is not otherwise prohibited by the law of the jurisdiction in which the proceedings were held from receiving or possessing any firearms.

National Firearms Act. 26 U.S.C. Chapter 53.

NICS. The National Instant Criminal Background Check System established by the Attorney General pursuant to 18 U.S.C. 922(t).

Nonimmigrant alien. An alien in the United States in a nonimmigrant classification as defined by section 101(a)(15) of the Immigration and Nationality Act (8 U.S.C. 1101(a)(15)).

Pawnbroker. Any person whose business or occupation includes the taking or receiving, by way of pledge or pawn, of any firearm as security for the payment or repayment of money. The term shall include any person who engages in such business on a part-time basis.

Permanently inoperable. A firearm which is incapable of discharging a shot by means of an explosive and incapable of being readily restored to a firing condition. An acceptable method of rendering most firearms permanently inoperable is to fusion weld the chamber closed and fusion weld the barrel solidly to the frame. Certain unusual firearms require other methods to render the firearm permanently inoperable. Contact ATF for instructions.

Person. Any individual, corporation, company, association, firm, partnership, society, or joint stock company.

Pistol. A weapon originally designed, made, and intended to fire a projectile (bullet) from one or more barrels when held in one hand, and having **(a)** a chamber(s) as an integral part(s) of, or permanently aligned with, the bore(s); and **(b)** a short stock designed to be gripped by one hand and at an angle to and extending below the line of the bore(s).

Principal objective of livelihood and profit. The intent underlying the sale or disposition of firearms is predominantly one of obtaining livelihood and pecuniary gain, as opposed to other intents such as improving or liquidating a personal firearms collection: **Provided**, That proof of profit shall not be required as to a person who engages in the regular and repetitive purchase and disposition of firearms for criminal purposes or terrorism. For purposes of this part, the term "terrorism" means activity, directed against United States persons, which—

(a) Is committed by an individual who is not a national or permanent resident alien of the United States;

(b) Involves violent acts or acts dangerous to human life which would be a criminal violation if committed within the jurisdiction of the United States; and

(c) Is intended—

(1) To intimidate or coerce a civilian population;

(2) To influence the policy of a government by intimidation or coercion; or

(3) To affect the conduct of a government by assassination or kidnapping.

Published ordinance. A published law of any political subdivision of a State which the Director determines to be relevant to the enforcement of this part and which is contained on a list compiled by the Director, which list is incorporated by reference in the FEDERAL REGISTER, revised annually, and furnished to licensees under this part.

Renounced U.S. citizenship. (a) A person has renounced his U.S. citizenship if the person, having been a citizen of the United States, has renounced citizenship either—

(1) Before a diplomatic or consular officer of the United States in a foreign state pursuant to 8 U.S.C. 1481(a)(5); or

(2) Before an officer designated by the Attorney General when the United States is in a state of war pursuant to 8 U.S.C. 1481(a)(6).

(b) The term shall not include any renunciation of citizenship that has been reversed as a result of administrative or judicial appeal.

Revolver. A projectile weapon, of the pistol type, having a breechloading chambered cylinder so arranged that the cocking of the hammer or movement of the trigger rotates it and brings the next cartridge in line with the barrel for firing.

Rifle. A weapon designed or redesigned, made or remade, and intended to be fired from the shoulder, and designed or redesigned and made or remade to use the energy of the explosive in a fixed metallic cartridge to fire only a single projectile through a rifled bore for each single pull of the trigger.

Semiautomatic assault weapon.

(a) Any of the firearms, or copies or duplicates of the firearms in any caliber, known as:

(1) Norinco, Mitchell, and Poly Technologies Avtomat Kalashnikovs (all models),

(2) Action Arms Israeli Military Industries UZI and Galil,

(3) Beretta Ar70 (SC–70),

(4) Colt AR–15,

(5) Fabrique National FN/FAL, FN/LAR, and FNC,

(6) SWD M–10, M–11, M–11/9, and M–12,

(7) Steyr AUG,

(8) INTRATEC TEC–9, TEC–DC9 and TEC–22, and

(9) Revolving cylinder shotguns, such as (or similar to) the Street Sweeper and Striker 12;

(b) A semiautomatic rifle that has an ability to accept a detachable magazine and has at least 2 of—

(1) A folding or telescoping stock,

(2) A pistol grip that protrudes conspicuously beneath the action of the weapon,

(3) A bayonet mount,

(4) A flash suppressor or threaded barrel designed to accommodate a flash suppressor, and

(5) A grenade launcher;

(c) A semiautomatic pistol that has an ability to accept a detachable magazine and has at least 2 of—

(1) An ammunition magazine that attaches to the pistol outside of the pistol grip,

(2) A threaded barrel capable of accepting a barrel extender, flash suppressor, forward handgrip, or silencer,

(3) A shroud that is attached to, or partially or completely encircles, the barrel and that permits the shooter to hold the firearm with the nontrigger hand without being burned,

(4) A manufactured weight of 50 ounces or more when the pistol is unloaded, and

(5) A semiautomatic version of an automatic firearm; and

(d) A semiautomatic shotgun that has at least 2 of—

(1) A folding or telescoping stock,

(2) A pistol grip that protrudes conspicuously beneath the action of the weapon,

(3) A fixed magazine capacity in excess of 5 rounds, and

(4) An ability to accept a detachable magazine.

Semiautomatic pistol. Any repeating pistol which utilizes a portion of the energy of a firing cartridge to extract the fired cartridge case and chamber the next round, and which requires a separate pull of the trigger to fire each cartridge.

Semiautomatic rifle. Any repeating rifle which utilizes a portion of the energy of a firing cartridge to extract the fired cartridge case and chamber the next round, and which requires a separate pull of the trigger to fire each cartridge.

Semiautomatic shot gun. Any repeating shotgun which utilizes a portion of the energy of a firing cartridge to extract the fired cartridge case and chamber the next round, and which requires a separate pull of the trigger to fire each cartridge.

Short-barreled rifle. A rifle having one or more barrels less than 16 inches in length, and any weapon made from a rifle, whether by alteration, modification, or otherwise, if such weapon, as modified, has an overall length of less than 26 inches.

Short-barreled shotgun. A shotgun having one or more barrels less than 18 inches in length, and any weapon made from a shotgun, whether by alteration, modification, or otherwise, if such weapon as modified has an overall length of less than 26 inches.

Shotgun. A weapon designed or redesigned, made or remade, and intended to be fired from the shoulder, and designed or redesigned and made or remade to use the energy of the explosive in a fixed shotgun shell to fire through a smooth bore either a number of ball shot or a single projectile for each single pull of the trigger.

State. A State of the United States. The term shall include the District of Columbia, the Commonwealth of Puerto Rico, and the possessions of the United States (not including the Canal Zone).

State of residence. The State in which an individual resides. An individual resides in a State if he or she is present in a State with the intention of making a home in that State. If an individual is on active duty as a member of the Armed Forces, the individual's State of residence is the State in which his or her permanent duty station is located. An alien who is legally in the United States shall be considered to be a resident of a State only if the alien is residing in the State and has resided in the State for a period of at least 90 days prior to the date of sale or delivery of a firearm. The following are examples that illustrate this definition:

Example 1. A maintains a home in State X. A travels to State Y on a hunting, fishing, business, or other type of trip. A does not become a resident of State Y by reason of such trip.

Example 2. A is a U.S. citizen and maintains a home in State X and a home in State Y. A resides in State X except for weekends or the summer months of the year and in State Y for the weekends or the summer months of the year. During the time that A actually resides in State X, A is a resident of State X, and during the time that A actually resides in State Y, A is a resident of State Y.

Example 3. A, an alien, travels on vacation or on a business trip to State X. Regardless of the length of time A spends in State X, A does not have a State of residence in State X. This is because A does not have a home in State X at which he has resided for at least 90 days.

Unlawful user of or addicted to any controlled substance. A person who uses a controlled substance and has lost the power of self-control with reference to the use of controlled substance; and any person who is a current user of a controlled substance in a manner other than as prescribed by a licensed physician. Such use is not limited to the use of drugs on a particular day, or within a matter of days or weeks before, but rather that the unlawful use has occurred recently enough to indicate that the individual is actively engaged in such conduct. A person may be an unlawful current user of a controlled substance even though the substance is not being used at the precise time the person seeks to acquire a firearm or receives or possesses a firearm. An inference of current use may be drawn from evidence of a recent use or possession of a controlled substance or a pattern of use or possession that reasonably covers the present time, e.g., a conviction for use or possession of a controlled substance within the past year; multiple arrests for such offenses within the past 5 years if the most recent arrest occurred within the past year; or persons found through a drug test to use a controlled substance unlawfully, provided that the test was administered within the past year. For a current or former member of the Armed Forces, an inference of current use may be drawn from recent disciplinary or other administrative action based on confirmed drug use, e.g., court-martial conviction, nonjudicial punishment, or an administrative discharge based on drug use or drug rehabilitation failure.

Unserviceable firearm. A firearm which is incapable of discharging a shot by means of an explosive and is incapable of being readily restored to a firing condition.

U.S.C. The United States Code.

Subpart C—Administrative and Miscellaneous Provisions

§ 478.21 Forms prescribed.

(a) The Director is authorized to prescribe all forms required by this part. All of the information called for in each form shall be furnished as indicated by the headings on the form and the instructions on or pertaining to the form. In addition, information called for in each form shall be furnished as required by this part.

(b) Requests for forms should be mailed to the ATF Distribution Center, 7943 Angus Court, Springfield, Virginia 22153.

§ 478.22 Alternate methods or procedures; emergency variations from requirements.

(a) Alternate methods or procedures. The licensee, on specific approval by the Director as provided in this paragraph, may use an alternate method or procedure in lieu of a method or procedure specifically prescribed in this part. The Director may approve an alternate method or procedure, subject to stated conditions, when it is found that:

(1) Good cause is shown for the use of the alternate method or procedure;

(2) The alternate method or procedure is within the purpose of, and consistent with the effect intended by, the specifically prescribed method or procedure and that the alternate method or procedure is substantially equivalent to that specifically prescribed method or procedure; and

(3) The alternate method or procedure will not be contrary to any provision of law and will not result in an increase in cost to the Government or hinder the effective administration of this part. Where the licensee desires to employ an alternate method or procedure, a written application shall be submitted to the appropriate Director of Industry Operations, for transmittal to the Director. The application shall specifically describe the proposed alternate method or procedure and shall set forth the reasons for it. Alternate methods or procedures may not be employed until the application is approved by the Director. The licensee shall, during the period of authorization of an alternate method or procedure, comply with the terms of the approved application. Authorization of any alternate method or procedure may be withdrawn whenever, in the judgment of the Director, the effective administration of this part is hindered by the continuation of the authorization.

(b) Emergency variations from requirements. The Director may approve a method of operation other than as specified in this part, where it is found that an emergency exists and the proposed variation from the specified requirements are necessary and the proposed variations (1) will not hinder the effective administration of this part, and (2) will not be contrary to any provisions of law. Variations from requirements granted under this paragraph are conditioned on compliance with the procedures, conditions, and limitations set forth in the approval of the application. Failure to comply in good faith with the procedures, conditions, and limitations shall automatically terminate the authority for the variations, and the licensee shall fully comply with the prescribed requirements of regulations from which the variations were authorized. Authority for any variation may be withdrawn whenever, in the judgment of the Director, the effective administration of this part is hindered by the continuation of the variation. Where the licensee desires to employ an emergency variation, a written application shall be submitted to the appropriate Director of Industry Operations for transmittal to the Director. The application shall describe the proposed variation and set forth the reasons for it. Variations may not be employed until the application is approved.

(c) Retention of approved variations. The licensee shall retain, as part of the licensee's records, available for examination by ATF officers, any application approved by the Director under this section.

§ 478.23 Right of entry and examination.

(a) Except as provided in paragraph (b), any ATF officer, when there is reasonable cause to believe a violation of the Act has occurred and that evidence of the violation may be found on the premises of any licensed manufacturer, licensed importer, licensed dealer, or licensed collector, may, upon demonstrating such cause before a Federal magistrate and obtaining from the magistrate a warrant authorizing entry, enter during business hours (or, in the case of a licensed collector, the hours of operation) the premises, including places of storage, of any such licensee for the purpose of inspecting or examining:

(1) Any records or documents required to be kept by such licensee under this part and

(2) Any inventory of firearms or ammunition kept or stored by any licensed manufacturer, licensed importer, or licensed dealer at such premises or any firearms curios or relics or ammunition kept or stored by any licensed collector at such premises.

(b) Any ATF officer, without having reasonable cause to believe a violation of the Act has occurred or that evidence of the violation may be found and without demonstrating such cause before a Federal magistrate or obtaining from the magistrate a warrant authorizing entry, may enter during business hours the premises, including places of storage, of any licensed manufacturer, licensed importer, or licensed dealer for the purpose of inspecting or examining the records, documents, ammunition and firearms referred to in paragraph (a) of this section:

(1) In the course of a reasonable inquiry during the course of a criminal investigation of a person or persons other than the licensee,

(2) For insuring compliance with the recordkeeping requirements of this part:

(i) Not more than once during any 12-month period, or

(ii) At any time with respect to records relating to a firearm involved in a criminal investigation that is traced to the licensee, or

(3) When such inspection or examination may be required for determining the disposition of one or more particular firearms in the course of a bona fide criminal investigation.

(c) Any ATF officer, without having reasonable cause to believe a violation of the Act has occurred or that evidence of the violation may be found and without demonstrating such cause before a Federal magistrate or obtaining from the magistrate a warrant authorizing entry, may enter during hours of operation the premises, including places of storage, of any licensed collector for the purpose of inspecting or examining the records, documents, firearms, and ammunition referred to in paragraph (a) of this section **(1)** for ensuring compliance with the recordkeeping requirements of this part not more than once during any 12-month period or **(2)** when such inspection or examination may be required for determining the disposition of one or more particular firearms in the course of a bona fide criminal investigation. At the election of the licensed collector, the annual inspection permitted by this paragraph shall be performed at the ATF office responsible for conducting such inspection in closest proximity to the collectors premises.

(d) The inspections and examinations provided by this section do not authorize an ATF officer to seize any records or documents other than those records or documents constituting material evidence of a violation of law. If an ATF officer seizes such records or documents, copies shall be provided to the licensee within a reasonable time.

§ 478.24 Compilation of State laws and published ordinances.

(a) The Director shall annually revise and furnish Federal firearms licensees with a compilation of State laws and published ordinances which are relevant to the enforcement of this part. The Director annually revises the compilation and publishes it as **"State Laws and Published Ordinances—Firearms"** which is furnished free of charge to licensees under this part. Where the compilation has previously been furnished to licensees, the Director need only furnish amendments of the relevant laws and ordinances to such licensees.

(b) "State Laws and Published Ordinances—Firearms" is incorporated by reference in this part. It is ATF Publication 5300.5, revised yearly. The current edition is available from the Superintendent of Documents, U.S. Government Printing Office, Washington, DC 20402. It is also available for inspection at the Office of the Federal Register, Room 8401, 1100 L Street, N.W., Washington D.C. This incorporation by reference was approved by the Director of the Federal Register.

§ 478.25 Disclosure of information.

The Director of Industry Operations may make available to any Federal, State or local law enforcement agency any information which is obtained by reason of the provisions of the Act with respect to the identification of persons prohibited from purchasing or receiving firearms or ammunition who have purchased or received firearms or ammunition, together with a description of such firearms or ammunition. Upon the request of any Federal, State or local law enforcement agency, the Director of Industry Operations may provide such agency any information contained in the records required to be maintained by the Act or this part.

§ 478.25a Responses to requests for information.

Each licensee shall respond immediately to, and in no event later than 24 hours after the receipt of, a request by an ATF officer at the National Tracing Center for information contained in the records required to be kept by this part for determining the disposition of one or more firearms in the course of a bona fide criminal investigation. The requested information shall be provided orally to the ATF officer within the 24-hour period. Verification of the identity and employment of National Tracing Center personnel requesting information may be established at the time the requested information is provided by telephoning the toll-free number 1–800–788–7132 or using the toll-free facsimile (FAX) number 1–800–578–7223.

(Approved by the Office of Management and Budget under control number 1512–0387)

§ 478.26 Curio and relic determination.

Any person who desires to obtain a determination whether a particular firearm is a curio or relic shall submit a written request, in duplicate, for a ruling thereon to the Director. Each such request shall be executed under the penalties of perjury and shall contain a complete and accurate description of the firearm, and such photographs, diagrams, or drawings as may be necessary to enable the Director to make a determination. The Director may require the submission of the firearm for examination and evaluation. If the submission of the firearm is impractical, the person requesting the determination shall so advise the Director and designate the place where the firearm will be available for examination and evaluation.

§ 478.27 Destructive device determination.

The Director shall determine in accordance with 18 U.S.C. 921(a)(4) whether a device is excluded from the definition of a destructive device. A person who desires to obtain a determination under that provision of law for any device which he believes is not likely to be used as a weapon shall submit a written request, in triplicate, for a ruling thereon to the Director. Each such request shall be executed under the penalties of perjury and contain a complete and accurate description of the device, the name and address of the manufacturer or importer thereof, the purpose of and use for which it is intended, and such photographs, diagrams, or drawings as may be necessary to enable the Director to make his determination. The Director may require the submission to him, of a sample of such device for examination and evaluation. If the submission of such device is impracticable, the person requesting the ruling shall so advise the Director and designate the place where the device will be available for examination and evaluation.

§ 478.28 Transportation of destructive devices and certain firearms.

(a) The Director may authorize a person to transport in interstate or foreign commerce any destructive device, machine gun, short-barreled shotgun, or short-barreled rifle, if he finds that such transportation is reasonably necessary and is consistent with public safety and applicable State and local law. A person who desires to transport in interstate or foreign commerce any such device or weapon shall submit a written request so to do, in duplicate, to the Director. The request shall contain:

(1) A complete description and identification of the device or weapon to be transported;

(2) A statement whether such transportation involves a transfer of title;

(3) The need for such transportation;

(4) The approximate date such transportation is to take place;

(5) The present location of such device or weapon and the place to which it is to be transported;

(6) The mode of transportation to be used (including, if by common or contract carrier, the name and address of such carrier); and

(7) Evidence that the transportation or possession of such device or weapon is not inconsistent with the laws at the place of destination.

(b) No person shall transport any destructive device, machine gun, short-barreled shotgun, or short-barreled rifle in interstate or foreign commerce under the provisions of this section until he has received specific authorization so to do from the Director. Authorization granted under this section does not carry or import relief from any other statutory or regulatory provision relating to firearms.

(c) This section shall not be construed as requiring licensees to obtain authorization to transport destructive devices, machine guns, short-barreled shotguns, and short-barreled rifles in interstate or foreign commerce: **Provided,** That in the case of a licensed importer, licensed manufacturer, or licensed dealer, such a licensee is qualified under the National Firearms Act (see also Part 479 of this chapter) and this part to engage in the business with respect to the device or weapon to be transported, and that in the case of a licensed collector, the device or weapon to be transported is a curio or relic.

§ 478.29 Out-of-State acquisition of firearms by nonlicensees.

No person, other than a licensed importer, licensed manufacturer, licensed dealer, or licensed collector, shall transport into or receive in the State where the person resides (or if a corporation or other business entity, where it maintains a place of business) any firearm purchased or otherwise obtained by such person outside that State: **Provided,** That the provisions of this section:

(a) Shall not preclude any person who lawfully acquires a firearm by bequest or intestate succession in a State other than his State of residence from transporting the firearm into or receiving it in that State, if it is lawful for such person to purchase or possess such firearm in that State,

(b) Shall not apply to the transportation or receipt of a rifle or shotgun obtained from a licensed manufacturer, licensed importer, licensed dealer, or licensed collector in a State other than the transferee's State of residence in an over-the-counter transaction at the licensee's premises obtained in conformity with the provisions of § 478.96(c) and

(c) Shall not apply to the transportation or receipt of a firearm obtained in conformity with the provisions of §§ 478.30 and 478.97.

§ 478.29a Acquisition of firearms by nonresidents.

No person, other than a licensed importer, licensed manufacturer, licensed dealer, or licensed collector, who does not reside in any State shall receive any firearms unless such receipt is for lawful sporting purposes.

§ 478.30 Out-of-State disposition of firearms by nonlicensees.

No nonlicensee shall transfer, sell, trade, give, transport, or deliver any firearm to any other nonlicensee, who the transferor knows or has reasonable cause to believe does not reside in (or if the person is a corporation or other business entity, does not maintain a place of business in) the State in which the transferor resides: **Provided,** That the provisions of this section:

(a) shall not apply to the transfer, transportation, or delivery of a firearm made to carry out a bequest of a firearm to, or any acquisition by intestate succession of a firearm by, a person who is permitted to acquire or possess a firearm under the laws of the State of his residence; and

(b) shall not apply to the loan or rental of a firearm to any person for temporary use for lawful sporting purposes.

§ 478.31 Delivery by common or contract carrier.

(a) No person shall knowingly deliver or cause to be delivered to any common or contract carrier for transportation or shipment in interstate or foreign commerce to any person other than a licensed importer, licensed manufacturer, licensed dealer, or licensed collector, any package or other container in which there is any firearm or ammunition without written notice to the carrier that such firearm or ammunition is being transported or shipped:

Provided, that any passenger who owns or legally possesses a firearm or ammunition being transported aboard any common or contract carrier for movement with the passenger in interstate or foreign commerce may deliver said firearm or ammunition into the custody of the pilot, captain, conductor or operator of such common or contract carrier for the duration of that trip without violating any provision of this part.

(b) No common or contract carrier shall require or cause any label, tag, or other written notice to be placed on the outside of any package, luggage, or other container indicating that such package, luggage, or other container contains a firearm.

(c) No common or contract carrier shall transport or deliver in interstate or foreign commerce any firearm or ammunition with knowledge or reasonable cause to believe that the shipment, transportation, or receipt thereof would be in violation of any provision of this part:

Provided, however, that the provisions of this paragraph shall not apply in respect to the transportation of firearms or ammunition in in-bond shipment under Customs laws and regulations.

(d) No common or contract carrier shall knowingly deliver in interstate or foreign commerce any firearm without obtaining written acknowledgement of receipt from the recipient of the package or other container in which there is a firearm: **Provided,** That this paragraph shall not apply with respect to the return of a firearm to a passenger who places firearms in the carrier's custody for the duration of the trip.

§ 478.32 Prohibited shi pment, transportation, p ossession, or receipt of firearms and ammunition by certain persons.

(a) No person may ship or transport any firearm or ammunition in interstate or foreign commerce, or receive any firearm or ammunition which has been shipped or transported in interstate or foreign commerce, or possess any firearm or ammunition in or affecting commerce, who:

(1) Has been convicted in any court of a crime punishable by imprisonment for a term exceeding 1 year,

(2) Is a fugitive from justice,

(3) Is an unlawful user of or addicted to any controlled substance (as defined in section 102 of the Controlled Substances Act, 21 U.S.C. 802),

(4) Has been adjudicated as a mental defective or has been committed to a mental institution,

(5) Being an alien —

(i) Is illegally or unlawfully in the United States; or

(ii) Except as provided in paragraph (f) of this section, is a nonimmigrant alien: Provided, That the provisions of this paragraph (a)(5)(ii) do not apply to any nonimmigrant alien if that alien is-

(A) Admitted to the United States for lawful hunting or sporting purposes or is in possession of a hunting license or permit lawfully issued in the United States;

(B) An official representative of a foreign government who is either accredited to the United States Government or the Government's mission to an international organization having its headquarters in the United States or is en route to or from another country to which that alien is accredited. This exception only applies if the firearm or ammunition is shipped, transported, possessed, or received in the representative's official capacity;

(C) An official of a foreign government or a distinguished foreign visitor who has been so designated by the Department of State. This exception only applies if the firearm or ammunition is shipped, transported, possessed, or received in the official's or visitor's official capacity, except if the visitor is a private individual who does not have an official capacity; or

(D) A foreign law enforcement officer of a friendly foreign government entering the United States on official law enforcement business,

(6) Has been discharged from the Armed Forces under dishonorable conditions,

(7) Having been a citizen of the United States, has renounced citizenship,

(8) Is subject to a court order that—

(i) Was issued after a hearing of which such person received actual notice, and at which such person had an opportunity to participate;

(ii) Restrains such person from harassing, stalking, or threatening an intimate partner of such person or child of such intimate partner or person, or engaging in other conduct that would place an intimate partner in reasonable fear of bodily injury to the partner or child; and

(iii) (A) Includes a finding that such person represents a credible threat to the physical safety of such intimate partner or child; or

(B) By its terms explicitly prohibits the use, attempted use, or threatened use of physical force against such intimate partner or child that would reasonably be expected to cause bodily injury, or

(9) Has been convicted of a misdemeanor crime of domestic violence.

(b) No person who is under indictment for a crime punishable by imprisonment for a term exceeding one year may ship or transport any firearm or ammunition in interstate or foreign commerce or receive any firearm or ammunition which has been shipped or transported in interstate or foreign commerce.

(c) Any individual, who to that individual's knowledge and while being employed by any person described in paragraph (a) of this section, may not in the course of such employment receive, possess, or transport any firearm or ammunition in commerce or affecting commerce or receive any firearm or ammunition which has been shipped or transported in interstate or foreign commerce.

(d) No person may sell or otherwise dispose of any firearm or ammunition to any person knowing or having reasonable cause to believe that such person:

(1) Is under indictment for, or has been convicted in any court of, a crime punishable by imprisonment for a term exceeding 1 year,

(2) Is a fugitive from justice,

(3) Is an unlawful user of or addicted to any controlled substance (as defined in section 102 of the Controlled Substances Act, 21 U.S.C. 802),

(4) Has been adjudicated as a mental defective or has been committed to a mental institution,

(5) Being an alien —

(i) Is illegally or unlawfully in the United States; or

(ii) Except as provided in paragraph (f) of this section, is a nonimmigrant alien: Provided, That the provisions of this paragraph (d)(5)(ii) do not apply to any nonimmigrant alien if that alien is-

(A) Admitted to the United States for lawful hunting or sporting purposes or is in possession of a hunting license or permit lawfully issued in the United States;

(B) An official representative of a foreign government who is either accredited to the United States Government or the Government's mission to an international organization having its headquarters in the United States or en route to or from another country to which that alien is accredited. This exception only applies if the firearm or ammunition is shipped, transported, possessed, or received in the representative's official capacity;

(C) An official of a foreign government or a distinguished foreign visitor who has been so designated by the Department of State. This exception only applies if the firearm or ammunition is shipped, transported, possessed, or received in the official's or visitor's official capacity, except if the visitor is a private individual who does not have an official capacity; or

(D) A foreign law enforcement officer of a friendly foreign government entering the United States on official law enforcement business,

(6) Has been discharged from the Armed Forces under dishonorable conditions,

(7) Having been a citizen of the United States, has renounced citizenship,

(8) Is subject to a court order that restrains such person from harassing, stalking, or threatening an intimate partner of such person or child of such intimate partner or person, or engaging in other conduct that would place an intimate partner in reasonable fear of bodily injury to the partner or child: **Provided,** That the provisions of this paragraph shall only apply to a court order that—

(i) Was issued after a hearing of which such person received actual notice, and at which such person had the opportunity to participate; and

(ii)(A) Includes a finding that such person represents a credible threat to the physical safety of such intimate partner or child; or

(B) By its terms explicitly prohibits the use, attempted use, or threatened use of physical force against such intimate partner or child that would reasonably be expected to cause bodily injury, or

(9) Has been convicted of a misdemeanor crime of domestic violence.

(e) The actual notice required by paragraphs (a)(8)(i) and (d)(8)(i) of this section is notice expressly and actually given, and brought home to the party directly, including service of process personally served on the party and service by mail. Actual notice also includes proof of facts and circumstances that raise the inference that the party received notice including, but not limited to, proof that notice was left at the party's dwelling house or usual place of abode with some person of suitable age and discretion residing therein; or proof that the party signed a return receipt for a hearing notice which had been mailed to the party. It does not include notice published in a newspaper.

(f) Pursuant to 18 U.S.C. 922(y)(3), any nonimmigrant alien may receive a waiver from the prohibition contained in paragraph (a)(5)(ii) of this section, if the Attorney General approves a petition for the waiver.

§ 478.33 Stolen firearms and ammunition.

No person shall transport or ship in interstate or foreign commerce any stolen firearm or stolen ammunition knowing or having reasonable cause to believe that the firearm or ammunition was stolen, and no person shall receive, possess, conceal, store, barter, sell, or dispose of any stolen firearm or stolen ammunition, or pledge or accept as security for a loan any stolen firearm or stolen ammunition, which is moving as, which is a part of, which constitutes, or which has been shipped or transported in, interstate or foreign commerce, either before or after it was stolen, knowing or having reasonable cause to believe that the firearm or ammunition was stolen.

§ 478.33a Theft of firearms.

No person shall steal or unlawfully take or carry away from the person or the premises of a person who is licensed to engage in the business of importing, manufacturing, or dealing in firearms, any firearm in the licensee's business inventory that has been shipped or transported in interstate or foreign commerce.

§ 478.34 Re moved, obliterated, or altered serial number.

No person shall knowingly transport, ship, or receive in interstate or foreign commerce any firearm which has had the importer's or manufacturer's serial number removed, obliterated, or altered, or possess or receive any firearm which has had the importer's or manufacturer's serial number removed, obliterated, or altered and has, at any time, been shipped or transported in interstate or foreign commerce.

§ 478.35 Skeet, trap, target, and similar shooting activities.

Licensing and recordkeeping requirements, including permissible alternate records, for skeet, trap, target, and similar organized activities shall be determined by the Director of Industry Operations on a case by case basis.

§ 478.36 Transfer or possession of machine guns.

No person shall transfer or possess a machine gun except:

(a) A transfer to or by, or possession by or under the authority of, the United States, or any department or agency thereof, or a State, or a department, agency, or political subdivision thereof. (See Part 479 of this chapter); or

(b) Any lawful transfer or lawful possession of a machine gun that was lawfully possessed before May 19, 1986 (See Part 479 of this chapter).

§ 478.37 Manufacture, importation and sale of armor piercing ammunition.

No person shall manufacture or import, and no manufacturer or importer shall sell or deliver, armor piercing ammunition, except:

(a) The manufacture or importation, or the sale or delivery by any manufacturer or importer, of armor piercing ammunition for the use of the United States or any department or agency thereof or any State or any department, agency or political subdivision thereof;

(b) The manufacture, or the sale or delivery by a manufacturer or importer, of armor piercing ammunition for the purpose of exportation; or

(c) The sale or delivery by a manufacturer or importer of armor piercing ammunition for the purposes of testing or experimentation as authorized by the Director under the provisions of § 478.149.

§ 478.38 Transportation of firearms.

Notwithstanding any other provision of any law or any rule or regulation of a State or any political subdivision thereof, any person who is not otherwise prohibited by this chapter from transporting, shipping, or receiving a firearm shall be entitled to transport a firearm for any lawful purpose from any place where such person may lawfully possess and carry such firearm to any other place where such person may lawfully possess and carry such firearm if, during such transportation the firearm is unloaded, and neither the firearm nor any ammunition being transported is readily accessible or is directly accessible from the passenger compartment of such transporting vehicle: **Provided,** That in the case of a vehicle without a compartment separate from the driver's compartment the firearm or ammunition shall be contained in a locked container other than the glove compartment or console.

§ 478.39 Assembly of semiautomatic rifles or shotguns.

(a) No person shall assemble a semiautomatic rifle or any shotgun using more than 10 of the imported parts listed in paragraph (c) of this section if the assembled firearm is prohibited from importation under section 925(d)(3) as not being particularly suitable for or readily adaptable to sporting purposes.

(b) The provisions of this section shall not apply to:

(1) The assembly of such rifle or shotgun for sale or distribution by a licensed manufacturer to the United States or any department or agency thereof or to any State or any department, agency, or political subdivision thereof; or

(2) The assembly of such rifle or shotgun for the purposes of testing or experimentation authorized by the Director under the provisions of § 478.151; or

(3) The repair of any rifle or shotgun which had been imported into or assembled in the United States prior to November 30, 1990, or the replacement of any part of such firearm.

(c) For purposes of this section, the term **imported parts** are:

(1) Frames, receivers, receiver castings, forgings or stampings

(2) Barrels

(3) Barrel extensions

(4) Mounting blocks (trunions)

(5) Muzzle attachments

(6) Bolts

(7) Bolt carriers

(8) Operating rods

(9) Gas pistons

(10) Trigger housings

(11) Triggers

(12) Hammers

(13) Sears

(14) Disconnectors

(15) Buttstocks

(16) Pistol grips

(17) Forearms, handguards

(18) Magazine bodies

(19) Followers

(20) Floorplates

§ 478.39a Reporting theft or loss of firearms.

Each licensee shall report the theft or loss of a firearm from the licensee's inventory (including any firearm which has been transferred from the licensee's inventory to a personal collection and held as a personal firearm for at least 1 year), or from the collection of a licensed collector, within 48 hours after the theft or loss is discovered. Licensees shall report thefts or losses by telephoning 1-800-800-3855 (nationwide toll free number) and by preparing ATF Form 3310.11, Federal Firearms Licensee Theft/Loss Report, in accordance with the instructions on the form. The original of the report shall be forwarded to the office specified thereon, and Copy 1 shall be retained by the licensee as part of the licensee's permanent records. Theft or loss of any firearm shall also be reported to the appropriate local authorities.

(Approved by the Office of Management and Budget under control number 1512–0524)

§ 478.40 Manufacture, transfer, and possession of semiautomatic assault weapons.

(a) Prohibition. No person shall manufacture, transfer, or possess a semiautomatic assault weapon.

(b) Exceptions. The provisions of paragraph (a) of this section shall not apply to:

(1) The possession or transfer of any semiautomatic assault weapon otherwise lawfully possessed in the United States under Federal law on September 13, 1994;

(2) Any of the firearms, or replicas or duplicates of the firearms, specified in 18 U.S.C. 922, Appendix A, as such firearms existed on October 1, 1993;

(3) Any firearm that—

(i) Is manually operated by bolt, pump, lever, or slide action;

(ii) Has been rendered permanently inoperable; or

(iii) Is an antique firearm;

(4) Any semiautomatic rifle that cannot accept a detachable magazine that holds more than 5 rounds of ammunition;

(5) Any semiautomatic shotgun that cannot hold more than 5 rounds of ammunition in a fixed or detachable magazine;

(6) The manufacture for, transfer to, or possession by the United States or a department or agency of the United States or a State or a department, agency, or political subdivision of a State, or a transfer to or possession by a law enforcement officer employed by such an entity for purposes of law enforcement;

(7) The transfer to a licensee under title I of the Atomic Energy Act of 1954 (42 U.S.C. 2011 et seq.) for purposes of establishing and maintaining an on-site physical protection system and security organization required by Federal law, or possession by an employee or contractor of such licensee on-site for such purposes or off-site for purposes of licensee-authorized training or transportation of nuclear materials;

(8) The possession, by an individual who is retired from service with a law enforcement agency and is not otherwise prohibited from receiving a firearm, of a semiautomatic assault weapon transferred to the individual by the agency upon such retirement;

(9) The manufacture, transfer, or possession of a semiautomatic assault weapon by a licensed manufacturer or licensed importer for the purposes of testing or experimentation as authorized by the Director under the provisions of § 478.153; or

(10) The manufacture, transfer, or possession of a semiautomatic assault weapon by a licensed manufacturer, licensed importer, or licensed dealer for the purpose of exportation in compliance with the Arms Export Control Act (22 U.S.C. 2778).

(c) Manufacture and dealing in semiautomatic assault weapons. Subject to compliance with the provisions of this part, licensed manufacturers and licensed dealers in semiautomatic assault weapons may manufacture and deal in such weapons manufactured after September 13, 1994: **Provided**, The licensee obtains evidence that the weapons will be disposed of in accordance with paragraph (b) of this section. Examples of acceptable evidence include the following:

(1) Contracts between the manufacturer and dealers stating that the weapons may only be sold to law enforcement agencies, law enforcement officers, or other purchasers specified in paragraph (b) of this section;

(2) Copies of purchase orders submitted to the manufacturer or dealer by law enforcement agencies or other purchasers specified in paragraph (b) of this section;

(3) Copies of letters submitted to the manufacturer or dealer by government agencies, law enforcement officers, or other purchasers specified in paragraph (b) of this section expressing an interest in purchasing the semiautomatic assault weapons;

(4) Letters from dealers to the manufacturer stating that sales will only be made to law enforcement agencies, law enforcement officers, or other purchasers specified in paragraph (b) of this section; and

(5) Letters from law enforcement officers purchasing in accordance with paragraph (b)(6) of this section and § 478.132.

(Paragraph (c) approved by the Office of Management and Budget under control number 1512–0526)

§ 478.40a Transfer and possession of large capacity ammunition feeding devices.

(a) Prohibition. No person shall transfer or possess a large capacity ammunition feeding device.

(b) Exceptions. The provisions of paragraph (a) of this section shall not apply to:

(1) The possession or transfer of any large capacity ammunition feeding device otherwise lawfully possessed on September 13, 1994;

(2) The manufacture for, transfer to, or possession by the United States or a department or agency of the United States or a State or a department, agency, or political subdivision of a State, or a transfer to or possession by a law enforcement officer employed by such an entity for purposes of law enforcement;

(3) The transfer to a licensee under title I of the Atomic Energy Act of 1954 for purposes of establishing and maintaining an on-site physical protection system and security organization required by Federal law, or possession by an employee or contractor of such licensee on-site for such purposes or off-site for purposes of licensee-authorized training or transportation of nuclear materials;

(4) The possession, by an individual who is retired from service with a law enforcement agency and is not otherwise prohibited from receiving ammunition, of a large capacity ammunition feeding device transferred to the individual by the agency upon such retirement;

(5) The manufacture, transfer, or possession of any large capacity ammunition feeding device by a manufacturer or importer for the purposes of testing or experimentation in accordance with § 478.153; or

(6) The manufacture, transfer, or possession of any large capacity ammunition feeding device by a manufacturer or importer for the purpose of exportation in accordance with the Arms Export Control Act (22 U.S.C. 2778).

(c) Importation, manufacture, and dealing in large capacity ammunition feeding devices. Possession and transfer of large capacity ammunition feeding devices by persons who manufacture, import, or deal in such devices will be presumed to be lawful if such persons maintain evidence establishing that the devices are possessed and transferred for sale to purchasers specified in paragraph (b) of this section. Examples of acceptable evidence include the following:

(1) Contracts between persons who import or manufacture such devices and persons who deal in such devices stating that the devices may only be sold to law enforcement agencies or other purchasers specified in paragraph (b) of this section;

(2) Copies of purchase orders submitted to persons who manufacture, import, or deal in such devices by law enforcement agencies or other purchasers specified in paragraph (b) of this section;

(3) Copies of letters submitted to persons who manufacture, import, or deal in such devices by government agencies or other purchasers specified in paragraph (b) of this section expressing an interest in purchasing the devices;

(4) Letters from persons who deal in such devices to persons who import or manufacture such devices stating that sales will only be made to law enforcement agencies or other purchasers specified in paragraph (b) of this section; and

(5) Letters from law enforcement officers purchasing in accordance with paragraph (b)(2) of this section and § 478.132.

(Paragraph (c) approved by the Office of Management and Budget under control number 1512–0526)

Subpart D—Licenses

§ 478.41 General.

(a) Each person intending to engage in business as an importer or manufacturer of firearms or ammunition, or a dealer in firearms shall, before commencing such business, obtain the license required by this subpart for the business to be operated. Each person who desires to obtain a license as a collector of curios or relics may obtain such a license under the provisions of this subpart.

(b) Each person intending to engage in business as a firearms or ammunition importer or manufacturer, or dealer in firearms shall file an application, with the required fee (see § 478.42), with ATF in accordance with the instructions on the form (see § 478.44), and, pursuant to § 478.47, receive the license required for such business from the Chief, National Licensing Center. Except as provided in § 478.50, a license must be obtained for each business and each place at which the applicant is to do business. A license as an importer or manufacturer of firearms or ammunition, or a dealer in firearms shall, subject to the provisions of the Act and other applicable provisions of law, entitle the licensee to transport, ship, and receive firearms and ammunition covered by such license in interstate or foreign commerce and to engage in the business specified by the license, at the location described on the license, and for the period stated on the license. However, it shall not be necessary for a licensed importer or a licensed manufacturer to also obtain a dealer's license in order to engage in business on the licensed premises as a dealer in the same type of firearms authorized by the license to be imported or manufactured. Payment of the license fee as an importer or manufacturer of destructive devices, ammunition for destructive devices or armor piercing ammunition or as a dealer in destructive devices includes the privilege of importing or manufacturing firearms other than destructive devices and ammunition for other than destructive devices or ammunition other than armor piercing ammunition, or dealing in firearms other than destructive devices, as the case may be, by such a licensee at the licensed premises.

(c) Each person seeking the privileges of a collector licensed under this part shall file an application, with the required fee (see § 478.42), with ATF in accordance with the instructions on the form (see § 478.44), and pursuant to § 478.47, receive from the Chief, National Licensing Center, the license covering the collection of curios and relics. A separate license may be obtained for each collection premises, and such license shall, subject to the provisions of the Act and other applicable provisions of law, entitle the licensee to transport, ship, receive, and acquire curios and relics in interstate or foreign commerce, and to make disposition of curios and relics in interstate or foreign commerce, to any other person licensed under the provisions of this part, for the period stated on the license.

(d) The collector license provided by this part shall apply only to transactions related to a collector's activity in acquiring, holding or disposing of curios and relics. A collector's license does not authorize the collector to engage in a business required to be licensed under the Act or this part. Therefore, if the acquisitions and dispositions of curios and relics by a collector bring the collector within the definition of a manufacturer, importer, or dealer under this part, he shall qualify as such. (See also § 478.93 of this part.)

§ 478.42 License fees.

Each applicant shall pay a fee for obtaining a firearms license or ammunition license, a separate fee being required for each business or collecting activity at each place of such business or activity, as follows:

(a) For a manufacturer:

(1) Of destructive devices, ammunition for destructive devices or armor piercing ammunition—$1,000 per year.

(2) Of firearms other than destructive devices—$50 per year.

(3) Of ammunition for firearms other than ammunition for destructive devices or armor piercing ammunition—$10 per year.

(b) For an importer:

(1) Of destructive devices, ammunition for destructive devices or armor piercing ammunition—$1,000 per year.

(2) Of firearms other than destructive devices or ammunition for firearms other than destructive devices or ammunition other than armor piercing ammunition—$50 per year.

(c) For a dealer:

(1) In destructive devices—$1,000 per year.

(2) Who is not a dealer in destructive devices—$200 for 3 years, except that the fee for renewal of a valid license shall be $90 for 3 years.

(d) For a c ollector of curios and relics—$10 per year.

§ 478.43 License fee not refundable.

No refund of any part of the amount paid as a license fee shall be made where the operations of the licensee are, for any reason, discontinued during the period of an issued license. However, the license fee submitted with an application for a license shall be refunded if that application is denied or withdrawn by the applicant prior to being acted upon.

§ 478.44 Original license.

(a) (1) Any person who intends to engage in business as a firearms or ammunition importer or manufacturer, or firearms dealer, or who has not previously been licensed under the provisions of this part to so engage in business, or who has not timely submitted an application for renewal of the previous license issued under this part, must file an application for license, ATF Form 7 (Firearms), in duplicate, with ATF in accordance with the instructions on the form. The application must:

(i) Be executed under the penalties of perjury and the penalties imposed by 18 U.S.C. 924;

(ii) Include a photograph and fingerprints as required in the instructions on the form;

(iii) If the applicant (including, in the case of a corporation, partnership, or association, any individual possessing, directly or indirectly, the power to direct or cause the direction of the management and policies of the corporation, partnership, or association) is a nonimmigrant alien, applicable documentation demonstrating that the nonimmigrant alien falls within an exception to or has obtained a waiver from the nonimmigrant alien provision (e.g., a hunting license or permit lawfully issued in the United States; waiver);

(iv) Be accompanied by a completed ATF Form 5300.37 (Certification of Compliance with State and Local Law) and ATF Form 5300.36 (Notification of Intent to Apply for a Federal Firearms License); and

(v) Include the appropriate fee in the form of money order or check made payable to the Bureau of Alcohol, Tobacco and Firearms.

(2) ATF Forms 7 (Firearms), ATF Forms 5300.37, and ATF Forms 5300.36 may be obtained by contacting any ATF office.

(b) Any person who desires to obtain a license as a collector under the Act and this part, or who has not timely submitted an application for renewal of the previous license issued under this part, shall file an application, ATF Form 7CR (Curios and Relics), with ATF in accordance with the instructions on the form. If the applicant (including, in the case of a corporation, partnership, or association, any individual possessing, directly or indirectly, the power to direct or cause the direction of the management and policies of the corporation, partnership, or association) is a nonimmigrant alien, the application must include applicable documentation demonstrating that the nonimmigrant alien falls within an exception to or has obtained a waiver from the nonimmigrant alien provision (e.g., a hunting license or permit lawfully issued in the United States; waiver). The application must be executed under the penalties of perjury and the penalties imposed by 18 U.S.C. 924. The application shall be accompanied by a completed ATF Form 5300.37 and ATF Form 5300.36 and shall include the appropriate fee in the form of a money order or check made payable to the Bureau of Alcohol, Tobacco and Firearms. ATF Forms 7CR (Curios and Relics), ATF Forms 5300.37, and ATF Forms 5300.36 may be obtained by contacting any ATF office.

(Paragraphs (a) and (b) approved by the Office of Management and Budget under control number 1512–0570)

§ 478.45 Renewal of license.

If a licensee intends to continue the business or activity described on a license issued under this part during any portion of the ensuing year, the licensee shall, unless otherwise notified in writing by the Chief, National Licensing Center, execute and file with ATF prior to the expiration of the license an application for a license renewal, ATF Form 8 Part II, accompanied by a completed ATF Form 5300.37 and ATF Form 5300.36, in accordance with the instructions on the forms, and the required fee. If the applicant is a nonimmigrant alien, the application must include applicable documentation demonstrating that the nonimmigrant alien falls within an exception to or has obtained a waiver from the nonimmigrant alien provision (e.g., a hunting license or permit lawfully issued in the United States; waiver). The Chief, National Licensing Center, may, in writing, require the applicant for license renewal to also file completed ATF Form 7 or ATF Form 7CR in the manner required by § 478.44. In the event the licensee does not timely file an ATF Form 8 Part II, the licensee must file an ATF Form 7 or ATF Form 7CR as required by § 478.44, and obtain the required license before continuing business or collecting activity. If an ATF Form 8 Part II is not timely received through the mails, the licensee should so notify the Chief, National Licensing Center.

(Approved by the Office of Management and Budget under control number 1512–0570)

§ 478.46 Insufficient fee.

If an application is filed with an insufficient fee, the application and any fee submitted will be returned to the applicant.

§ 478.47 Issuance of license.

(a) Upon receipt of a properly executed application for a license on ATF Form 7, ATF Form 7CR, or ATF Form 8 Part II, the Chief, National Licensing Center, shall, upon finding through further inquiry or investigation, or otherwise, that the applicant is qualified, issue the appropriate license. Each license shall bear a serial number and such number may be assigned to the licensee to whom issued for so long as the licensee maintains continuity of renewal in the same location (State).

(b) The Chief, National Licensing Center, shall approve a properly executed application for license on ATF Form 7, ATF Form 7CR, or ATF Form 8 Part II, if:

(1) The applicant is 21 years of age or over;

(2) The applicant (including, in the case of a corporation, partnership, or association, any individual possessing, directly or indirectly, the power to direct or cause the direction of the management and policies of the corporation, partnership, or association) is not prohibited under the provisions of the Act from shipping or transporting in interstate or foreign commerce, or possessing in or affecting commerce, any firearm or ammunition, or from receiving any firearm or ammunition which has been shipped or transported in interstate or foreign commerce;

(3) The applicant has not willfully violated any of the provisions of the Act or this part;

(4) The applicant has not willfully failed to disclose any material information required, or has not made any false statement as to any material fact, in connection with his application;

(5) The applicant has in a State

(i) premises from which he conducts business subject to license under the Act or from which he intends to conduct such business within a reasonable period of time, or

(ii) in the case of a collector, premises from which he conducts his collecting subject to license under the Act or from which he intends to conduct such collecting within a reasonable period of time; and

(6) The applicant has filed an ATF Form 5300.37 (Certification of Compliance with State and Local Law) with ATF in accordance with the instructions on the form certifying under the penalties of perjury that—

(i) The business to be conducted under the license is not prohibited by State or local law in the place where the licensed premises are located;

(ii) Within 30 days after the application is approved the business will comply with the requirements of State and local law applicable to the conduct of business;

(iii) The business will not be conducted under the license until the requirements of State and local law applicable to the business have been met; and

(iv) The applicant has completed and sent or delivered ATF F 5300.36 (Notification of Intent to Apply for a Federal Firearms License) to the chief law enforcement officer of the locality in which the premises are located, which indicates that the applicant intends to apply for a Federal firearms license. For purposes of this paragraph, the **"chief law enforcement officer"** is the chief of police, the sheriff, or an equivalent officer.

(c) The Chief, National Licensing Center, shall approve or the Director of Industry Operations shall deny an application for a license within the 60-day period beginning on the date the properly executed application was received: **Provided,** That when an applicant for license renewal is a person who is, pursuant to the provisions of § 478.78, § 478.143, or § 478.144, conducting business or collecting activity under a previously issued license, action regarding the application will be held in abeyance pending the completion of the proceedings against the applicant's existing license or license application, final determination of the applicant's criminal case, or final action by the Director on an application for relief submitted pursuant to § 478.144, as the case may be.

(d) When the Director of Industry Operations or the Chief, National Licensing Center fails to act on an application for a license within the 60-day period prescribed by paragraph (c) of this section, the applicant may file an action under section 1361 of title 28, United States Code, to compel ATF to act upon the application.

(Paragraph (b)(6) approved by the Office of Management and Budget under control numbers 1512–0522 and 1512–0523)

§ 478.48 Correction of error on license.

(a) Upon receipt of a license issued under the provisions of this part, each licensee shall examine same to ensure that the information contained thereon is accurate. If the license is incorrect, the licensee shall return the license to the Chief, National Licensing Center, with a statement showing the nature of the error. The Chief, National Licensing Center, shall correct the error, if the error was made in his office, and return the license. However, if the error resulted from information contained in the licensee's application for the license, the Chief, National Licensing Center, shall require the licensee to file an amended application setting forth the correct information and a statement explaining the error contained in the application. Upon receipt of the amended application and a satisfactory explanation of the error, the Chief, National Licensing Center, shall make the correction on the license and return same to the licensee.

(b) When the Chief, National Licensing Center, finds through any means other than notice from the licensee that an incorrect license has been issued, the Chief, National Licensing Center, may require the holder of the incorrect license to:

(1) return the license for correction, and

(2) if the error resulted from information contained in the licensee's application for the license, the Chief, National Licensing Center, shall require the licensee to file an amended application setting forth the correct information, and a statement explaining the error contained in the application. The Chief, National Licensing Center, then shall make the correction on the license and return same to the licensee.

§ 478.49 Duration of license.

The license entitles the person to whom issued to engage in the business or activity specified on the license, within the limitations of the Act and the regulations contained in this part, for a three-year period, unless terminated sooner.

§ 478.50 Locations covered by license.

The license covers the class of business or the activity specified in the license at the address specified therein. A separate license must be obtained for each location at which a firearms or ammunition business or activity requiring a license under this part is conducted except:

(a) No license is required to cover a separate warehouse used by the licensee solely for storage of firearms or ammunition if the records required by this part are maintained at the licensed premises served by such warehouse;

(b) A licensed collector may acquire curios and relics at any location, and dispose of curios or relics to any licensee or to other persons who are residents of the State where the collector's license is held and the disposition is made;

(c) A licensee may conduct business at a gun show pursuant to the provisions of § 478.100; or

(d) A licensed importer, manufacturer, or dealer may engage in the business of dealing in curio or relic firearms with another licensee at any location pursuant to the provisions of § 478.100.

§ 478.51 License not transferable.

Licenses issued under this part are not transferable. In the event of the lease, sale, or other transfer of the operations authorized by the license, the successor must obtain the license required by this part prior to commencing such operations. However, for rules on right of succession, see § 478.56.

§ 478.52 Change of address.

(a) Licensees may during the term of their current license remove their business or activity to a new location at which they intend regularly to carry on such business or activity by filing an Application for an Amended Federal Firearms License, ATF Form 5300.38, in duplicate, not less than 30 days prior to such removal with the Chief, National Licensing Center. The ATF Form 5300.38 shall be completed in accordance with the instructions on the form. The application must be executed under the penalties of perjury and penalties imposed by 18 U.S.C. 924. The application shall be accompanied by the licensee's original license. The Chief, National Licensing Center, may, in writing, require the applicant for an amended license to also file completed ATF Form 7 or ATF Form 7CR, or portions thereof, in the manner required by § 478.44.

(b) Upon receipt of a properly executed application for an amended license, the Chief, National Licensing Center, shall, upon finding through further inquiry or investigation, or otherwise, that the applicant is qualified at the new location, issue the amended license, and return it to the applicant. The license shall be valid for the remainder of the term of the original license. The Chief, National Licensing Center, shall, if the applicant is not qualified, refer the application for amended license to the Director of Industry Operations for denial in accordance with § 478.71.

(Approved by the Office of Management and Budget under control number 1512–0525)

§ 478.53 Change in trade name.

A licensee continuing to conduct business at the location shown on his license is not required to obtain a new license by reason of a mere change in trade name under which he conducts his business: **Provided,** That such licensee furnishes his license for endorsement of such change to the Chief, National Licensing Center, within 30 days from the date the licensee begins his business under the new trade name.

§ 478.54 Change of control.

In the case of a corporation or association holding a license under this part, if actual or legal control of the corporation or association changes, directly or indirectly, whether by reason of change in stock ownership or control (in the licensed corporation or in any other corporation), by operations of law, or in any other manner, the licensee shall, within 30 days of such change, give written notification thereof, executed under the penalties of perjury, to the Chief, National Licensing Center. Upon expiration of the license, the corporation or association must file a Form 7 (Firearms) as required by § 478.44.

§ 478.55 Continuing partnerships.

Where, under the laws of the particular State, the partnership is not terminated on death or insolvency of a partner, but continues until the winding up of the partnership affairs is completed, and the surviving partner has the exclusive right to the control and possession of the partnership assets for the purpose of liquidation and settlement, such surviving partner may continue to operate the business under the license of the partnership. If such surviving partner acquires the business on completion of the settlement of the partnership, he shall obtain a license in his own name from the date of acquisition, as provided in § 478.44. The rule set forth in this section shall also apply where there is more than one surviving partner.

§ 478.56 Right of su ccession by certain persons.

(a) Certain persons other than the licensee may secure the right to carry on the same firearms or ammunition business at the same address shown on, and for the remainder of the term of, a current license. Such persons are:

(1) The surviving spouse or child, or executor, administrator, or other legal representative of a deceased licensee; and

(2) A receiver or trustee in bankruptcy, or an assignee for benefit of creditors.

(b) In order to secure the right provided by this section, the person or persons continuing the business shall furnish the license for that business for endorsement of such succession to the Chief, National Licensing Center, within 30 days from the date on which the successor begins to carry on the business.

§ 478.57 Discontinuance of business.

(a) Where a firearm or ammunition business is either discontinued or succeeded by a new owner, the owner of the business discontinued or succeeded shall within 30 days thereof furnish to the Chief, National Licensing Center notification of the discontinuance or succession. (See also § 478.127.)

(b) Since section 922(v), Title 18, U.S.C., makes it unlawful to transfer or possess a semiautomatic assault weapon, except as provided in the law, any licensed manufacturer, licensed importer, or licensed dealer intending to discontinue business shall, prior to going out of business, transfer in compliance with the provisions of this part any semiautomatic assault weapon manufactured or imported after September 13, 1994, to a person specified in § 478.40(b), or, subject to the provisions of §§ 478.40(c) and 478.132, a licensed manufacturer, a licensed importer, or a licensed dealer.

(c) Since section 922(w), Title 18, U.S.C., makes it unlawful to transfer or possess a large capacity ammunition feeding device, except as provided in the law, any person who manufactures, imports, or deals in such devices and who intends to discontinue business shall, prior to going out of business, transfer in compliance with the provisions of this part any large capacity ammunition feeding device manufactured after September 13, 1994, to a person specified in § 478.40a(b), or, subject to the provisions of §§ 478.40a(c) and 478.132, a person who manufactures, imports, or deals in such devices.

§ 478.58 State or other law.

A license issued under this part confers no right or privilege to conduct business or activity contrary to State or other law. The holder of such a license is not by reason of the rights and privileges granted by that license immune from punishment for operating a firearm or ammunition business or activity in violation of the provisions of any State or other law. Similarly, compliance with the provisions of any State or other law affords no immunity under Federal law or regulations.

§ 478.59 Abandoned application.

Upon receipt of an incomplete or improperly executed application on ATF form 7 (5310.12), or ATF Form 8 (5310.11) Part II, the applicant shall be notified of the deficiency in the application. If the application is not corrected and returned within 30 days following the date of notification, the application shall be considered as having been abandoned and the license fee returned.

§ 478.60 Certain contin uances of business.

A licensee who furnishes his license to the Chief, National Licensing Center for correction or endorsement in compliance with the provisions contained in this subpart may continue his operations while awaiting its return.

Subpart E—License Proceedings

§ 478.71 Den ial of an application for license.

Whenever the Director of Industry Operations has reason to believe that an applicant is not qualified to receive a license under the provisions of § 478.47, he may issue a notice of denial, on Form 4498, to the applicant. The notice shall set forth the matters of fact and law relied upon in determining that the application should be denied, and shall afford the applicant 15 days from the date of receipt of the notice in which to request a hearing to review the denial. If no request for a hearing is filed within such time, the application shall be disapproved and a copy, so marked, shall be returned to the applicant.

§ 478.72 Hear ing after appli cation denial.

If the applicant for an original or renewal license desires a hearing to review the denial of his application, he shall file a request therefor, in duplicate, with the Director of Industry Operations within 15 days after receipt of the notice of denial. The request should include a statement of the reasons therefor. On receipt of the request, the Director of Industry Operations shall, as expeditiously as possible, make the necessary arrangements for the hearing and advise the applicant of the date, time, location, and the name of the officer before whom the hearing will be held. Such notification shall be made not less than 10 days in advance of the date set for the hearing. On conclusion of the hearing and consideration of all relevant facts and circumstances presented by the applicant or his representative, the Director of Industry Operations shall render his decision confirming or reversing the denial of the application. If the decision is that the denial should stand, a certified copy of the Director of Industry Operations findings and conclusions shall be furnished to the applicant with a final notice of denial, Form 4501. A copy of the application, marked "Disapproved," will be returned to the applicant. If the decision is that the license applied for should be issued, the applicant shall be so notified, in writing, and the license shall be issued as provided by § 478.47.

§ 478.73 Notice of revocation, suspension, or imposition of civil fine.

(a) Basis for action. Whenever the Director of Industry Operations has reason to believe that a licensee has willfully violated any provision of the Act or this part, a notice of revocation of the license, ATF Form 4500, may be issued. In addition, a notice of revocation, suspension, or imposition of a civil fine may be issued on ATF Form 4500 whenever the Director of Industry Operations has reason to believe that a licensee has knowingly transferred a firearm to an unlicensed person and knowingly failed to comply with the requirements of 18 U.S.C. 922(t)(1) with respect to the transfer and, at the time that the transferee most recently proposed the transfer, the national instant criminal background check system was operating and information was available to the system demonstrating that the transferee's receipt of a firearm would violate 18 U.S.C. 922(g) or 922(n) or State law.

(b) Issuance of notice. The notice shall set forth the matters of fact constituting the violations specified, dates, places, and the sections of law and regulations violated. The Director of Industry Operations shall afford the licensee 15 days from the date of receipt of the notice in which to request a hearing prior to suspension or revocation of the license, or imposition of a civil fine. If the licensee does not file a timely request for a hearing, the Director of Industry Operations shall issue a final notice of suspension or revocation and/or imposition of a civil fine on ATF Form 4501, as provided in § 478.74.

§ 478.74 Request for hearing after notice of suspension, revocation, or imposition of civil fine.

If a licensee desires a hearing after receipt of a notice of suspension or revocation of a license, or imposition of a civil fine, the licensee shall file a request, in duplicate, with the Director of Industry Operations within 15 days after receipt of the notice of suspension or revocation of a license, or imposition of a civil fine. On receipt of such request, the Director of Industry Operations shall, as expeditiously as possible, make necessary arrangements for the hearing and advise the licensee of the date, time, location and the name of the officer before whom the hearing will be held. Such notification shall be made no less than 10 days in advance of the date set for the hearing. On conclusion of the hearing and consideration of all the relevant presentations made by the licensee or the licensee's representative, the Director of Industry Operations shall render a decision and shall prepare a brief summary of the findings and conclusions on which the decision is based. If the decision is that the license should be revoked, or, in actions under 18 U.S.C. 922(t)(5), that the license should be revoked or suspended, and/or that a civil fine should be imposed, a certified copy of the summary shall be furnished to the licensee with the final notice of revocation, suspension, or imposition of a civil fine on ATF Form 4501. If the decision is that the license should not be revoked, or in actions under 18 U.S.C. 922(t)(5), that the license should not be revoked or suspended, and a civil fine should not be imposed, the licensee shall be notified in writing.

§ 478.75 Service on applicant or licensee.

All notices and other documents required to be served on an applicant or licensee under this subpart shall be served by certified mail or by personal delivery. Where service is by certified mail, a signed duplicate original copy of the formal document shall be mailed, with return receipt requested, to the applicant or licensee at the address stated in his application or license, or at his last known address. Where service is by personal delivery, a signed duplicate original copy of the formal document shall be delivered to the applicant or licensee, or, in the case of a corporation, partnership, or association, by delivering it to an officer, manager, or general agent thereof, or to its attorney of record.

§ 478.76 Representation at a hearing.

An applicant or licensee may be represented by an attorney, certified public accountant, or other person recognized to practice before the Bureau of Alcohol, Tobacco and Firearms, as provided in 31 CFR Part 8 (Practice Before the Bureau of Alcohol, Tobacco and Firearms), if he has otherwise complied with the applicable requirements of 26 CFR 601.521 through 601.527 (conference and practice requirements for Alcohol, Tobacco and Firearms activities) of this chapter. The Director of Industry Operations may be represented in proceedings by an attorney in the office of the Assistant Chief Counsel or Division Counsel who is authorized to execute and file motions, briefs and other papers in the proceeding, on behalf of the Director of Industry Operations, in his own name as "Attorney for the Government."

§ 478.77 Designated place of hearing.

The designated place of the hearing shall be a location convenient to the aggrieved party.

§ 478.78 Operations by licensee after notice.

In any case where denial, suspension, or revocation proceedings are pending before the Bureau of Alcohol, Tobacco and Firearms, or notice of denial, suspension, or revocation has been served on the licensee and he has filed timely request for a hearing, the license in the possession of the licensee shall remain in effect even though such license has expired, or the suspension or revocation date specified in the notice of revocation on Form 4500 served on the licensee has passed: **Provided,** That with respect to a license that has expired, the licensee has timely filed an application for the renewal of his license. If a licensee is dissatisfied with a posthearing decision revoking or suspending the license or denying the application or imposing a civil fine, as the case may be, he may, pursuant to 18 U.S.C. 923(f)(3), within 60 days after receipt of the final notice denying the application or revoking or suspending the license or imposing a civil fine, file a petition for judicial review of such action. Such petition should be filed with the U.S. district court for the district in which the applicant or licensee resides or has his principal place of business. In such case, when the Director of Industry Operations finds that justice so requires, he may postpone the effective date of suspension or revocation of a license or authorize continued operations under the expired license, as applicable, pending judicial review.

Subpart F—Conduct of Business

§ 478.91 Posting of license.

Any license issued under this part shall be kept posted and kept available for inspection on the premises covered by the license.

§ 478.92 How must licensed manufacturers and licensed importers identify firearms, armor piercing ammunition, and large capacity ammunition feeding devices?

(a) (1) Firearms. You, as a licensed manufacturer or licensed importer of firearms, must legibly identify each firearm manufactured or imported as follows:

(i) By engraving, casting, stamping (impressing), or otherwise conspicuously placing or causing to be engraved, cast, stamped (impressed) or placed on the frame or receiver thereof an individual serial number. The serial number must be placed in a manner not susceptible of being readily obliterated, altered, or removed, and must not duplicate any serial number placed by you on any other firearm. For firearms manufactured or imported on and after January 30, 2002, the engraving, casting, or stamping (impressing) of the serial number must be to a minimum depth of .003 inch and in a print size no smaller than 1/16 inch; and

(ii) By engraving, casting, stamping (impressing), or otherwise conspicuously placing or causing to be engraved, cast, stamped (impressed) or placed on the frame, receiver, or barrel thereof certain additional information. This information must be placed in a manner not susceptible of being readily obliterated, altered, or removed. For firearms manufactured or imported on and after January 30, 2002, the engraving, casting, or stamping (impressing) of this information must be to a minimum depth of

.003 inch. The additional information includes:

(A) The model, if such designation has been made;

(B) The caliber or gauge;

(C) Your name (or recognized abbreviation) and also, when applicable, the name of the foreign manufacturer;

(D) In the case of a domestically made firearm, the city and State (or recognized abbreviation thereof) where you as the manufacturer maintain your place of business; and

(E) In the case of an imported firearm, the name of the country in which it was manufactured and the city and State (or recognized abbreviation thereof) where you as the importer maintain your place of business. For additional requirements relating to imported firearms, see Customs regulations at 19 CFR part 134.

(2) Firearm frames or receivers. A firearm frame or receiver that is not a component part of a complete weapon at the time it is sold, shipped, or otherwise disposed of by you must be identified as required by this section.

(3) Special markings for semiautomatic assault weapons, effective July 5, 1995. In the case of any semiautomatic assault weapon manufactured after September 13, 1994, you must mark the frame or receiver **"RESTRICTED LAW ENFORCEMENT/GOVERNMENT USE ONLY"** or, in the case of weapons manufactured for export, **"FOR EXPORT ONLY,"** in a manner not susceptible of being readily obliterated, altered, or removed. For weapons manufactured or imported on and after January 30, 2002, the engraving, casting, or stamping (impressing) of the special markings prescribed in this paragraph (a)(3) must be to a minimum depth of .003 inch.

(4) Exceptions.

(i) Alternate means of identification. The Director may authorize other means of identification upon receipt of a letter application from you, submitted in duplicate, showing that such other identification is reasonable and will not hinder the effective administration of this part.

(ii) Destructive devices. In the case of a destructive device, the Director may authorize other means of identifying that weapon upon receipt of a letter application from you, submitted in duplicate, showing that engraving, casting, or stamping (impressing) such a weapon would be dangerous or impracticable.

(iii) Machine guns, silencers, and parts. Any part defined as a machine gun, firearm muffler, or firearm silencer in § 478.11, that is not a component part of a complete weapon at the time it is sold, shipped, or otherwise disposed of by you, must be identified as required by this section. The Director may authorize other means of identification of parts defined as machine guns other than frames or receivers and parts defined as mufflers or silencers upon receipt of a letter application from you, submitted in duplicate, showing that such other identification is reasonable and will not hinder the effective administration of this part.

(5) Measurement of height and depth of markings. The depth of all markings required by this section will be measured from the flat surface of the metal and not the peaks or ridges. The height of serial numbers required by paragraph (a)(1)(i) of this section will be measured as the distance between the latitudinal ends of the character impression bottoms (bases).

(b) Armor piercing ammunition.

(1) Marking of ammunition. Each licensed manufacturer or licensed importer of armor piercing ammunition shall identify such ammunition by means of painting, staining or dying the exterior of the projectile with an opaque black coloring. This coloring must completely cover the point of the projectile and at least 50 percent of that portion of the projectile which is visible when the projectile is loaded into a cartridge case.

(2) Labeling of packages. Each licensed manufacturer or licensed importer of armor piercing ammunition shall clearly and conspicuously label each package in which armor piercing ammunition is contained, e.g., each box, carton, case, or other container. The label shall include the words **"ARMOR PIERCING"** in block letters at least 1/4 inch in height. The lettering shall be located on the exterior surface of the package which contains information concerning the caliber or gauge of the ammunition. There shall also be placed on the same surface of the package in block lettering at least 1/8 inch in height the words **"FOR GOVERNMENTAL ENTITIES OR EXPORTATION ONLY."** The statements required by this subparagraph shall be on a contrasting background.

(c) Large capacity ammunition feeding devices manufactured after September 13, 1994. (1) Each person who manufactures or imports any large capacity ammunition feeding device manufactured after September 13, 1994, shall legibly identify each such device with a serial number. Such person may use the same serial number for all large capacity ammunition feeding devices produced.

(i) Additionally, in the case of a domestically made large capacity ammunition feeding device, such device shall be marked with the name, city and State (or recognized abbreviation thereof) of the manufacturer;

(ii) And in the case of an imported large capacity ammunition feeding device, such device shall be marked:

(A) With the name of the manufacturer, country of origin, and,

(B) Effective July 5, 1995, the name, city and State (or recognized abbreviation thereof) of the importer.

(iii) Further, large capacity ammunition feeding devices manufactured after September 13, 1994, shall be marked **"RESTRICTED LAW ENFORCEMENT/GOVERNMENT USE ONLY"** or, in the case of devices manufactured or imported for export, effective July 5, 1995, "FOR EXPORT ONLY."

(2) All markings required by this paragraph (c) shall be cast, stamped, or engraved on the exterior of the device. In the case of a magazine, the markings shall be placed on the magazine body.

(3) Exceptions—(i) Metallic links. Persons who manufacture or import metallic links for use in the assembly of belted ammunition are only required to place the identification marks prescribed in paragraph (c)(1) of this section on the containers used for the packaging of the links.

(ii) Alternate means of identification. The Director may authorize other means of identifying large capacity ammunition feeding devices upon receipt of a letter application, in duplicate, from the manufacturer or importer showing that such other identification is reasonable and will not hinder the effective administration of this part.

(Approved by the Office of Management and Budget under control number 1512–0550)

§ 478.93 A uthorized o perations by a licensed collector.

The license issued to a collector of curios or relics under the provisions of this part shall cover only transactions by the licensed collector in curios and relics. The collector's license is of no force or effect and a licensed collector is of the same status under the Act and this part as a nonlicensee with respect to **(a)** any acquisition or disposition of firearms other than curios or relics, or any transportation, shipment, or receipt of firearms other than curios or relics in interstate or foreign commerce, and **(b)** any transaction with a nonlicensee involving any firearm other than a curio or relic. (See also § 478.50.) A collectors license is not necessary to receive or dispose of ammunition, and a licensed collector is not precluded by law from receiving or disposing of armor piercing ammunition. However, a licensed collector may not dispose of any ammunition to a person prohibited from receiving or possessing ammunition (see § 478.99(c)). Any licensed collector who disposes of armor piercing ammunition must record the disposition as required by § 478.125 (a) and (b).

§ 478.94 Sales or deliveries bet ween licensees.

A licensed importer, licensed manufacturer, or licensed dealer selling or otherwise disposing of firearms, and a licensed collector selling or otherwise disposing of curios or relics, to another licensee shall verify the identity and licensed status of the transferee prior to making the transaction. Verification shall be established by the transferee furnishing to the transferor a certified copy of the transferee's license and by such other means as the transferor deems necessary: **Provided,** That it shall not be required **(a)** for a transferee who has furnished a certified copy of its license to a transferor to again furnish such certified copy to that transferor during the term of the transferee's current license, **(b)** for a licensee to furnish a certified copy of its license to another licensee if a firearm is being returned either directly or through another licensee to such licensee and **(c)** for licensees of multi-licensed business organizations to furnish certified copies of their licenses to other licensed locations operated by such organization: **Provided further,** That a multi-licensed business organization may furnish to a transferor, in lieu of a certified copy of each license, a list, certified to be true, correct and complete, containing the name, address, license number, and the date of license expiration of each licensed location operated by such organization, and the transferor may sell or otherwise dispose of firearms as provided by this section to any licensee appearing on such list without requiring a certified copy of a license therefrom. A transferor licensee who has the certified information required by this section may sell or dispose of firearms to a licensee for not more than 45 days following the expiration date of the transferee's license.

(Approved by the Office of Management and Budget under control number 1512–0387)

§ 478.95 Certified copy of license.

The license furnished to each person licensed under the provisions of this part contains a purchasing certification statement. This original license may be reproduced and the reproduction then certified by the licensee for use pursuant to § 478.94. If the licensee desires an additional copy of the license for certification (instead of making a reproduction of the original license), the licensee may submit a request, in writing, for a certified copy or copies of the license to the Chief, National Licensing Center. The request must set forth the name, trade name (if any) and address of the licensee, and the number of license copies desired. There is a charge of $1 for each copy. The fee paid for copies of the license must accompany the request for copies. The fee may be paid by **(a)** cash, or **(b)** money order or check made payable to the Bureau of Alcohol, Tobacco and Firearms.

(Approved by the Office of Management and Budget under control number 1512–0387)

§ 478.96 Out-of-State and mail order sales.

(a) The provisions of this section shall apply when a firearm is purchased by or delivered to a person not otherwise prohibited by the Act from purchasing or receiving it.

(b) A licensed importer, licensed manufacturer, or licensed dealer may sell a firearm that is not subject to the provisions of § 478.102(a) to a nonlicensee who does not appear in person at the licensee's business premises if the nonlicensee is a resident of the same State in which the licensee's business premises are located, and the nonlicensee furnishes to the licensee the firearms transaction record, Form 4473, required by § 478.424. The nonlicensee shall attach to such record a true copy of any permit or other information required pursuant to any statute of the State and published ordinance applicable to the locality in which he resides. The licensee shall prior to shipment or delivery of the firearm, forward by registered or certified mail (return receipt requested) a copy of the record, Form 4473, to the chief law enforcement officer named on such record, and delay shipment or delivery of the firearm for a period of at least 7 days following receipt by the licensee of the return receipt evidencing delivery of the copy of the record to such chief law enforcement officer, or the return of the copy of the record to him due to the refusal of such chief law enforcement officer to accept same in accordance with U.S. Postal Service regulations. The original Form 4473, and evidence of receipt or rejection of delivery of the copy of the Form 4473 sent to the chief law enforcement officer shall be retained by the licensee as a part of the records required of him to be kept under the provisions of subpart H of this part.

(c) **(1)** A licensed importer, licensed manufacturer, or licensed dealer may sell or deliver a rifle or shotgun, and a licensed collector may sell or deliver a rifle or shotgun that is a curio or relic to a nonlicensed resident of a State other than the State in which the licensee's place of business is located if—

(i) The purchaser meets with the licensee in person at the licensee's premises to accomplish the transfer, sale, and delivery of the rifle or shotgun;

(ii) The licensed importer, licensed manufacturer, or licensed dealer complies with the provisions of § 478.102;

(iii) The purchaser furnishes to the licensed importer, licensed manufacturer, or licensed dealer the firearms transaction record, Form 4473, required by § 478.124; and

(iv) The sale, delivery, and receipt of the rifle or shotgun fully comply with the legal conditions of sale in both such States.

(2) For purposes of paragraph (c) of this section, any licensed manufacturer, licensed importer, or licensed dealer is presumed, in the absence of evidence to the contrary, to have had actual knowledge of the State laws and published ordinances of both such States.

(Approved by the Office of Management and Budget under control number 1512–0130)

§ 478.97 Loan or rental of firearms.

(a) A licensee may lend or rent a firearm to any person for temporary use off the premises of the licensee for lawful sporting purposes: **Provided,** That the delivery of the firearm to such person is not prohibited by § 478.99(b) or § 478.99(c), the licensee complies with the requirements of § 478.102, and the licensee records such loan or rental in the records required to be kept by him under Subpart H of this part.

(b) A club, association, or similar organization temporarily furnishing firearms (whether by loan, rental, or otherwise) to participants in a skeet, trap, target, or similar shooting activity for use at the time and place such activity is held does not, unattended by other circumstances, cause such club, association, or similar organization to be engaged in the business of a dealer in firearms or as engaging in firearms transactions. Therefore, licensing and recordkeeping requirements contained in this part pertaining to firearms transactions would not apply to this temporary furnishing of firearms for use on premises on which such an activity is conducted.

§ 478.98 Sales or deliveries of destructive devices and certain firearms.

The sale or delivery by a licensee of any destructive device, machine gun, short-barreled shotgun, or short-barreled rifle, to any person other than another licensee who is licensed under this part to deal in such device or firearm, is prohibited unless the person to receive such device or firearm furnishes to the licensee a sworn statement setting forth

(a) The reasons why there is a reasonable necessity for such person to purchase or otherwise acquire the device or weapon; and

(b) That such person's receipt or possession of the device or weapon would be consistent with public safety. Such sworn statement shall be made on the application to transfer and register the firearm required by Part 479 of this chapter. The sale or delivery of the device or weapon shall not be made until the application for transfer is approved by the Director and returned to the licensee (transferor) as provided in Part 479 of this chapter.

§ 478.99 Certain prohibited sales or deliveries.

(a) Interstate sales or deliveries. A licensed importer, licensed manufacturer, licensed dealer, or licensed collector shall not sell or deliver any firearm to any person not licensed under this part and who the licensee knows or has reasonable cause to believe does not reside in (or if a corporation or other business entity, does not maintain a place of business in) the State in which the licensee's place of business or activity is located: **Provided,** that the foregoing provisions of this paragraph **(1)** shall not apply to the sale or delivery of a rifle or shotgun (curio or relic, in the case of a licensed collector) to a resident of a State other than the State in which the licensee's place of business or collection premises is located if the requirements of § 478.96(c) are fully met, and **(2)** shall not apply to the loan or rental of a firearm to any person for temporary use for lawful sporting purposes (see § 478.97).

(b) Sales or deliveries to underaged persons. A licensed importer, licensed manufacturer, licensed dealer, or licensed collector shall not sell or deliver **(1)** any firearm or ammunition to any individual who the importer, manufacturer, dealer, or collector knows or has reasonable cause to believe is less than 18 years of age, and, if the firearm, or ammunition, is other than a shotgun or rifle, or ammunition for a shotgun or rifle, to any individual who the importer, manufacturer, dealer, or collector knows or has reasonable cause to believe is less than 21 years of age, or **(2)** any firearm to any person in any State where the purchase or possession by such person of such firearm would be in violation of any State law or any published ordinance applicable at the place of sale, delivery, or other disposition, unless the importer, manufacturer, dealer, or collector knows or has reasonable cause to believe that the purchase or possession would not be in violation of such State law or such published ordinance.

(c) Sales or deliveries to prohibited categories of persons. A licensed manufacturer, licensed importer, licensed dealer, or licensed collector shall not sell or otherwise dispose of any firearm or ammunition to any person knowing or having reasonable cause to believe that such person:

(1) Is, except as provided by § 478.143, under indictment for, or, except as provided by § 478.144, has been convicted in any court of a crime punishable by imprisonment for a term exceeding 1 year;

(2) Is a fugitive from justice;

(3) Is an unlawful user of or addicted to any controlled substance (as defined in section 102 of the Controlled Substance Act, 21 U.S.C. 802);

(4) Has been adjudicated as a mental defective or has been committed to any mental institution;

(5) Is an alien illegally or unlawfully in the United States or, except as provided in § 478.32(f), is a nonimmigrant alien: Provided, That the provisions of this paragraph (c)(5) do not apply to any nonimmigrant alien if that alien is—

(i) Admitted to the United States for lawful hunting or sporting purposes or is in possession of a hunting license or permit lawfully issued in the United States;

(ii) An official representative of a foreign government who is either accredited to the United States Government or the Government's mission to an international organization having its headquarters in the United States or en route to or from another country to which that alien is accredited. This exception only applies if the firearm or ammunition is shipped, transported, possessed, or received in the representative's official capacity;

(iii) An official of a foreign government or a distinguished foreign visitor who has been so designated by the Department of State. This exception only applies if the firearm or ammunition is shipped, transported, possessed, or received in the official's or visitor's official capacity, except if the visitor is a private individual who does not have an official capacity; or

(iv) A foreign law enforcement officer of a friendly foreign government entering the United States on official law enforcement business;

(6) Has been discharged from the Armed Forces under dishonorable conditions;

(7) Who, having been a citizen of the United States, has renounced citizenship;

(8) Is subject to a court order that restrains such person from harassing, stalking, or threatening an intimate partner of such person or child of such intimate partner or person, or engaging in other conduct that would place an intimate partner in reasonable fear of bodily injury to the partner or child, except that this paragraph shall only apply to a court order that—

(i) Was issued after a hearing of which such person received actual notice, and at which such person had the opportunity to participate; and

(ii) **(A)** Includes a finding that such person represents a credible threat to the physical safety of such intimate partner or child; or

(B) By its terms explicitly prohibits the use, attempted use, or threatened use of physical force against such intimate partner or child that would reasonably be expected to cause bodily injury, or

(9) Has been convicted of a misdemeanor crime of domestic violence.

(d) Manufacture, importation, and sale of armor piercing ammunition by licensed importers and licensed manufacturers. A licensed importer or licensed manufacturer shall not import or manufacture armor piercing ammunition or sell or deliver such ammunition, except:

(1) For use of the United States or any department or agency thereof or any State or any department, agency, or political subdivision thereof;

(2) For the purpose of exportation; or

(3) For the purpose of testing or experimentation authorized by the Director under the provisions of § 478.149.

(e) Transfer of armor piercing ammunition by licensed dealers. A licensed dealer shall not willfully transfer armor piercing ammunition: **Provided,** That armor piercing ammunition received and maintained by the licensed dealer as business inventory prior to August 28, 1986, may be transferred to any department or agency of the United States or any State or political subdivision thereof if a record of such ammunition is maintained in the form and manner prescribed by § 478.125(c). Any licensed dealer who violates this paragraph is subject to license revocation. See subpart E of this part. For purposes of this paragraph, the Director shall furnish each licensed dealer information defining which projectiles are considered armor piercing. Such information may not be all-inclusive for purposes of the prohibition on manufacture, importation, or sale or delivery by a manufacturer or importer of such ammunition or 18 U.S.C. 929 relating to criminal misuse of such ammunition.

§ 478.100 Conduct of business away from licensed premises.

(a) (1) A licensee may conduct business temporarily at a gun show or event as defined in paragraph (b) if the gun show or event is located in the same State specified on the license: **Provided,** that such business shall not be conducted from any motorized or towed vehicle. The premises of the gun show or event at which the licensee conducts business shall be considered part of the licensed premises. Accordingly, no separate fee or license is required for the gun show or event locations. However, licensees shall comply with the provisions of § 478.91 relating to posting of licenses (or a copy thereof) while conducting business at the gun show or event.

(2) A licensed importer, manufacturer, or dealer may engage in the business of dealing in curio or relic firearms with another licensee at any location.

(b) A gun show or an event is a function sponsored by any national, State, or local organization, devoted to the collection, competitive use, or other sporting use of firearms, or an organization or association that sponsors functions devoted to the collection, competitive use, or other sporting use of firearms in the community.

(c) Licensees conducting business at locations other than the premises specified on their license under the provisions of paragraph (a) of this section shall maintain firearms records in the form and manner prescribed by subpart H of this part. In addition, records of firearms transactions conducted at such locations shall include the location of the sale or other disposition, be entered in the acquisition and disposition records of the licensee, and retained on the premises specified on the license.

§ 478.101 Record of transactions.

Every licensee shall maintain firearms and armor piercing ammunition records in such form and manner as is prescribed by subpart H of this part.

§ 478.102 Sales or deliveries of firearms on and after November 30, 1998.

(a) Background check. Except as provided in paragraph (d) of this section, a licensed importer, licensed manufacturer, or licensed dealer (the licensee) shall not sell, deliver, or transfer a firearm to any other person who is not licensed under this part unless the licensee meets the following requirements:

(1) Before the completion of the transfer, the licensee has contacted NICS;

(2)(i) NICS informs the licensee that it has no information that receipt of the firearm by the transferee would be in violation of Federal or State law and provides the licensee with a unique identification number; or

(ii) Three business days (meaning days on which State offices are open) have elapsed from the date the licensee contacted NICS and NICS has not notified the licensee that receipt of the firearm by the transferee would be in violation of law; and

(3) The licensee verifies the identity of the transferee by examining the identification document presented in accordance with the provisions of § 478.124(c).

Example for paragraph (a). A licensee contacts NICS on Thursday, and gets a "delayed" response. The licensee does not get a further response from NICS. If State offices are not open on Saturday and Sunday, 3 business days would have elapsed on the following Tuesday. The licensee may transfer the firearm on the next day, Wednesday.

(b) Transaction number. In any transaction for which a licensee receives a transaction number from NICS (which shall include either a NICS transaction number or, in States where the State is recognized as a point of contact for NICS checks, a State transaction number), such number shall be recorded on a firearms transaction record, Form 4473, which shall be retained in the records of the licensee in accordance with the provisions of § 478.129. This applies regardless of whether the transaction is approved or denied by NICS, and regardless of whether the firearm is actually transferred.

(c) Time limitation on NICS checks. A NICS check conducted in accordance with paragraph (a) of this section may be relied upon by the licensee only for use in a single transaction, and for a period not to exceed 30 calendar days from the date that NICS was initially contacted. If the transaction is not completed within the 30-day period, the licensee shall initiate a new NICS check prior to completion of the transfer.

Example 1 for paragraph (c). A purchaser completes the Form 4473 on December 15, 1998, and a NICS check is initiated by the licensee on that date. The licensee is informed by NICS that the information available to the system does not indicate that receipt of the firearm by the transferee would be in violation of law, and a unique identification number is provided. However, the State imposes a 7-day waiting period on all firearms transactions, and the purchaser does not return to pick up the firearm until January 22, 1999. The licensee must conduct another NICS check before transferring the firearm to the purchaser.

Example 2 for paragraph (c). A purchaser completes the Form 4473 on January 25, 1999, and arranges for the purchase of a single firearm. A NICS check is initiated by the licensee on that date. The licensee is informed by NICS that the information available to the system does not indicate that receipt of the firearm by the transferee would be in violation of law, and a unique identification number is provided. The State imposes a 7-day waiting period on all firearms transactions, and the purchaser returns to pick up the firearm on February 15, 1999. Before the licensee executes the Form 4473, and the firearm is transferred, the purchaser decides to purchase an additional firearm. The transfer of these two firearms is considered a single transaction; accordingly, the licensee may add the second firearm to the Form 4473, and transfer that firearm without conducting another NICS check.

Example 3 f or para graph (c). A purchaser completes a Form 4473 on February 15, 1999. The licensee receives a unique identification number from NICS on that date, the Form 4473 is executed by the licensee, and the firearm is transferred. On February 20, 1999, the purchaser returns to the licensee's premises and wishes to purchase a second firearm. The purchase of the second firearm is a separate transaction; thus, a new NICS check must be initiated by the licensee.

(d) Exceptions to NICS check. The provisions of paragraph (a) of this section shall not apply if—

(1) The transferee has presented to the licensee a valid permit or license that—

(i) Allows the transferee to possess, acquire, or carry a firearm;

(ii) Was issued not more than 5 years earlier by the State in which the transfer is to take place; and

(iii) The law of the State provides that such a permit or license is to be issued only after an authorized government official has verified that the information available to such official does not indicate that possession of a firearm by the transferee would be in violation of Federal, State, or local law: **Provided,** That on and after November 30, 1998, the information available to such official includes the NICS;

(2) The firearm is subject to the provisions of the National Firearms Act and has been approved for transfer under 27 CFR part 479; or

(3) On application of the licensee, in accordance with the provisions of § 478.150, the Director has certified that compliance with paragraph (a)(1) of this section is impracticable.

(e) The document referred to in paragraph (d)(1) of this section (or a copy thereof) shall be retained or the required information from the document shall be recorded on the firearms transaction record in accordance with the provisions of § 478.131.

(Approved by the Office of Management and Budget under control number 1512–0544)

§ 478.103 Posting of signs and written notification to purchasers of handguns.

(a) Each licensed importer, manufacturer, dealer, or collector who delivers a handgun to a nonlicensee shall provide such nonlicensee with written notification as described in paragraph (b) of this section.

(b) The written notification (ATF I 5300.2) required by paragraph (a) of this section shall state as follows:

(1) The misuse of handguns is a leading contributor to juvenile violence and fatalities.

(2) Safely storing and securing firearms away from children will help prevent the unlawful possession of handguns by juveniles, stop accidents, and save lives.

(3) Federal law prohibits, except in certain limited circumstances, anyone under 18 years of age from knowingly possessing a handgun, or any person from transferring a handgun to a person under 18.

(4) A knowing violation of the prohibition against selling, delivering, or otherwise transferring a handgun to a person under the age of 18 is, under certain circumstances, punishable by up to 10 years in prison.

FEDERAL LAW

The Gun Control Act of 1968, 18 U.S.C. Chapter 44, provides in pertinent part as follows:

18 U.S.C. 922(x)

(x)(1) It shall be unlawful for a person to sell, deliver, or otherwise transfer to a person who the transferor knows or has reasonable cause to believe is a juvenile—

(A) a handgun; or

(B) ammunition that is suitable for use only in a handgun.

(2) It shall be unlawful for any person who is a juvenile to knowingly possess—

(A) a handgun; or

(B) ammunition that is suitable for use only in a handgun.

(3) This subsection does not apply to—

(A) a temporary transfer of a handgun or ammunition to a juvenile or to the possession or use of a handgun or ammunition by a juvenile if the handgun and ammunition are possessed and used by the juvenile—

(i) in the course of employment, in the course of ranching or farming related to activities at the residence of the juvenile (or on property used for ranching or farming at which the juvenile, with the permission of the property owner or lessee, is performing activities related to the operation of the farm or ranch), target practice, hunting, or a course of instruction in the safe and lawful use of a handgun;

(ii) with the prior written consent of the juvenile's parent or guardian who is not prohibited by Federal, State, or local law from possessing a firearm, except—

(I) during transportation by the juvenile of an unloaded handgun in a locked container directly from the place of transfer to a place at which an activity described in clause (i) is to take place and transportation by the juvenile of that handgun, unloaded and in a locked container, directly from the place at which such an activity took place to the transferor; or

(II) with respect to ranching or farming activities as described in clause (i) a juvenile may possess and use a handgun or ammunition with the prior written approval of the juvenile's parent or legal guardian and at the direction of an adult who is not prohibited by Federal, State, or local law from possessing a firearm;

(iii) the juvenile has the prior written consent in the juvenile's possession at all times when a handgun is in the possession of the juvenile; and

(iv) in accordance with State and local law;

(B) a juvenile who is a member of the Armed Forces of the United States or the National Guard who possesses or is armed with a handgun in the line of duty;

(C) a transfer by inheritance of title (but not possession) of a handgun or ammunition to a juvenile; or

(D) the possession of a handgun or ammunition by a juvenile taken in defense of the juvenile or other persons against an intruder into the residence of the juvenile or a residence in which the juvenile is an invited guest.

(4) A handgun or ammunition, the possession of which is transferred to a juvenile in circumstances in which the transferor is not in violation of this subsection shall not be subject to permanent confiscation by the Government if its possession by the juvenile subsequently becomes unlawful because of the conduct of the juvenile, but shall be returned to the lawful owner when such handgun or ammunition is no longer required by the Government for the purposes of investigation or prosecution.

(5) For purposes of this subsection, the term "juvenile" means a person who is less than 18 years of age.

(6)(A) In a prosecution of a violation of this subsection, the court shall require the presence of a juvenile defendant's parent or legal guardian at all proceedings.

(B) The court may use the contempt power to enforce subparagraph (A).

(C) The court may excuse attendance of a parent or legal guardian of a juvenile defendant at a proceeding in a prosecution of a violation of this subsection for good cause shown.

18 U.S.C. 924(a)(6)

(6)(A)(i) A juvenile who violates section 922(x) shall be fined under this title, imprisoned not more than 1 year, or both, except that a juvenile described in clause (ii) shall be sentenced to probation on appropriate conditions and shall not be incarcerated unless the juvenile fails to comply with a condition of probation.

(ii) A juvenile is described in this clause if—

(I) the offense of which the juvenile is charged is possession of a handgun or ammunition in violation of section 922(x)(2); and

(II) the juvenile has not been convicted in any court of an offense (including an offense under section 922(x) or a similar State law, but not including any other offense consisting of conduct that if engaged in by an adult would not constitute an offense) or adjudicated as a juvenile delinquent for conduct that if engaged in by an adult would constitute an offense.

(B) A person other than a juvenile who knowingly violates section 922(x)—

(i) shall be fined under this title, imprisoned not more than 1 year, or both; and

(ii) if the person sold, delivered, or otherwise transferred a handgun or ammunition to a juvenile knowing or having reasonable cause to know that the juvenile intended to carry or otherwise possess or discharge or otherwise use the handgun or ammunition in the commission of a crime of violence, shall be fined under this title, imprisoned not more than 10 years, or both.

(c) This written notification shall be delivered to the nonlicensee on ATF I 5300.2, or in the alternative, the same written notification may be delivered to the nonlicensee on another type of written notification, such as a manufacturer's or importer's brochure accompanying the handgun; a manufacturer's or importer's operational manual accompanying the handgun; or a sales receipt or invoice applied to the handgun package or container delivered to a nonlicensee. Any written notification delivered to a nonlicensee other than on ATF I 5300.2 shall include the language set forth in paragraph (b) of this section in its entirety. Any written notification other than ATF I 5300.2 shall be legible, clear, and conspicuous, and the required language shall appear in type size no smaller than 10-point type.

(d) Except as provided in paragraph (f) of this section, each licensed importer, manufacturer, or dealer who delivers a handgun to a nonlicensee shall display at its licensed premises (including temporary business locations at gun shows) a sign as described in paragraph (e) of this section. The sign shall be displayed where customers can readily see it. Licensed importers, manufacturers, and dealers will be provided with such signs by ATF. Replacement signs may be requested from the ATF Distribution Center.

(e) The sign (ATF I 5300.1) required by paragraph (d) of this section shall state as follows:

(1) The misuse of handguns is a leading contributor to juvenile violence and fatalities.

(2) Safely storing and securing firearms away from children will help prevent the unlawful possession of handguns by juveniles, stop accidents, and save lives.

(3) Federal law prohibits, except in certain limited circumstances, anyone under 18 years of age from knowingly possessing a handgun, or any person from transferring a handgun to a person under 18.

(4) A knowing violation of the prohibition against selling, delivering, or otherwise transferring a handgun to a person under the age of 18 is, under certain circumstances, punishable by up to 10 years in prison.

Editor's Note:

ATF I 5300.2 provides the complete language of the statutory prohibitions and exceptions provided in 18 U.S.C. 922(x) and the penalty provisions of 18 U.S.C. 924(a)(6). The Federal firearms licensee posting this sign will provide customers with a copy of this publication upon request. Requests for additional copies of ATF I 5300.2 should be mailed to the ATF Distribution Center, P.O. Box 5950, Springfield, Virginia 22150–5950.

(f) The sign required by paragraph (d) of this section need not be posted on the premises of any licensed importer, manufacturer, or dealer whose only dispositions of handguns to nonlicensees are to nonlicensees who do not appear at the licensed premises and the dispositions otherwise comply with the provisions of this part.

Subpart G—Importation

§ 478.111 General.

(a) Section 922(a)(3) of the Act makes it unlawful, with certain exceptions not pertinent here, for any person other than a licensee to transport into or receive in the State where the person resides any firearm purchased or otherwise obtained by the person outside of that State. However, section 925(a)(4) provides a limited exception for the transportation, shipment, receipt or importation of certain firearms and ammunition by certain members of the United States Armed Forces. Section 922(1) of the Act makes it unlawful for any person knowingly to import or bring into the United States or any possession thereof any firearm or ammunition except as provided by section 925(d) of the Act, which section provides standards for importing or bringing firearms or ammunition into the United States. Section 925(d) also provides standards for importing or bringing firearm barrels into the United States. Accordingly, no firearm, firearm barrel, or ammunition may be imported or brought into the United States except as provided by this part.

(b) Where a firearm, firearm barrel, or ammunition is imported and the authorization for importation required by this subpart has not been obtained by the person importing same, such person shall:

(1) Store, at the person's expense, such firearm, firearm barrel, or ammunition at a facility designated by U.S Customs or the Director of Industry Operations to await the issuance of the required authorization or other disposition; or

(2) Abandon such firearm, firearm barrel, or ammunition to the U.S. Government; or

(3) Export such firearm, firearm barrel, or ammunition.

(c) Any inquiry relative to the provisions or procedures under this subpart, other than that pertaining to the payment of customs duties or the release from Customs custody of firearms, firearm barrels, or ammunition authorized by the Director to be imported, shall be directed to the Director of Industry Operations for reply.

§ 478.112 Importation by a licensed importer.

(a) No firearm, firearm barrel, or ammunition shall be imported or brought into the United States by a licensed importer (as defined in § 478.11) unless the Director has authorized the importation of the firearm, firearm barrel, or ammunition.

(b) (1) An application for a permit, ATF Form 6—Part I, to import or bring a firearm, firearm barrel, or ammunition into the United States or a possession thereof under this section must be filed, in triplicate, with the Director. The application must be signed and dated and must contain the information requested on the form, including:

(i) The name, address, telephone number, and license number (including expiration date) of the importer;

(ii) The country from which the firearm, firearm barrel, or ammunition is to be imported;

(iii) The name and address of the foreign seller and foreign shipper;

(iv) A description of the firearm, firearm barrel, or ammunition to be imported, including:

(A) The name and address of the manufacturer;

(B) The type (e.g., rifle, shotgun, pistol, revolver and, in the case of ammunition only, ball, wadcutter, shot, etc.);

(C) The caliber, gauge, or size;

(D) The model;

(E) The barrel length, if a firearm or firearm barrel (in inches);

(F) The overall length, if a firearm (in inches);

(G) The serial number, if known;

(H) Whether the firearm is new or used;

(I) The quantity;

(J) The unit cost of the firearm, firearm barrel, or ammunition to be imported;

(v) The specific purpose of importation, including final recipient information if different from the importer;

(vi) Verification that if a firearm, it will be identified as required by this part; and

(vii)(A) If a firearm or ammunition imported or brought in for scientific or research purposes, a statement describing such purpose; or

(B) If a firearm or ammunition for use in connection with competition or training pursuant to Chapter 401 of Title 10, U.S.C., a statement describing such intended use; or

(C) If an unserviceable firearm (other than a machine gun) being imported as a curio or museum piece, a description of how it was rendered unserviceable and an explanation of why it is a curio or museum piece; or

(D) If a firearm other than a surplus military firearm, of a type that does not fall within the definition of a firearm under section 5845(a) of the Internal Revenue Code of 1954, and is for sporting purposes, an explanation of why the firearm is generally recognized as particularly suitable for or readily adaptable to sporting purposes; or

(E) If ammunition being imported for sporting purposes, a statement why the ammunition is particularly suitable for or readily adaptable to sporting purposes; or

(F) If a firearm barrel for a handgun, an explanation why the handgun is generally recognized as particularly suitable for or readily adaptable to sporting purposes.

(2)(i) If the Director approves the application, such approved application will serve as the permit to import the firearm, firearm barrel, or ammunition described therein, and importation of such firearms, firearm barrels, or ammunition may continue to be made by the licensed importer under the approved application (permit) during the period specified thereon. The Director will furnish the approved application (permit) to the applicant and retain two copies thereof for administrative use.

(ii) If the Director disapproves the application, the licensed importer will be notified of the basis for the disapproval.

(c) A firearm, firearm barrel, or ammunition imported or brought into the United States or a possession thereof under the provisions of this section by a licensed importer may be released from Customs custody to the licensed importer upon showing that the importer has obtained a permit from the Director for the importation of the firearm, firearm barrel, or ammunition to be released. The importer will also submit to Customs a copy of the export license authorizing the export of the firearm, firearm barrel, or ammunition from the exporting country. If the exporting country does not require issuance of an export license, the importer must submit a certification, under penalty of perjury, to that effect.

(1) In obtaining the release from Customs custody of a firearm, firearm barrel, or ammunition authorized by this section to be imported through the use of a permit, the licensed importer will prepare ATF Form 6A, in duplicate, and furnish the original ATF Form 6A to the Customs officer releasing the firearm, firearm barrel, or ammunition. The Customs officer will, after certification, forward the ATF Form 6A to the address specified on the form.

(2) The ATF Form 6A must contain the information requested on the form, including:

(i) The name, address, and license number of the importer;

(ii) The name of the manufacturer of the firearm, firearm barrel, or ammunition;

(iii) The country of manufacture;

(iv) The type;

(v) The model;

(vi) The caliber, gauge, or size;

(vii) The serial number in the case of firearms, if known; and

(viii) The number of firearms, firearm barrels, or rounds of ammunition released.

(d) Within 15 days of the date of release from Customs custody, the licensed importer must:

(1) Forward to the address specified on the form a copy of ATF Form 6A on which must be reported any error or discrepancy appearing on the ATF Form 6A certified by Customs and serial numbers if not previously provided on ATF Form 6A;

(2) Pursuant to § 478.92, place all required identification data on each imported firearm if same did not bear such identification data at the time of its release from Customs custody; and

(3) Post in the records required to be maintained by the importer under subpart H of this part all required information regarding the importation.

(Paragraph (b) approved by the Office of Management and Budget under control number 1512–0017; paragraphs (c) and (d) approved by the Office of Management and Budget under control number 1512–0019)

§ 478.113 Im portation by other licensees.

(a) No person other than a licensed importer (as defined in § 478.11) shall engage in the business of importing firearms or ammunition. Therefore, no firearm or ammunition shall be imported or brought into the United States or a possession thereof by any licensee other than a licensed importer unless the Director issues a permit authorizing the importation of the firearm or ammunition. No barrel for a handgun not generally recognized as particularly suitable for or readily adaptable to sporting purposes shall be imported or brought into the United States or a possession thereof by any person. Therefore, no firearm barrel shall be imported or brought into the United States or possession thereof by any licensee other than a licensed importer unless the Director issues a permit authorizing the importation of the firearm barrel.

(b) (1) An application for a permit, ATF Form 6—Part I, to import or bring a firearm, firearm barrel, or ammunition into the United States or a possession thereof by a licensee, other than a licensed importer, must be filed, in triplicate, with the Director. The application must be signed and dated and must contain the information requested on the form, including:

(i) The name, address, telephone number, and license number (including expiration date) of the applicant;

(ii) The country from which the firearm, firearm barrel, or ammunition is to be imported;

(iii) The name and address of the foreign seller and foreign shipper;

(iv) A description of the firearm, firearm barrel, or ammunition to be imported, including:

(A) The name and address of the manufacturer;

(B) The type (e.g., rifle, shotgun, pistol, revolver and, in the case of ammunition only, ball, wadcutter, shot, etc.);

(C) The caliber, gauge, or size;

(D) The model;

(E) The barrel length, if a firearm or firearm barrel (in inches);

(F) The overall length, if a firearm (in inches);

(G) The serial number, if known;

(H) Whether the firearm is new or used;

(I) The quantity;

(J) The unit cost of the firearm, firearm barrel, or ammunition to be imported;

(v) The specific purpose of importation, including final recipient information if different from the applicant; and

(vi)(A) If a firearm or ammunition imported or brought in for scientific or research purposes, a statement describing such purpose; or

(B) If a firearm or ammunition for use in connection with competition or training pursuant to Chapter 401 of Title 10, U.S.C., a statement describing such intended use; or

(C) If an unserviceable firearm (other than a machine gun) being imported as a curio or museum piece, a description of how it was rendered unserviceable and an explanation of why it is a curio or museum piece; or

(D) If a firearm other than a surplus military firearm, of a type that does not fall within the definition of a firearm under section 5845(a) of the Internal Revenue Code of 1986, and is for sporting purposes, an explanation of why the firearm is generally recognized as particularly suitable for or readily adaptable to sporting purposes; or

(E) If ammunition being imported for sporting purposes, a statement why the ammunition is particularly suitable for or readily adaptable to sporting purposes; or

(F) If a firearm barrel, for a handgun, an explanation why the handgun is generally recognized as particularly suitable for or readily adaptable to sporting purposes.

(2)(i) If the Director approves the application, such approved application will serve as the permit to import the firearm, firearm barrel, or ammunition described therein, and importation of such firearms, firearm barrels, or ammunition may continue to be made by the applicant under the approved application (permit) during the period specified thereon. The Director will furnish the approved application (permit) to the applicant and retain two copies thereof for administrative use.

(ii) If the Director disapproves the application, the applicant will be notified of the basis for the disapproval.

(c) A firearm, firearm barrel, or ammunition imported or brought into the United States or a possession thereof under the provisions of this section may be released from Customs custody to the licensee upon showing that the licensee has obtained a permit from the Director for the importation of the firearm, firearm barrel, or ammunition to be released.

(1) In obtaining the release from Customs custody of a firearm, firearm barrel, or ammunition authorized by this section to be imported through the use of a permit, the licensee will prepare ATF Form 6A, in duplicate, and furnish the original ATF Form 6A to the Customs officer releasing the firearm, firearm barrel, or ammunition. The Customs officer will, after certification, forward the ATF Form 6A to the address specified on the form.

(2) The ATF Form 6A must contain the information requested on the form, including:

(i) The name, address, and license number of the licensee;

(ii) The name of the manufacturer of the firearm, firearm barrel, or ammunition;

(iii) The country of manufacture;

(iv) The type;

(v) The model;

(vi) The caliber, gauge, or size;

(vii) The serial number in the case of firearms; and

(viii) The number of firearms, firearm barrels, or rounds of ammunition released.

(Paragraph (b) approved by the Office of Management and Budget under control number 1512–0017; paragraph (c) approved by the Office of Management and Budget under control number 1512–0019)

§ 478.113a Importation of firearm barrels by nonlicensees.

(a) A permit will not be issued for a firearm barrel for a handgun not generally recognized as particularly suitable for or readily adaptable to sporting purposes. No firearm barrel shall be imported or brought into the United States or possession thereof by any nonlicensee unless the Director issues a permit authorizing the importation of the firearm barrel.

(b)(1) An application for a permit, ATF Form 6—Part I, to import or bring a firearm barrel into the United States or a possession thereof under this section must be filed, in triplicate, with the Director. The application must be signed and dated and must contain the information requested on the form, including:

(i) The name, address, and telephone number of the applicant;

(ii) The country from which the firearm barrel is to be imported;

(iii) The name and address of the foreign seller and foreign shipper;

(iv) A description of the firearm barrel to be imported, including:

(A) The name and address of the manufacturer;

(B) The type (e.g., rifle, shotgun, pistol, revolver);

(C) The caliber, gauge, or size;

(D) The model;

(E) The barrel length (in inches);

(F) The quantity;

(G) The unit cost of the firearm barrel;

(v) The specific purpose of importation, including final recipient information if different from the importer; and

(vi) If a handgun barrel, an explanation of why the barrel is for a handgun that is generally recognized as particularly suitable for or readily adaptable to sporting purposes.

(2)(i) If the Director approves the application, such approved application will serve as the permit to import the firearm barrel, and importation of such firearm barrels may continue to be made by the applicant under the approved application (permit) during the period specified thereon. The Director will furnish the approved application (permit) to the applicant and retain two copies thereof for administrative use.

(ii) If the Director disapproves the application, the applicant will be notified of the basis for the disapproval.

(c) A firearm barrel imported or brought into the United States or a possession thereof under the provisions of this section may be released from Customs custody to the person importing the firearm barrel upon showing that the person has obtained a permit from the Director for the importation of the firearm barrel to be released.

(1) In obtaining the release from Customs custody of a firearm barrel authorized by this section to be imported through the use of a permit, the person importing the firearm barrel will prepare ATF Form 6A, in duplicate, and furnish the original ATF Form 6A to the Customs officer releasing the firearm barrel. The Customs officer will, after certification, forward the ATF Form 6A to the address specified on the form.

(2) The ATF Form 6A must contain the information requested on the form, including:

(i) The name and address of the person importing the firearm barrel;

(ii) The name of the manufacturer of the firearm barrel;

(iii) The country of manufacture;

(iv) The type;

(v) The model;

(vi) The caliber or gauge of the firearm barrel so released; and

(vii) The number of firearm barrels released.

(Paragraph (b) approved by the Office of Management and Budget under control number 1512–0017; paragraph (c) approved by the Office of Management and Budget under control number 1512–0019)

§ 478.114 Importation by members of the U.S. Armed Forces.

(a) The Director may issue a permit authorizing the importation of a firearm or ammunition into the United States to the place of residence of any military member of the U.S. Armed Forces who is on active duty outside the United States, or who has been on active duty outside the United States within the 60-day period immediately preceding the intended importation: **Provided,** That such firearm or ammunition is generally recognized as particularly suitable for or readily adaptable to sporting purposes and is intended for the personal use of such member.

(1) An application for a permit, ATF Form 6—Part II, to import a firearm or ammunition into the United States under this section must be filed, in triplicate, with the Director. The application must be signed and dated and must contain the information requested on the form, including:

(i) The name, current address, and telephone number of the applicant;

(ii) Certification that the transportation, receipt, or possession of the firearm or ammunition to be imported would not constitute a violation of any provision of the Act or of any State law or local ordinance at the place of the applicant's residence;

(iii) The country from which the firearm or ammunition is to be imported;

(iv) The name and address of the foreign seller and foreign shipper;

(v) A description of the firearm or ammunition to be imported, including:

(A) The name and address of the manufacturer;

(B) The type (e.g., rifle, shotgun, pistol, revolver and, in the case of ammunition only, ball, wadcutter, shot, etc.);

(C) The caliber, gauge, or size;

(D) The model;

(E) The barrel length, if a firearm (in inches);

(F) The overall length, if a firearm (in inches);

(G) The serial number;

(H) Whether the firearm is new or used;

(I) The quantity;

(J) The unit cost of the firearm or ammunition to imported;

(vi) The specific purpose of importation, that is —

(A) That the firearm or ammunition being imported is for the personal use of the applicant; and

(B) If a firearm, a statement that it is not a surplus military firearm, that it does not fall within the definition of a firearm under section 5845(a) of the Internal Revenue Code of 1986, and an explanation of why the firearm is generally recognized as particularly suitable for or readily adaptable to sporting purposes; or

(C) If ammunition, a statement why it is generally recognized as particularly suitable for or readily adaptable to sporting purposes; and

(vii) The applicant's date of birth;

(viii) The applicant's rank or grade;

(ix) The applicant's place of residence;

(x) The applicant's present foreign duty station or last foreign duty station, as the case may be;

(xi) The date of the applicant's reassignment to a duty station within the United States, if applicable; and

(xii) The military branch of which the applicant is a member.

(2)(i) If the Director approves the application, such approved application will serve as the permit to import the firearm or ammunition described therein. The Director will furnish the approved application (permit) to the applicant and retain two copies thereof for administrative use.

(ii) If the Director disapproves the application, the applicant will be notified of the basis for the disapproval.

(b) Except as provided in paragraph (b)(3) of this section, a firearm or ammunition imported into the United States under the provisions of this section by the applicant may be released from Customs custody to the applicant upon showing that the applicant has obtained a permit from the Director for the importation of the firearm or ammunition to be released.

(1) In obtaining the release from Customs custody of a firearm or ammunition authorized by this section to be imported through the use of a permit, the military member of the U.S. Armed Forces will prepare ATF Form 6A and furnish the completed form to the Customs officer releasing the firearm or ammunition. The Customs officer will, after certification, forward the ATF Form 6A to the address specified on the form.

(2) The ATF Form 6A must contain the information requested on the form, including:

(i) The name and address of the military member;

(ii) The name of the manufacturer of the firearm or ammunition;

(iii) The country of manufacture;

(iv) The type;

(v) The model;

(vi) The caliber, gauge, or size;

(vii) The serial number in the case of firearms; and

(viii) If applicable, the number of firearms or rounds of ammunition released.

(3) When such military member is on active duty outside the United States, the military member may appoint, in writing, an agent to obtain the release of the firearm or ammunition from Customs custody for such member. Such agent will present sufficient identification of the agent and the written authorization to act on behalf of such military member to the Customs officer who is to release the firearm or ammunition.

(c) Firearms determined by the Department of Defense to be war souvenirs may be imported into the United States by the military members of the U.S. Armed Forces under such provisions and procedures as the Department of Defense may issue.

Paragraph (a) approved by the Office of Management and Budget under control number 1512–0018; paragraph (b) approved by the Office of Management and Budget under control number 1512–0019.)

§ 478.115 Exempt importation.

(a) Firearms and ammunition may be brought into the United States or any possession thereof by any person who can establish to the satisfaction of Customs that such firearm or ammunition was previously taken out of the United States or any possession thereof by such person. Registration on Customs Form 4457 or on any other registration document available for this purpose may be completed before departure from the United States at any U.S. customhouse or any office of a Director of Industry Operations. A bill of sale or other commercial document showing transfer of the firearm or ammunition in the United States to such person also may be used to establish proof that the firearm or ammunition was taken out of the United States by such person. Firearms and ammunition furnished under the provisions of section 925(a)(3) of the Act to military members of the U.S. Armed Forces on active duty outside of the United States also may be imported into the United States or any possession thereof by such military members upon establishing to the satisfaction of Customs that such firearms and ammunition were so obtained.

(b) Firearms, firearm barrels, and ammunition may be imported or brought into the United States by or for the United States or any department or agency thereof, or any State or any department, agency, or political subdivision thereof. A firearm, firearm barrel or ammunition imported or brought into the United States under this paragraph may be released from Customs custody upon a showing that the firearm, firearm barrel or ammunition is being imported or brought into the United States by or for such a governmental entity.

(c) The provisions of this subpart shall not apply with respect to the importation into the United States of any antique firearm.

(d) Firearms and ammunition are not imported into the United States, and the provisions of this subpart shall not apply, when such firearms and ammunition are brought into the United States by:

(1) A nonresident of the United States for legitimate hunting or lawful sporting purposes, and such firearms and such ammunition as remains following such shooting activity are to be taken back out of the territorial limits of the United States by such person upon conclusion of the shooting activity;

(2) Foreign military personnel on official assignment to the United States who bring such firearms or ammunition into the United States for their exclusive use while on official duty in the United States, and such firearms and unexpended ammunition are taken back out of the territorial limits of the United States by such foreign military personnel when they leave the United States;

(3) Official representatives of foreign governments who are accredited to the U.S. Government or are en route to or from other countries to which accredited, and such firearms and unexpended ammunition are taken back out of the territorial limits of the United States by such official representatives of foreign governments when they leave the United States;

(4) Officials of foreign governments and distinguished foreign visitors who have been so designated by the Department of State, and such firearms and unexpended ammunition are taken back out of the territorial limits of the United States by such officials of foreign governments and distinguished foreign visitors when they leave the United States; and

(5) Foreign law enforcement officers of friendly foreign governments entering the United States on official law enforcement business, and such firearms and unexpended ammunition are taken back out of the territorial limits of the United States by such foreign law enforcement officers when they leave the United States.

(e) Notwithstanding the provisions of paragraphs (d) (1), (2), (3), (4) and (5) of this section, the Secretary of the Treasury or his delegate may in the interest of public safety and necessity require a permit for the importation or bringing into the United States of any firearms or ammunition.

§ 478.116 Conditional importation.

The Director shall permit the conditional importation or bringing into the United States or any possession thereof of any firearm, firearm barrel, ammunition, or ammunition feeding device as defined in § 478.119(b) for the purpose of examining and testing the firearm, firearm barrel, ammunition, or ammunition feeding device in connection with making a determination as to whether the importation or bringing in of such firearm, firearm barrel, ammunition, or ammunition feeding device will be authorized under this part. An application on ATF Form 6 for such conditional importation shall be filed, in duplicate, with the Director. The Director may impose conditions upon any importation under this section including a requirement that the firearm, firearm barrel, ammunition, or ammunition feeding device be shipped directly from Customs custody to the Director and that the person importing or bringing in the firearm, firearm barrel, ammunition, or ammunition feeding device must agree to either export the firearm, firearm barrel, ammunition, or ammunition feeding device or destroy same if a determination is made that the firearm, firearm barrel, ammunition, or ammunition feeding device may not be imported or brought in under this part. A firearm, firearm barrel, ammunition, or ammunition feeding device imported or brought into the United States or any possession thereof under the provisions of this section shall be released from Customs custody upon the payment of customs duties, if applicable, and in the manner prescribed in the conditional authorization issued by the Director.

Editor's Note:

The references to "ammunition feeding device" in section 478.116 are not applicable on or after September 13, 2004.

§ 478.117 Function outside a customs territory.

In the insular possessions of the United States outside customs territory, the functions performed by U.S. Customs officers under this subpart within a customs territory may be performed by the appropriate authorities of a territorial government or other officers of the United States who have been designated to perform such functions. For the purpose of this subpart, the term customs territory means the United States, the District of Columbia, and the Commonwealth of Puerto Rico.

§ 478.118 Importation of certain firearms classified as curios or relics.

Notwithstanding any other provision of this part, a licensed importer may import all rifles and shotguns classified by the Director as curios or relics, and all handguns classified by the Director as curios or relics that are determined to be generally recognized as particularly suitable for or readily adaptable to sporting purposes. The importation of such curio or relic firearms must be in accordance with the applicable importation provisions of this part and the importation provisions of 27 CFR part 447. Curios or relics which fall within the definition of "firearm" under 26 U.S.C. 5845(a) must also meet the importation provisions of 27 CFR part 479 before they may be imported.

§ 478.119 Importation of ammunition feeding devices.

(a) No ammunition feeding device shall be imported or brought into the United States unless the Director has authorized the importation of such device.

(b) For purposes of this section, an **"ammunition feeding device"** is a magazine, belt, drum, feed strip, or similar device for a firearm that has a capacity of, or that can be readily restored or converted to accept, more than 10 rounds of ammunition. The term does not include an attached tubular device designed to accept, and capable of operating only with, .22 caliber rimfire ammunition, or a fixed device for a manually operated firearm, or a fixed device for a firearm listed in 18 U.S.C. 922, Appendix A.

(c) An application for a permit, ATF Form 6, to import or bring an ammunition feeding device into the United States or a possession thereof under this section shall be filed, in triplicate, with the Director. The application shall contain:

(1) The name and address of the person importing the device,

(2) A description of the device to be imported, including type and cartridge capacity, model and caliber of firearm for which the device was made, country of manufacture, and name of the manufacturer if known,

(3) The unit cost of the device to be imported,

(4) The country from which to be imported,

(5) The name and address of the foreign seller and the foreign shipper,

(6) Verification that such device will be marked as required by this part, and

(7) A statement by the importer that the device is being imported for sale to purchasers specified in § 478.40a(b) or physical or reasonable documentary evidence establishing that the magazine was manufactured on or before September 13, 1994. Any one of the following examples, which are not meant to be exhaustive, may be sufficient to establish the time of manufacture:

(i) Permanent markings or physical characteristics which establish that the magazine was manufactured on or before September 13, 1994;

(ii) A certification from the importer, under penalty of perjury, that the importer maintained continuous custody beginning on a date prior to September 14, 1994, and continuing until the date of the certification. Such certification shall also be supported by reasonable documentary evidence, such as commercial records;

(iii) A certification from the importer, under penalty of perjury, that the magazines sought to be imported were in the custody and control of a foreign Government on or before September 13, 1994, along with reasonable documentary evidence to support the certification; or

(iv) A certification from the importer, under penalty of perjury, that the magazine was in the possession of a foreign arms supplier on or before September 13, 1994, along with reasonable documentary evidence to support the certification.

(d) The Director shall act upon applications to import ammunition feeding devices as expeditiously as possible. If the Director approves the application, such approved application shall serve as the permit to import the device described therein, and importation of such devices may continue to be made by the person importing such devices under the approved application (permit) during the period specified thereon. The Director shall furnish the approved application (permit) to the applicant and retain two copies thereof for administrative use. If the Director disapproves the application, the person importing such devices shall be notified of the basis for the disapproval.

(e) An ammunition feeding device imported or brought into the United States by a person importing such a device may be released from Customs custody to the person importing such a device upon showing that such person has obtained a permit from the Director for the importation of the device to be released. In obtaining the release from Customs custody of such a device authorized by this section to be imported through use of a permit, the person importing such a device shall prepare ATF Form 6A, in duplicate, and furnish the original ATF Form 6A to the Customs officer releasing the device. The Customs officer shall, after certification, forward the ATF Form 6A to the address specified on the form. The ATF Form 6A shall show the name and address of the person importing the device, the name of the manufacturer of the device, the country of manufacture, the type, model, caliber, size, and the number of devices released.

(f) Within 15 days of the date of release from Customs custody, the person importing such a device shall:

(1) Forward to the address specified on the form a copy of ATF Form 6A on which shall be reported any error or discrepancy appearing on the ATF Form 6A certified by Customs, and

(2) Pursuant to § 478.92, place all required identification data on each imported device manufactured after September 13, 1994, if same did not bear such identification data at the time of its release from Customs custody.

(g) The Director may authorize the conditional importation of an ammunition feeding device as provided in § 478.116.

(Paragraphs (a), (c), and (d) approved by the Office of Management and Budget under control numbers 1512–0017 and 1512–0018; paragraphs (e) and (f) approved by the Office of Management and Budget under control number 1512–0019)

Editor's Note:

An import permit is still needed for ammunition feeding devices pursuant to the Arms Export Control Act – see 27 CFR 447.41(a).

§ 478.120 Firearms or ammunition imported by or for a nonimmigrant alien.

(a) Any nonimmigrant alien who completes an ATF Form 6 to import firearms or ammunition into the United States, or any licensee who completes an ATF Form 6 to import firearms or ammunition for a nonimmigrant alien, must attach applicable documentation to the Form 6 (e.g., a hunting license or permit lawfully issued in the United States; waiver) establishing the nonimmigrant alien falls within an exception to or has obtained a waiver from the nonimmigrant alien prohibition.

(b) Nonimmigrant aliens importing or bringing firearms or ammunition into the United States must provide the United States Customs Service with applicable documentation (e.g., a hunting license or permit lawfully issued in the United States; waiver) establishing the nonimmigrant alien falls within an exception to or has obtained a waiver from the nonimmigrant alien prohibition before the firearm or ammunition may be imported. This provision applies in all cases, whether or not a Form 6 is needed to bring the firearms or ammunition into the United States.

(Approved by the Office of Management and Budget under control number 1512–0570)

Subpart H—Records

§ 478.121 General.

(a) The records pertaining to firearms transactions prescribed by this part shall be retained on the licensed premises in the manner prescribed by this subpart and for the length of time prescribed by § 478.129. The records pertaining to ammunition prescribed by this part shall be retained on the licensed premises in the manner prescribed by § 478.125.

(b) ATF officers may, for the purposes and under the conditions prescribed in § 478.23, enter the premises of any licensed importer, licensed manufacturer, licensed dealer, or licensed collector for the purpose of examining or inspecting any record or document required by or obtained under this part. Section 923(g) of the Act requires licensed importers, licensed manufacturers, licensed dealers, and licensed collectors to make such records available for such examination or inspection during business hours or, in the case of licensed collectors, hours of operation, as provided in § 478.23.

(c) Each licensed importer, licensed manufacturer, licensed dealer, and licensed collector shall maintain such records of importation, production, shipment, receipt, sale, or other disposition, whether temporary or permanent, of firearms and such records of the disposition of ammunition as the regulations contained in this part prescribe. Section 922(m) of the Act makes it unlawful for any licensed importer, licensed manufacturer, licensed dealer, or licensed collector knowingly to make any false entry in, to fail to make appropriate entry in, or to fail to properly maintain any such record.

(d) For recordkeeping requirements for sales by licensees at gun shows see § 478.100(c).

(Information collection requirements in paragraph (a) approved by the Office of Management and Budget under control number 1512–0129; information collection requirements in paragraphs (b) and (c) approved by the Office of Management and Budget under control number 1512–0387)

§ 478.122 Records maintained by importers.

(a) Each licensed importer shall, within 15 days of the date of importation or other acquisition, record the type, model, caliber or gauge, manufacturer, country of manufacture, and the serial number of each firearm imported or otherwise acquired, and the date such importation or other acquisition was made.

(b) A record of firearms disposed of by a licensed importer to another licensee and a separate record of armor piercing ammunition dispositions to governmental entities, for exportation, or for testing or experimentation authorized under the provisions of § 478.149 shall be maintained by the licensed importer on the licensed premises. For firearms, the record shall show the quantity, type, manufacturer, country of manufacture, caliber or gauge, model, serial number of the firearms so transferred, the name and license number of the licensee to whom the firearms were transferred, and the date of the transaction. For armor piercing ammunition, the record shall show the date of the transaction, manufacturer, caliber or gauge, quantity of projectiles, and the name and address of the purchaser. The information required by this paragraph shall be entered in the proper record book not later than the seventh day following the date of the transaction, and such information shall be recorded under the following formats (See Tables 1 & 2 below):

(c) Notwithstanding the provisions of paragraph (b) of this section, the Director of Industry Operations may authorize alternate records to be maintained by a licensed importer to record the disposal of firearms and armor piercing ammunition when it is shown by the licensed importer that such alternate records will accurately and readily disclose the information required by paragraph (b) of this section. A licensed importer who proposes to use alternate records shall submit a letter application, in duplicate, to the Director of Industry Operations and shall describe the proposed alternate records and the need therefor. Such alternate records shall not be employed by the licensed importer until approval in such regard is received from the Director of Industry Operations.

(d) Each licensed importer shall maintain separate records of the sales or other dispositions made of firearms to nonlicensees. Such records shall be maintained in the form and manner as prescribed by §§ 478.124 and 478.125 in regard to firearms transaction records and records of acquisition and disposition of firearms.

(Approved by the Office of Management and Budget under control number 1512–0387)

§ 478.123 Records maintained by manufacturers.

(a) Each licensed manufacturer shall record the type, model, caliber or gauge, and serial number of each complete firearm manufactured or otherwise acquired, and the date such manufacture or other acquisition was made. The information required by this paragraph shall be recorded not later than the seventh day following the date such manufacture or other acquisition was made.

(b) A record of firearms disposed of by a manufacturer to another licensee and a separate record of armor piercing ammunition dispositions to governmental entities, for exportation, or for testing or experimentation authorized under the provision of § 478.149 shall be maintained by the licensed manufacturer on the licensed premises. For firearms, the record shall show the quantity, type, model, manufacturer, caliber, size or gauge, serial number of the firearms so transferred, the name and license number of the licensee to whom the firearms were transferred, and the date of the transaction. For armor piercing ammunition, the record shall show the manufacturer, caliber or gauge, quantity, the name and address of the transferee to whom the armor piercing ammunition was transferred, and the date of the transaction. The information required by this paragraph shall be entered in the proper record book not later than the seventh day following the date of the transaction, and such information shall be recorded under the format prescribed by § 478.122, except that the name of the manufacturer of a firearm or armor piercing ammunition need not be recorded if the firearm or armor piercing ammunition is of the manufacturer's own manufacture.

(c) Notwithstanding the provisions of paragraph (b) of this section, the Director of Industry Operations may authorize alternate records to be maintained by a licensed manufacturer to record the disposal of firearms and armor piercing ammunition when it is shown by the licensed manufacturer that such alternate records will accurately and readily disclose the information required by paragraph (b) of this section. A licensed manufacturer who proposes to use alternate records shall submit a letter application, in duplicate, to the Director of Industry Operations and shall describe the proposed alternate record and the need therefor. Such alternate records shall not be employed by the licensed manufacturer until approval in such regard is received from the Director of Industry Operations.

(d) Each licensed manufacturer shall maintain separate records of the sales or other dispositions made of firearms to nonlicensees. Such records shall be maintained in the form and manner as prescribed by § 478.124 and § 478.125 in regard to firearms transaction records and records of acquisition and disposition of firearms.

(Approved by the Office of Management and Budget under control number 1512–0369)

§ 478.124 Firearms transaction record.

(a) A licensed importer, licensed manufacturer, or licensed dealer shall not sell or otherwise dispose, temporarily or permanently, of any firearm to any person, other than another licensee, unless the licensee records the transaction on a firearms transaction record, Form 4473: **Provided,** That a firearms transaction record, Form 4473, shall not be required to record the disposition made of a firearm delivered to a licensee for the sole purpose of repair or customizing when such firearm or a replacement firearm is returned to the person from whom received.

(b) A licensed manufacturer, licensed importer, or licensed dealer shall retain in alphabetical (by name of purchaser), chronological (by date of disposition), or numerical (by transaction serial number) order, and as a part of the required records, each Form 4473 obtained in the course of transferring custody of the firearms.

(c) (1) Prior to making an over-the-counter transfer of a firearm to a nonlicen-

TABLE 1: Importer's Firearms Disposition Record

Quantity	Type	Manufacturer	Country of manufacture	Caliber or gauge	Model	Serial No.	Name and license No. of licensee to whom transferred	Date of the transaction

TABLE 2: Importer's Armor Piercing Ammunition Disposition Record

Date	Manufacturer	Caliber or gauge	Quantity of projectiles	Purchaser – Name and address

see who is a resident of the State in which the licensee's business premises is located, the licensed importer, licensed manufacturer, or licensed dealer so transferring the firearm shall obtain a Form 4473 from the transferee showing the transferee's name, sex, residence address (including county or similar political subdivision), date and place of birth; height, weight and race of the transferee; the transferee's country of citizenship; the transferee's INS-issued alien number or admission number; the transferee's State of residence; and certification by the transferee that the transferee is not prohibited by the Act from transporting or shipping a firearm in interstate or foreign commerce or receiving a firearm which has been shipped or transported in interstate or foreign commerce or possessing a firearm in or affecting commerce.

(2) In order to facilitate the transfer of a firearm and enable NICS to verify the identity of the person acquiring the firearm, ATF Form 4473 also requests certain optional information. This information includes the transferee's social security number. Such information may help avoid the possibility of the transferee being misidentified as a felon or other prohibited person.

(3) After the transferee has executed the Form 4473, the licensee:

(i) Shall verify the identity of the transferee by examining the identification document (as defined in § 478.11) presented, and shall note on the Form 4473 the type of identification used;

(ii) Shall, in the case of a transferee who is an alien legally in the United States, cause the transferee to present documentation establishing that the transferee is a resident of the State (as defined in § 478.11) in which the licensee's business premises is located, and shall note on the form the documentation used. Examples of acceptable documentation include utility bills or a lease agreement which show that the transferee has resided in the State continuously for at least 90 days prior to the transfer of the firearm; and

(iii) Must, in the case of a transferee who is a nonimmigrant alien who states that he or she falls within an exception to, or has a waiver from, the nonimmigrant alien prohibition, have the transferee present applicable documentation establishing the exception or waiver, note on the Form 4473 the type of documentation provided, and attach a copy of the documentation to the Form 4473.

(iv) Shall comply with the requirements of § 478.102 and record on the form the date on which the licensee contacted the NICS, as well as any response provided by the system, including any identification number provided by the system.

(4) The licensee shall identify the firearm to be transferred by listing on the Form 4473 the name of the manufacturer, the name of the importer (if any), the type, model, caliber or gauge, and the serial number of the firearm.

(5) The licensee shall sign and date the form if the licensee does not know or have reasonable cause to believe that the transferee is disqualified by law from receiving the firearm and transfer the firearm described on the Form 4473.

(d) Prior to making an over-the-counter transfer of a shotgun or rifle under the provisions contained in § 478.96(c) to a nonlicensee who is not a resident of the State in which the licensee's business premises is located, the licensee so transferring the shotgun or rifle, and such transferee, shall comply with the requirements of paragraph (c) of this section: **Provided,** That in the case of a transferee who is an alien legally in the United States, the documentation required by paragraph (c)(3)(ii) of this section need only establish that the transferee is a resident of any State and has resided in such State continuously for at least 90 days prior to the transfer of the firearm. Examples of acceptable documentation include utility bills or a lease agreement. The licensee shall note on the form the documentation used.

(e) Prior to making a transfer of a firearm to any nonlicensee who is not a resident of the State in which the licensee's business premises is located, and such nonlicensee is acquiring the firearm by loan or rental from the licensee for temporary use for lawful sporting purposes, the licensed importer, licensed manufacturer, or licensed dealer so furnishing the firearm, and such transferee, shall comply with the provisions of paragraph (c) of this section, except for the provisions of paragraph (c)(3)(ii).

(f) Form 4473 shall be submitted, in duplicate, to a licensed importer, licensed manufacturer, or licensed dealer by a transferee who is purchasing or otherwise acquiring a firearm by other than an over-the-counter transaction, who is not subject to the provisions of § 478.102(a), and who is a resident of the State in which the licensee's business premises are located. The Form 4473 shall show the name, address, date and place of birth, height, weight, and race of the transferee; and the title, name, and address of the principal law enforcement officer of the locality to which the firearm will be delivered. The transferee also must date and execute the sworn statement contained on the form showing, in case the firearm to be transferred is a firearm other than a shotgun or rifle, the transferee is 21 years or more of age; in case the firearm to be transferred is a shotgun or rifle, the transferee is 18 years or more of age; whether the transferee is a citizen of the United States; the transferee's State of residence, and in the case of a transferee who is an alien legally in the United States, the transferee has resided in that State continuously for at least 90 days prior to the transfer of the firearm; the transferee is not prohibited by the provisions of the Act from shipping or transporting a firearm in interstate or foreign commerce or receiving a firearm which has been shipped or transported in interstate or foreign commerce or possessing a firearm in or affecting commerce; and the transferee's receipt of the firearm would not be in violation of any statute of the State or published ordinance applicable to the locality in which the transferee resides. Upon receipt of such Forms 4473, the licensee shall identify the firearm to be transferred by listing in the Forms 4473 the name of the manufacturer, the name of the importer if any), the type, model, caliber or gauge, and the serial number of the firearm to be transferred. The licensee shall prior to shipment or delivery of the firearm to such transferee, forward by registered or certified mail (return receipt requested) a copy of the Form 4473 to the principal law enforcement officer named in the Form 4473 by the transferee, and shall delay shipment or delivery of the firearm to the transferee for a period of at least 7 days following receipt by the licensee of the return receipt evidencing delivery of the copy of the Form 4473 to such principal law enforcement officer, or the return of the copy of the Form 4473 to the licensee due to the refusal of such principal law enforcement officer to accept same in accordance with U.S. Postal Service regulations. The original Form 4473, and evidence of receipt or rejection of delivery of the copy of the Form 4473 sent to the principal law enforcement officer, shall be retained by the licensee as a part of the records required to be kept under this subpart.

(g) A licensee who sells or otherwise disposes of a firearm to a nonlicensee who is other than an individual, shall obtain from the transferee the information required by this section from an individual authorized to act on behalf of the transferee. In addition, the licensee shall obtain from the individual acting on behalf of the transferee a written statement, executed under the penalties of perjury, that the firearm is being acquired for the use of and will be the property of the transferee, and showing the name and address of that transferee.

(h) The requirements of this section shall be in addition to any other recordkeeping requirement contained in this part.

(i) A licensee may obtain, upon request, an emergency supply of Forms 4473 from any Director of Industry Operations. For normal usage, a licensee should request a year's supply from the ATF Distribution Center, 7943 Angus Court, Springfield, Virginia 22153.

(Paragraph (c) approved by the Office of Management and Budget under control numbers 1512–0544, 1512–0129, and 1512–0570; paragraph (f) approved by the Office of Management and Budget under control number 1512–0130; all other recordkeeping approved by the Office of Management and Budget under control number 1512–0129)

§ 478.124a Firearms transaction record in lieu of record of receipt and disposition.

(a) A licensed dealer acquiring firearms after August 1, 1988 and contemplating the disposition of not more than 50 firearms within a succeeding 12-month period to licensees or nonlicensees may maintain a record of the acquisition and disposition of such firearms on a firearms transaction record, Form 4473(LV), Part I or II, in lieu of the records prescribed by § 478.125. Such 12-month period shall commence from the date the licensed dealer first records the purchase or other acquisition of a firearm on Form 4473(LV) pursuant to this section. A licensed dealer who maintains records pursuant to this section, but whose firearms dispositions exceed 50 firearms within such 12-month period, shall make and maintain the acquisition and disposition records required by § 478.125 with respect to each firearm exceeding 50.

(b) Each licensed dealer maintaining firearms acquisition and disposition records pursuant to this section shall record the purchase or other acquisition of a firearm on Form 4473(LV), Part I or II, in accordance with the instructions on the form not later than the close of the next business day following the date of such purchase or acquisition. However, when disposition is made of a firearm before the close of the next business day after the receipt of that firearm, the licensed dealer making such disposition shall enter all required acquisition information regarding the firearm on the Form 4473(LV) at the time such transfer or disposition is made. The record on Form 4473(LV) shall show the date of receipt, the name and address or the name and license number of the person from whom received, the name of the manufacturer and importer (if any), the model, serial number, type, and caliber or gauge of the firearm.

(c) Each licensed dealer maintaining firearms acquisition and disposition records pursuant to this section shall retain Form 4473(LV), Part I or II, reflecting firearms possessed by such business in chronological (by date of receipt) or numerical (by transaction serial number) order. Forms 4473(LV) reflecting the licensee's sale or disposition of firearms shall be retained in alphabetical (by name of purchaser), chronological (by date of disposition) or numerical (by transaction serial number) order.

(d) A licensed dealer maintaining records pursuant to this section shall record the sale or other disposition of a firearm to another licensee by entering on the Form 4473(LV), Part I, associated with such firearm, the name and license number of the person to whom transferred and by signing and dating the form.

(e) A licensed dealer shall obtain the Form 4473(LV), Part I, associated with the firearm in lieu of a Form 4473 and comply with the requirements specified in § 478.124(c) prior to making an over-the-counter transfer of a firearm to a nonlicensee:

(1) Who is a resident of the State in which the licensee's business premises is located,

(2) Who is not a resident of the State in which the licensee's business premises is located and the firearm is a shotgun or rifle and the transfer is under the provisions of § 478.96(c), or

(3) Who is not a resident of the State in which the licensee's business premises is located and who is acquiring the firearm by loan or rental for temporary use for lawful sporting purposes.

(f) A licensed dealer shall obtain the Form 4473(LV), Part II, associated with the firearm in lieu of a Form 4473 and comply with the requirements specified in § 478.124(f) prior to making a disposition of a firearm to a nonlicensee who is purchasing or otherwise acquiring a firearm by other than an over-the-counter transaction and who is a resident of the State in which the licensee's business premises is located. If the licensee's record of the acquisition of the firearm is, at the time of the disposition, being maintained on a Form 4473(LV), Part I, for over-the-counter transactions, the licensee shall transfer the information relative to the receipt of the firearm, as required by paragraph (b) of this section, to Form 4473(LV), Part II. The corresponding form 4473(LV), Part I, may then be destroyed.

§ 478.125 Record of receipt and disposition.

(a) Armor piercing ammunition sales by licensed collectors to nonlicensees. The sale or other disposition of armor piercing ammunition by licensed collectors shall be recorded in a bound record at the time a transaction is made. The bound record shall be maintained in chronological order by date of sale or disposition of the armor piercing ammunition, and shall be retained on the licensed premises of the licensee for a period not less than two years following the date of the recorded sale or disposition of the armor piercing ammunition. The bound record entry shall show:

(1) The date of the transaction;

(2) The name of the manufacturer;

(3) The caliber or gauge;

(4) The quantity of projectiles;

(5) The name, address, and date of birth of the nonlicensee; and

(6) The method used to establish the identity of the armor piercing ammunition purchaser. The format required for the bound record is as follows:
(Table 3)

However, when a commercial record is made at the time a transaction is made, a licensee may delay making an entry into the bound record if the provisions of paragraph (d) of this section are complied with.

TABLE 3: Disposition Record of Armor Piercing Ammunition

Date	Manufacturer	Caliber or gauge	Quantity of projectiles	Purchaser		Enter a (x) in the "known" column if purchaser is personally known to you. Otherwise, establish the purchaser's identification		
				Name and address	Date of birth	Known	Driver's License	Other type (specify)

(b) Armor piercing ammunition sales by licensed collectors to licensees. Sales or other dispositions of armor piercing ammunition from a licensed collector to another licensee shall be recorded and maintained in the manner prescribed in § 478.122(b) for importers: **Provided**, That the license number of the transferee may be recorded in lieu of the transferee's address.

(c) Armor piercing ammunition sales by licensed dealers to governmental entities. A record of armor piercing ammunition disposed of by a licensed dealer to a governmental entity pursuant to § 478.99(e) shall be maintained by the licensed dealer on the licensed premises and shall show the name of the manufacturer, the caliber or gauge, the quantity, the name and address of the entity to which the armor piercing ammunition was transferred, and the date of the transaction. Such information shall be recorded under the format prescribed by § 478.122(b). Each licensed dealer disposing of armor piercing ammunition pursuant to § 478.99(e) shall also maintain a record showing the date of acquisition of such ammunition which shall be filed in an orderly manner separate from other commercial records maintained and be readily available for inspection. The records required by this paragraph shall be retained on the licensed premises of the licensee for a period not less than two years following the date of the recorded sale or disposition of the armor piercing ammunition.

(d) Commercial records of armor piercing ammunition transactions. When a commercial record is made at the time of sale or other disposition of armor piercing ammunition, and such record contains all information required by the bound record prescribed by paragraph (a) of this section, the licensed collector transferring the armor piercing ammunition may, for a period not exceeding 7 days following the date of such transfer, delay making the required entry into such bound record: **Provided,** That the commercial record pertaining to the transfer is:

(1) Maintained by the licensed collector separate from other commercial documents maintained by such licensee, and

(2) Is readily available for inspection on the licensed premises until such time as the required entry into the bound record is made.

(e) Firearms receipt and disposition by dealers. Except as provided in § 478.124a with respect to alternate records for the receipt and disposition of firearms by dealers, each licensed dealer shall enter into a record each receipt and disposition of firearms. In addition, before commencing or continuing a firearms business, each licensed dealer shall inventory the firearms possessed for such business and shall record same in the record required by this paragraph. The record required by this paragraph shall be maintained in bound form under the format prescribed below. The purchase or other acquisition of a firearm shall, except as provided in paragraph (g) of this section, be recorded not later than the close of the next business day following the date of such purchase or acquisition. The record shall show the date of receipt, the name and address or the name and license number of the person from whom received, the name of the manufacturer and importer (if any), the model, serial number, type, and the caliber or gauge of the firearm. The sale or other disposition of a firearm shall be recorded by the licensed dealer not later than 7 days following the date of such transaction. When such disposition is made to a nonlicensee, the firearms transaction record, Form 4473, obtained by the licensed dealer shall be retained, until the transaction is recorded, separate from the licensee's Form 4473 file and be readily available for inspection. When such disposition is made to a licensee, the commercial record of the transaction shall be retained, until the transaction is recorded, separate from other commercial documents maintained by the licensed dealer, and be readily available for inspection. The record shall show the date of the sale or other disposition of each firearm, the name and address of the person to whom the firearm is transferred, or the name and license number of the person to whom transferred if such person is a licensee, or the firearms transaction record, Form 4473, serial number if the licensed dealer transferring the firearm serially numbers the Forms 4473 and files them numerically. The format required for the record of receipt and disposition of firearms is as follows: **(Table 4)**

(f) Firearms receipt and disposition by licensed collectors. Each licensed collector shall enter into a record each receipt and disposition of firearms curios or relics. The record required by this paragraph shall be maintained in bound form under the format prescribed below. The purchase or other acquisition of a curio or relic shall, except as provided in paragraph (g) of this section, be recorded not later than the close of the next business day following the date of such purchase or other acquisition. The record shall show the date of receipt, the name and address or the name and license number of the person from whom received, the name of the manufacturer and importer (if any), the model, serial number, type, and the caliber or gauge of the firearm curio or relic. The sale or other disposition of a curio or relic shall be recorded by the licensed collector not later than 7 days following the date of such transaction. When such disposition is made to a licensee, the commercial record of the transaction shall be retained, until the transaction is recorded, separate from other commercial documents maintained by the licensee, and be readily available for inspection. The record shall show the date of the sale or other disposition of each firearm curio or relic, the name and address of the person to whom the firearm curio or relic is transferred, or the name and license number of the person to whom transferred if such person is a licensee, and the date of birth of the transferee if other than a licensee. In addition, the licensee shall—

(1) Cause the transferee, if other than a licensee, to be identified in any manner customarily used in commercial transactions (e.g., a driver's license), and note on the record the method used, and

(2) In the case of a transferee who is an alien legally in the United States and who is other than a licensee—

(i) Verify the identity of the transferee by examining an identification document (as defined in § 478.11), and

TABLE 4: Firearms Acquisition and Disposition Record

Description of firearm					Receipt		Disposition		
Manufacturer and/or importer	Model	Serial No.	Type	Caliber or gauge	Date	Name and address or name and license No.	Date	Name	Address or license No. if licensee, or Form 4473 Serial No. if Forms 4733 filed numerically

(ii) Cause the transferee to present documentation establishing that the transferee is a resident of the State (as defined in § 478.11) in which the licensee's business premises is located if the firearm curio or relic is other than a shotgun or rifle, and note on the record the documentation used or is a resident of any State and has resided in such State continuously for at least 90 days prior to the transfer of the firearm if the firearm curio or relic is a shotgun or rifle and shall note on the record the documentation used. Examples of acceptable documentation include utility bills or a lease agreement which show that the transferee has resided in the State continuously for at least 90 days prior to the transfer of the firearm curio or relic.

(3) The format required for the record of receipt and disposition of firearms by collectors is as follows: **(See Table 5)**

(g) Commercial record s o f firearms received. When a commercial record is held by a licensed dealer or licensed collector showing the acquisition of a firearm or firearm curio or relic, and such record contains all acquisition information required by the bound record prescribed by paragraphs (e) and (f) of this section, the licensed dealer or licensed collector acquiring such firearm or curio or relic, may, for a period not exceeding 7 days following the date of such acquisition, delay making the required entry into such bound record: **Provided,** That the commercial record is, until such time as the required entry into the bound record is made, **(1)** maintained by the licensed dealer or licensed collector separate from other commercial documents maintained by such licensee, and **(2)** readily available for inspection on the licensed premises: **Provided further,** That when disposition is made of a firearm or firearm curio or relic not entered in the bound record under the provisions of this paragraph, the licensed dealer or licensed collector making such disposition shall enter all required acquisition information regarding the firearm or firearm curio or relic in the bound record at the time such transfer or disposition is made.

(h) Alternate records. Notwithstanding the provisions of paragraphs (a), (e), and (f) of this section, the Director of Industry Operations may authorize alternate records to be maintained by a licensed dealer or licensed collector to record the acquisition and disposition of firearms or curios or relics and the disposition of armor piercing ammunition when it is shown by the licensed dealer or the licensed collector that such alternate records will accurately and readily disclose the required information. A licensed dealer or licensed collector who proposes to use alternate records shall submit a letter application, in duplicate, to the Director of Industry Operations and shall describe the proposed alternate records and the need therefor. Such alternate records shall not be employed by the licensed dealer or licensed collector until approval in such regard is received from the Director of Industry Operations.

(i) Requirements for imp orters an d manufacturers. Each licensed importer and licensed manufacturer selling or otherwise disposing of firearms or armor piercing ammunition to nonlicensees shall maintain such records of such transactions as are required of licensed dealers by this section.

(Approved by the Office of Management and Budget under control number 1512–0387)

§ 478.125a Personal firearms collection.

(a) Notwithstanding any other provision of this subpart, a licensed manufacturer, licensed importer, or licensed dealer is not required to comply with the provisions of § 478.102 or record on a firearms transaction record, Form 4473, the sale or other disposition of a firearm maintained as part of the licensee's personal firearms collection: **Provided**, That **(1)** The licensee has maintained the firearm as part of such collection for 1 year from the date the firearm was transferred from the business inventory into the personal collection or otherwise acquired as a personal firearm, **(2)** The licensee recorded in the bound record prescribed by § 478.125(e) the receipt of the firearm into the business inventory or other acquisition, **(3)** The licensee recorded the firearm as a disposition in the bound record prescribed by § 478.125(e) when the firearm was transferred from the business inventory into the personal firearms collection or otherwise acquired as a personal firearm, and **(4)** The licensee enters the sale or other disposition of the firearm from the personal firearms collection into a bound record, under the format prescribed below, identifying the firearm transferred by recording the name of the manufacturer and importer (if any), the model, serial number, type, and the caliber or gauge, and showing the date of the sale or other disposition, the name and address of the transferee, or the name and business address of the transferee if such person is a licensee, and the date of birth of the transferee if other than a licensee. In addition, the licensee shall cause the transferee, if other than a licensee, to be identified in any manner customarily used in commercial transactions (e.g., a driver's license). The format required for the disposition record of personal firearms is as follows: **(See Table 6 on page 67)**

(b) Any licensed manufacturer, licensed importer, or licensed dealer selling or otherwise disposing of a firearm from the licensee's personal firearms collection under this section shall be subject to the restrictions imposed by the Act and this part on the dispositions of firearms by persons other than licensed manufacturers, licensed importers, and licensed dealers.

(Approved by the Office of Management and Budget under control number 1512–0387)

§ 478.126 Furnishing transaction information.

(a) Each licensee shall, when required by letter issued by the Director of Industry Operations, and until notified to the contrary in writing by such officer, submit on Form 4483, Report of Firearms Transac-

TABLE 5: Firearms Collectors Acquisition and Disposition Record

Description of firearm					Receipt		Disposition				
Manufacturer and/or importer	Model	Serial No.	Type	Caliber or gauge	Date	Name and address or name and license No.	Date	Name and address or name and license No.	Date of birth if nonlicensee	Driver's license No. or other identification if nonlicensee	For transfers to aliens, documentation used to establish residency

tions, for the periods and at the times specified in the letter issued by the Director of Industry Operations, all record information required by this subpart, or such lesser record information as the Director of Industry Operations in his letter may specify.

(b) The Director of Industry Operations may authorize the information to be submitted in a manner other than that prescribed in paragraph (a) of this section when it is shown by a licensee that an alternate method of reporting is reasonably necessary and will not unduly hinder the effective administration of this part. A licensee who proposes to use an alternate method of reporting shall submit a letter application, in duplicate, to the Director of Industry Operations and shall describe the proposed alternate method of reporting and the need therefor. An alternate method of reporting shall not be employed by the licensee until approval in such regard is received from the Director of Industry Operations.

(Approved by the Office of Management and Budget under control number 1512–0387)

§ 478.126a Reporting multiple sales or other disposition of pistols and revolvers.

Each licensee shall prepare a report of multiple sales or other disposition whenever the licensee sells or otherwise disposes of, at one time or during any five consecutive business days, two or more pistols, or revolvers, or any combination of pistols and revolvers totaling two or more, to an unlicensed person: **Provided,** That a report need not be made where pistols or revolvers, or any combination thereof, are returned to the same person from whom they were received. The report shall be prepared on Form 3310.4, Report of Multiple Sale or Other Disposition of Pistols and Revolvers. Not later than the close of business on the day that the multiple sale or other disposition occurs, the licensee shall forward two copies of Form 3310.4 to the ATF office specified thereon and one copy to the State police or to the local law enforcement agency in which the sale or other disposition took place. Where the State or local law enforcement officials have notified the licensee that a particular official has been designated to receive Forms 3310.4, the licensee shall forward such forms to that designated official. The licensee shall retain one copy of Form 3310.4 and attach it to the firearms transaction record, Form 4473, executed upon delivery of the pistols or revolvers.

Example. 1. A licensee sells a pistol and revolver in a single transaction to an unlicensed person. This is a multiple sale and must be reported not later than the close of business on the date of the transaction.

Example. 2. A licensee sells a pistol on Monday and sells a revolver on the following Friday to the same unlicensed person. This is a multiple sale and must be reported not later than the close of business on Friday. If the licensee sells the same unlicensed person another pistol or revolver on the following Monday, this would constitute an additional multiple sale and must also be reported.

Example 3. A licensee maintaining business hours on Monday through Saturday sells a revolver to an unlicensed person on Monday and sells another revolver to the same person on the following Saturday. This does not constitute a multiple sale and need not be reported since the sales did not occur during five consecutive business days.

(Approved by the Office of Management and Budget under control number 1512–0006)

§ 478.127 Discontinuance of business.

Where a licensed business is discontinued and succeeded by a new licensee, the records prescribed by this subpart shall appropriately reflect such facts and shall be delivered to the successor. Where discontinuance of the business is absolute, the records shall be delivered within 30 days following the business discontinuance to the ATF Out-of-Business Records Center, Spring Mills Office Park, 2029 Stonewall Jackson Drive, Falling Waters, West Virginia 25419, or to any ATF office in the division in which the business was located: **Provided, however,** Where State law or local ordinance requires the delivery of records to other responsible authority, the Chief, National Licensing Center may arrange for the delivery of the records required by this subpart to such authority: **Provided further,** That where a licensed business is discontinued and succeeded by a new licensee, the records may be delivered within 30 days following the business discontinuance to the ATF Out-of-Business Records Center or to any ATF office in the division in which the business was located.

§ 478.128 False statement or representation.

(a) Any person who knowingly makes any false statement or representation in applying for any license or exemption or relief from disability, under the provisions of the Act, shall be fined not more than $5,000 or imprisoned not more than 5 years, or both.

(b) Any person other than a licensed manufacturer, licensed importer, licensed dealer, or licensed collector who knowingly makes any false statement or representation with respect to any information required by the provisions of the Act or this part to be kept in the records of a person licensed under the Act or this part shall be fined not more than $5,000 or imprisoned not more than 5 years, or both.

(c) Any licensed manufacturer, licensed importer, licensed dealer, or licensed collector who knowingly makes any false statement or representation with respect to any information required by the provisions of the Act or this part to be kept in the records of a person licensed under the Act or this part shall be fined not more than $1,000 or imprisoned not more than 1 year, or both.

§ 478.129 Record retention.

(a) Records prior to Act. Licensed importers and licensed manufacturers may dispose of records of sale or other disposition of firearms prior to December 16, 1968. Licensed dealers and licensed collectors may dispose of all records of firearms transactions that occurred prior to December 16, 1968.

(b) Firearms transaction record. Licensees shall retain each Form 4473 and Form 4473(LV) for a period of not less than 20 years after the date of sale or disposition. Where a licensee has initiated a NICS check for a proposed firearms transaction, but the sale, delivery, or transfer of the firearm is not made, the licensee shall record any transaction number on the Form 4473, and retain the Form 4473 for a period of not less than 5 years after the date of the NICS inquiry. Forms 4473 shall be retained in the licensee's records as provided in § 478.124(b): **Provided,** That Forms 4473 with respect to which a

Table 6: Disposition Record of Personal Firearms

Description of firearm					Disposition		
Manufacturer and/or importer	Model	Serial No.	Type	Caliber or gauge	Date	Name and address (business address if licensee)	Date of birth of nonlicensee

sale, delivery or transfer did not take place shall be separately retained in alphabetical (by name of transferee) or chronological (by date of transferee's certification) order.

(c) Statement of intent to obtain a handgun, reports of multiple sales or other disposition of pistols and revolvers, and reports of theft or loss of firearms. Licensees shall retain each Form 5300.35 (Statement of Intent to Obtain a Handgun(s)) for a period of not less than 5 years after notice of the intent to obtain the handgun was forwarded to the chief law enforcement officer, as defined in § 478.150(c). Licensees shall retain each copy of Form 3310.4 (Report of Multiple Sale or Other Disposition of Pistols and Revolvers) for a period of not less than 5 years after the date of sale or other disposition. Licensees shall retain each copy of Form 3310.11 (Federal Firearms Licensee Theft/Loss Report) for a period of not less than 5 years after the date the theft or loss was reported to ATF.

(d) Records of importation and manufacture. Licensees will maintain permanent records of the importation, manufacture, or other acquisition of firearms, including ATF Forms 6 and 6A as required by subpart G of this part. Licensed importers' records and licensed manufacturers' records of the sale or other disposition of firearms after December 15, 1968, shall be retained through December 15, 1988, after which records of transactions over 20 years of age may be discarded.

(e) Records of dealers and collectors under the Act. The records prepared by licensed dealers and licensed collectors under the Act of the sale or other disposition of firearms and the corresponding record of receipt of such firearms shall be retained through December 15, 1988, after which records of transactions over 20 years of age may be discarded.

(f) Retention of records of transactions in semiautomatic assault weapons. The documentation required by §§ 478.40(c) and 478.132 shall be retained in the licensee's permanent records for a period of not less than 5 years after the date of sale or other disposition.

(Paragraph (b) approved by the Office of Management and Budget under control number 1512–0544; Paragraph (c) approved by the Office of Management and Budget under control numbers 1512–0520, 1512–0006, and 1512–0524; Paragraph (f) approved by the Office of Management and Budget under control number 1512–0526; all other recordkeeping approved by the Office of Management and Budget under control number 1512–0129)

§ 478.131 Firearms transactions not subject to a NICS check.

(a) **(1)** A licensed importer, licensed manufacturer, or licensed dealer whose sale, delivery, or transfer of a firearm is made pursuant to the alternative provisions of § 478.102(d) and is not subject to the NICS check prescribed by § 478.102(a) shall maintain the records required by paragraph (a) of this section.

(2) If the transfer is pursuant to a permit or license in accordance with § 478.102(d)(1), the licensee shall either retain a copy of the purchaser's permit or license and attach it to the firearms transaction record, Form 4473, or record on the firearms transaction record, Form 4473, any identifying number, the date of issuance, and the expiration date (if provided) from the permit or license.

(3) If the transfer is pursuant to a certification by ATF in accordance with §§ 478.102(d)(3) and 478.150, the licensee shall maintain the certification as part of the records required to be kept under this subpart and for the period prescribed for the retention of Form 5300.35 in § 478.129(c).

(b) The requirements of this section shall be in addition to any other recordkeeping requirements contained in this part.

(Approved by the Office of Management and Budget under control number 1512–0544)

§ 478.132 Dispositions of semiautomatic assault weapons and large capacity ammunition feeding devices to law enforcement officers for official use and to employees or contractors of nuclear facilities.

Licensed manufacturers, licensed importers, and licensed dealers in semiautomatic assault weapons, as well as persons who manufacture, import, or deal in large capacity ammunition feeding devices, may transfer such weapons and devices manufactured after September 13, 1994, to law enforcement officers and to employees or contractors of nuclear facilities with the following documentation:

(a) Law enforcement officers. **(1)** A written statement from the purchasing officer, under penalty of perjury, stating that the weapon or device is being purchased for use in performing official duties and that the weapon or device is not being acquired for personal use or for purposes of transfer or resale; and

(2) A written statement from a supervisor of the purchasing officer, on agency letterhead, under penalty of perjury, stating that the purchasing officer is acquiring the weapon or device for use in official duties, that the firearm is suitable for use in performing official duties, and that the weapon or device is not being acquired for personal use or for purposes of transfer or resale.

(b) Employees or contractors of nuclear facilities. **(1)** Evidence that the employee is employed by a nuclear facility licensed pursuant to 42 U.S.C. 2133 or evidence that the contractor has a valid contract with such a facility.

(2) A written statement from the purchasing employee or contractor under penalty of perjury, stating that the weapon or device is being purchased for one of the purposes authorized in § 478.40(b)(7) and § 478.40(b)(3), i.e., on-site physical protection, on-site or off-site training, or off-site transportation of nuclear materials.

(3) A written statement from a supervisor of the purchasing employee or contractor, on agency or company letterhead, under penalty of perjury, stating that the purchasing employee or contractor is acquiring the weapon or device for use in official duties, and that the weapon or device is not being acquired for personal use or for purposes of transfer or resale.

(Approved by the Office of Management and Budget under control number 1512–0526)

§ 478.133 Records of transactions in semiautomatic assault weapons.

The evidence specified in § 478.40(c), relating to transactions in semiautomatic assault weapons, shall be retained in the permanent records of the manufacturer or dealer and in the records of the licensee to whom the weapons are transferred.

(Approved by the Office of Management and Budget under control number 1512–0526)

§ 478.134 Sale of firearms to law enforcement officers.

(a) Law enforcement officers purchasing firearms for official use who provide the licensee with a certification on agency letterhead, signed by a person in authority within the agency (other than the officer purchasing the firearm), stating that the officer will use the firearm in official duties and that a records check reveals that the purchasing officer has no convictions for misdemeanor crimes of domestic violence are not required to complete Form 4473 or Form 5300.35. The law enforcement officer purchasing the firearm may purchase a firearm from a licensee in another State, regardless of where the officer resides or where the agency is located.

(b) **(1)** The following individuals are considered to have sufficient authority to certify that law enforcement officers purchasing firearms will use the firearms in the performance of official duties:

(i) In a city or county police department, the director of public safety or the chief or commissioner of police.

(ii) In a sheriff's office, the sheriff.

(iii) In a State police or highway patrol department, the superintendent or the supervisor in charge of the office to which the State officer or employee is assigned.

(iv) In Federal law enforcement offices, the supervisor in charge of the office to which the Federal officer or employee is assigned.

(2) An individual signing on behalf of the person in authority is acceptable, provided there is a proper delegation of authority.

(c) Licensees are not required to prepare a Form 4473 or Form 5300.35 covering sales of firearm made in accordance with paragraph (a) of this section to law enforcement officers for official use. However, disposition to the officer must be entered into the licensee's permanent records, and the certification letter must be retained in the licensee's files.

Subpart I—Exemptions, Seizures, and Forfeitures

§ 478.141 General.

With the exception of §§ 478.32(a)(9) and (d)(9) and 478.99(c)(9), the provisions of this part shall not apply with respect to:

(a) The transportation, shipment, receipt, possession, or importation of any firearm or ammunition imported for, sold or shipped to, or issued for the use of, the United States or any department or agency thereof or any State or any department, agency, or political subdivision thereof.

(b) The shipment or receipt of firearms or ammunition when sold or issued by the Secretary of the Army pursuant to section 4308 of Title 10, U.S.C., and the transportation of any such firearm or ammunition carried out to enable a person, who lawfully received such firearm or ammunition from the Secretary of the Army, to engage in military training or in competitions.

(c) The shipment, unless otherwise prohibited by the Act or any other Federal law, by a licensed importer, licensed manufacturer, or licensed dealer to a member of the U.S. Armed Forces on active duty outside the United States, or to clubs recognized by the Department of Defense whose entire membership is composed of such members of the U.S. Armed Forces, and such members or clubs may receive a firearm or ammunition determined by the Director to be generally recognized as particularly suitable for sporting purposes and intended for the personal use of such member or club. Before making a shipment of firearms or ammunition under the provisions of this paragraph, a licensed importer, licensed manufacturer, or licensed dealer may submit a written request, in duplicate, to the Director for a determination by the Director whether such shipment would constitute a violation of the Act or any other Federal law, or whether the firearm or ammunition is considered by the Director to be generally recognized as particularly suitable for sporting purposes.

(d) The transportation, shipment, receipt, possession, or importation of any antique firearm.

§ 478.142 Effect of pard ons and e xpunctions of convictions.

(a) A pardon granted by the President of the United States regarding a Federal conviction for a crime punishable by imprisonment for a term exceeding 1 year shall remove any disability which otherwise would be imposed by the provisions of this part with respect to that conviction.

(b) A pardon granted by the Governor of a State or other State pardoning authority or by the pardoning authority of a foreign jurisdiction with respect to a conviction, or any expunction, reversal, setting aside of a conviction, or other proceeding rendering a conviction nugatory, or a restoration of civil rights shall remove any disability which otherwise would be imposed by the provisions of this part with respect to the conviction, unless:

(1) The pardon, expunction, setting aside, or other proceeding rendering a conviction nugatory, or restoration of civil rights, expressly provides that the person may not ship, transport, possess or receive firearms; or

(2) The pardon, expunction, setting aside, or other proceeding rendering a conviction nugatory, or restoration of civil rights, did not fully restore the rights of the person to possess or receive firearms under the law of the jurisdiction where the conviction occurred.

§ 478.143 Relief from disabilities incurred by indictment.

A licensed importer, licensed manufacturer, licensed dealer, or licensed collector who is indicted for a crime punishable by imprisonment for a term exceeding 1 year may, notwithstanding any other provision of the Act, continue operations pursuant to his existing license during the term of such indictment and until any conviction pursuant to the indictment becomes final: **Provided,** That if the term of the license expires during the period between the date of the indictment and the date the conviction thereunder becomes final, such importer, manufacturer, dealer, or collector must file a timely application for the renewal of his license in order to continue operations. Such application shall show that the applicant is under indictment for a crime punishable by imprisonment for a term exceeding 1 year.

§ 478.144 Relief from disa bilities under the Act.

(a) Any person may make application for relief from the disabilities under section 922 (g) and (n) of the Act (see § 478.32).

(b) An application for such relief shall be filed, in triplicate, with the Director. It shall include the information required by this section and such other supporting data as the Director and the applicant deem appropriate.

(c) Any record or document of a court or other government entity or official required by this paragraph to be furnished by an applicant in support of an application for relief shall be certified by the court or other government entity or official as a true copy. An application shall include:

(1) In the case of an applicant who is an individual, a written statement from each of 3 references, who are not related to the applicant by blood or marriage and have known the applicant for at least 3 years, recommending the granting of relief;

(2) Written consent to examine and obtain copies of records and to receive statements and information regarding the applicant's background, including records, statements and other information concerning employment, medical history, military service, and criminal record;

(3) In the case of an applicant under indictment, a copy of the indictment or information;

(4) In the case of an applicant having been convicted of a crime punishable by imprisonment for a term exceeding 1 year, a copy of the indictment or information on which the applicant was convicted, the judgment of conviction or record of any plea of *nolo contendere* or plea of guilty or finding of guilt by the court, and any pardon, expunction, setting aside or other record purporting to show that the conviction was rendered nugatory or that civil rights were restored;

(5) In the case of an applicant who has been adjudicated a mental defective or committed to a mental institution, a copy of the order of a court, board, commission, or other lawful authority that made the adjudication or ordered the commitment, any petition that sought to have the applicant so adjudicated or committed, any medical records reflecting the reasons for commitment and diagnoses of the applicant, and any court order or finding of a court, board, commission, or other lawful authority showing the applicant's discharge from commitment, restoration of mental competency and the restoration of rights;

(6) In the case of an applicant who has been discharged from the Armed Forces under dishonorable conditions, a copy of the applicant's summary of service record (Department of Defense Form 214), charge sheet (Department of Defense Form 458), and final court martial order;

(7) In the case of an applicant who, having been a citizen of the United States, has renounced his or her citizenship, a copy of the formal renunciation of nationality before a diplomatic or consular officer of the United States in a foreign state or before an officer designated by the Attorney General when the United States was in a state of war (see 8 U.S.C. 1481(a) (5) and (6)); and

(8) In the case of an applicant who has been convicted of a misdemeanor crime of domestic violence, a copy of the indictment or information on which the applicant was convicted, the judgment of conviction or record of any plea of *nolo contendere* or plea of guilty or finding of guilt by the court, and any pardon, expunction, setting aside or other record purporting to show that the conviction was rendered nugatory or that civil rights were restored.

(d) The Director may grant relief to an applicant if it is established to the satisfaction of the Director that the circumstances regarding the disability, and the applicant's record and reputation, are such that the applicant will not be likely to act in a manner dangerous to public safety, and that the granting of the relief would not be contrary to the public interest. The Director will not ordinarily grant relief if the applicant has not been discharged from parole or probation for a period of at least 2 years. Relief will not be granted to an applicant who is prohibited from possessing all types of firearms by the law of the State where such applicant resides.

(e) In addition to meeting the requirements of paragraph (d) of this section, an applicant who has been adjudicated a mental defective or committed to a mental institution will not be granted relief unless the applicant was subsequently determined by a court, board, commission, or other lawful authority to have been restored to mental competency, to be no longer suffering from a mental disorder, and to have had all rights restored.

(f) Upon receipt of an incomplete or improperly executed application for relief, the applicant shall be notified of the deficiency in the application. If the application is not corrected and returned within 30 days following the date of notification, the application shall be considered as having been abandoned.

(g) Whenever the Director grants relief to any person pursuant to this section, a notice of such action shall be promptly published in the FEDERAL REGISTER, together with the reasons therefor.

(h) A person who has been granted relief under this section shall be relieved of any disabilities imposed by the Act with respect to the acquisition, receipt, transfer, shipment, transportation, or possession of firearms or ammunition and incurred by reason of such disability.

(i) **(1)** A licensee who incurs disabilities under the Act (see § 478.32(a)) during the term of a current license or while the licensee has pending a license renewal application, and who files an application for removal of such disabilities, shall not be barred from licensed operations for 30 days following the date on which the applicant was first subject to such disabilities (or 30 days after the date upon which the conviction for a crime punishable by imprisonment for a term exceeding 1 year becomes final), and if the licensee files the application for relief as provided by this section within such 30-day period, the licensee may further continue licensed operations during the pendency of the application. A licensee who does not file such application within such 30-day period shall not continue licensed operations beyond 30 days following the date on which the licensee was first subject to such disabilities (or 30 days from the date the conviction for a crime punishable by imprisonment for a term exceeding 1 year becomes final).

(2) In the event the term of a license of a person expires during the 30-day period specified in paragraph (i)(1) of this section, or during the pendency of the application for relief, a timely application for renewal of the license must be filed in order to continue licensed operations. Such license application shall show that the applicant is subject to Federal firearms disabilities, shall describe the event giving rise to such disabilities, and shall state when the disabilities were incurred.

(3) A licensee shall not continue licensed operations beyond 30 days following the date the Director issues notification that the licensee's applications for removal of disabilities has been denied.

(4) When as provided in this paragraph a licensee may no longer continue licensed operations, any application for renewal of license filed by the licensee during the pendency of the application for removal of disabilities shall be denied by the Director of Industry Operations.

§ 478.145 Research organizations.

The provisions of § 478.98 with respect to the sale or delivery of destructive devices, machine guns, short-barreled shotguns, and short-barreled rifles shall not apply to the sale or delivery of such devices and weapons to any research organization designated by the Director to receive same. A research organization desiring such designation shall submit a letter application, in duplicate, to the Director. Such application shall contain the name and address of the research organization, the names and addresses of the persons directing or controlling, directly or indirectly, the policies and management of such organization, the nature and purpose of the research being conducted, a description of the devices and weapons to be received, and the identity of the person or persons from whom such devices and weapons are to be received.

§ 478.146 Deliveries by mail to certain persons.

The provisions of this part shall not be construed as prohibiting a licensed importer, licensed manufacturer, or licensed dealer from depositing a firearm for conveyance in the mails to any officer, employee, agent, or watchman who, pursuant to the provisions of section 1715 of title 18, U.S.C., is eligible to receive through the mails pistols, revolvers, and other firearms capable of being concealed on the person, for use in connection with his official duties.

§ 478.147 Return of firearm.

A person not otherwise prohibited by Federal, State or local law may ship a firearm to a licensed importer, licensed manufacturer, or licensed dealer for any lawful purpose, and, notwithstanding any other provision of this part, the licensed manufacturer, licensed importer, or licensed dealer may return in interstate or foreign commerce to that person the firearm or a replacement firearm of the same kind and type. See § 478.124(a) for requirements of a Form 4473 prior to return. A person not otherwise prohibited by Federal, State or local law may ship a firearm curio or relic to a licensed collector for any lawful purpose, and, notwithstanding any other provision of this part, the licensed collector may return in interstate or foreign commerce to that person the firearm curio or relic.

§ 478.148 Armor piercing ammunition intended for sporting or industrial purposes.

The Director may exempt certain armor piercing ammunition from the requirements of this part. A person who desires to obtain an exemption under this section for any such ammunition which is primarily intended for sporting purposes or intended for industrial purposes, including charges used in oil and gas well perforating devices, shall submit a written request to the Director. Each request shall be executed under the penalties of perjury and contain a complete and accurate description of the ammunition, the name and address of the manufacturer or importer, the purpose of and use for which it is designed and intended, and any photographs, diagrams, or drawings as may be necessary to enable the Director to make a determination. The Director may require that a sample of the ammunition be submitted for examination and evaluation.

§ 478.149 Armor piercing ammunition manufactured or imported for the purpose of testing or experimentation.

The provisions of §§ 478.37 and 478.99(d) with respect to the manufacture or importation of armor piercing ammunition and the sale or delivery of armor piercing ammunition by manufacturers and importers shall not apply to the manufacture, importation, sale or delivery of armor piercing ammunition for the purpose of testing or experimentation as authorized by the Director. A person desiring such authorization to receive armor piercing ammunition shall submit a letter application, in duplicate, to the Director. Such application shall contain the name and addresses of the persons directing or controlling, directly or indirectly, the policies and management of the applicant, the nature or purpose of the testing or experimentation, a description of the armor piercing ammunition to be received, and the identity of the manufacturer or importer from whom such ammunition is to be received. The approved application shall be submitted to the manufacturer or importer who shall retain a copy as part of the records required by subpart H of this part.

§ 478.150 Alternative to NICS in certain geographical locations.

(a) The provisions of § 478.102(d)(3) shall be applicable when the Director has certified that compliance with the provisions of § 478.102(a)(1) is impracticable because:

(1) The ratio of the number of law enforcement officers of the State in which the transfer is to occur to the number of square miles of land area of the State does not exceed 0.0025;

(2) The business premises of the licensee at which the transfer is to occur are extremely remote in relation to the chief law enforcement officer; and

(3) There is an absence of telecommunications facilities in the geographical area in which the business premises are located.

(b) A licensee who desires to obtain a certification under this section shall submit a written request to the Director. Each request shall be executed under the penalties of perjury and contain information sufficient for the Director to make such certification. Such information shall include statistical data, official reports, or other statements of government agencies pertaining to the ratio of law enforcement officers to the number of square miles of land area of a State and statements of government agencies and private utility companies regarding the absence of telecommunications facilities in the geographical area in which the licensee's business premises are located.

(c) For purposes of this section and § 478.129(c), the **"chief law enforcement officer"** means the chief of police, the sheriff, or an equivalent officer or the designee of any such individual.

(Approved by the Office of Management and Budget under control number 1512–0544)

§ 478.151 Semiautomatic rifles or shotguns for testing or experimentation.

(a) The provisions of § 478.39 shall not apply to the assembly of semiautomatic rifles or shotguns for the purpose of testing or experimentation as authorized by the Director.

(b) A person desiring authorization to assemble nonsporting semiautomatic rifles or shotguns shall submit a written request, in duplicate, to the Director. Each such request shall be executed under the penalties of perjury and shall contain a complete and accurate description of the firearm to be assembled, and such diagrams or drawings as may be necessary to enable the Director to make a determination. The Director may require the submission of the firearm parts for examination and evaluation. If the submission of the firearm parts is impractical, the person requesting the authorization shall so advise the Director and designate the place where the firearm parts will be available for examination and evaluation.

§ 478.152 Seizure and forfeiture.

(a) Any firearm or ammunition involved in or used in any knowing violation of subsections (a)(4), (a)(6), (f), (g), (h), (i), (j), or (k) of section 922 of the Act, or knowing importation or bringing into the United States or any possession thereof any firearm or ammunition in violation of section 922(l) of the Act, or knowing violation of section 924 of the Act, or willful violation of any other provision of the Act or of this part, or any violation of any other criminal law of the United States, or any firearm or ammunition intended to be used in any offense referred to in paragraph (c) of this section, where such intent is demonstrated by clear and convincing evidence, shall be subject to seizure and forfeiture, and all provisions of the Internal Revenue Code of 1986 relating to the seizure, forfeiture, and disposition of firearms, as defined in section 5845(a) of that Code, shall, so far as applicable, extend to seizures and forfeitures under the provisions of the Act: **Provided,** That upon acquittal of the owner or possessor, or dismissal of the charges against such person other than upon motion of the Government prior to trial, or lapse of or court termination of the restraining order to which he is subject, the seized or relinquished firearms or ammunition shall be returned forthwith to the owner or possessor or to a person delegated by the owner or possessor unless the return of the firearms or ammunition would place the owner or possessor or the delegate of the owner or possessor in violation of law. Any action or proceeding for the forfeiture of firearms or ammunition shall be commenced within 120 days of such seizure.

(b) Only those firearms or quantities of ammunition particularly named and individually identified as involved in or used in any violation of the provisions of the Act or this part, or any other criminal law of the United States or as intended to be used in any offense referred to in paragraph (c) of this section, where such intent is demonstrated by clear and convincing evidence, shall be subject to seizure, forfeiture and disposition.

(c) The offenses referred to in paragraphs (a) and (b) of this section for which firearms and ammunition intended to be used in such offenses are subject to seizure and forfeiture are:

(1) Any crime of violence, as that term is defined in section 924(c)(3) of the Act;

(2) Any offense punishable under the Controlled Substances Act (21 U.S.C. 801 et seq.) or the Controlled Substances Import and Export Act (21 U.S.C. 951 et seq.);

(3) Any offense described in section 922(a)(1), 922(a)(3), 922(a)(5), or 922(b)(3) of the Act, where the firearm or ammunition intended to be used in such offense is involved in a pattern of activities which includes a violation of any offense described in section 922(a)(1), 922(a)(3), 922(a)(5), or 922(b)(3) of the Act;

(4) Any offense described in section 922(d) of the Act where the firearm or ammunition is intended to be used in such offense by the transferor of such firearm or ammunition;

(5) Any offense described in section 922(i), 922(j), 922(l), 922(n), or 924(b) of the Act; and

(6) Any offense which may be prosecuted in a court of the United States which involves the exportation of firearms or ammunition.

§ 478.153 Semiautomatic assault weapons and large capacity ammunition feeding devices manufactured or imported for the purposes of testing or experimentation.

The provisions of § 478.40 with respect to the manufacture, transfer, or possession of a semiautomatic assault weapon, and § 478.40a with respect to large capacity ammunition feeding devices, shall not apply to the manufacture, transfer, or possession of such weapons or devices by a manufacturer or importer for the purposes of testing or experimentation as authorized by the Director. A person desiring such authorization shall submit a letter application, in duplicate, to the Director. Such application shall contain the name and addresses of the persons directing or controlling, directly or indirectly, the policies and management of the applicant, the nature or purpose of the testing or experimentation, a description of the weapons or devices to be manufactured or imported, and the source of the weapons or devices. The approved application shall be retained as part of the records required by subpart H of this part.

Subpart J—[Reserved]

Subpart K—Exportation

§ 478.171 Exportation.

Firearms and ammunition shall be exported in accordance with the applicable provisions of section 38 of the Arms Export Control Act (22 U.S.C. 2778) and regulations thereunder. However, licensed manufacturers, licensed importers, and licensed dealers exporting firearms shall maintain records showing the manufacture or acquisition of the firearms as required by this part and records showing the name and address of the foreign consignee of the firearms and the date the firearms were exported. Licensed manufacturers and licensed importers exporting armor piercing ammunition and semiautomatic assault weapons manufactured after September 13, 1994, shall maintain records showing the name and address of the foreign consignee and the date the armor piercing ammunition or semiautomatic assault weapons were exported.

Editor's Note:

The references to "semiautomatic assault weapons" in section 478.171 are not applicable on or after September 13, 2004.

THE NATIONAL FIREARMS ACT

TITLE 26, UNITED STATES CODE, CHAPTER 53

INTERNAL REVENUE CODE

Editor's Note:

The National Firearms Act (NFA) is part of the Internal Revenue Code of 1986. The Internal Revenue Code, with the exception of the NFA, is administered and enforced by the Secretary of the Treasury. When ATF transferred to the Department of Justice under the Homeland Security Act of 2002, all its authorities, including the authority to administer and enforce the NFA, were transferred to the Attorney General. In order to keep all the references throughout the Internal Revenue Code consistent, references to the Secretary of the Treasury in the NFA were left unchanged by the Homeland Security Act. However, section 7801(a)(2), Title 26, U.S.C., provides that references to the term "Secretary" or "Secretary of the Treasury" in the NFA shall mean the Attorney General.

CHAPTER 53 – MACHINE GUNS, DESTRUCTIVE DEVICES, AND CERTAIN OTHER FIREARMS

SUBCHAPTER A — TAXES

Part I – Special Occupational Taxes

§ 5801 Imposition of tax.

(a) General rule. On first engaging in business and thereafter on or before July 1 of each year, every importer, manufacturer, and dealer in firearms shall pay a special (occupational) tax for each place of business at the following rates:

(1) Importers and manufacturers: $1,000 a year or fraction thereof.

(2) Dealers: $500 a year or fraction thereof.

(b) Reduced rates of tax for small importers and manufacturers.

(1) In general. Paragraph (1) of subsection (a) shall be applied by substituting "$500" for "$1,000" with respect to any taxpayer the gross receipts of which (for the most recent taxable year ending before the 1st day of the taxable period to which the tax imposed by subsection (a) relates) are less than $500,000.

(2) Controlled group rules. All persons treated as 1 taxpayer under section 5061(e)(3) shall be treated as 1 taxpayer for purposes of paragraph (1).

(3) Certain rules to apply. For purposes of paragraph (1), rules similar to the rules of subparagraphs (B) and (C) of section 448(c)(3) shall apply.

§ 5802 Registration of importers, manufacturers, and dealers.

On first engaging in business and thereafter on or before the first day of July of each year, each importer, manufacturer, and dealer in firearms shall register with the Secretary in each internal revenue district in which such business is to be carried on, his name, including any trade name, and the address of each location in the district where he will conduct such business. An individual required to register under this section shall include a photograph and fingerprints of the individual with the initial application. Where there is a change during the taxable year in the location of, or the trade name used in, such business, the importer, manufacturer, or dealer shall file an application with the Secretary to amend his registration. Firearms operations of an importer, manufacturer, or dealer may not be commenced at the new location or under a new trade name prior to approval of the application by the Secretary.

Part II – Tax on Transferring Firearms.

§ 5811 Transfer tax.

(a) Rate. There shall be levied, collected, and paid on firearms transferred a tax at the rate of $200 for each firearm transferred, except, the transfer tax on any firearm classified as any other weapon under section 5845(e) shall be at the rate of $5 for each such firearm transferred.

(b) By whom paid. The tax imposed

by subsection (a) of this section shall be paid by the transferor.

(c) Payment. The tax imposed by subsection (a) of this section shall be payable by the appropriate stamps prescribed for payment by the Secretary.

§ 5812 Transfers.

(a) Application. A firearm shall not be transferred unless **(1)** the transferor of the firearm has filed with the Secretary a written application, in duplicate, for the transfer and registration of the firearm to the transferee on the application form prescribed by the Secretary; **(2)** any tax payable on the transfer is paid as evidenced by the proper stamp affixed to the original application form; **(3)** the transferee is identified in the application form in such manner as the Secretary may by regulations prescribe, except that, if such person is an individual, the identification must include his fingerprints and his photograph; **(4)** the transferor of the firearm is identified in the application form in such manner as the Secretary may by regulations prescribe; **(5)** the firearm is identified in the application form in such manner as the Secretary may by regulations prescribe; and **(6)** the application form shows that the Secretary has approved the transfer and the registration of the firearm to the transferee. Applications shall be denied if the transfer, receipt, or possession of the firearm would place the transferee in violation of law.

(b) Transfer of possession. The transferee of a firearm shall not take possession of the firearm unless the Secretary has approved the transfer and registration of the firearm to the transferee as required by subsection (a) of this section.

Part III – Tax on Making Firearms.

§ 5821 Making tax.

(a) Rate. There shall be levied, collected, and paid upon the making of a firearm a tax at the rate of $200 for each firearm made.

(b) By whom paid. The tax imposed by subsection (a) of this section shall be paid by the person making the firearm.

(c) Payment. The tax imposed by subsection (a) of this section shall be payable by the stamp prescribed for payment by the Secretary.

§ 5822 Making.

No person shall make a firearm unless he has (a) filed with the Secretary a written application, in duplicate, to make and register the firearm on the form prescribed by the Secretary; (b) paid any tax payable on the making and such payment is evidenced by the proper stamp affixed to the original application form; (c) identified the firearm to be made in the application form in such manner as the Secretary may by regulations prescribe; (d) identified himself in the application form in such manner as the Secretary may by regulations prescribe, except that, if such person is an individual, the identification must include his fingerprints and his photograph; and (e) obtained the approval of the Secretary to make and register the firearm and the application form shows such approval. Applications shall be denied if the making or possession of the firearm would place the person making the firearm in violation of law.

SUBCHAPTER B — GENERAL PROVISIONS AND EXEMPTIONS

Part I – General Provisions.

§ 5841 Registration of firearms.

(a) Central registry. The Secretary shall maintain a central registry of all firearms in the United States which are not in the possession or under the control of the United States. This registry shall be known as the National Firearms Registration and Transfer Record. The registry shall include—

(1) identification of the firearm;

(2) date of registration; and

(3) identification and address of person entitled to possession of the firearm.

(b) By whom registered. Each manufacturer, importer, and maker shall register each firearm he manufactures, imports, or makes. Each firearm transferred shall be registered to the transferee by the transferor.

(c) How registered. Each manufacturer shall notify the Secretary of the manufacture of a firearm in such manner as may by regulations be prescribed and such notification shall effect the registration of the firearm required by this section. Each importer, maker, and transferor of a firearm shall, prior to importing, making, or transferring a firearm, obtain authorization in such manner as required by this chapter or regulations issued thereunder to import, make, or transfer the firearm, and such authorization shall effect the registration of the firearm required by this section.

(d) Firearms registered on effective date of this act. A person shown as possessing a firearm by the records maintained by the Secretary pursuant to the National Firearms Act in force on the day immediately prior to the effective date of the National Firearms Act of 1968 shall be considered to have registered under this section the firearms in his possession which are disclosed by that record as being in his possession.

(e) Proof of registration. A person possessing a firearm registered as required by this section shall retain proof of registration which shall be made available to the Secretary upon request.

§ 5842 Identification of firearms.

(a) Identification of firearms other than destructive devices. Each manufacturer and importer and anyone making a firearm shall identify each firearm, other than a destructive device, manufactured, imported, or made by a serial number which may not be readily removed, obliterated, or altered, the name of the manufacturer, importer, or maker, and such other identification as the Secretary may by regulations prescribe.

(b) Firearms without serial number. Any person who possesses a firearm, other than a destructive device, which does not bear the serial number and other information required by subsection (a) of this section shall identify the firearm with a serial number assigned by the Secretary and any other information the Secretary may by regulations prescribe.

(c) Identification of destructive device. Any firearm classified as a destructive device shall be identified in such manner as the Secretary may by regulations prescribe.

§ 5843 Records and returns.

Importers, manufacturers, and dealers shall keep such records of, and render such returns in relation to, the importation, manufacture, making, receipt, and sale, or other disposition, of firearms as the Secretary may by regulations prescribe.

§ 5844 Importation.

No firearm shall be imported or brought into the United States or any territory under its control or jurisdiction unless the importer establishes, under regulations as may be prescribed by the Secretary, that the firearm to be imported or brought in is—

(1) being imported or brought in for the use of the United States or any department, independent establish-ment, or agency thereof or any State or possession or any political subdivision thereof; or

(2) being imported or brought in for scientific or research purposes; or

(3) being imported or brought in solely for testing or use as a model by a registered manufacturer or solely for use as a sample by a registered importer or registered dealer;

except that, the Secretary may permit the conditional importation or bringing in of a firearm for examination and testing in connection with classifying the firearm.

§ 5845 Definitions.

For the purpose of this chapter--

(a) Firearm. The term **'firearm'** means **(1)** a shotgun having a barrel or barrels of less than 18 inches in length; **(2)** a weapon made from a shotgun if such weapon as modified has an overall length of less than 26 inches or a barrel or barrels of less than 18 inches in length; **(3)** a rifle having a barrel or barrels of less than 16 inches in length; **(4)** a weapon made from a rifle if such weapon as modified has an overall length of less than 26 inches or a barrel or barrels of less than 16 inches in length; **(5)** any other weapon, as defined in subsection (e); **(6)** a machinegun; **(7)** any silencer (as defined in section 921 of title 18, United States Code); and **(8)** a destructive device. The term **'firearm'** shall not include an antique firearm or any device (other than a machinegun or destructive device) which, although designed as a weapon, the Secretary finds by reason of the date of its manufacture, value, design, and other characteristics is primarily a collector's item and is not likely to be used as a weapon.

(b) Machinegun. The term **'machinegun'** means any weapon which shoots, is designed to shoot, or can be readily restored to shoot, automatically more than one shot, without manual reloading, by a single function of the trigger. The term shall also include the frame or receiver of any such weapon, any part designed and intended solely and exclusively, or combination of parts designed and intended, for use in converting a weapon into a machinegun, and any combination of parts from which a machinegun can be assembled if such parts are in the possession or under the control of a person.

(c) Rifle. The term **'rifle'** means a weapon designed or redesigned, made or remade, and intended to be fired from the shoulder and designed or redesigned and made or remade to use the energy of the explosive in a fixed cartridge to fire only a single projectile through a rifled bore for each single pull of the trigger, and shall include any such weapon which may be readily restored to fire a fixed cartridge.

(d) Shotgun. The term **'shotgun'** means a weapon designed or redesigned, made or remade, and intended to be fired from the shoulder and designed or redesigned and made or remade to use the energy of the explosive in a fixed shotgun shell to fire through a smooth bore either a number of projectiles (ball shot) or a single projectile for each pull of the trigger, and shall include any such weapon which may be readily restored to fire a fixed shotgun shell.

(e) Any other weapon. The term **'any other w eapon'** means any weapon or device capable of being concealed on the person from which a shot can be discharged through the energy of an explosive, a pistol or revolver having a barrel with a smooth bore designed or redesigned to fire a fixed shotgun shell, weapons with combination shotgun and rifle barrels 12 inches or more, less than 18 inches in length, from which only a single discharge can be made from either barrel without manual reloading, and shall include any such weapon which may be readily restored to fire. Such term shall not include a pistol or a revolver having a rifled bore, or rifled bores, or weapons designed, made, or intended to be fired from the shoulder and not capable of firing fixed ammunition.

(f) Destructive de vice. The term **'destructive de vice'** means **(1)** any explosive, incendiary, or poison gas (A) bomb, (B) grenade, (C) rocket having a propellant charge of more than four ounces, (D) missile having an explosive or incendiary charge of more than one-quarter ounce, (E) mine, or (F) similar device; **(2)** any type of weapon by whatever name known which will, or which may be readily converted to, expel a projectile by the action of an explosive or other propellant, the barrel or barrels of which have a bore of more than one-half inch in diameter, except a shotgun or shotgun shell which the Secretary finds is generally recognized as particularly suitable for sporting purposes; and **(3)** any combination of parts either designed or intended for use in converting any device into a destructive device as defined in subparagraphs (1) and (2) and from which a destructive device may be readily assembled. The term **'destructive device'** shall not include any device which is neither designed nor redesigned for use as a weapon; any device, although originally designed for use as a weapon, which is redesigned for use as a signaling, pyrotechnic, line throwing, safety, or similar device; surplus ordnance sold, loaned, or given by the Secretary of the Army pursuant to the provisions of section 4684(2), 4685, or 4686 of title 10 of the United States Code; or any other device which the Secretary finds is not likely to be used as a weapon, or is an antique or is a rifle which the owner intends to use solely for sporting purposes.

(g) Antique firearm. The term **'antique firearm'** means any firearm not designed or redesigned for using rim fire or conventional center fire ignition with fixed ammunition and manufactured in or before 1898 (including any matchlock, flintlock, percussion cap, or similar type of ignition system or replica thereof, whether actually manufactured before or after the year 1898) and also any firearm using fixed ammunition manufactured in or before 1898, for which ammunition is no longer manufactured in the United States and is not readily available in the ordinary channels of commercial trade.

(h) Unserviceable firearm . The term **'unserviceable firearm'** means a firearm which is incapable of discharging a shot by means of an explosive and incapable of being readily restored to a firing condition.

(i) Make. The term **'make'**, and the various derivatives of such word, shall include manufacturing (other than by one qualified to engage in such business under this chapter), putting together, altering, any combination of these, or otherwise producing a firearm.

(j) Transfer. The term **'transfer'** and the various derivatives of such word, shall include selling, assigning, pledging, leasing, loaning, giving away, or otherwise disposing of.

(k) Dealer. The term **'dealer'** means any person, not a manufacturer or importer, engaged in the business of selling, renting, leasing, or loaning firearms and shall include pawnbrokers who accept firearms as collateral for loans.

(l) Importer. The term **'importer'** means any person who is engaged in the business of importing or bringing firearms into the United States.

(m) Manufacturer. The term **'manufacturer'** means any person who is engaged in the business of manufacturing firearms.

§ 5846 Other laws applicable.

All provisions of law relating to special taxes imposed by chapter 51 and to engraving, issuance, sale, accountability, cancellation, and distribution of stamps for tax payment shall, insofar as not inconsistent with the provisions of this chapter, be applicable with respect to the taxes imposed by sections 5801, 5811, and 5821.

§ 5847 Effect on other laws.

Nothing in this chapter shall be construed as modifying or affecting the requirements of section 414 of the Mutual Security Act of 1954, as amended, with respect to the manufacture, exportation, and importation of arms, ammunition, and implements of war.

§ 5848 Restrictive use of information.

(a) General Rule. No information or evidence obtained from an application, registration, or records required to be submitted or retained by a natural person in order to comply with any provision of this chapter or regulations issued thereunder, shall, except as provided in subsection (b) of this section, be used, directly or indirectly, as evidence against that person in a criminal proceeding with respect to a violation of law occurring prior to or concurrently with the filing of the application or registration, or the compiling of the records containing the information or evidence.

(b) Furnishing false information. Subsection (a) of this section shall not preclude the use of any such information or evidence in a prosecution or other action under any applicable provision of law with respect to the furnishing of false information.

§ 5849 Citation of chapter.

This chapter may be cited as the 'National Firearms Act' and any reference in any other provision of law to the 'National Firearms Act' shall be held to refer to the provisions of this chapter.

Part II – Exemptions.

§ 5851 Special (occupational) tax exemption.

(a) Business with United States. Any person required to pay special (occupational) tax under section 5801 shall be relieved from payment of that tax if he establishes to the satisfaction of the Secretary that his business is conducted exclusively with, or on behalf of, the United States or any department, independent establishment, or agency thereof. The Secretary may relieve any person manufacturing firearms for, or on behalf of, the United States from compliance with any provision of this chapter in the conduct of such business.

(b) Application. The exemption provided for in subsection (a) of this section may be obtained by filing with the Secretary an application on such form and containing such information as may by regulations be prescribed. The exemptions must thereafter be renewed on or before July 1 of each year. Approval of the application by the Secretary shall entitle the applicant to the exemptions stated on the approved application.

§ 5852 General transfer and making tax exemption.

(a) Transfer. Any firearm may be transferred to the United States or any department, independent establishment, or agency thereof, without payment of the transfer tax imposed by section 5811.

(b) Making by a person other than a qualified manufacturer. Any firearm may be made by, or on behalf of, the United States, or any department, independent establishment, or agency thereof, without payment of the making tax imposed by section 5821.

(c) Making by a qualified manufacturer. A manufacturer qualified under this chapter to engage in such business may make the type of firearm which he is qualified to manufacture without payment of the making tax imposed by section 5821.

(d) Transfers between special (occupational) taxpayers. A firearm registered to a person qualified under this chapter to engage in business as an importer, manufacturer, or dealer may be transferred by that person without payment of the transfer tax imposed by section 5811 to any other person qualified under this chapter to manufacture, import, or deal in that type of firearm.

(e) Unserviceable firearm. An unserviceable firearm may be transferred as a curio or ornament without payment of the transfer tax imposed by section 5811, under such requirements as the Secretary may by regulations prescribe.

(f) Right to exemption. No firearm may be transferred or made exempt from tax under the provisions of this section unless the transfer or making is performed pursuant to an application in such form and manner as the Secretary may by regulations prescribe.

§ 5853 Transfer and making tax exemption available to certain governmental entities.

(a) Transfer. A firearm may be transferred without the payment of the transfer tax imposed by section 5811 to any State, possession of the United States, any political subdivision thereof, or any official police organization of such a government entity engaged in criminal investigations.

(b) Making. A firearm may be made without payment of the making tax imposed by section 5821 by, or on behalf of, any State, or possession of the United States, any political subdivision thereof, or any official police organization of such a government entity engaged in criminal investigations.

(c) Right to exemption. No firearm may be transferred or made exempt from tax under this section unless the transfer or making is performed pursuant to an application in such form and manner as the Secretary may by regulations prescribe.

§ 5854 Exportation of firearms exempt from transfer tax.

A firearm may be exported without payment of the transfer tax imposed under section 5811 provided that proof of the exportation is furnished in such form and manner as the Secretary may by regulations prescribe.

SUBCHAPTER C — PROHIBITED ACTS

§ 5861 Prohibited acts.

It shall be unlawful for any person—

(a) to engage in business as a manufacturer or importer of, or dealer in, firearms without having paid the special (occupational) tax required by section 5801 for his business or having registered as required by section 5802; or

(b) to receive or possess a firearm transferred to him in violation of the provisions of this chapter; or

(c) to receive or possess a firearm made in violation of the provisions of this chapter; or

(d) to receive or possess a firearm

which is not registered to him in the National Firearms Registration and Transfer Record; or

(e) to transfer a firearm in violation of the provisions of this chapter; or

(f) to make a firearm in violation of the provisions of this chapter; or

(g) to obliterate, remove, change, or alter the serial number or other identification of a firearm required by this chapter; or

(h) to receive or possess a firearm having the serial number or other identification required by this chapter obliterated, removed, changed, or altered; or

(i) to receive or possess a firearm which is not identified by a serial number as required by this chapter; or

(j) to transport, deliver, or receive any firearm in interstate commerce which has not been registered as required by this chapter; or

(k) to receive or possess a firearm which has been imported or brought into the United States in violation of section 5844; or

(l) to make, or cause the making of, a false entry on any application, return, or record required by this chapter, knowing such entry to be false.

SUBCHAPTER D — PENALTIES AND FORFEITURES

§ 5871 Penalties.

Any person who violates or fails to comply with any provision of this chapter shall, upon conviction, be fined not more than $10,000, or be imprisoned not more than ten years, or both.

§ 5872 Forfeitures.

(a) Laws applicable. Any firearm involved in any violation of the provisions of this chapter shall be subject to seizure and forfeiture, and (except as provided in subsection (b)) all the provisions of internal revenue laws relating to searches, seizures, and forfeitures of unstamped articles are extended to and made to apply to the articles taxed under this chapter, and the persons to whom this chapter applies.

(b) Disposal. In the case of the forfeiture of any firearm by reason of a violation of this chapter, no notice of public sale shall be required; no such firearm shall be sold at public sale; if such firearm is forfeited for a violation of this chapter and there is no remission or mitigation of forfeiture thereof, it shall be delivered by the Secretary to the Administrator of General Services, General Services Administration, who may order such firearm destroyed or may sell it to any State, or possession, or political subdivision thereof, or at the request of the Secretary, may authorize its retention for official use of the Treasury Department, or may transfer it without charge to any executive department or independent establishment of the Government for use by it.

TITLE 27 CFR CHAPTER II

PART 479—MACHINE GUNS, DESTRUCTIVE DEVICES, AND CERTAIN OTHER FIREARMS

(This part was formerly designated as Part 179)

Editor's Note:

Effective January 24, 2003, the Homeland Security Act transferred the Bureau of Alcohol, Tobacco and Firearms from the Department of the Treasury to the Department of Justice. In addition, the agency's name was changed to the Bureau of Alcohol, Tobacco, Firearms and Explosives. The regulations, as printed in this publication, do not yet reflect this change. The regulations will be amended to change the references from the "Bureau of Alcohol, Tobacco and Firearms," the "Department of the Treasury," and the "Secretary of the Treasury" to the "Bureau of Alcohol, Tobacco, Firearms and Explosives," the "Department of Justice" and the "Attorney General," respectively.

Subpart A—Scope of Regulations

§ 479.1 General.

This part contains the procedural and substantive requirements relative to the importation, manufacture, making, exportation, identification and registration of, and the dealing in, machine guns, destructive devices and certain other firearms under the provisions of the National Firearms Act (26 U.S.C. Chapter 53).

Subpart B—Definitions

§ 479.11 Meaning of terms.

When used in this part and in forms prescribed under this part, where not otherwise distinctly expressed or manifestly incompatible with the intent thereof, terms shall have the meanings ascribed in this section. Words in the plural form shall include the singular, and vice versa, and words importing the masculine gender shall include the feminine. The terms "includes" and "including" do not exclude other things not enumerated which are in the same general class or are otherwise within the scope thereof.

Antique firearm. Any firearm not designed or redesigned for using rim fire or conventional center fire ignition with fixed ammunition and manufactured in or before 1898 (including any matchlock, flintlock, percussion cap, or similar type of ignition system or replica thereof, whether actually manufactured before or after the year 1898) and also any firearm using fixed ammunition manufactured in or before 1898, for which ammunition is no longer manufactured in the United States and is not readily available in the ordinary channels of commercial trade.

Any other weapon. Any weapon or device capable of being concealed on the person from which a shot can be discharged through the energy of an explosive, a pistol or revolver having a barrel with a smooth bore designed or redesigned to fire a fixed shotgun shell, weapons with combination shotgun and rifle barrels 12 inches or more, less than 18 inches in length, from which only a single discharge can be made from either barrel without manual reloading, and shall include any such weapon which may be readily restored to fire. Such term shall not include a pistol or a revolver having a rifled bore, or rifled bores, or weapons designed, made, or intended to be fired from the shoulder and not capable of firing fixed ammunition.

ATF officer. An officer or employee of the Bureau of Alcohol, Tobacco and Firearms (ATF) authorized to perform any function relating to the administration or enforcement of this part.

Customs officer. Any officer of the Customs Service or any commissioned, warrant, or petty officer of the Coast Guard, or any agent or other person authorized by law or designated by the Secretary of the Treasury to perform any duties of an officer of the Customs Service.

Dealer. Any person, not a manufacturer or importer, engaged in the business of selling, renting, leasing, or loaning firearms and shall include pawnbrokers who accept firearms as collateral for loans.

Destructive device. **(a)** Any explosive, incendiary, or poison gas **(1)** bomb, **(2)** grenade, **(3)** rocket having a propellent charge of more than 4 ounces, **(4)** missile having an explosive or incendiary charge of more than one-quarter ounce, **(5)** mine, or **(6)** similar device; **(b)** any type of weapon by whatever name known which will, or which may be readily converted to, expel a projectile by the action of an explosive or other propellant, the barrel or barrels of which have a bore of more than one-half inch in diameter, except a shotgun or shotgun shell which the Director finds is generally recognized as particularly suitable for sporting purposes; and **(c)** any combination of parts either designed or intended for use in converting any device into a destructive device as described in paragraphs (a) and (b) of this definition and from which a destructive device may be readily assembled. The term shall not include any device which is neither designed or redesigned for use as a weapon; any device, although originally designed for use as a weapon, which is redesigned for use as a signaling, pyrotechnic, line throwing, safety, or similar device; surplus ordnance sold, loaned, or given by the Secretary of the Army under 10 U.S.C. 4684(2), 4685, or 4686, or any device which the Director finds is not likely to be used as a weapon, or is an antique or is a rifle which the owner intends to use solely for sporting purposes.

Director. The Director, Bureau of Alcohol, Tobacco and Firearms, the Department of the Treasury, Washington, DC.

Director of the Service Center. A director of an Internal Revenue Service Center in an internal revenue region.

District director. A district director of the Internal Revenue Service in an internal revenue district.

Executed under penalties of perjury. Signed with the prescribed declaration under the penalties of perjury as provided on or with respect to the return, form, or other document or, where no form of declaration is prescribed, with the declaration:

> **"I declare und er the penaltie s of perjury that this—(inse rt type of do cument, suc h a s, statement, a pplication, r e-quest, c ertificate), includin g the doc uments submitted in support thereo f, has been ex - amined by me and, to the b est of my knowledge and belief, is true, correct, and complete."**

Exportation. The severance of goods from the mass of things belonging to this country with the intention of uniting them to the mass of things belonging to some foreign country.

Exporter. Any person who exports firearms from the United States.

Firearm. **(a)** A shotgun having a barrel or barrels of less than 18 inches in length; **(b)** a weapon made from a shotgun if such weapon as modified has an overall length of less than 26 inches or a barrel or barrels of less than 18 inches in length; **(c)** a rifle having a barrel or barrels of less than 16 inches in length; **(d)** a weapon made from a rifle if such weapon as modified has an overall length of less than 26 inches or a barrel or barrels of less than 16 inches in length; **(e)** any other weapon, as defined in this subpart; **(f)** a machine gun; **(g)** a muffler or a silencer for any firearm whether or not such firearm is included within this definition; and **(h)** a destructive device. The term shall not include an antique firearm or any device (other than a machine gun or destructive device) which, although designed as a weapon, the Director finds by reason of the date of its manufacture, value, design, and other characteristics is primarily a collector's item and is not likely to be used as a weapon. For purposes of this definition, the length of the barrel having an integral chamber(s) on a shotgun or rifle shall be determined by measuring the distance between the muzzle and the face of the bolt, breech, or breech block when closed and when the shotgun or rifle is cocked. The overall length of a weapon made from a shotgun or rifle is the distance between the extreme ends of the weapon measured along a line parallel to the center line of the bore.

Fixed ammunition. That self-contained unit consisting of the case, primer, propellant charge, and projectile or projectiles.

Frame or receiver. That part of a firearm which provides housing for the hammer, bolt or breechblock and firing mechanism, and which is usually threaded at its forward portion to receive the barrel.

Importation. The bringing of a firearm within the limits of the United States or any territory under its control or jurisdiction, from a place outside thereof (whether such place be a foreign country or territory subject to the jurisdiction of the United States), with intent to unlade. Except that, bringing a firearm from a foreign country or a territory subject to the jurisdiction of the United States into a foreign trade zone for storage pending shipment to a foreign country or subsequent importation into this country, under Title 26 of the United States Code, and this part, shall not be deemed importation.

Importer. Any person who is engaged in the business of importing or bringing firearms into the United States.

Machine gun. Any weapon which shoots, is designed to shoot, or can be readily restored to shoot, automatically more than one shot, without manual reloading, by a single function of the trigger. The term shall also include the frame or receiver of any such weapon, any part designed and intended solely and exclusively, or combination of parts designed and intended, for use in converting a weapon into a machine gun, and any combination of parts from which a machine gun can be assembled if such parts are in the possession or under the control of a person.

Make. This term and the various derivatives thereof shall include manufacturing (other than by one qualified to engage in such business under this part), putting together, altering, any combination of these, or otherwise producing a firearm.

Manual reloading. The inserting of a cartridge or shell into the chamber of a firearm either with the hands or by means of a mechanical device controlled and energized by the hands.

Manufacturer. Any person who is engaged in the business of manufacturing firearms.

Muffler or silencer. Any device for silencing, muffling, or diminishing the report of a portable firearm, including any combination of parts, designed or redesigned, and intended for the use in assembling or fabricating a firearm silencer or firearm muffler, and any part intended only for use in such assembly or fabrication.

Person. A partnership, company, association, trust, estate, or corporation, as well as a natural person.

Pistol. A weapon originally designed, made, and intended to fire a projectile (bullet) from one or more barrels when held in one hand, and having **(a)** a chamber(s) as an integral part(s) of, or permanently aligned with, the bore(s); and **(b)** a short stock designed to be gripped by one hand and at an angle to and extending below the line of the bore(s).

Regional director (compliance). The principal ATF regional official responsible for administering regulations in this part.

Revolver. A projectile weapon, of the pistol type, having a breechloading chambered cylinder so arranged that the cocking of the hammer or movement of the trigger rotates it and brings the next cartridge in line with the barrel for firing.

Rifle. A weapon designed or redesigned, made or remade, and intended to be fired from the shoulder and designed or redesigned and made or remade to use the energy of the explosive in a fixed cartridge to fire only a single projectile through a rifled bore for each single pull of the trigger, and shall include any such weapon which may be readily restored to fire a fixed cartridge.

Shotgun. A weapon designed or redesigned, made or remade, and intended to be fired from the shoulder and designed or redesigned and made or remade to use the energy of the explosive in a fixed shotgun shell to fire through a smooth bore either a number of projectiles (ball shot) or a single projectile for each pull of the trigger, and shall include any such weapon which may be readily restored to fire a fixed shotgun shell.

Transfer. This term and the various derivatives thereof shall include selling, assigning, pledging, leasing, loaning, giving away, or otherwise disposing of.

United States. The States and the District of Columbia.

U.S.C. The United States Code.

Unserviceable firearm. A firearm which is incapable of discharging a shot by means of an explosive and incapable of being readily restored to a firing condition.

Subpart C—Administrative and Miscellaneous Provisions

§ 479.21 Forms prescribed.

(a) The Director is authorized to prescribe all forms required by this part. All of the information called for in each form shall be furnished as indicated by the headings on the form and the instructions on or pertaining to the form. In addition, information called for in each form shall be furnished as required by this part. Each form requiring that it be executed under penalties of perjury shall be executed under penalties of perjury.

(b) Requests for forms should be

mailed to the ATF Distribution Center, 7943 Angus Court, Springfield, Virginia 22153.

§ 479.22 Right of entry and examination.

Any ATF officer or employee of the Bureau of Alcohol, Tobacco and Firearms duly authorized to perform any function relating to the administration or enforcement of this part may enter during business hours the premises (including places of storage) of any importer or manufacturer of or dealer in firearms, to examine any books, papers, or records required to be kept pursuant to this part, and any firearms kept by such importer, manufacturer or dealer on such premises, and may require the production of any books, papers, or records necessary to determine any liability for tax under 26 U.S.C. Chapter 53, or the observance of 26 U.S.C. Chapter 53, and this part.

§ 479.23 Restrictive use of required information.

No information or evidence obtained from an application, registration, or record required to be submitted or retained by a natural person in order to comply with any provision of 26 U.S.C. Chapter 53, or this part or section 207 of the Gun Control Act of 1968 shall be used, directly or indirectly, as evidence against that person in a criminal proceeding with respect to a violation of law occurring prior to or concurrently with the filing of the application or registration, or the compiling of the record containing the information or evidence: **Provided, however,** that the provisions of this section shall not preclude the use of any such information or evidence in a prosecution or other action under any applicable provision of law with respect to the furnishing of false information.

§ 479.24 Destructive device determination.

The Director shall determine in accordance with 26 U.S.C. 5845(f) whether a device is excluded from the definition of a destructive device. A person who desires to obtain a determination under that provision of law for any device which he believes is not likely to be used as a weapon shall submit a written request, in triplicate, for a ruling thereon to the Director. Each such request shall be executed under the penalties of perjury and contain a complete and accurate description of the device, the name and address of the manufacturer or importer thereof, the purpose of and use for which it is intended, and such photographs, diagrams, or drawings as may be necessary to enable the Director to make his determination. The Director may require the submission to him, of a sample of such device for examination and evaluation. If the submission of such device is impracticable, the person requesting the ruling shall so advise the Director and designate the place where the device will be available for examination and evaluation.

§ 479.25 Collector's items.

The Director shall determine in accordance with 26 U.S.C. 5845(a) whether a firearm or device, which although originally designed as a weapon, is by reason of the date of its manufacture, value, design, and other characteristics primarily a collector's item and is not likely to be used as a weapon. A person who desires to obtain a determination under that provision of law shall follow the procedures prescribed in § 479.24 relating to destructive device determinations, and shall include information as to date of manufacture, value, design and other characteristics which would sustain a finding that the firearm or device is primarily a collector's item and is not likely to be used as a weapon.

§ 479.26 Alternate methods or procedures; emergency variations from requirements.

(a) Alternate methods or procedures. Any person subject to the provisions of this part, on specific approval by the Director as provided in this paragraph, may use an alternate method or procedure in lieu of a method or procedure specifically prescribed in this part. The Director may approve an alternate method or procedure, subject to stated conditions, when it is found that:

(1) Good cause is shown for the use of the alternate method or procedure;

(2) The alternate method or procedure is within the purpose of, and consistent with the effect intended by, the specifically prescribed method or procedure and that the alternate method or procedure is substantially equivalent to that specifically prescribed method or procedure; and

(3) The alternate method or procedure will not be contrary to any provision of law and will not result in an increase in cost to the Government or hinder the effective administration of this part. Where such person desires to employ an alternate method or procedure, a written application shall be submitted to the appropriate regional director (compliance), for transmittal to the Director. The application shall specifically describe the proposed alternate method or procedure and shall set forth the reasons for it. Alternate methods or procedures may not be employed until the application is approved by the Director. Such person shall, during the period of authorization of an alternate method or procedure, comply with the terms of the approved application. Authorization of any alternate method or procedure may be withdrawn whenever, in the judgment of the Director, the effective administration of this part is hindered by the continuation of the authorization.

(b) Emergency variations from requirements. The Director may approve a method of operation other than as specified in this part, where it is found that an emergency exists and the proposed variation from the specified requirements are necessary and the proposed variations **(1)** will not hinder the effective administration of this part, and **(2)** will not be contrary to any provisions of law. Variations from requirements granted under this paragraph are conditioned on compliance with the procedures, conditions, and limitations set forth in the approval of the application. Failure to comply in good faith with the procedures, conditions, and limitations shall automatically terminate the authority for the variations, and the person granted the variance shall fully comply with the prescribed requirements of regulations from which the variations were authorized. Authority for any variation may be withdrawn whenever, in the judgment of the Director, the effective administration of this part is hindered by the continuation of the variation. Where a person desires to employ an emergency variation, a written application shall be submitted to the appropriate regional director (compliance) for transmittal to the Director. The application shall describe the proposed variation and set forth the reasons for it. Variations may not be employed until the application is approved.

(c) Retention of approved variations. The person granted the variance shall retain and make available for examination by ATF officers any application approved by the Director under this section.

Subpart D—Special (Occupational) Taxes

§ 479.31 Liability for tax.

(a) General. Every person who engages in the business of importing, manufacturing, or dealing in (including pawnbrokers) firearms in the United States shall pay a special (occupational) tax at a rate specified by § 479.32. The tax shall be paid on or before the date of commencing the taxable business, and thereafter every year on or before July 1. Special (occupational) tax shall not be prorated. The tax shall be computed for the entire tax year (July 1 through June 30), regardless of the portion of the year during which the taxpayer engages in business. Persons commencing business at any time after July 1 in any year are liable for the special (occupational) tax for the entire tax year.

(b) Each place of business taxable. An importer, manufacturer, or dealer in firearms incurs special tax liability at each place of business where an occupation subject to special tax is conducted. A place of business means the entire office, plant or area of the business in any one location under the same proprietorship. Passageways, streets, highways, rail crossings, waterways, or partitions dividing the premises are not sufficient separation to require additional special tax, if the divisions of the premises are otherwise contiguous. See also §§ 479.38–479.39.

§ 479.32 Special (occupational) tax rates.

(a) Prior to January 1, 1988, the special (occupational) tax rates were as follows:

	Per year or fraction thereof
Class 1 – Importer of firearms	**$500.**
Class 2 – Manufacturer of firearms	**$500.**
Class 3 – Dealer in firearms.	**$200.**
Class 4 – Importer only of weapons classified as "any other weapon"	**$25.**
Class 5 – Manufacturer only of weapons classified as "any other weapon"	**$25.**
Class 6 – Dealer only in weapons classified as "any other weapon"	**$10.**

(b) Except as provided in § 479.32a, the special (occupational) tax rates effective January 1, 1988, are as follows:

	Per year or fraction thereof
Class 1 – Importer of firearms (including an importer only of weapons classified as "any other weapon")	**$1000.**
Class 2 – Manufacturer of firearms (including a manufacturer only of weapons classified as "any other weapon")	**$1000.**
Class 3 – Dealer in firearms (including a dealer only of weapons classified as "any other weapon")	**$500.**

(c) A taxpayer who was engaged in a business on January 1, 1988, for which a special (occupational) tax was paid for a taxable period which began before January 1, 1988, and included that date, shall pay an increased special tax for the period January 1, 1988, through June 30, 1988. The increased tax shall not exceed one-half the excess (if any) of **(1)** the rate of special tax in effect on January 1, 1988, over **(2)** the rate of such tax in effect on December 31, 1987. The increased special tax shall be paid on or before April 1, 1988.

§ 479.32a Reduced rate of tax for small importers and manufacturers.

(a) General. Effective January 1, 1988, 26 U.S.C. 5801(b) provides for a reduced rate of special tax with respect to any importer or manufacturer whose gross receipts (for the most recent taxable year ending before the first day of the taxable period to which the special tax imposed by § 479.32 relates) are less than $500,000. The rate of tax for such an importer or manufacturer is $500 per year or fraction thereof. The **"taxable year"** to be used for determining gross receipts is the taxpayer's income tax year. All gross receipts of the taxpayer shall be included, not just the gross receipts of the business subject to special tax. Proprietors of new businesses that have not yet begun a taxable year, as well as proprietors of existing businesses that have not yet ended a taxable year, who commence a new activity subject to special tax, quality for the reduced special (occupational) tax rate, unless the business is a member of a **"controlled group;"** in that case, the rules of paragraph (b) of this section shall apply.

(b) Controlled group. All persons treated as one taxpayer under 26 U.S.C. 5061(e)(3) shall be treated as one taxpayer for the purpose of determining gross receipts under paragraph (a) of this section. **"Controlled group"** means a controlled group of corporations, as defined in 26 U.S.C. 1563 and implementing regulations in 26 CFR 1.1563–1 through 1.1563–4, except that the words "at least 80 percent" shall be replaced by the words "more than 50 percent" in each place they appear in subsection (a) of 26 U.S.C. 1563, as well as in the implementing regulations. Also, the rules for a **"controlled gro up of corporations"** apply in a similar fashion to groups which include partnerships and/or sole proprietorships. If one entity maintains more than 50% control over a group consisting of corporations and one, or more, partnerships and/or sole proprietorships, all of the members of the controlled group are one taxpayer for the purpose of this section.

(c) Short taxable year. Gross receipts for any taxable year of less than 12 months shall be annualized by multiplying the gross receipts for the short period by 12 and dividing the result by the number of months in the short period, as required by 26 U.S.C. 448(c)(3).

(d) Returns and allowances. Gross receipts for any taxable year shall be reduced by returns and allowances made during that year under 26 U.S.C. 448(c)(3).

§ 479.33 Special exemption.

(a) Any person required to pay special (occupational) tax under this part shall be relieved from payment of that tax if he establishes to the satisfaction of the Director that his business is conducted exclusively with, or on behalf of, the United States or any department, independent establishment, or agency thereof. The Director may relieve any person manufacturing firearms for or on behalf of the United States from compliance with any provision of this part in the conduct of the business with respect to such firearms.

(b) The exemption in this section may be obtained by filing with the Director an application, in letter form, setting out the manner in which the applicant conducts his business, the type of firearm to be manufactured, and proof satisfactory to the Director of the existence of the contract with the United States, department, independent establishment, or agency thereof, under which the applicant intends to operate.

§ 479.34 Spe cial tax registration and return.

(a) General. Special tax shall be paid by return. The prescribed return is ATF Form 5630.7, Special Tax Registration and Return. Special tax returns, with payment of tax, shall be filed with ATF in accordance with instructions on the form. Properly completing, signing, and timely filing of a return (Form 5630.7) constitutes compliance with 26 U.S.C. 5802.

(b) Preparation of ATF Form 5630.7. All of the information called for on Form 5630.7 shall be provided, including:

(1) The true name of the taxpayer.

(2) The trade name(s) (if any) of the business(es) subject to special tax.

(3) The employer identification number (see § 479.35).

(4) The exact location of the place of business, by name and number of building or street, or if these do not exist, by some description in addition to the post office address. In the case of one return for two or more locations, the address to be shown shall be the taxpayer's principal place of business (or principal office, in the case of a corporate taxpayer).

(5) The class(es) of special tax to which the taxpayer is subject.

(6) Ownership and control information: That is, the name, position, and residence address of every owner of the business and of every person having power to control its management and policies with respect to the activity subject to special tax. "Owner of the business" shall include every partner, if the taxpayer is a partnership, and every person owning 10% or more of its stock, if the taxpayer is a corporation. However, the ownership and control information required by this paragraph need not be stated if the same information has been previously provided to ATF in connection with a license application under Part 478 of this chapter, and if the information previously provided is still current.

(c) Multiple locations and/or classes of tax. A taxpayer subject to special tax for the same period at more than one location or for more than one class of tax shall—

(1) File one special tax return, ATF Form 5630.7, with payment of tax, to cover all such locations and classes of tax; and

(2) Prepare, in duplicate, a list identified with the taxpayer's name, address (as shown on ATF Form 5630.7), employer identification number, and period covered by the return. The list shall show, by States, the name, address, and tax class of each location for which special tax is being paid. The original of the list shall be filed with ATF in accordance with instructions on the return, and the copy shall be retained at the taxpayer's principal place of business (or principal office, in the case of a corporate taxpayer) for not less than 3 years.

(d) Signing of ATF Forms 5630.7— (1) Ordinary returns. The return of an individual proprietor shall be signed by the individual. The return of a partnership shall be signed by a general partner. The return of a corporation shall be signed by any officer. In each case, the person signing the return shall designate his or her capacity as "individual owner," "member of firm," or, in the case of a corporation, the title of the officer.

(2) Fiduciaries. Receivers, trustees, assignees, executors, administrators, and other legal representatives who continue the business of a bankrupt, insolvent, deceased person, etc., shall indicate the fiduciary capacity in which they act.

(3) Agent or attorney in fact. If a return is signed by an agent or attorney in fact, the signature shall be preceded by the name of the principal and followed by the title of the agent or attorney in fact. A return signed by a person as agent will not be accepted unless there is filed, with the ATF office with which the return is required to be filed, a power of attorney authorizing the agent to perform the act.

(4) Perjury statement. ATF Forms 5630.7 shall contain or be verified by a written declaration that the return has been executed under the penalties of perjury.

(e) Identification of taxpayer. If the taxpayer is an individual, with the initial return such person shall securely attach to Form 5630.7 a photograph of the individual 2 × 2 inches in size, clearly showing a full front view of the features of the individual with head bare, with the distance from the top of the head to the point of the chin approximately 1 1/4 inches, and which shall have been taken within 6 months prior to the date of completion of the return. The individual shall also attach to the return a properly completed FBI Form FD–258 (Fingerprint Card). The fingerprints must be clear for accurate classification and should be taken by someone properly equipped to take them: **Provided,** That the provisions of this paragraph shall not apply to individuals who have filed with ATF a properly executed Application for License under 18 U.S.C. Chapter 44, Firearms, ATF Form 7 (5310.12) (12–93 edition), as specified in § 478.44(a).

§ 479.35 Employer identification number.

(a) Requirement. The employer identification number (defined in 26 CFR 301.7701–12) of the taxpayer who has been assigned such a number shall be shown on each special tax return, including amended returns, filed under this subpart. Failure of the taxpayer to include the employer identification number may result in the imposition of the penalty specified in § 70.113 of this chapter.

(b) Application for employer identification number. Each taxpayer who files a special tax return, who has not already been assigned an employer identification number, shall file IRS Form SS–4 to apply for one. The taxpayer shall apply for and be assigned only one employer identification number, regardless of the number of places of business for which the taxpayer is required to file a special tax return. The employer identification number shall be applied for no later than 7 days after the filing of the taxpayer's first special tax return. IRS Form SS–4 may be obtained from the director of an IRS service center or from any IRS district director.

(c) Preparation and filing of IRS Form SS–4. The taxpayer shall prepare and file IRS Form SS–4, together with any supplementary statement, in accordance with the instructions on the form or issued in respect to it.

§ 479.36 The special tax stamp, receipt for special (occupational) taxes.

Upon filing a properly completed and executed return (Form 5630.7) accompanied by remittance of the full amount due, the taxpayer will be issued a special tax stamp as evidence of payment of the special (occupational) tax.

§ 479.37 Certificates in lieu of stamps lost or destroyed.

When a special tax stamp has been lost or destroyed, such fact should be reported immediately to the regional director (compliance) who issued the stamp. A certificate in lieu of the lost or destroyed stamp will be issued to the taxpayer upon the submission of an affidavit showing to the satisfaction of the regional director (compliance) that the stamp was lost or destroyed.

§ 479.38 Engaging in business at more than one location.

A person shall pay the special (occupational) tax for each location where he engages in any business taxable under 26 U.S.C. 5801. However, a person paying a special (occupational) tax covering his principal place of business may utilize other locations solely for storage of firearms without incurring special (occupational) tax liability at such locations. A manufacturer, upon the single payment of the appropriate special (occupational) tax, may sell firearms, if such firearms are of his own manufacture, at the place of manufacture and at his principal office or place of business if no such firearms, except samples, are kept at such office or place of business. When a person changes the location of a business for which he has paid the special (occupational) tax, he will be liable for another such tax unless the change is properly registered with the regional director (compliance) for the region in which the special tax stamp was issued, as provided in § 479.46.

§ 479.39 Engaging in more than one business at the same location.

If more than one business taxable under 26 U.S.C. 5801 is carried on at the same location during a taxable year, the special (occupational) tax imposed on each such business must be paid. This section does not require a qualified manufacturer or importer to qualify as a dealer if such manufacturer or importer also engages in business on his qualified premises as a dealer. However, a qualified manufacturer who engages in business as an importer must also qualify as an importer. Further, a qualified dealer is not entitled to engage in business as a manufacturer or importer.

§ 479.40 Partnership liability.

Any number of persons doing business in partnership at any one location shall be required to pay but one special (occupational) tax.

§ 479.41 Single sale.

A single sale, unattended by circumstances showing the one making the sale to be engaged in business, does not create special (occupational) tax liability.

Change of Ownership

§ 479.42 Changes through death of owner.

Whenever any person who has paid special (occupational) tax dies, the surviving spouse or child, or executors or administrators, or other legal representatives, may carry on this business for the remainder of the term for which tax has been paid and at the place (or places) for which the tax was paid, without any additional payment, subject to the following conditions. If the surviving spouse or child, or executor or administrator, or other legal representative of the deceased taxpayer continues the business, such person shall, within 30 days after the date on which the successor begins to carry on the business, file a new return, Form 5630.7, with ATF in accordance with the instructions on the form. The return thus executed shall show the name of the original taxpayer, together with the basis of the succession. (As to liability in case of failure to register, see § 479.49.)

§ 479.43 Changes through bankruptcy of owner.

A receiver or referee in bankruptcy may continue the business under the stamp issued to the taxpayer at the place and for the period for which the tax was paid. An assignee for the benefit of creditors may continue business under his assignor's special tax stamp without incurring additional special (occupational) tax liability. In such cases, the change shall be registered with ATF in a manner similar to that required by § 479.42.

§ 479.44 Change in partnership or unincorporated association.

When one or more members withdraw from a partnership or an unincorporated association, the remaining member, or members, may, without incurring additional special (occupational) tax liability, carry on the same business at the same location for the balance of the taxable period for which special (occupational) tax was paid, provided any such change shall be registered in the same manner as required by § 479.42. Where new member(s) are taken into a partnership or an unincorporated association, the new firm so constituted may not carry on business under the special tax stamp of the old firm. The new firm must file a return, pay the special (occupational) tax and register in the same manner as a person who first engages in business is required to do under § 479.34 even though the name of the new firm may be the same as that of the old. Where the members of a partnership or an unincorporated association, which has paid special (occupational) tax, form a corporation to continue the business, a new special tax stamp must be taken out in the name of the corporation.

§ 479.45 Changes in corporation.

Additional special (occupational) tax is not required by reason of a mere change of name or increase in the capital stock of a corporation if the laws of the State of incorporation provide for such change or increase without the formation of a new corporation. A stockholder in a corporation, who after its dissolution continues the business, incurs new special (occupational) tax liability.

Change of Business Location

§ 479.46 Notice by taxpayer.

Whenever during the taxable year a taxpayer intends to remove his business to a location other than specified in his last special (occupational) tax return (see § 479.34), he shall file with ATF **(a)** a return, Form 5630.7, bearing the notation "Removal Registry," and showing the new address intended to be used, **(b)** his current special tax stamp, and **(c)** a letter application requesting the amendment of his registration. The regional director (compliance), upon approval of the application, shall return the special tax stamp, amended to show the new business location. Firearms operations shall not be commenced at the new business location by the taxpayer prior to the required approval of his application to so change his business location.

Change of Trade Name

§ 479.47 Notice by taxpayer.

Whenever during the taxable year a taxpayer intends to change the name of his business, he shall file with ATF **(a)** a return, Form 5630.7, bearing the notation "Amended," and showing the trade name intended to be used, **(b)** his current special tax stamp, and **(c)** a letter application requesting the amendment of his registration. The regional director (compliance), upon approval of the application, shall return the special tax stamp, amended to show the new trade name. Firearms operations shall not be commenced under the new trade name by the taxpayer prior to the required approval of his application to so change the trade name.

Penalties and Interest

§ 479.48 Failure to pay special (occupational) tax.

Any person who engages in a business taxable under 26 U.S.C. 5801, without timely payment of the tax imposed with respect to such business (see § 479.34) shall be liable for such tax, plus the interest and penalties thereon (see 26 U.S.C. 6601 and 6651). In addition, such person may be liable for criminal penalties under 26 U.S.C. 5871.

§ 479.49 Failure to register change or removal.

Any person succeeding to and carrying on a business for which special (occupational) tax has been paid without registering such change within 30 days thereafter, and any taxpayer removing his business with respect to which special (occupational) tax has been paid to a place other than that for which tax was paid without obtaining approval therefor (see § 479.46), will incur liability to an additional payment of the tax, addition to tax and interest, as provided in sections 5801, 6651, and 6601, respectively, I.R.C., for failure to make return (see § 479.50) or pay tax, as well as criminal penalties for carrying on business without payment of special (occupational) tax (see section 5871 I.R.C.).

§ 479.50 Delinquency.

Any person liable for special (occupational) tax under section 5801, I.R.C., who fails to file a return (Form 5630.7), as prescribed, will be liable for a delinquency penalty computed on the amount of tax due unless a return (Form 5630.7) is later filed and failure to file the return timely is shown to the satisfaction of the regional director (compliance), to be due to reasonable cause. The delinquency penalty to be added to the tax is 5 percent if the failure is for not more than 1 month, with an additional 5 percent for each additional month or fraction thereof during which failure continues, not to exceed 25 percent in the aggregate (section 6651, I.R.C.). However, no delinquency penalty is assessed where the 50 percent addition to tax is assessed for fraud (see § 479.51).

§ 479.51 Fraudulent return.

If any part of any underpayment of tax required to be shown on a return is due to fraud, there shall be added to the tax an amount equal to 50 percent of the underpayment, but no delinquency penalty shall be assessed with respect to the same underpayment (section 6653, I.R.C.).

Application of State Laws

§ 479.52 State regulations.

Special tax stamps are merely receipts for the tax. Payment of tax under Federal law confers no privilege to act contrary to State law. One to whom a special tax stamp has been issued may still be punishable under a State law prohibiting or controlling the manufacture, possession or transfer of firearms. On the other hand, compliance with State law confers no immunity under Federal law. Persons who engage in the business of importing, manufacturing or dealing in firearms, in violation of the law of a State, are nevertheless required to pay special (occupational) tax as imposed under the internal revenue laws of the United States. For provisions relating to restrictive use of information furnished to comply with the provisions of this part see § 479.23.

Subpart E—Tax on Making Firearms

§ 479.61 Rate of tax.

Except as provided in this subpart, there shall be levied, collected, and paid upon the making of a firearm a tax at the rate of $200 for each firearm made. This tax shall be paid by the person making the firearm. Payment of the tax on the making of a firearm shall be represented by a $200 adhesive stamp bearing the words "National Firearms Act." The stamps are maintained by the Director.

Application to Make a Firearm

§ 479.62 Application to make.

No person shall make a firearm unless the person has filed with the Director a written application on Form 1 (Firearms), Application to Make and Register a Firearm, in duplicate, executed under the penalties of perjury, to make and register the firearm and has received the approval of the Director to make the firearm which approval shall effectuate registration of the weapon to the applicant. The application shall identify the firearm to be made by serial number, type, model, caliber or gauge, length of barrel, other marks of identification, and the name and address of original manufacturer (if the applicant is not the original manufacturer). The applicant must be identified on the Form 1 (Firearms) by name and address and, if other than a natural person, the name and address of the principal officer or authorized representative and the employer identification number and, if an individual, the identification must include the date and place of birth and the information prescribed in § 479.63. Each applicant shall identify the Federal firearms license and special (occupational) tax stamp issued to the applicant, if any. The applicant shall also show required information evidencing that making or possession of the firearm would not be in violation of law. If the making is taxable, a remittance in the amount of $200 shall be submitted with the application in accordance with the instructions on the form. If the making is taxable and the application is approved, the Director will affix a National Firearms Act stamp to the original application in the space provided therefor and properly cancel the stamp (see § 479.67). The approved application will be returned to the applicant. If the making of the firearm is tax exempt under this part, an explanation of the basis of the exemption shall be attached to the Form 1 (Firearms).

§ 479.63 Identification of applicant.

If the applicant is an individual, the applicant shall securely attach to each copy of the Form 1 (Firearms), in the space provided on the form, a photograph of the applicant 2 × 2 inches in size, clearly showing a full front view of the features of the applicant with head bare, with the distance from the top of the head to the point of the chin approximately 1 1/4 inches, and which shall have been taken within 1 year prior to the date of the application. The applicant shall attach two properly completed FBI Forms FD–258 (Fingerprint Card) to the application. The fingerprints must be clear for accurate classification and should be taken by someone properly equipped to take them. A certificate of the local chief of police, sheriff of the county, head of the State police, State or local district attorney or prosecutor, or such other person whose certificate may in a particular case be acceptable to the Director, shall be completed on each copy of the Form 1 (Firearms). The certificate shall state that the certifying official is satisfied that the fingerprints and photograph accompanying the application are those of the applicant and that the certifying official has no information indicating that possession of the firearm by the maker would be in violation of State or local law or that the maker will use the firearm for other than lawful purposes.

§ 479.64 Procedure for approval of application.

The application to make a firearm, Form 1 (Firearms), must be forwarded directly, in duplicate, by the maker of the firearm to the Director in accordance with the instructions on the form. The Director will consider the application for approval or disapproval. If the application is approved, the Director will return the original thereof to the maker of the firearm and retain the duplicate. Upon receipt of the approved application, the maker is authorized to make the firearm described therein. The maker of the firearm shall not, under any circumstances, make the firearm until the application, satisfactorily executed, has been forwarded to the Director and has been approved and returned by the Director with the National Firearms Act stamp affixed. If the application is disapproved, the original Form 1 (Firearms) and the remittance submitted by the applicant for the purchase of the stamp will be returned to the applicant with the reason for disapproval stated on the form.

§ 479.65 Denial of application.

An application to make a firearm shall not be approved by the Director if the making or possession of the firearm would place the person making the firearm in violation of law.

§ 479.66 Subsequent transfer of firearms.

Where a firearm which has been made in compliance with 26 U.S.C. 5821, and the regulations contained in this part, is to be transferred subsequently, the transfer provisions of the firearms laws and regulations must be complied with. (See subpart F of this part).

§ 479.67 Cancellation of stamp.

The person affixing to a Form 1 (Firearms) a "National Firearms Act" stamp shall cancel it by writing or stamping thereon, in ink, his initials, and the day, month and year, in such manner as to render it unfit for reuse. The cancellation shall not so deface the stamp as to prevent its denomination and genuineness from being readily determined.

Exceptions to Tax on Making Firearms

§ 479.68 Qualified manufacturer.

A manufacturer qualified under this part to engage in such business may make firearms without payment of the making tax. However, such manufacturer shall report and register each firearm made in the manner prescribed by this part.

§ 479.69 Making a firearm for the United States.

A firearm may be made by, or on behalf of, the United States or any department, independent establishment, or agency thereof without payment of the making tax. However, if a firearm is to be made on behalf of the United States, the maker must file an application, in duplicate, on Form 1 (Firearms) and obtain the approval of the Director in the manner prescribed in § 479.62.

§ 479.70 Certain gov ernment entities.

A firearm may be made without payment of the making tax by, or on behalf of, any State, or possession of the United States, any political subdivision thereof, or any official police organization of such a government entity engaged in criminal investigations. Any person making a firearm under this exemption shall first file an application, in duplicate, on Form 1 (Firearms) and obtain the approval of the Director as prescribed in § 479.62.

Registration

§ 479.71 Proof of registration.

The approval by the Director of an application, Form 1 (Firearms), to make a firearm under this subpart shall effectuate registration of the firearm described in the Form 1 (Firearms) to the person making the firearm. The original Form 1 (Firearms) showing approval by the Director shall be retained by the maker to establish proof of his registration of the firearm described therein, and shall be made available to any ATF officer on request.

Subpart F—Transfer Tax

§ 479.81 Scope of tax.

Except as otherwise provided in this part, each transfer of a firearm in the United States is subject to a tax to be represented by an adhesive stamp of the proper denomination bearing the words "National Firearms Act" to be affixed to the Form 4 (Firearms), Application for Transfer and Registration of Firearm, as provided in this subpart.

§ 479.82 Rate of tax.

The transfer tax imposed with respect to firearms transferred within the United States is at the rate of $200 for each firearm transferred, except that the transfer tax on any firearm classified as "any other weapon" shall be at the rate of $5 for each such firearm transferred. The tax imposed on the transfer of the firearm shall be paid by the transferor.

§ 479.83 Transfer tax in addition to import duty.

The transfer tax imposed by section 5811, I.R.C., is in addition to any import duty.

Application and Order for Transfer of Firearm

§ 479.84 Application to transfer.

Except as otherwise provided in this subpart, no firearm may be transferred in the United States unless an application, Form 4 (Firearms), Application for Transfer and Registration of Firearm, in duplicate, executed under the penalties of perjury to transfer the firearm and register it to the transferee has been filed with and approved by the Director. The application, Form 4 (Firearms), shall be filed by the transferor and shall identify the firearm to be transferred by type; serial number; name and address of the manufacturer and importer, if known; model; caliber, gauge or size; in the case of a short-barreled shotgun or a short-barreled rifle, the length of the barrel; in the case of a weapon made from a rifle or shotgun, the overall length of the weapon and the length of the barrel; and any other identifying marks on the firearm. In the event the firearm does not bear a serial number, the applicant shall obtain a serial number from the Regional director (compliance) and shall stamp (impress) or otherwise conspicuously place such serial number on the firearm in a manner not susceptible of being readily obliterated, altered or removed. The application, Form 4 (Firearms), shall identify the transferor by name and address; shall identify the transferor's Federal firearms license and special (occupational) Chapter tax stamp, if any; and if the transferor is other than a natural person, shall show the title or status of the person executing the application. The application also shall identify the transferee by name and address, and, if the transferee is a natural person not qualified as a manufacturer, importer or dealer under this part, he shall be further identified in the manner prescribed in § 479.85. The application also shall identify the special (occupational) tax stamp and Federal firearms license of the transferee, if any. Any tax payable on the transfer must be represented by an adhesive stamp of proper denomination being affixed to the application, Form 4 (Firearms), properly cancelled.

§ 479.85 Identification of transferee.

If the transferee is an individual, such person shall securely attach to each copy of the application, Form 4 (Firearms), in the space provided on the form, a photograph of the applicant 2 × 2 inches in size, clearly showing a full front view of the features of the applicant with head bare, with the distance from the top of the head to the point of the chin approximately 1 1/4 inches, and which shall have been taken within 1 year prior to the date of the application. The transferee shall attach two properly completed FBI Forms FD–258 (Fingerprint Card) to the application. The fingerprints must be clear for accurate classification and should be taken by someone properly equipped to take them. A certificate of the local chief of police, sheriff of the county, head of the State police, State or local district attorney or prosecutor, or such other person whose certificate may in a particular case be acceptable to the Director, shall be completed on each copy of the Form 4 (Firearms). The certificate shall state that the certifying official is satisfied that the fingerprints and photograph accompanying the application are those of the applicant and that the certifying official has no information indicating that the receipt or possession of the firearm would place the transferee in violation of State or local law or that the transferee will use the firearm for other than lawful purposes.

§ 479.86 Action on application.

The Director will consider a completed and properly executed application, Form 4 (Firearms), to transfer a firearm. If the application is approved, the Director will affix the appropriate National Firearms Act stamp, cancel it, and return the original application showing approval to the transferor who may then transfer the firearm to the transferee along with the approved application. The approval of an application, Form 4 (Firearms), by the Director will effectuate registration of the firearm to the transferee. The transferee shall not take possession of a firearm until the application, Form 4 (Firearms), for the transfer filed by the transferor has been approved by the Director and registration of the firearm is effectuated to the transferee. The transferee shall retain the approved application as proof that the firearm described therein is registered to the transferee, and shall make the approved Form 4 (Firearms) available to any ATF officer on request. If the application, Form 4 (Firearms), to transfer a firearm is disapproved by the Director, the original application and the remittance for purchase of the stamp will be returned to the transferor with reasons for the disapproval stated on the application. An application, Form 4 (Firearms), to transfer a firearm shall be denied if the transfer, receipt, or possession of a firearm would place the transferee in violation of law. In addition to any other records checks that may be conducted to determine whether the transfer, receipt, or possession of a firearm would place the transferee in violation of law, the Director shall contact the National Instant Criminal Background Check System.

§ 479.87 Cancellation of stamp.

The method of cancellation of the stamp required by this subpart as prescribed in § 479.67 shall be used.

Exemptions Relating to Transfers of Firearms

§ 479.88 Special (occupational) taxpayers.

(a) A firearm registered to a person qualified under this part to engage in business as an importer, manufacturer, or dealer may be transferred by that person without payment of the transfer tax to any other person qualified under this part to manufacture, import, or deal in firearms.

(b) The exemption provided in paragraph (a) of this section shall be obtained by the transferor of the firearm filing with the Director an application, Form 3 (Firearms), Application for Tax-exempt Transfer of Firearm and Registration to Special (Occupational) Taxpayer, in duplicate, executed under the penalties of perjury. The application, Form 3 (Firearms), shall **(1)** show the name and address of the transferor and of the transferee, **(2)** identify the Federal firearms license and special (occupational) tax stamp of the transferor and of the transferee, **(3)** show the name and address of the manufacturer and the importer of the firearm, if known, **(4)** show the type, model, overall length (if applicable), length of barrel, caliber, gauge or size, serial number, and other marks of identification of the firearm, and **(5)** contain a statement by the transferor that he is entitled to the exemption because the transferee is a person qualified under this part to manufacture, import, or deal in firearms. If the Director approves an application, Form 3 (Firearms), he shall return the original Form 3 (Firearms) to the transferor with the approval noted thereon. Approval of an application, Form 3 (Firearms), by the Director shall remove registration of the firearm reported thereon from the transferor and shall effectuate the registration of that firearm to the transferee. Upon receipt of the approved Form 3 (Firearms), the transferor shall deliver same with the firearm to the transferee. The transferor shall not transfer the firearm to the transferee until his application, Form 3 (Firearms), has been approved by the Director and the original thereof has been returned to the transferor. If the Director disapproves the application, Form 3 (Firearms), he shall return the original Form 3 (Firearms) to the transferor with the reasons for the disapproval stated thereon.

(c) The transferor shall be responsible for establishing the exempt status of the transferee before making a transfer under the provisions of this section. Therefore, before engaging in transfer negotiations with the transferee, the transferor should satisfy himself as to the claimed exempt status of the transferee and the bona fides of the transaction. If not fully satisfied, the transferor should communicate with the Director, report all circumstances regarding the proposed transfer, and await the Director's advice before making application for the transfer. An unapproved transfer or a transfer to an unauthorized person may subject the transferor to civil and criminal liabilities. (See 26 U.S.C. 5852, 5861, and 5871.)

§ 479.89 Transfers to the United States.

A firearm may be transferred to the United States or any department, independent establishment or agency thereof without payment of the transfer tax. However, the procedures for the transfer of a firearm as provided in § 479.90 shall be followed in a tax-exempt transfer of a firearm under this section, unless the transferor is relieved of such requirement under other provisions of this part.

§ 479.90 Certain government entities.

(a) A firearm may be transferred without payment of the transfer tax to or from any State, possession of the United States, any political subdivision thereof, or any official police organization of such a governmental entity engaged in criminal investigations.

(b) The exemption provided in paragraph (a) of this section shall be obtained by the transferor of the firearm filing with the Director an application, Form 5 (Firearms), Application for Tax-exempt Transfer and Registration of Firearm, in duplicate, executed under the penalties of perjury. The application shall **(1)** show the name and address of the transferor and of the transferee, **(2)** identify the Federal firearms license and special (occupational) tax stamp, if any, of the transferor and of the transferee, **(3)** show the name and address of the manufacturer and the importer of the firearm, if known, **(4)** show the type, model, overall length (if applicable), length of barrel, caliber, gauge or size, serial number, and other marks of identification of the firearm, and **(5)** contain a statement by the transferor that the transferor is entitled to the exemption because either the transferor or the transferee is a governmental entity coming within the purview of paragraph (a) of this section. In the case of a transfer of a firearm by a governmental entity to a transferee who is a natural person not qualified as a manufacturer, importer, or dealer under this part, the transferee shall be further identified in the manner prescribed in § 479.85. If the Director approves an application, Form 5 (Firearms), the original Form 5 (Firearms) shall be returned to the transferor with the approval noted thereon. Approval of an application, Form 5 (Firearms), by the Director shall effectuate the registration of that firearm to the transferee. Upon receipt of the approved Form 5 (Firearms), the transferor shall deliver same with the firearm to the transferee. The transferor shall not transfer the firearm to the transferee until the application, Form 5 (Firearms), has been approved by the Director and the original thereof has been returned to the transferor. If the Director disapproves the application, Form 5 (Firearms), the original Form 5 (Firearms) shall be returned to the transferor with the reasons for the disapproval stated thereon. An application by a governmental entity to transfer a firearm shall be denied if the transfer, receipt, or possession of a firearm would place the transferee in violation of law.

(c) The transferor shall be responsible for establishing the exempt status of the transferee before making a transfer under the provisions of this section. Therefore, before engaging in transfer negotiations with the transferee, the transferor should satisfy himself of the claimed exempt status of the transferee and the bona fides of the transaction. If not fully satisfied, the transferor should communicate with the Director, report all circumstances regarding the proposed transfer, and await the Director's advice before making application for transfer. An unapproved transfer or a transfer to an unauthorized person may subject the transferor to civil and criminal liabilities. (See 26 U.S.C. 5852, 5861, and 5871.)

§ 479.91 Unserviceable firearms.

An unserviceable firearm may be transferred as a curio or ornament without payment of the transfer tax. However, the procedures for the transfer of a firearm as provided in § 479.90 shall be followed in a tax-exempt transfer of a firearm under this section, except a statement shall be entered on the transfer application, Form 5 (Firearms), by the transferor that he is entitled to the exemption because the firearm to be transferred is unservicable and is being transferred as a curio or ornament. An unapproved transfer, the transfer of a firearm under the provisions of this section which is in fact not an unserviceable firearm, or the transfer of an unserviceable firearm as something other than a curio or ornament, may subject the transferor to civil and criminal liabilities. (See 26 U.S.C. 5811, 5852, 5861, and 5871.)

§ 479.92 Transp ortation of firearms to effect transfer.

Notwithstanding any provision of § 478.28 of this chapter, it shall not be required that authorization be obtained from the Director for the transportation in interstate or foreign commerce of a firearm in order to effect the transfer of a firearm authorized under the provisions of this subpart.

Other Provisions

§ 479.93 Tra nsfers of firearms to certain persons.

Where the transfer of a destructive device, machine gun, short-barreled shotgun, or short-barreled rifle is to be made by a person licensed under the provisions of Title I of the Gun Control Act of 1968 (82 Stat. 1213) to a person not so licensed, the sworn statement required by § 478.98 of this chapter shall be attached to and accompany the transfer application required by this subpart.

Subpart G—Registration and Identification of Firearms

§ 479.101 Registration of firearms.

(a) The Director shall maintain a central registry of all firearms in the United States which are not in the possession of or under the control of the United States. This registry shall be known as the National Firearms Registration and Transfer Record and shall include:

(1) Identification of the firearm as required by this part;

(2) Date of registration; and

(3) Identification and address of person entitled to possession of the firearm as required by this part.

(b) Each manufacturer, importer, and maker shall register each firearm he manufactures, imports, or makes in the manner prescribed by this part. Each firearm transferred shall be registered to the transferee by the transferor in the manner prescribed by this part. No firearm may be registered by a person unlawfully in possession of the firearm except during an amnesty period established under section 207 of the Gun Control Act of 1968 (82 Stat. 1235).

(c) A person shown as possessing firearms by the records maintained by the Director pursuant to the National Firearms Act (26 U.S.C. Chapter 53) in force on October 31, 1968, shall be considered to have registered the firearms in his possession which are disclosed by that record as being in his possession on October 31, 1968.

(d) The National Firearms Registration and Transfer Record shall include firearms registered to the possessors thereof under the provisions of section 207 of the Gun Control Act of 1968.

(e) A person possessing a firearm registered to him shall retain proof of registration which shall be made available to any ATF officer upon request.

(f) A firearm not identified as required by this part shall not be registered.

§ 479.102 How must firearms be identified?

(a) You, as a manufacturer, importer, or maker of a firearm, must legibly identify the firearm as follows:

(1) By engraving, casting, stamping (impressing), or otherwise conspicuously placing or causing to be engraved, cast, stamped (impressed) or placed on the frame or receiver thereof an individual serial number. The serial number must be placed in a manner not susceptible of being readily obliterated, altered, or removed, and must not duplicate any serial number placed by you on any other firearm. For firearms manufactured, imported, or made on and after January 30, 2002, the engraving, casting, or stamping (impressing) of the serial number must be to a minimum depth of .003 inch and in a print size no smaller than 1/16 inch; and

(2) By engraving, casting, stamping (impressing), or otherwise conspicuously placing or causing to be engraved, cast, stamped (impressed), or placed on the frame, receiver, or barrel thereof certain additional information. This information must be placed in a manner not susceptible of being readily obliterated, altered or removed. For firearms manufactured, imported, or made on and after January 30, 2002, the engraving, casting, or stamping (impressing) of this information must be to a minimum depth of .003 inch. The additional information includes:

(i) The model, if such designation has been made;

(ii) The caliber or gauge;

(iii) Your name (or recognized abbreviation) and also, when applicable, the name of the foreign manufacturer or maker;

(iv) In the case of a domestically made firearm, the city and State (or recognized abbreviation thereof) where you as the manufacturer maintain your place of business, or where you, as the maker, made the firearm; and

(v) In the case of an imported firearm, the name of the country in which it was manufactured and the city and State (or recognized abbreviation thereof) where you as the importer maintain your place of business. For additional require-ments relating to imported firearms, see Customs regulations at 19 CFR part 134.

(b) The depth of all markings required by this section will be measured from the flat surface of the metal and not the peaks or ridges. The height of serial numbers required by paragraph (a)(1) of this section will be measured as the distance between the latitudinal ends of the character impression bottoms (bases).

(c) The Director may authorize other means of identification upon receipt of a letter application from you, submitted in duplicate, showing that such other identification is reasonable and will not hinder the effective administration of this part.

(d) In the case of a destructive device, the Director may authorize other means of identifying that weapon upon receipt of a letter application from you, submitted in duplicate, showing that engraving, casting, or stamping (impressing) such a weapon would be dangerous or impracticable.

(e) A firearm frame or receiver that is not a component part of a complete weapon at the time it is sold, shipped, or otherwise disposed of by you must be identified as required by this section.

(f)(1) Any part defined as a machine gun, muffler, or silencer for the purposes of this part that is not a component part of a complete firearm at the time it is sold, shipped, or otherwise disposed of by you must be identified as required by this section.

(2) The Director may authorize other means of identification of parts defined as machine guns other than frames or receivers and parts defined as mufflers or silencers upon receipt of a letter application from you, submitted in duplicate, showing that such other identification is reasonable and will not hinder the effective administration of this part.

(Approved by the Office of Management and Budget under control number 1512–0550)

§ 479.103 Registration of firearms manufactured.

Each manufacturer qualified under this part shall file with the Director an accurate notice on Form 2 (Firearms), Notice of Firearms Manufactured or Imported, executed under the penalties of perjury, to show his manufacture of firearms. The notice shall set forth the name and address of the manufacturer, identify his special (occupational) tax stamp and Federal firearms license, and show the date of manufacture, the type, model, length of barrel, overall length, caliber, gauge or size, serial numbers, and other marks of identification of the firearms he manufactures, and the place where the manufactured firearms will be kept. All firearms manufactured by him during a single day shall be included on one notice, Form 2 (Firearms), filed by the manufacturer no later than the close of the next business day. The manufacturer shall prepare the notice, Form 2 (Firearms), in duplicate, file the original notice as prescribed herein and keep the copy with the records required by subpart I of this part at the premises covered by his special (occupational) tax stamp. Receipt of the notice, Form 2 (Firearms), by the Director shall effectuate the registration of the firearms listed on that notice. The requirements of this part relating to the transfer of a firearm are applicable to transfers by qualified manufacturers.

§ 479.104 Registration of firearms by certain go vernmental entities.

Any State, any political subdivision thereof, or any official police organization of such a government entity engaged in criminal investigations, which acquires for official use a firearm not registered to it, such as by abandonment or by forfeiture, will register such firearm with the Director by filing Form 10 (Firearms), Registration of Firearms Acquired by Certain Governmental Entities, and such registration shall become a part of the National Firearms Registration and Transfer Record. The application shall identify the applicant, describe each firearm covered by the application, show the location where each firearm usually will be kept, and, if the firearm is unserviceable, the application shall show how the firearm was made unserviceable. This section shall not apply to a firearm merely being held for use as evidence in a criminal proceeding. The Form 10 (Firearms) shall be executed in duplicate in accordance with the instructions thereon. Upon registering the firearm, the Director shall return the original Form 10 (Firearms) to the registrant with notification thereon that registration of the firearm has been made. The registration of any firearm under this section is for official use only and a subsequent transfer will be approved only to other governmental entities for official use.

Machine Guns

§ 479.105 Transfer and possession of machine guns.

(a) General. As provided by 26 U.S.C. 5812 and 26 U.S.C. 5822, an application to make or transfer a firearm shall be denied if the making, transfer, receipt, or possession of the firearm would place the maker or transferee in violation of law. Section 922(o), Title 18, U.S.C., makes it unlawful for any person to transfer or possess a machine gun, except a transfer to or by, or possession by or under the authority of, the United States or any department or agency thereof or a State, or a department, agency, or political subdivision thereof; or any lawful transfer or lawful possession of a machine gun that was lawfully possessed before May 19, 1986. Therefore, notwithstanding any other provision of this part, no application to make, transfer, or import a machine gun will be approved except as provided by this section.

(b) Machine guns lawfully possessed prior to May 19, 1986. A machine gun possessed in compliance with the provisions of this part prior to May 19, 1986, may continue to be lawfully possessed by the person to whom the machine gun is registered and may, upon compliance with the provisions of this part, be lawfully transferred to and possessed by the transferee.

(c) Importation and manufacture. Subject to compliance with the provisions of this part, importers and manufacturers qualified under this part may import and manufacture machine guns on or after May 19, 1986, for sale or distribution to any department or agency of the United States or any State or political subdivision thereof, or for use by dealers qualified under this part as sales samples as provided in paragraph (d) of this section. The registration of such machine guns under this part and their subsequent transfer shall be conditioned upon and restricted to the sale or distribution of such weapons for the official use of Federal, State or local governmental entities. Subject to compliance with the provisions of this part, manufacturers qualified under this part may manufacture machine guns on or after May 19, 1986, for exportation in compliance with the Arms Export Control Act (22 U.S.C. 2778) and regulations prescribed thereunder by the Department of State.

(d) Dealer sales samples. Subject to compliance with the provisions of this part, applications to transfer and register a machine gun manufactured or imported on or after May 19, 1986, to dealers qualified under this part will be approved if it is established by specific information the expected governmental customers who would require a demonstration of the weapon, information as to the availability of the machine gun to fill subsequent orders, and letters from governmental entities expressing a need for a particular model or interest in seeing a demonstration of a particular weapon. Applications to transfer more than one machine gun of a particular model to a dealer must also establish the dealer's need for the quantity of samples sought to be transferred.

(e) The making of machine guns on or after May 19, 1986. Subject to compliance with the provisions of this part, applications to make and register machine guns on or after May 19, 1986, for the benefit of a Federal, State or local governmental entity (e.g., an invention for possible future use of a governmental entity or the making of a weapon in connection with research and development on behalf of such an entity) will be approved if it is established by specific information that the machine gun is particularly suitable for use by Federal, State or local governmental entities and that the making of the weapon is at the request and on behalf of such an entity.

(f) Discontinuance of business. Since section 922(o), Title 18, U.S.C., makes it unlawful to transfer or possess a machine gun except as provided in the law, any qualified manufacturer, importer, or dealer intending to discontinue business shall, prior to going out of business, transfer in compliance with the provisions of this part any machine gun manufactured or imported after May 19, 1986, to a Federal, State or local governmental entity, qualified manufacturer, qualified importer, or, subject to the provisions of paragraph (d) of this section, dealer qualified to possess such, machine gun.

Subpart H—Importation and Exportation

Importation

§ 479.111 Procedure.

(a) No firearm shall be imported or brought into the United States or any territory under its control or jurisdiction unless the person importing or bringing in the firearm establishes to the satisfaction of the Director that the firearm to be imported or brought in is being imported or brought in for:

(1) The use of the United States or any department, independent establishment, or agency thereof or any State or possession or any political subdivision thereof; or

(2) Scientific or research purposes; or

(3) Testing or use as a model by a registered manufacturer or solely for use as a sample by a registered importer or registered dealer.

The burden of proof is affirmatively on any person importing or bringing the firearm into the United States or any territory under its control or jurisdiction to show that the firearm is being imported or brought in under one of the above paragraphs. Any person desiring to import or bring a firearm into the United States under this paragraph shall file with the Director an application on Form 6 (Firearms), Application and Permit for Importation of Firearms, Ammunition and Implements of War, in triplicate, executed under the penalties of perjury. The application shall show the information required by subpart G of Part 478 of this chapter. A detailed explanation of why the importation of the firearm falls within the standards set out in this paragraph shall be attached to the application. The person seeking to import or bring in the firearm will be notified of the approval or disapproval of his application. If the application is approved, the original Form 6 (Firearms) will be returned to the applicant showing such approval and he will present the approved application, Form 6 (Firearms), to the Customs officer at the port of importation. The approval of an application to import a firearm shall be automatically terminated at the expiration of one year from the date of approval unless, upon request, it is further extended by the Director. If the firearm described in the approved application is not imported prior to the expiration of the approval, the Director shall be so notified. Customs officers will not permit release of a firearm from Customs custody, except for exportation, unless covered by an application which has been approved by the Director and which is currently effective. The importation or bringing in of a firearm not covered by an approved application may subject the person responsible to civil and criminal liabilities. (26 U.S.C. 5861, 5871, and 5872.)

(b) Part 478 of this chapter also contains requirements and procedures for the importation of firearms into the United States. A firearm may not be imported into the United States under this part unless those requirements and procedures are also complied with by the person importing the firearm.

(c) The provisions of this subpart shall not be construed as prohibiting the return to the United States or any territory under its control or jurisdiction of a firearm by a person who can establish to the satisfaction of Customs that **(1)** the firearm was taken out of the United States or any territory under its control or jurisdiction by such person, **(2)** the firearm is registered to that person, and **(3)** if appropriate, the authorization required by Part 478 of this chapter for the transportation of such a firearm in interstate or foreign commerce has been obtained by such person.

§ 479.112 Registration of imported firearms.

(a) Each importer shall file with the Director an accurate notice on Form 2 (Firearms), Notice of Firearms Manufactured or Imported, executed under the penalties of perjury, showing the importation of a firearm. The notice shall set forth the name and address of the importer, identify the importer's special (occupational) tax stamp and Federal firearms license, and show the import permit number, the date of release from Customs custody, the type, model, length of barrel, overall length, caliber, gauge or size, serial number, and other marks of identification of the firearm imported, and the place where the imported firearm will be kept. The Form 2 (Firearms) covering an imported firearm shall be filed by the importer no later than fifteen (15) days from the date the firearm was released from Customs custody. The importer shall prepare the notice, Form 2 (Firearms), in duplicate, file the original return as prescribed herein, and keep the copy with the records required by subpart I of this part at the premises covered by the special (occupational) tax stamp. The timely receipt by the Director of the notice, Form 2 (Firearms), and the timely receipt by the Director of the copy of Form 6A (Firearms), Release and Receipt of Imported Firearms, Ammunition and Implements of War, required by § 478.112 of this chapter, covering the weapon reported on the Form 2 (Firearms) by the qualified importer, shall effectuate the registration of the firearm to the importer.

(b) The requirements of this part relating to the transfer of a firearm are applicable to the transfer of imported firearms by a qualified importer or any other person.

(c) Subject to compliance with the provisions of this part, an application, Form 6 (Firearms), to import a firearm by an importer or dealer qualified under this part, for use as a sample in connection with sales of such firearms to Federal, State or local governmental entities, will be approved if it is established by specific information attached to the application that the firearm is suitable or potentially suitable for use by such entities. Such information must show why a sales sample of a particular firearm is suitable for such use and the expected governmental customers who would require a demonstration of the firearm. Information as to the availability of the firearm to fill subsequent orders and letters from governmental entities expressing a need for a particular model or interest in seeing a demonstration of a particular firearm would establish suitability for governmental use. Applications to import more than one firearm of a particular model for use as a sample by an importer or dealer must also establish the importer's or dealer's need for the quantity of samples sought to be imported.

(d) Subject to compliance with the provisions of this part, an application, Form 6 (Firearms), to import a firearm by an importer or dealer qualified under this part, for use as a sample in connection with sales of such firearms to Federal, State or local governmental entities, will be approved if it is established by specific information attached to the application that the firearm is particularly suitable for use by such entities. Such information must show why a sales sample of a particular firearm is suitable for such use and the expected governmental customers who would require a demonstration of the firearm. Information as to the availability of the firearm to fill subsequent orders and letters from governmental entities expressing a need for a particular model or interest in seeing a demonstration of a particular firearm would establish suitability for governmental use. Applications to import more than one firearm of a particular model for use as a sample by an importer or dealer must also establish the importer's or dealer's need for the quantity of samples sought to be imported.

§ 479.113 Conditional importation.

The Director shall permit the conditional importation or bringing into the United States of any firearm for the purpose of examining and testing the firearm in connection with making a determination as to whether the importation or bringing in of such firearm will be authorized under this subpart. An application under this section shall be filed on Form 6 (Firearms), in triplicate, with the Director. The Director may impose conditions upon any importation under this section including a requirement that the firearm be shipped directly from Customs custody to the Director and that the person importing or bringing in the firearm must agree to either export the weapon or destroy it if a final determination is made that it may not be imported or brought in under this subpart. A firearm so imported or brought into the United States may be released from Customs custody in the manner prescribed by the conditional authorization of the

Director.

Exportation

§ 479.114 Application and permit for exportation of firearms.

Any person desiring to export a firearm without payment of the transfer tax must file with the Director an application on Form 9 (Firearms), Application and Permit for Exportation of Firearms, in quadruplicate, for a permit providing for deferment of tax liability. Part 1 of the application shall show the name and address of the foreign consignee, number of firearms covered by the application, the intended port of exportation, a complete description of each firearm to be exported, the name, address, State Department license number (or date of application if not issued), and identification of the special (occupational) tax stamp of the transferor. Part 1 of the application shall be executed under the penalties of perjury by the transferor and shall be supported by a certified copy of a written order or contract of sale or other evidence showing that the firearm is to be shipped to a foreign designation. Where it is desired to make a transfer free of tax to another person who in turn will export the firearm, the transferor shall likewise file an application supported by evidence that the transfer will start the firearm in course of exportation, except, however, that where such transferor and exporter are registered special-taxpayers the transferor will not be required to file an application on Form 9 (Firearms).

§ 479.115 Action by Director.

If the application is acceptable, the Director will execute the permit, Part 2 of Form 9 (Firearms), to export the firearm described on the form and return three copies thereof to the applicant. Issuance of the permit by the Director will suspend assertion of tax liability for a period of six (6) months from the date of issuance. If the application is disapproved, the Director will indicate thereon the reason for such action and return the forms to the applicant.

§ 479.116 Procedure by exporter.

Shipment may not be made until the permit, Form 9 (Firearms), is received from the Director. If exportation is to be made by means other than by parcel post, two copies of the form must be addressed to the District Director of Customs at the port of exportation, and must precede or accompany the shipment in order to permit appropriate inspection prior to lading. If exportation is to be made by parcel post, one copy of the form must be presented to the postmaster at the office receiving the parcel who will execute Part 4 of such form and return the form to the exporter for transmittal to the Director. In the event exportation is not effected, all copies of the form must be immediately returned to the Director for cancellation.

§ 479.117 Action by Customs.

Upon receipt of a permit, Form 9 (Firearms), in duplicate, authorizing the exportation of firearms, the District Director of Customs may order such inspection as deemed necessary prior to lading of the merchandise. If satisfied that the shipment is proper and the information contained in the permit to export is in agreement with information shown in the shipper's export declaration, the District Director of Customs will, after the merchandise has been duly exported, execute the certificate of exportation (Part 3 of Form 9 (Firearms)). One copy of the form will be retained with the shipper's export declaration and the remaining copy thereof will be transmitted to the Director.

§ 479.118 Proof of exportation.

Within a six-month's period from date of issuance of the permit to export firearms, the exporter shall furnish or cause to be furnished to the Director **(a)** the certificate of exportation (Part 3 of Form 9 (Firearms)) executed by the District Director of Customs as provided in § 479.117, or **(b)** the certificate of mailing by parcel post (Part 4 of Form 9 (Firearms)) executed by the postmaster of the post office receiving the parcel containing the firearm, or **(c)** a certificate of lading executed by a Customs officer of the foreign country to which the firearm is exported, or **(d)** a sworn statement of the foreign consignee covering the receipt of the firearm, or **(e)** the return receipt, or a reproduced copy thereof, signed by the addressee or his agent, where the shipment of a firearm was made by insured or registered parcel post. Issuance of a permit to export a firearm and furnishing of evidence establishing such exportation under this section will relieve the actual exporter and the person selling to the exporter for exportation from transfer tax liability. Where satisfactory evidence of exportation of a firearm is not furnished within the stated period, the transfer tax will be assessed.

§ 479.119 Transp ortation of firearms to effect exportation.

Notwithstanding any provision of § 478.28 of this chapter, it shall not be required that authorization be obtained from the Director for the transportation in interstate or foreign commerce of a firearm in order to effect the exportation of a firearm authorized under the provisions of this subpart.

§ 479.120 Refunds.

Where, after payment of tax by the manufacturer, a firearm is exported, and satisfactory proof of exportation (see § 479.118) is furnished, a claim for refund may be submitted on Form 843 (see § 479.172). If the manufacturer waives all claim for the amount to be refunded, the refund shall be made to the exporter. A claim for refund by an exporter of tax paid by a manufacturer should be accompanied by waiver of the manufacturer and proof of tax payment by the latter.

§ 479.121 Insular possessions.

Transfers of firearms to persons in the insular possessions of the United States are exempt from transfer tax, provided title in cases involving change of title (and custody or control, in cases not involving change of title), does not pass to the transferee or his agent in the United States. However, such exempt transactions must be covered by approved permits and supporting documents corresponding to those required in the case of firearms exported to foreign countries (see §§ 479.114 and 479.115), except that the Director may vary the requirements herein set forth in accordance with the requirements of the governing authority of the insular possession. Shipments to the insular possessions will not be authorized without compliance with the requirements of the governing authorities thereof. In the case of a nontaxable transfer to a person in such insular possession, the exemption extends only to such transfer and not to prior transfers.

Arms Export Control Act

§ 479.122 Requirements.

(a) Persons engaged in the business of importing firearms are required by the Arms Export Control Act (22 U.S.C. 2778) to register with the Director. (See Part 447 of this chapter.)

(b) Persons engaged in the business of exporting firearms caliber .22 or larger are subject to the requirements of a license issued by the Secretary of State. Application for such license should be made to the Office of Munitions Control, Department of State, Washington, DC 20502, prior to exporting firearms.

Subpart I—Records and Returns

§ 479.131 Records.

For the purposes of this part, each manufacturer, importer, and dealer in firearms shall keep and maintain such records regarding the manufacture, importation, acquisition (whether by making, transfer, or otherwise), receipt, and disposition of firearms as are prescribed, and in the manner and place required, by part 478 of this chapter. In addition, each manufacturer, importer, and dealer shall maintain, in chronological order, at his place of business a separate record consisting of the documents required by this part showing the registration of any firearm to him. If firearms owned or possessed by a manufacturer, importer, or dealer are stored or kept on premises other than the place of business shown on his special (occupational) tax stamp, the record establishing registration shall show where such firearms are stored or kept. The records required by this part shall be readily accessible for inspection at all reasonable times by ATF officers.

(Approved by the Office of Management and Budget under control number 1512–0387)

Subpart J—Stolen or Lost Firearms or Documents

§ 479.141 Stolen or lost firearms.

Whenever any registered firearm is stolen or lost, the person losing possession thereof will, immediately upon discovery of such theft or loss, make a report to the Director showing the following: **(a)** Name and address of the person in whose name the firearm is registered, **(b)** kind of firearm, **(c)** serial number, **(d)** model, **(e)** caliber, **(f)** manufacturer of the firearm, **(g)** date and place of theft or loss, and **(h)** complete statement of facts and circumstances surrounding such theft or loss.

§ 479.142 Stolen or lost documents.

When any Form 1, 2, 3, 4, 5, 6A, or 10 (Firearms) evidencing possession of a firearm is stolen, lost, or destroyed, the person losing possession will immediately upon discovery of the theft, loss, or destruction report the matter to the Director. The report will show in detail the circumstances of the theft, loss, or destruction and will include all known facts which may serve to identify the document. Upon receipt of the report, the Director will make such investigation as appears appropriate and may issue a duplicate document upon such conditions as the circumstances warrant.

Subpart K—Examination of Books and Records

§ 479.151 Failure to make returns: Substitute returns.

If any person required by this part to make returns shall fail or refuse to make any such return within the time prescribed by this part or designated by the Director, then the return shall be made by an ATF officer upon inspection of the books, but the making of such return by an ATF officer shall not relieve the person from any default or penalty incurred by reason of failure to make such return.

§ 479.152 Penalties (records and returns).

Any person failing to keep records or make returns, or making, or causing the making of, a false entry on any application, return or record, knowing such entry to be false, is liable to fine and imprisonment as provided in section 5871, I.R.C.

Subpart L—Distribution and Sale of Stamps

§ 479.161 National Firearms Act stamps.

"National Firearms Act" stamps evidencing payment of the transfer tax or tax on the making of a firearm are maintained by the Director. The remittance for purchase of the appropriate tax stamp shall be submitted with the application. Upon approval of the application, the Director will cause the appropriate tax to be paid by affixing the appropriate stamp to the application.

§ 479.162 Stamps authorized.

Adhesive stamps of the $5 and $200 denomination, bearing the words "National Firearms Act," have been prepared and only such stamps shall be used for the payment of the transfer tax and for the tax on the making of a firearm.

§ 479.163 Reuse of stamps prohibited.

A stamp once affixed to one document cannot lawfully be removed and affixed to another. Any person willfully reusing such a stamp shall be subject to the penalty prescribed by 26 U.S.C. 7208.

Subpart M—Redemption of or Allowance for Stamps or Refunds

§ 479.171 Redemption of or allowance for stamps.

Where a National Firearms Act stamp is destroyed, mutilated or rendered useless after purchase, and before liability has been incurred, such stamp may be redeemed by giving another stamp in lieu thereof. Claim for redemption of the stamp should be filed on ATF Form 2635 (5620.8) with the Director. Such claim shall be accompanied by the stamp or by a satisfactory explanation of the reasons why the stamp cannot be returned, and shall be filed within 3 years after the purchase of the stamp.

§ 479.172 Refunds.

As indicated in this part, the transfer tax or tax on the making of a firearm is ordinarily paid by the purchase and affixing of stamps, while special tax stamps are issued in payment of special (occupational) taxes. However, in exceptional cases, transfer tax, tax on the making of firearms, and/or special (occupational) tax may be paid pursuant to assessment. Claims for refunds of such taxes, paid pursuant to assessment, shall be filed on ATF Form 2635 (5620.8) within 3 years next after payment of the taxes. Such claims shall be filed with the regional director (compliance) serving the region in which the tax was paid. (For provisions relating to hand-carried documents and manner of filing, see 26 CFR 301.6091–1(b) and 301.6402–2(a).) When an applicant to make or transfer a firearm wishes a refund of the tax paid on an approved application where the firearm was not made pursuant to an approved Form 1 (Firearms) or transfer of the firearm did not take place pursuant to an approved Form 4 (Firearms), the applicant shall file a claim for refund of the tax on ATF Form 2635 (5620.8) with the Director. The claim shall be accompanied by the approved application bearing the stamp and an explanation why the tax liability was not incurred. Such claim shall be filed within 3 years next after payment of the tax.

Subpart N—Penalties and Forfeitures

§ 479.181 Penalties.

Any person who violates or fails to comply with the requirements of 26 U.S.C. Chapter 53 shall, upon conviction, be subject to the penalties imposed under 26 U.S.C. 5871.

§ 479.182 Forfeitures.

Any firearm involved in any violation of the provisions of 26 U.S.C. Chapter 53, shall be subject to seizure, and forfeiture under the internal revenue laws: **Provided, however,** That the disposition of forfeited firearms shall be in conformance with the requirements of 26 U.S.C. 5872. In addition, any vessel, vehicle or aircraft used to transport, carry, convey or conceal or possess any firearm with respect to which there has been committed any violation of any provision of 26 U.S.C. Chapter 53, or the regulations in this part issued pursuant thereto, shall be subject to seizure and forfeiture under the Customs laws, as provided by the act of August 9, 1939 (49 U.S.C. App., Chapter 11).

Subpart O—Other Laws Applicable

§ 479.191 Applicability of other provisions of internal revenue laws.

All of the provisions of the internal revenue laws not inconsistent with the provisions of 26 U.S.C. Chapter 53 shall be applicable with respect to the taxes imposed by 26 U.S.C. 5801, 5811, and 5821 (see 26 U.S.C. 5846).

§ 479.192 Commerce in firearms and ammunition.

For provisions relating to commerce in firearms and ammunition, including the movement of destructive devices, machine guns, short-barreled shotguns, or short-barreled rifles, see 18 U.S.C. Chapter 44, and Part 478 of this chapter issued pursuant thereto.

§ 479.193 Arms Export Control Act.

For provisions relating to the registration and licensing of persons engaged in the business of manufacturing, importing or exporting arms, ammunition, or implements of war, see the Arms Export Control Act (22 U.S.C. 2778), and the regulations issued pursuant thereto. (See also Part 447 of this chapter.)

ARMS EXPORT CONTROL ACT, SEC. 38

TITLE 22, UNITED STATES CODE, § 2778

Editor's Note:

With respect to Section 38 of the Arms Export Control Act of 1976 (22 U.S.C. 2778), only the importation provisions are administered by ATF. Export provisions are administered by the Department of State. Importation regulations issued under this law are in 27 CFR Part 447, and are included in this publication.

§ 2778. Control of arms exports and imports

(a) Presidential control of exports and imports of defense articles and services, guidance of policy, etc.; designation of United States Munitions List; issuance of export licenses; condition for export; negotiations information.

(1) In furtherance of world peace and the security and foreign policy of the United States, the President is authorized to control the import and the export of defense articles and defense services and to provide foreign policy guidance to persons of the United States involved in the export and import of such articles and services. The President is authorized to designate those items which shall be considered as defense articles and defense services for the purposes of this section and to promulgate regulations for the import and export of such articles and services. The items so designated shall constitute the United States Munitions List.

(2) Decisions on issuing export licenses under this section shall take into account whether the export of an article would contribute to an arms race, aid in the development of weapons of mass destruction, support international terrorism, increase the possibility of outbreak or escalation of conflict, or prejudice the development of bilateral or multilateral arms control or nonproliferation agreements or other arrangements.

(3) In exercising the authorities conferred by this section, the President may require that any defense article or defense service be sold under this Act as a condition of its eligibility for export, and may require that persons engaged in the negotiation for the export of defense articles and services keep the President fully and currently informed of the progress and future prospects of such negotiations.

(b) Registration and licensing requirements for manufacturers, exporters, or importers of designated defense articles and defense services; exceptions.

(1) (A) (i) As prescribed in regulations issued under this section, every person (other than an officer or employee of the United States Government acting in an official capacity) who engages in the business of manufacturing, exporting, or importing any defense articles or defense services designated by the President under subsection (a)(1) shall register with the United States Government agency charged with the administration of this section, and shall pay a registration fee which shall be prescribed by such regulations. Such regulations shall prohibit the return to the United States for sale in the United States (other than for the Armed Forces of the United States and its allies or for any State or local law enforcement agency) of any military firearms or ammunition of United States manufacture furnished to foreign governments by the United States under this Act or any other foreign assistance or sales program of the United States, whether or not enhanced in value or improved in condition in a foreign country. This prohibition shall not extend to similar firearms that have been so substantially transformed as to become, in effect, articles of foreign manufacture.

(ii) (I) As prescribed in regulations issued under this section, every person (other than an officer or employee of the United States Government acting in official capacity) who engages in the business of brokering activities with respect to the manufacture, export, import, or transfer of any defense article or defense service designated by the President under subsection (a)(1) or in the business of brokering activities with respect to the manufacture, export, import, or transfer of any foreign defense article or defense service (as defined in subclause (IV)), shall register with the United States Government agency charged with the administration of this section, and shall pay a registration fee which shall be prescribed by such regulations.

(II) Such brokering activities shall include the financing, transportation, freight forwarding, or taking of any other action that facilitates the manufacture, export, or import of a defense article or defense service.

(III) No person may engage in the business of brokering activities described in subclause (I) without a license, issued in accordance with this Act, except that no license shall be required for such activities undertaken by or for an agency of the United States Government—

(aa) for use by an agency of the United States Government; or

(bb) for carrying out any foreign assistance or sales program authorized by law and subject to the control of the President by other means.

(IV) For purposes of this clause, the term "foreign defense article or defense service" includes any non-United States defense article or defense service of a nature described on the United States Munitions List regardless of whether such article or service is of United States origin or whether such article or service contains United States origin components.

(B) The prohibition under such regulations required by the second sentence of subparagraph (A) shall not extend to any military firearms (or ammunition, components, parts, accessories, and attachments for such firearms) of United States manufacture furnished to any foreign government by the United States under this Act or any other foreign assistance or sales program of the United States if—

(i) such firearms are among those firearms that the Secretary of the Treasury is, or was at any time, required to authorize the importation of by reason of the provisions of section 925(e) of title 18, United States Code (including the requirement for the listing of such firearms as curios or relics under section 921(a)(13) of that title); and

(ii) such foreign government certifies to the United States Government that such firearms are owned by such foreign government.

(C) A copy of each registration made under this paragraph shall be transmitted to the Secretary of the Treasury for review regarding law enforcement concerns. The Secretary shall report to the President regarding such concerns as necessary.

(2) Except as otherwise specifically provided in regulations issued under subsection (a)(1), no defense articles or defense services designated by the President under subsection (a)(1) may be exported or imported without a license for such export or import, issued in accordance with this Act and regulations issued under this Act, except that no license shall be required for exports or imports made by or for an agency of the United States Government **(A)** for official use by a department or agency of the United States Government, or **(B)** for carrying out any foreign assistance or sales program authorized by law and subject to the control of the President by other means.

(3) (A) For each of the fiscal years 1988 and 1989, $250,000 of registration fees collected pursuant to paragraph (1) shall be credited to a Department of State account, to be available without fiscal year limitation. Fees credited to that account shall be available only for the payment of expenses incurred for—

(i) contract personnel to assist in the evaluation of munitions control license applications, reduce processing time for license applications, and improve monitoring of compliance with the terms of licenses; and

(ii) the automation of munitions control functions and the processing of munitions control license applications, including the development, procurement, and utilization of computer equipment and related software.

(B) The authority of this paragraph may be exercised only to such extent or in such amounts as are provided in advance in appropriation Acts.

(c) Criminal violations; punishment.

Any person who willfully violates any provision of this section or section 39, or any rule or regulation issued under either section, or who willfully, in a registration or license application or required report, makes any untrue statement of a material fact or omits to state a material fact required to be stated therein or necessary to make the statements therein not misleading, shall upon conviction be fined for each violation not more than $1,000,000 or imprisoned not more than ten years, or both.

(d) [Repealed]

(e) Enforcement powers of President.

In carrying out functions under this section with respect to the export of defense articles and defense services, the President is authorized to exercise the same powers concerning violations and enforcement which are conferred upon departments, agencies and officials by subsections (c), (d), (e), and (g) of section 11 of the Export Administration Act of 1979 [50 USCS Appx § 2410], and by subsections (a) and (c) of section 12 of such Act [50 USCS Appx § 2411(a) and (c)], subject to the same terms and conditions as are applicable to such powers under such Act, except that section 11(c)(2)(B) of such Act shall not apply, and instead, as prescribed in regulations issued under this section, the Secretary of State may assess civil penalties for violations of this Act and regulations prescribed thereunder and further may commence a civil action to recover such civil penalties, and except further that the names of the countries and the types and quantities of defense articles for which licenses are issued under this section shall not be withheld from public disclosure unless the President determines that the release of such information would be contrary to the national interest. Nothing in this subsection shall be construed as authorizing the withholding of information from the Congress. Notwithstanding section 11(c) of the Export Administration Act of 1979, the civil penalty for each violation involving controls imposed on the export of defense articles and defense services under this section may not exceed $500,000.

(f) Periodic review of items on the munitions list; notification regarding exemption from licensing requirements for export of defense items.

(1) The President shall periodically review the items on the United States Munitions List to determine what items, if any, no longer warrant export controls under this section. The results of such reviews shall be reported to the Speaker of the House of Representatives and to the Committee on Foreign Relations and the Committee on Banking, Housing, and Urban Affairs of the Senate. The President may not remove any item from the Munitions List until 30 days after the date on which the President has provided notice of the proposed removal to the Committee on International Relations of the House of Representatives and to the Committee on Foreign Relations of the Senate in accordance with the procedures applicable to reprogramming notifications under section 634A(a) of the Foreign Assistance Act of 1961. Such notice shall describe the nature of any controls to be imposed on that item under any other provision of law.

(2) The President may not authorize an exemption for a foreign country from the licensing requirements of this Act for the export of defense items under subsection (j) or any other provision of this Act until 30 days after the date on which the President has transmitted to the Committee on International Relations of the House of Representatives and the Committee on Foreign Relations of the Senate a notification that includes—

(A) a description of the scope of the exemption, including a detailed summary of the defense articles, defense services, and related technical data covered by the exemption; and

(B) a determination by the Attorney General that the bilateral agreement concluded under subsection (j) requires the compilation and maintenance of sufficient documentation relating to the export of United States defense articles, defense services, and related technical data to facilitate law enforcement efforts to detect, prevent, and prosecute criminal violations of any provision of this Act, including the efforts on the part of countries and factions engaged in international terrorism to illicitly acquire sophisticated United States defense items.

(3) Paragraph (2) shall not apply with respect to an exemption for Canada from the licensing requirements of this Act for the export of defense items.

TITLE 27 CFR CHAPTER II

PART 447—IMPORTATION OF ARMS, AMMUNITION AND IMPLEMENTS OF WAR

(This section was formerly designated as Part 47)

Editor's Note:

Effective January 24, 2003, the Homeland Security Act transferred the Bureau of Alcohol, Tobacco and Firearms from the Department of the Treasury to the Department of Justice. In addition, the agency's name was changed to the Bureau of Alcohol, Tobacco, Firearms and Explosives. The regulations, as printed in this publication, do not yet reflect this change. The regulations will be amended to change the references from the "Bureau of Alcohol, Tobacco and Firearms," the "Department of the Treasury," and the "Secretary of the Treasury" to the "Bureau of Alcohol, Tobacco, Firearms and Explosives," the "Department of Justice" and the "Attorney General," respectively.

Subpart A—Scope

§ 447.1 General.

The regulations in this part relate to that portion of Section 38, Arms Export Control Act of 1976, as amended, which is concerned with the importation of arms, ammunition and implements of war. This part contains the U.S. Munitions Import List and includes procedural and administrative requirements and provisions relating to registration of importers, permits, articles in transit, import certification, delivery verification, import restrictions applicable to certain countries, exemptions, U.S. military firearms or ammunition, penalties, seizures, and forfeitures. All designations and changes in designation of articles subject to import control under Section 414 of the Mutual Security Act of 1954, as amended, have the concurrence of the Secretary of State and the Secretary of Defense.

§ 447.2 Relation to ot her laws and regulations.

(a) All of those items on the U.S. Munitions Import List (see § 447.21) which are **"firearms"** or **"ammunition"** as defined in 18 U.S.C. 921(a) are subject to the interstate and foreign commerce controls contained in Chapter 44 of Title 18 U.S.C. and 27 CFR Part 478 and if they are **"firearms"** within the definition set out in 26 U.S.C. 5845(a) are also subject to the provisions of 27 CFR Part 479. Any person engaged in the business of importing firearms or ammunition as defined in 18 U.S.C. 921(a) must obtain a license under the provisions of 27 CFR Part 478, and if he imports firearms which fall within the definition of 26 U.S.C. 5845(a) must also register and pay special tax pursuant to the provisions of 27 CFR Part 479. Such licensing, registration and special tax requirements are in addition to registration under subpart D of this part.

(b) The permit procedures of subpart E of this part are applicable to all importations of articles on the U.S. Munitions Import List not subject to controls under 27 CFR Part 478 or 479. U.S. Munitions Import List articles subject to controls under 27 CFR Part 478 or 27 CFR Part 479 are subject to the import permit procedures of those regulations if imported into the United States (within the meaning of 27 CFR Parts 478 and 479).

(c) Articles on the U.S. Munitions Import List imported for the United States or any State or political subdivision thereof are exempt from the import controls of 27 CFR Part 478 but are not exempt from control under Section 38, Arms Export Control Act of 1976, unless imported by the United States or any agency thereof. All such importations not imported by the United States or any agency thereof shall be subject to the import permit procedures of subpart E of this part.

(d) For provisions requiring the registration of persons engaged in the business of brokering activities with respect to the importation of any defense article or defense service, see Department of State regulations in 22 CFR Part 129.

Subpart B—Definitions

§ 447.11 Meaning of terms.

When used in this part and in forms prescribed under this part, where not otherwise distinctly expressed or manifestly incompatible with the intent thereof, terms shall have the meanings ascribed in this section. Words in the plural form shall include the singular, and vice versa, and words imparting the masculine gender shall include the feminine. The terms "includes" and "including" do not exclude other things not enumerated which are in the same general class or are otherwise within the scope thereof.

Appropriate ATF officer. An officer or employee of the Bureau of Alcohol, Tobacco, and Firearms (ATF) specified by ATF Order 1130.34, Delegation of the Director's Authorities in 27 CFR Part 447, Importation of Arms, Ammunition and Implements of War.

Article. Any of the arms, ammunition, and implements of war enumerated in the U.S. Munitions Import List.

Bureau. Bureau of Alcohol, Tobacco and Firearms, the Department of the Treasury.

Carbine. A short-barrelled rifle whose barrel is generally not longer than 22 inches and is characterized by light weight.

CFR. The Code of Federal Regulations.

Chemical agent. A substance useful in war which, by its ordinary and direct chemical action, produces a powerful physiological effect.

Defense articles. Any item designated in § 447.21 or § 447.22. This term includes models, mockups, and other such items which reveal technical data directly relating to § 447.21 or § 447.22. For purposes of Category XXII, any item enumerated on the U.S. Munitions List (22 CFR Part 121).

Defense services. (a) The furnishing of assistance, including training, to foreign persons in the design, engineering, development, production, processing, manufacture, use, operation, overhaul, repair, maintenance, modification, or reconstruction of defense articles, whether in the United States or abroad; or

(b) The furnishing to foreign persons of any technical data, whether in the United States or abroad.

Director. The Director, Bureau of Alcohol, Tobacco and Firearms, the Department of the Treasury, Washington, DC 20226.

Executed under the penalties of perjury. Signed with the prescribed declaration under the penalties of perjury as provided on or with respect to the application, form, or other document or, where no form of declaration is prescribed, with the declaration: **"I declare under the penalties of perjury that this ____ __ (insert type of document such as statement, certificate, application, or other document), including the documents submitted in support thereof, has been examined by me and, to be st of my knowledge and belief, is true, correct, and complete."**

Firearms. A weapon, and all components and parts therefor, not over .50 caliber which will or is designed to or may be readily converted to expel a projectile by the action of an explosive, but shall not include BB and pellet guns, and muzzle loading (black powder) firearms (including any firearm with a matchlock, flintlock, percussion cap, or similar type of ignition system) or firearms covered by Category I(a) established to have been manufactured in or before 1898.

Import or importation. Bringing into the United States from a foreign country any of the articles on the Import List, but shall not include intransit, temporary import or temporary export transactions subject to Department of State controls under Title 22, Code of Federal Regulations.

Import List. The list of articles contained in § 447.21 and identified therein as "The U.S. Munitions Import List".

Machinegun. A **"machinegun," "machine pistol," "submachinegun,"** or **"automatic rifle"** is a firearm originally designed to fire, or capable of being fired fully automatically by a single pull of the trigger.

Permit. The same as "license" for purposes of 22 U.S.C. 1934(c).

Person. A partnership, company, association, or corporation, as well as a natural person.

Pistol. A hand-operated firearm having a chamber integral with, or permanently aligned with, the bore.

Revolver. A hand-operated firearm with a revolving cylinder containing chambers for individual cartridges.

Rifle. A shoulder firearm discharging bullets through a rifled barrel at least 16 inches in length, including combination and drilling guns.

Sporting type sight including optical. A telescopic sight suitable for daylight use on a rifle, shotgun, pistol, or revolver for hunting or target shooting.

This chapter. Title 27, Code of Federal Regulations, Chapter II (27 CFR Chapter II).

United States. When used in the geographical sense, includes the several States, the Commonwealth of Puerto Rico, the insular possessions of the United States, the District of Columbia, and any territory over which the United States exercises any powers of administration, legislation, and jurisdiction. (26 U.S.C. 7805 (68A Stat. 917), 27 U.S.C. 205 (49 Stat. 981 as amended), 18 U.S.C. 926 (82 Stat. 959), and sec. 38, Arms Export Control Act (22 U.S.C. 2778, 90 Stat. 744)).

Subpart C—The U.S. Munitions Import List

§ 447.21 The U.S. Munitions Import List.

The U.S. Munitions List compiled by the Department of State, Office of Defense Trade Controls, and published at 22 CFR 121.1, with the deletions indicated, has been adopted as an enumeration of the defense articles subject to controls under this part. The expurgated list, set out below, shall, for the purposes of this part, be known as the U.S. Munitions Import List:

THE U.S. MUNITIONS IMPORT LIST

CATEGORY I—FIREARMS

(a) Nonautomatic and semi-automatic firearms, to caliber .50 inclusive, combat shotguns, and shotguns with barrels less than 18 inches in length, and all components and parts for such firearms.

(b) Automatic firearms and all components and parts for such firearms to caliber .50 inclusive.

(c) Insurgency-counterinsurgency type firearms of other weapons having a special military application (e.g. close assault weapons systems) regardless of caliber and all components and parts for such firearms.

(d) Firearms silencers and suppressors, including flash suppressors.

(e) Riflescopes manufactured to military specifications and specifically designed or modified components therefor.

NOTE: Rifles, carbines, revolvers, and pistols, to caliber .50 inclusive, combat shotguns, and shotguns with barrels less than 18 inches in length are included under Category 1(a). Machineguns, submachineguns, machine pistols and fully automatic rifles to caliber .50 inclusive are included under Category 1(b).

CATEGORY II—ARTILLERY PROJECTORS

(a) Guns over caliber .50, howitzers, mortars, and recoiless rifles.

(b) Military flamethrowers and projectors.

(c) Components, parts, accessories, and attachments for the articles in paragraphs (a) and (b) of this category, including but not limited to mounts and carriages for these articles.

CATEGORY III—AMMUNITION

(a) Ammunition for the arms in Categories I and II of this section.

(b) Components, parts, accessories, and attachments for articles in paragraph (a) of this category, including but not limited to cartridge cases, powder bags, bullets, jackets, cores, shells (excluding shotgun shells), projectiles, boosters, fuzes and components therefor, primers, and other detonating devices for such ammunition.

(c) Ammunition belting and linking machines.

(d) Ammunition manufacturing machines and ammunition loading machines (except handloading ones).

NOTE: Cartridge and shell casings are included under Category III unless, prior to their importation, they have been rendered useless beyond the possibility of restoration for use as a cartridge or shell casing by means of heating, flame treatment, mangling, crushing, cutting, or popping.

CATEGORY IV—LAUNCH VEHICLES, GUIDED MISSILES, BALLISTIC MISSILES, ROCKETS, TORPEDOES, BOMBS AND MINES

(a) Rockets (including but not limited to meteorological and other sounding rockets), bombs, grenades, torpedoes, depth charges, land and naval mines, as well as launchers for such defense articles, and demolition blocks and blasting caps.

(b) Launch vehicles and missile and anti-missile systems including but not limited to guided, tactical and strategic missiles, launchers, and systems.

(c) Apparatus, devices, and materials for the handling, control, activation, monitoring, detection, protection, discharge, or detonation of the articles in paragraphs (a) and (b) of this category. Articles in this category include, but are not limited to, the following: Fuses and components for the items in this category, bomb racks and shackles, bomb shackle release units, bomb ejectors, torpedo tubes, torpedo and guided missile boosters, guidance system equipment and parts, launching racks and projectors, pistols (exploders), igniters, fuse arming devices, intervalometers, guided missile launchers and specialized handling equipment, and hardened missile launching facilities.

(d) Missile and space vehicle powerplants.

(e) Military explosive excavating devices.

(f) Ablative materials fabricated or semi-fabricated from advanced composites (e.g., silica, graphite, carbon, carbon/carbon, and boron filaments) for the articles in this category that are derived directly from or specifically developed or modified for defense articles.

(g) Non/nuclear warheads for rockets and guided missiles.

(h) All specifically designed components or modified components, parts, accessories, attachments, and associated equipment for the articles in this category.

NOTE: Military demolition blocks and blasting caps referred to in Category IV (a) do not include the following articles:

(a) Electric squibs.
(b) No. 6 and No. 8 blasting caps, including electric ones.
(c) Delay electric blasting caps (including No. 6 and No. 8 millisecond ones).
(d) Seismograph electric blasting caps (including SSS, Static-Master, Vibrocap SR, and SEISMO SR).
(e) Oil well perforating devices.

NOTE: Category V, "Munitions List," deleted as inapplicable to imports.

CATEGORY VI—VESSELS OF WAR AND SPECIAL NAVAL EQUIPMENT

(a) Warships, amphibious warfare vessels, landing craft, mine warfare vessels, patrol vessels, auxiliary vessels and service craft, experimental types of naval ships and any vessels specifically designed or modified for military purposes.

(b) Turrets and gun mounts, arresting gear, special weapons systems, protective systems, submarine storage batteries, catapults and other components, parts, attachments, and accessories specifically designed or modified for combatant vessels.

(c) Mine sweeping equipment, components, parts, attachments and accessories specifically designed or modified therefor.

(d) Harbor entrance detection devices (magnetic, pressure, and acoustic ones) and controls and components therefor.

(e) Naval nuclear propulsion plants, their land prototypes and special facilities for their construction, support and maintenance. This includes any machinery, device, component, or equipment specifically developed or designed or modified for use in such plants or facilities.

NOTE: The term "vessels of war" includes, but is not limited to the following:

(a) Combatant vessels:
(1) Warships (including nuclear-powered versions):
(i) Aircraft carriers (CV, CVN)
(ii) Battleships (BB)
(iii) Cruisers (CA, CG, CGN)
(iv) Destroyers (DD, DDG)
(v) Frigates (FF, FFG)
(vi) Submarines (SS, SSN, SSBN, SSG, SSAG).
(2) Other Combatant Classifications:
(i) Patrol Combatants (PC, PHM)
(ii) Amphibious Helicopter/Landing Craft Carriers (LHA, LPD, LPH)
(iii) Amphibious Landing Craft Carriers (LKA, LPA, LSD, LST)
(iv) Amphibious Command Ships (LCC)
(v) Mine Warfare Ships (MSO).

(b) Auxiliaries:
(1) Mobile Logistics Support:
(i) Under way Replenishment (AD, AF, AFS, AO, AOE, AOR)
(ii) Material Support (AD, AR, AS).
(2) Support Ships:
(i) Fleet Support Ships (ARS, ASR, ATA, ATF, ATS)
(ii) Other Auxiliaries (AG, AGDS, AGF, AGM, AGOR, AGOS, AGS, AH, AK, AKR, AOG, AOT, AP, APB, ARC, ARL, AVM, AVT).

(c) Combatant Craft:
(1) Patrol Craft:
(i) Coastal Patrol Combatants (PB, PCF, PCH, PTF)
(ii) River, Roadstead Craft (ATC, PBR).
(2) Amphibious Warfare Craft:
(i) Landing Craft (AALC, LCAC, LCM, LCPL, LCPR, LCU, LWT, SLWT)
(ii) Special Warfare Craft (LSSC, MSSC, SDV, SWCL, SWCM).
(3) Mine Warfare Craft:
(i) Mine Countermeasures Craft (MSB, MSD, MSI, MSM, MSR).

(d) Support and Service Craft:
(1) Tugs (YTB, YTL, YTM)
(2) Tankers (YO, YOG, Y)
(3) Lighters (YC, YCF, YCV, YF, YFN, YFNB, YFNX, YFR, YFRN, YFU, YG, YGN, YOGN, YON, YOS, YSR, YWN)
(4) Floating Dry Docks (AFDB, AFDL, AFDM, ARD, ARDM, YFD)
(5) Miscellaneous (APL, DSRV, DSV, IX, NR, YAG, YD, YDT, YFB, YFND, YEP,YFRT, YHLC, YM, YNG, YP, YPD, YR, YRB, YRBN, YRDH, YRDM, YRR, YRST, YSD).

(e) Coast Guard Patrol and Service Vessels and Craft:
(1) Coast Guard Cutters (CGC, WHEC, WMEC)
(2) Patrol Craft (WPB)
(3) Icebreakers (WAGB)
(4) Oceanography Vessels (WAGO)
(5) Special Vessels (WIX)
(6) Buoy Tenders (WLB, WLM, WLI, WLR, WLIC)
(7) Tugs (WYTM, WYTL)
(8) Light Ships (WLV).

CATEGORY VII—TANKS AND MILITARY VEHICLES

(a) Military type armed or armored vehicles, military railway trains, and vehicles specifically designed or modified to accommodate mountings for arms or other specialized military equipment or fitted with such items.

(b) Military tanks, combat engineer vehicles, bridge launching vehicles, halftracks and gun carriers.

(c) Self-propelled guns and howitzers.

NOTE: Category VII (d) and (e) of U.S. Munitions List" deleted as inappropriate to imports.

(f) Amphibious vehicles.

(g) Engines specifically designed or modified for the vehicles in paragraphs (a), (b), (c), and (f) of this category.

(h) All specifically designed or modified components and parts, accessories, attachments, and associated equipment for the articles in this category, including but not limited to military bridging and deep water fording kits.

NOTE: An "amphibious vehicle" in Category VII(f) is an automotive vehicle or chassis which embodies all-wheel drive, which is equipped to meet special military requirements, and which has sealed electrical systems and adaptation features for deep water fording.

CATEGORY VIII—AIRCRAFT, SPACECRAFT, AND ASSOCIATED EQUIPMENT

(a) Aircraft, including but not limited to helicopters, non-expansive balloons, drones and lighter-than-air aircraft, which are specifically designed, modified, or equipped for military purposes. This includes but is not limited to the following military purposes: gunnery, bombing, rocket or missile launching, electronic and other surveillance, reconnaissance, refueling, aerial mapping, military liaison, cargo carrying or dropping, personnel dropping, airborne warning and control, and military training.

NOTE: Category VIII (b) through (j) and Categories IX, X, XI, XII and XIII of "Munitions List" deleted as inapplicable to imports.

NOTE: In Category VIII, "aircraft" means aircraft designed, modified, or equipped for a military purpose, including aircraft described as "demilitarized." All aircraft bearing an original military designation are included in Category VIII. However, the following aircraft are not so included so long as they have not been specifically equipped, reequipped, or modified for military operations:

(a) Cargo aircraft bearing "C" designations and numbered C–45 through C–118 inclusive, and C–121 through C–125 inclusive, and C–131, using reciprocating engines only.
(b) Trainer aircraft bearing "T" designations and using reciprocating engines or turboprop engines with less than 600 horsepower (s.h.p.).
(c) Utility aircraft bearing "U" designations and using reciprocating engines only.
(d) All liaison aircraft bearing an "L" designation.
(e) All observation aircraft bearing "O" designations and using reciprocating engines.

CATEGORY XIV-TOXICOLOGICAL AGENTS AND EQUIPMENT AND RADIOLOGICAL EQUIPMENT

(a) Chemical agents, including but not limited to lung irritants, vesicants, lachrymators, and tear gases (except tear gas formulations containing 1% or less CN or CS), sternutators and irritant smoke, and nerve gases and incapacitating agents.

(b) Biological agents.

(c) Equipment for dissemination, detection, and identification of, and defense against, the articles in paragraphs (a) and (b) of this category.

(d) Nuclear radiation detection and measuring devices manufactured to military specification.

(e) Components, parts, accessories, attachments, and associated equipment specifically designed or modified for the articles in paragraphs (c) and (d) of this category.

NOTE: A chemical agent in Category XIV(a) is a substance having military application which by its ordinary and direct chemical action produces a powerful physiological effect. The term "chemical agent" includes, but is not limited to, the following chemical compounds:

(a) Lung irritants:
(1) Diphenylcyanoarsine (DC).
(2) Fluorine (but not fluorene).
(3) Trichloronitro methane (chloropicrin PS).

(b) Vesicants:
(1) B-Chlorovinyldichloroarsine (Lewisite, L).
(2) Bis(dichlorethyl) sulphide (Mustard Gas, HD or H).
(3) Ethyldichloroarsine (ED).
(4) Methyldichloroarsine (MD).

(c) Lachrymators and tear gases:
(1) A-Brombenzyl cyanide (BBC).
(2) Chloroacetophenone (CN).
(3) Dibromodimethyl ether.
(4) Dichlorodimethyl ether (ClCi).
(5) Ethyldibromoarsine.
(6) Phenylcarbylamine chloride.
(7) Tear gas solutions (CNB and CNS).
(8) Tear gas orthochlorobenzalmalononitrile (CS).

(d) Sternutators and irritant smokes:
(1) Diphenylamine chloroarsine (Adamsite, DM).
(2) Diphenylchloroarsine (BA).
(3) Liquid pepper.

(e) Nerve agents, gases, and aerosols. These are toxic compounds which affect the nervous system, such as:
(1) Dimethylaminoethoxycyanophosphine oxide (GA).
(2) Methylisopropoxyfluoro-phosphine oxide (GB).
(3) Methylpinacolyloxyfluori-phosphine oxide (GD).

(f) Antiplant chemicals, such as: Butyl 2-chloro-4-fluorophenoxyacetate (LNF).

CATEGORY XV [RESERVED]

CATEGORY XVI—NUCLEAR WEAPONS DESIGN AND TEST EQUIPMENT

(a) Any article, material, equipment, or device, which is specifically designed or modified for use in the design, development, or fabrication of nuclear weapons or nuclear explosive devices.

(b) Any article, material, equipment, or device, which is specifically designed or modified for use in the devising, carrying out, or evaluating of nuclear weapons tests or any other nuclear explosions, except such items as are in normal commercial use for other purposes.

NOTE: Categories XVII, XVIII, and XIX of "Munitions List" deleted as inapplicable to imports.

CATEGORY XX—SUBMERSIBLE VESSELS, OCEANOGRAPHIC AND ASSOCIATED EQUIPMENT

(a) Submersible vessels, manned and unmanned, designed or modified for military purposes or having independent capability to maneuver vertically or horizontally at depths below 1,000 feet, or powered by nuclear propulsion plants.

(b) Submersible vessels, manned or unmanned, designed or modified in whole or in part from technology developed by or for the U.S. Armed Forces.

(c) Any of the articles in Category VI and elsewhere in this part specifically designed or modified for use with submersible vessels, and oceanographic or associated equipment assigned a military designation.

(d) Equipment, components, parts, accessories, and attachments specifically designed for any of the articles in paragraphs (a) and (b) of this category.

CATEGORY XXI—MISCELLANEOUS ARTICLES

Any article not specifically enumerated in the other categories of the U.S. Munitions List which has substantial military applicability and which has been specifically designed or modified for military purposes. The decision on whether any article may be included in this category shall be made by the Director, Office of Defense Trade Controls, Department of State, with the concurrence of the Department of Defense.

CATEGORY XXII—SOUTH AFRICA

(a) Defense articles enumerated on the U.S. Munitions List (22 CFR Part 121).

(b) Technical data relating to defense articles enumerated on the U.S. Munitions List.

(1) Classified information relating to defense articles and defense services;

(2) Information covered by an invention secrecy order;

(3) Information which is directly related to the design, engineering, development, production, processing, manufacture, use, operation, overhaul, repair, maintenance, modification, or reconstruction of defense articles. This includes, for example, information in the form of blueprints, drawings, photographs, plans, instructions, computer software and documentation. This also includes information which advances the state of the art of articles on the U.S. Munitions List. This does not include information concerning general scientific, mathematical or engineering principles.

Editor's Note:

Category XXII is no longer a Munitions Import List category. The regulation will be amended to formally remove it.

§ 447.22 Forgings, castings, and machined bodies.

Articles on the U.S. Munitions Import List include articles in a partially completed state (such as forgings, castings, extrusions, and machined bodies) which have reached a stage in manufacture where they are clearly identifiable as defense articles. If the end-item is an article on the U.S. Munitions Import List, (including components, accessories, attachments and parts) then the particular forging, casting, extrusion, machined body, etc., is considered a defense article subject to the controls of this part, except for such items as are in normal commercial use.

Subpart D—Registration

§ 447.31 Registration requirement.

Persons engaged in the business, in the United States, of importing articles enumerated on the U.S. Munitions Import List must register by making an application on ATF Form 4587.

§ 447.32 Application for registration and refund of fee.

(a) Application for registration must be filed on ATF Form 4587, and must be accompanied by the registration fee at the rate prescribed in this section. The appropriate ATF officer will approve the application and return the original to the applicant.

(b) Registration may be effected for periods of from 1 to 5 years at the option of the registrant by identifying on Form 4587 the period of registration desired. The registration fees are as follows:

1 year	$250
2 years	$500
3 years	$700
4 years	$850
5 years	$1,000

(c) Fees paid in advance for whole future years of a multiple year registration will be refunded upon request if the registrant ceases to engage in importing articles on the U.S. Munitions Import List. A request for a refund must be submitted to the appropriate ATF officer at the Bureau of Alcohol, Tobacco and Firearms, Washington, DC 20226, prior to the beginning of any year for which a refund is claimed.

(Approved by the Office of Management and Budget under control number 1512–0021)

§ 447.33 Notification of changes in information furnished by registrants.

Registered persons shall notify the appropriate ATF officer in writing, in duplicate, of significant changes in the information set forth in their registration.

(Approved by the Office of Management and Budget under control number 1512–0021)

§ 447.34 Maintenance of records by persons required to register as importers of Import List articles.

(a) Registrants under this part engaged in the business of importing articles subject to controls under 27 CFR Parts 478 and 479 shall maintain records in accordance with the applicable provisions of those parts.

(b) Registrants under this part engaged in importing articles on the U.S. Munitions Import List subject to the permit procedures of subpart E of this part must maintain for a period of 6 years, records bearing on such articles imported, including records concerning their acquisition and disposition, including Forms 6 and 6A. The appropriate ATF officer may prescribe a longer or shorter period in individual cases as such officer deems necessary. See § 478.129 of this chapter for articles subject to import control under part 478 of this chapter.

(Approved by the Office of Management and Budget under control number 1512–0387)

§ 447.35 Forms prescribed.

(a) The appropriate ATF officer is authorized to prescribe all forms required by this part. All of the information called for in each form shall be furnished as indicated by the headings on the form and the instructions on or pertaining to the form. In addition, information called for in each form shall be furnished as required by this part. The form will be filed in accordance with the instructions for the form.

(b) Forms may be requested from the ATF Distribution Center, P.O. Box 5950, Springfield, Virginia 22150–5950, or by accessing the ATF Web site http://www.atf.treas.gov/.

Editor's Note:

The correct website is www.ATF.gov.

Subpart E—Permits

§ 447.41 Permit requirement.

(a) Articles on the U.S. Munitions Import List will not be imported into the United States except pursuant to a permit under this subpart. For articles subject to control under parts 478 or 479 of this chapter, a separate permit is not necessary.

(b) Articles on the U.S. Munitions Import List intended for the United States or any State or political subdivision thereof, or the District of Columbia, which are exempt from import controls of 27 CFR 478.115 shall not be imported into the United States, except by the United States or agency thereof, without first obtaining a permit under this subpart.

(c) A permit is not required for the importation of—

(1) (i) The U.S. Munitions Import List articles from Canada, except articles enumerated in Categories I, II, III, IV, VI(e), VIII(a), XVI, and XX; and

(ii) Nuclear weapons strategic delivery systems and all specifically designed components, parts, accessories, attachments, and associated equipment thereof (see Category XXI); or

(2) Minor components and parts for Category I(a) and I(b) firearms, except barrels, cylinders, receivers (frames) or complete breech mechanisms, when the total value does not exceed $100 wholesale in any single transaction.

§ 447.42 Application for permit.

(a)(1) Persons required to obtain a permit as provided in §447.41 must file a Form 6—Part I. The application must be signed and dated and must contain the information requested on the form, including:

(i) The name, address, telephone number, license and registration number, if any (including expiration date) of the importer;

(ii) The country from which the defense article is to be imported;

(iii) The name and address of the foreign seller and foreign shipper;

(iv) A description of the defense article to be imported, including—

(A) The name and address of the manufacturer;

(B) The type (e.g., rifle, shotgun, pistol, revolver, aircraft, vessel, and in the case of ammunition only, ball, wadcutter, shot, etc.);

(C) The caliber, gauge, or size;

(D) The model;

(E) The length of barrel, if any (in inches);

(F) The overall length, if a firearm (in inches);

(G) The serial number, if known;

(H) Whether the defense article is new or used;

(I) The quantity;

(J) The unit cost of the firearm, firearm barrel, ammunition, or other defense article to be imported;

(K) The category of U.S. Munitions Import List under which the article is regulated;

(v) The specific purpose of importation, including final recipient information if different from the importer; and

(vi) Certification of origin.

(2)(i) If the appropriate ATF officer approves the application, such approved application will serve as the permit to import the defense article described therein, and importation of such defense article may continue to be made by the licensed/registered importer (if applicable) under the approved application (permit) during the period specified thereon. The appropriate ATF officer will furnish the approved application (permit) to the applicant and retain two copies thereof for administrative use.

(ii) If the Director disapproves the application, the licensed/registered importer (if applicable) will be notified of the basis for the disapproval.

(b) For additional requirements relating to the importation of plastic explosives into the United States on or after April 24, 1997, see § 555.183 of this title.

(Approved by the Office of Management and Budget under control number 1512–0017)

§ 447.43 Terms of permit.

(a) Import permits issued under this subpart are valid for one year from their issuance date unless a different period of validity is stated thereon. They are not transferable.

(b) If shipment cannot be completed during the period of validity of the permit, another application must be submitted for permit to cover the unshipped balance. Such an application shall make reference to the previous permit and may include materials in addition to the unshipped balance.

(c) No amendments or alteration of a permit may be made, except by the appropriate ATF officer.

§ 447.44 Permit denial, revocation or suspension.

(a) Import permits under this subpart may be denied, revoked, suspended or revised without prior notice whenever the appropriate ATF officer finds the proposed importation to be inconsistent with the purpose or in violation of section 38, Arms Export Control Act of 1976 or the regulations in this part.

(b) Whenever, after appropriate consideration, a permit application is denied or an outstanding permit is revoked, suspended, or revised, the applicant or permittee shall be promptly advised in writing of the appropriate ATF officer's decision and the reasons therefor.

(c) Upon written request made within 30 days after receipt of an adverse decision, the applicant or permittee shall be accorded an opportunity to present additional information and to have a full review of his case by the appropriate ATF officer.

(d) Unused, expired, suspended, or revoked permits must be returned immediately to the appropriate ATF officer.

§ 447.45 Importation.

(a) Articles subject to the import permit procedures of this subpart imported into the United States may be released from Customs custody to the person authorized to import same upon his showing that he has a permit for the importation of the article or articles to be released. For articles in Categories I and III imported by a registered importer, the importer will also submit to Customs a copy of the export license authorizing the export of the article or articles from the exporting country. If the exporting country does not require issuance of an export license, the importer must submit a certification, under penalty of perjury, to that effect.

(1) In obtaining the release from Customs custody of an article imported pursuant to a permit, the permit holder will prepare and file Form 6A according to its instructions.

(2) The ATF Form 6A must contain the information requested on the form, including:

(i) The name, address, and license number (if any) of the importer;

(ii) The name of the manufacturer of the defense article;

(iii) The country of manufacture;

(iv) The type;

(v) The model;

(vi) The caliber, gauge, or size;

(vii) The serial number in the case of firearms, if known; and

(viii) The number of defense articles released.

(b) Within 15 days of the date of their release from Customs custody, the importer of the articles released will forward to the address specified on the form a copy of Form 6A on which will be reported any error or discrepancy appearing on the Form 6A certified by Customs and serial numbers if not previously provided on ATF Form 6A.

(Approved by the Office of Management and Budget under control number 1512–0019)

§ 447.46 Articles in transit.

Articles subject to the import permit procedures of this subpart which enter the United States for temporary deposit pending removal therefrom and such articles which are temporarily taken out of the United States for return thereto shall be regarded as in transit and will be considered neither imported nor exported under this part. Such transactions are subject to the Intransit or Temporary Export License procedures of the Department of State (see 22 CFR Part 123).

Subpart F—Miscellaneous Provisions

§ 447.51 Import certification and delivery verification.

Pursuant to agreement with the United States, certain foreign countries are entitled to request certification of legality of importation of articles on the U.S. Munitions Import List. Upon request of a foreign government, the appropriate ATF officer will certify the importation, on Form ITA–645P/ATF–4522/DSP53, for the U.S. importer. Normally, the U.S. importer will submit this form at the time he applies for an import permit. This document will serve as evidence to the government of the exporting company that the U.S. importer has complied with import regulations of the U.S. Government and is prohibited from diverting, transshipping, or re-exporting the material described therein without the approval of the U.S. Government. Foreign governments may also require documentation attesting to the delivery of the material into the United States. When such delivery certification is requested by a foreign government, the U.S. importer may obtain directly from the U.S. District Director of Customs the authenticated Delivery Verification Certificate (U.S. Department of Commerce Form ITA–647P) for this purpose.

(Approved by the Office of Management and Budget under control number 0625–0064)

§ 447.52 Import restrictions applicable to certain countries.

(a) It is the policy of the United States to deny licenses and other approvals with respect to defense articles and defense services originating in certain countries or areas. This policy applies to Cuba, Iran, Iraq, Libya, Mongolia, North Korea, Sudan, Syria, Vietnam, and some of the states that comprised the former Soviet Union (Armenia, Azerbaijan, Belarus, and Tajikistan). This policy applies to countries or areas with respect to which the United States maintains an arms embargo (e.g., Burma, China, the Federal Republic of Yugoslavia (Serbia and Montenegro), Haiti, Liberia, Rwanda, Somalia, Sudan, UNITA (Angola), and Zaire). It also applies when an import would not be in furtherance of world peace and the security and foreign policy of the United States.

NOTE: Changes in foreign policy may result in additions to and deletions from the above list of countries. The ATF will publish changes to this list in the Federal Register. Contact the Firearms and Explosives Imports Branch at (202) 927–8320 for current information.

(b) Notwithstanding paragraph (a) of this section, the appropriate ATF officer shall deny applications to import into the United States the following firearms and ammunition:

(1) Any firearm located or manufactured in Georgia, Kazakstan, Kyrgyzstan, Moldova, Russian Federation, Turkmenistan, Ukraine, or Uzbekistan, and any firearm previously manufactured in the Soviet Union, that is not one of the models listed below:

(i) Pistols/Revolvers:

(A) German Model P08 Pistol.

(B) IZH 34M, .22 caliber Target Pistol.

(C) IZH 35M, .22 caliber Target Pistol.

(D) Mauser Model 1896 Pistol.

(E) MC–57–1 Pistol.

(F) MC–1–5 Pistol.

(G) Polish Vis Model 35 Pistol.

(H) Soviet Nagant Revolver.

(I) TOZ 35, .22 caliber Target Pistol.

(ii) Rifles:

(A) BARS–4 Bolt Action Carbine.

(B) Biathlon Target Rifle, .22LR caliber.

(C) British Enfield Rifle.

(D) CM2, .22 caliber Target Rifle (also known as SM2, 22 caliber).

(E) German Model 98K Rifle.

(F) German Model G41 Rifle.

(G) German Model G43 Rifle.

(H) IZH–94.

(I) LOS–7 Bolt Action Rifle.

(J) MC–7–07.

(K) MC–18–3.

(L) MC–19–07.

(M) MC–105–01.

(N) MC–112–02.

(O) MC–113–02.

(P) MC–115–1.

(Q) MC–125/127.

(R) MC–126.

(S) MC–128.

(T) Saiga Rifle.

(U) Soviet Model 38 Carbine.

(V) Soviet Model 44 Carbine.

(W) Soviet Model 91/30 Rifle.

(X) TOZ 18, .22 caliber Bolt Action Rifle.

(Y) TOZ 55.

(Z) TOZ 78.

(AA) Ural Target Rifle, .22LR caliber.

(BB) VEPR Rifle.

(CC) Winchester Model 1895, Russian Model Rifle;

Editor's Note:

This list has been modified. The regulation will be amended to formally incorporate the revised list. The current list of affected firearms is available at www.ATF.gov.

(2) Ammunition located or

manufactured in Georgia, Kazakstan, Kyrgyzstan, Moldova, Russian Federation, Turkmenistan, Ukraine, or Uzbekistan, and ammunition previously manufactured in the Soviet Union, that is 7.62X25mm caliber (also known as 7.63X25mm caliber or .30 Mauser); or

(3) A type of firearm the manufacture of which began after February 9, 1996.

(c) The provisions of paragraph (b) of this section shall not affect the fulfillment of contracts with respect to firearms or ammunition entered or withdrawn from warehouse for consumption in the United States on or before February 9, 1996.

(d) A defense article authorized for importation under this part may not be shipped on a vessel, aircraft or other means or conveyance which is owned or operated by, or leased to or from, any of the countries or areas covered by paragraph (a) of this section.

(e) Applications for permits to import articles that were manufactured in, or have been in, a country or area proscribed under this section may be approved where the articles are covered by Category I(a) of the Import List (other than those subject to the provisions of 27 CFR Part 479), are importable as curios or relics under the provisions of 27 CFR 478.118, and meet the following criteria:

(1) The articles were manufactured in a proscribed country or area prior to the date, as established by the Department of State, the country or area became proscribed, or, were manufactured in a non-proscribed country or area; and

(2) The articles have been stored for the five year period immediately prior to importation in a non-proscribed country or area.

(f) Applicants desiring to import articles claimed to meet the criteria specified in paragraph (e) of this section shall explain, and certify to, how the firearms meet the criteria. The certification statement will be prepared in letter form, executed under the penalties of perjury, and should be submitted with the application for an import permit. The certification statement must be accompanied by documentary information on the country or area of original manufacture and on the country or area of storage for the five year period immediately prior to importation. Such information may, for example, include a verifiable statement in the English language of a government official or any other person having knowledge of the date and place of manufacture and/or the place of storage; a warehouse receipt or other document which provides the required history of storage; and any other document that the applicant believes substantiates the place and date of manufacture and the place of storage. The appropriate ATF officer, however, reserves the right to determine whether documentation is acceptable. Applicants shall, when required by the appropriate ATF officer, furnish additional documentation as may be necessary to determine whether an import permit application should be approved.

§ 447.53 Exemptions.

(a) The provisions of this part are not applicable to:

(1) Importations by the United States or any agency thereof;

(2) Importation of components for items being manufactured under contract for the Department of Defense; or

(3) Importation of articles (other than those which would be "firearms" as defined in 18 U.S.C. 921(a)(3) manufactured in foreign countries for persons in the United States pursuant to Department of State approval.

(b) Any person seeking to import articles on the U.S. Munitions Import List as exempt under paragraph (a)(2) or (3) of this section may obtain release of such articles from Customs custody by submitting, to the Customs officer with authority to release, a statement claiming the exemption accompanied by satisfactory proof of eligibility. Such proof may be in the form of a letter from the Department of Defense or State, as the case may be, confirming that the conditions of the exemption are met.

§ 447.54 Administrative procedures inapplicable.

The functions conferred under section 38, Arms Export Control Act of 1976, as amended, are excluded from the operation of Chapter 5, Title 5, United States Code, with respect to Rule Making and Adjudication, 5 U.S.C. 553 and 554.

§ 447.55 Departments of State and Defense consulted.

The administration of the provisions of this part will be subject to the guidance of the Secretaries of State and Defense on matters affecting world peace and the external security and foreign policy of the United States.

§ 447.56 Authority of Customs officers.

(a) Officers of the U.S. Customs Service are authorized to take appropriate action to assure compliance with this part and with 27 CFR Parts 478 and 479 as to the importation or attempted importation of articles on the U.S. Munitions Import List, whether or not authorized by permit.

(b) Upon the presentation to him of a permit or written approval authorizing importation of articles on the U.S. Munitions Import List, the Customs officer who has authority to release same may require, in addition to such documents as may be required by Customs regulations, the production of other relevant documents relating to the proposed importation, including, but not limited to, invoices, orders, packing lists, shipping documents, correspondence, and instructions.

§ 447.57 U.S. military defense articles.

(a) (1) Notwithstanding any other provision of this part or of parts 478 or 479 of this chapter, no military defense article of United States manufacture may be imported into the United States if such article was furnished to a foreign government under a foreign assistance or foreign military sales program of the United States.

(2) The restrictions in paragraph (a)(1) of this section cover defense articles which are advanced in value or improved in condition in a foreign country, but do not include those which have been substantially transformed as to become, in effect, articles of foreign manufacture.

(b) Paragraph (a) of this section will not apply if :

(1) The applicant submits with the ATF Form 6—Part I application written authorization from the Department of State to import the defense article; and

(2) In the case of firearms, such firearms are curios or relics under 18 U.S.C. 925(e) and the person seeking to import such firearms provides a certification of a foreign government that the firearms were furnished to such government under a foreign assistance or foreign military sales program of the United States and that the firearms are owned by such foreign government. (See § 478.118 of this chapter providing for the importation of certain curio or relic handguns, rifles and shotguns.)

(c) For the purpose of this section, the term **"military defense article"** includes all defense articles furnished to foreign governments under a foreign assistance or foreign military sales program of the United States as set forth in paragraph (a) of this section.

(Approved by the Office of Management and Budget under OMB Control No. 1512–0017)

§ 447.58 Delegations of the Director.

The regulatory authorities of the Director contained in this part are delegated to appropriate ATF officers. These ATF officers are specified in ATF O 1130.34, Delegation of the Director's Authorities in 27 CFR Part 447. ATF delegation orders, such as ATF O 1130.34, are available to any interested party by mailing a request to the ATF Distribution Center, PO Box 5950, Springfield, VA 22150–5950, or by accessing the ATF Web site http://www.atf.treas.gov/.

Editor's Note:

The correct website is www.ATF.gov.

Subpart G—Penalties, Seizures and Forfeitures

§ 447.61 Unlawful importation.

Any person who willfully:

(a) Imports articles on the U.S. Munitions Import List without a permit;

(b) Engages in the business of importing articles on the U.S. Munitions Import List without registering under this part; or

(c) Otherwise violates any provisions of this part;

Shall upon conviction be fined not more than $1,000,000 or imprisoned not more than 10 years, or both.

§ 447.62 False statements or concealment of facts.

Any person who willfully, in a registration or permit application, makes any untrue statement of a material fact or fails to state a material fact required to be stated therein or necessary to make the statements therein not misleading, shall upon conviction be fined not more than $1,000,000, or imprisoned not more than 10 years, or both.

§ 447.63 Seizure and forfeiture.

Whoever knowingly imports into the United States contrary to law any article on the U.S. Munitions Import List; or receives, conceals, buys, sells, or in any manner facilitates its transportation, concealment, or sale after importation, knowing the same to have been imported contrary to law, shall be fined not more than $10,000 or imprisoned not more than 5 years, or both; and the merchandise so imported, or the value thereof shall be forfeited to the United States.

TITLE 28 CFR CHAPTER I

PART 25 – DEPARTMENT OF JUSTICE INFORMATION SYSTEMS

Subpart A – The National Instant Criminal Background Check System

§ 25.1 Purpose and authority.

§ 25.2 Definitions.

§ 25.3 System information.

§ 25.4 Record source categories.

§ 25.5 Validation and data integrity of records in the system.

§ 25.6 Accessing records in the system.

§ 25.7 Querying records in the system.

§ 25.8 System safeguards.

§ 25.9 Retention and destruction of records in the system.

§ 25.10 Correction of erroneous system information.

§ 25.11 Prohibited activities and penalties.

§ 25.1 Purpose and authority.

The purpose of this subpart is to establish policies and procedures implementing the Brady Handgun Violence Prevention Act (Brady Act), Public Law 103-159, 107 Stat. 1536. The Brady Act requires the Attorney General to establish a National Instant Criminal Background Check System (NICS) to be contacted by any licensed importer, licensed manufacturer, or licensed dealer of firearms for information as to whether the transfer of a firearm to any person who is not licensed under 18 U.S.C. 923 would be in violation of Federal or State law. The regulations in this subpart are issued pursuant to section 103(h) of the Brady Act, 107 Stat. 1542 (18 U.S.C. 922 note), and include requirements to ensure the privacy and security of the NICS and appeals procedures for persons who have been denied the right to obtain a firearm as a result of a NICS background check performed by the Federal Bureau of Investigation (FBI) or a State or local law enforcement agency.

§ 25.2 Definitions.

Appeal means a formal procedure to challenge the denial of a firearm transfer.

ARI means a unique Agency Record Identifier assigned by the agency submitting records for inclusion in the NICS Index.

ATF means the Bureau of Alcohol, Tobacco, and Firearms of the Department of Treasury.

Audit log means a chronological record of system (computer) activities that enables the reconstruction and examination of the sequence of events and/or changes in an event.

Business day means a 24-hour day (beginning at 12:01 a.m.) on which state offices are open in the state in which the proposed firearm transaction is to take place.

Control Terminal Agency means a State or territorial criminal justice agency recognized by the FBI as the agency responsible for providing State- or territory-wide service to criminal justice users of NCIC data.

Data source means an agency that provided specific information to the NICS.

Delayed means the response given to the FFL indicating that the transaction is in an "Open" status and that more research is required prior to a NICS "Proceed" or "Denied" response. A "Delayed" response to the FFL indicates that it would be unlawful to transfer the firearm until receipt of a follow-up "Proceed" response from the NICS or the expiration of three business days, whichever occurs first.

Denied means denial of a firearm transfer based on a NICS response indicating one or more matching records were found providing information demonstrating that receipt of a firearm by a prospective transferee would violate 18 U.S.C. 922 or State law.

Denying agency means a POC or the NICS Operations Center, whichever determines that information in the NICS indicates that the transfer of a firearm to a person would violate Federal or State law, based on a background check.

Dial-up access means any routine access through commercial switched circuits on a continuous or temporary basis.

Federal agency means any authority of the United States that is an "Agency" under 44 U.S.C. 3502(1), other than those considered to be independent regulatory agencies, as defined in 44 U.S.C. 3502(10).

FFL (federal firearms licensee) means a person licensed by the ATF as a manufacturer, dealer, or importer of firearms.

Firearm has the same meaning as in 18 U.S.C. 921(a)(3).

Licensed dealer means any person defined in 27 CFR 478.11.

Licensed importer has the same meaning as in 27 CFR 478.11.

Licensed manufacturer has the same meaning as in 27 CFR 478.11.

NCIC (National Crime Information Center) means the nationwide computerized information system of criminal justice data established by the FBI as a service to local, State, and Federal criminal justice agencies.

NICS means the National Instant Criminal Background Check System, which an FFL must, with limited exceptions, contact for information on whether receipt of a firearm by a person who is not licensed under 18 U.S.C. 923 would violate Federal or State law.

NICS Index means the database, to

be managed by the FBI, containing information provided by Federal and State agencies about persons prohibited under Federal law from receiving or possessing a firearm. The NICS Index is separate and apart from the NCIC and the Interstate Identification Index (III).

NICS operational day means the period during which the NICS Operations Center has its daily regular business hours.

NICS Operations Center means the unit of the FBI that receives telephone or electronic inquiries from FFLs to perform background checks, makes a determination based upon available information as to whether the receipt or transfer of a firearm would be in violation of Federal or State law, researches criminal history records, tracks and finalizes appeals, and conducts audits of system use.

NICS Representative means a person who receives telephone inquiries to the NICS Operations Center from FFLs requesting background checks and provides a response as to whether the receipt or transfer of a firearm may proceed or is delayed.

NRI (NICS Record Identifier) means the system-generated unique number associated with each record in the NICS Index.

NTN (NICS Transaction Number) means the unique number that will be assigned to each valid background check inquiry received by the NICS. Its primary purpose will be to provide a means of associating inquiries to the NICS with the responses provided by the NICS to the FFLs.

Open means those non-canceled transactions where the FFL has not been notified of the final determination. In cases of "open" responses, the NICS continues researching potentially prohibiting records regarding the transferee and, if definitive information is obtained, communicates to the FFL the final determination that the check resulted in a proceed or a deny. An "open" response does not prohibit an FFL from transferring a firearm after three business days have elapsed since the FFL provided to the system the identifying information about the prospective transferee.

ORI (Originating Agency Identifier) means a nine-character identifier assigned by the FBI to an agency that has met the established qualifying criteria for ORI assignment to identify the agency in transactions on the NCIC System.

Originating Agency means an agency that provides a record to a database checked by the NICS.

POC (Point of Contact) means a State or local law enforcement agency serving as an intermediary between an FFL and the federal databases checked by the NICS. A POC will receive NICS background check requests from FFLs, check state or local record systems, perform NICS inquiries, determine whether matching records provide information demonstrating that an individual is disqualified from possessing a firearm under Federal or State law, and respond to FFLs with the results of a NICS background check. A POC will be an agency with express or implied authority to perform POC duties pursuant to State statute, regulation, or executive order.

Proceed means a NICS response indicating that the information available to the system at the time of the response did not demonstrate that transfer of the firearm would violate federal or State law. A "Proceed" response would not relieve an FFL from compliance with other provisions of Federal or State law that may be applicable to firearms transfers. For example, under 18 U.S.C. 922(d), an FFL may not lawfully transfer a firearm if he or she knows or has reasonable cause to believe that the prospective recipient is prohibited by law from receiving or possessing a firearm.

Record means any item, collection, or grouping of information about an individual that is maintained by an agency, including but not limited to information that disqualifies the individual from receiving a firearm, and that contains his or her name or other personal identifiers.

STN (State-Assigned Transaction Number) means a unique number that may be assigned by a POC to a valid background check inquiry.

System means the National Instant Criminal Background Check System (NICS).

§ 25.3 System information.

(a) There is established at the FBI a National Instant Criminal Background Check System.

(b) The system will be based at the Federal Bureau of Investigation, 1000 Custer Hollow Road, Clarksburg, West Virginia 26306-0147.

(c) The system manager and address are: Director, Federal Bureau of Investigation, J. Edgar Hoover F.B.I. Building, 935 Pennsylvania Avenue, NW, Washington, D.C. 20535.

§ 25.4 Record source categories.

It is anticipated that most records in the NICS Index will be obtained from Federal agencies. It is also anticipated that a limited number of authorized State and local law enforcement agencies will voluntarily contribute records to the NICS Index. Information in the NCIC and III systems that will be searched during a background check has been or will be contributed voluntarily by Federal, State, local, and international criminal justice agencies.

§ 25.5 Validation and data integrity of records in the system.

(a) The FBI will be responsible for maintaining data integrity during all NICS operations that are managed and carried out by the FBI. This responsibility includes:

(1) Ensuring the accurate adding, canceling, or modifying of NICS Index records supplied by Federal agencies;

(2) Automatically rejecting any attempted entry of records into the NICS Index that contain detectable invalid data elements;

(3) Automatic purging of records in the NICS Index after they are on file for a prescribed period of time; and

(4) Quality control checks in the form of periodic internal audits by FBI personnel to verify that the information provided to the NICS Index remains valid and correct.

(b) Each data source will be responsible for ensuring the accuracy and validity of the data it provides to the NICS Index and will immediately correct any record determined to be invalid or incorrect.

§ 25.6 Accessing records in the system.

(a) FFLs may initiate a NICS background check only in connection with a proposed firearm transfer as required by the Brady Act. FFLs are strictly prohibited from initiating a NICS background check for any other purpose. The process of accessing the NICS for the purpose of conducting a NICS background check is initiated by an FFL's contacting the FBI NICS Operations Center (by telephone or electronic dial-up access) or a POC. FFLs in each state will be advised by the ATF whether they are required to initiate NICS background checks with the NICS Operations Center or a POC and how they are to do so.

(b) Access to the NICS through the FBI NICS Operations Center. FFLs may contact the NICS Operations Center by use of a toll-free telephone number, only during its regular business hours. In addition to telephone access, toll-free electronic dial-up access to the NICS will be provided to FFLs after the beginning of the NICS operation. FFLs with electronic dial-up access will be able to contact the NICS 24 hours each day, excluding scheduled and unscheduled downtime.

(c)(1) The FBI NICS Operations Center, upon receiving an FFL telephone or electronic dial-up request for a background check, will:

(i) Verify the FFL Number and code word;

(ii) Assign a NICS Transaction Number (NTN) to a valid inquiry and provide the NTN to the FFL;

(iii) Search the relevant databases (i.e., NICS Index, NCIC, III) for any matching records; and

(iv) Provide the following NICS responses based upon the consolidated NICS search results to the FFL that requested the background check:

(A) "Proceed" response, if no disqualifying information was found in the NICS Index, NCIC, or III.

(B) "Delayed" response, if the NICS search finds a record that requires more research to determine whether the prospective transferee is disqualified from possessing a firearm by Federal or State law. A "Delayed" response to the FFL indicates that the firearm transfer should not proceed pending receipt of a follow-up "Proceed" response from the NICS or the expiration of three business days (exclusive of the day on which the query is made), whichever occurs first. (Example: An FFL requests a NICS check on a prospective firearm transferee at 9:00 a.m. on Friday and shortly thereafter receives a "Delayed" response from the NICS. If State offices in the state in which the FFL is located are closed on Saturday and Sunday and open the following Monday, Tuesday, and Wednesday, and the NICS has not yet responded with a "Proceed" or "Denied" response, the FFL may transfer the firearm at 12:01 a.m. Thursday.)

(C) "Denied" response, when at least one matching record is found in either the NICS Index, NCIC, or III that provides information demonstrating that receipt of a firearm by the prospective transferee would violate 18 U.S.C. 922 or State law. The "Denied" response will be provided to the requesting FFL by the NICS Operations Center during its regular business hours.

(2) None of the responses provided to the FFL under paragraph (c)(1) of this section will contain any of the underlying information in the records checked by the system.

(d) Access to the NICS through POCs. In states where a POC is designated to process background checks for the NICS, FFLs will contact the POC to initiate a NICS background check. Both ATF and the POC will notify FFLs in the POC's state of the means by which FFLs can contact the POC. The NICS will provide POCs with electronic access to the system virtually 24 hours each day through the NCIC communication network. Upon receiving a request for a background check from an FFL, a POC will:

(1) Verify the eligibility of the FFL either by verification of the FFL number or an alternative POC-verification system;

(2) Enter a purpose code indicating that the query of the system is for the purpose of performing a NICS background check in connection with the transfer of a firearm; and

(3) Transmit the request for a background check via the NCIC interface to the NICS.

(e) Upon receiving a request for a NICS background check, POCs may also conduct a search of available files in State and local law enforcement and other relevant record systems, and may provide a unique State-Assigned Transaction Number (STN) to a valid inquiry for a background check.

(f) When the NICS receives an inquiry from a POC, it will search the relevant databases (i.e., NICS Index, NCIC, III) for any matching record(s) and will provide an electronic response to the POC. This response will consolidate the search results of the relevant databases and will include the NTN. The following types of responses may be provided by the NICS to a state or local agency conducting a background check:

(1) No record response, if the NICS determines, through a complete search, that no matching record exists.

(2) Partial response, if the NICS has not completed the search of all of its records. This response will indicate the databases that have been searched (i.e., III, NCIC, and/or NICS Index) and the databases that have not been searched. It will also provide any potentially disqualifying information found in any of the databases searched. A follow-up response will be sent as soon as all the relevant databases have been searched. The follow-up response will provide the complete search results.

(3) Single matching record response, if all records in the relevant databases have been searched and one matching record was found.

(4) Multiple matching record response, if all records in the relevant databases have been searched and more than one matching record was found.

(g) Generally, based on the response(s) provided by the NICS, and other information available in the state and local record systems, a POC will:

(1) Confirm any matching records; and

(2) Notify the FFL that the transfer may proceed, is delayed pending further record analysis, or is denied. "Proceed" notifications made within three business days will be accompanied by the NTN or STN traceable to the NTN. The POC may or may not provide a transaction number

(NTN or STN) when notifying the FFL of a "Denied" response.

(h) POC Determination Messages. POCs shall transmit electronic NICS transaction determination messages to the FBI for the following transactions: open transactions that are not resolved before the end of the operational day on which the check is requested; denied transactions; transactions reported to the NICS as open and later changed to proceed; and denied transactions that have been overturned. The FBI shall provide POCs with an electronic capability to transmit this information. These electronic messages shall be provided to the NICS immediately upon communicating the POC determination to the FFL. For transactions where a determination has not been communicated to the FFL, the electronic messages shall be communicated no later than the end of the operational day on which the check was initiated. With the exception of permit checks, newly created POC NICS transactions that are not followed by a determination message (deny or open) before the end of the operational day on which they were initiated will be assumed to have resulted in a proceed notification to the FFL. The information provided in the POC determination messages will be maintained in the NICS Audit Log described in § 25.9(b). The NICS will destroy its records regarding POC determinations in accordance with the procedures detailed in § 25.9(b).

(i) Response recording. FFLs are required to record the system response, whether provided by the FBI NICS Operations Center or a POC, on the appropriate ATF form for audit and inspection purposes, under 27 CFR part 478 recordkeeping requirements. The FBI NICS Operations Center response will always include an NTN and associated "Proceed," "Delayed," or "Denied" determination. POC responses may vary as discussed in paragraph (g) of this section. In these instances, FFLs will record the POC response, including any transaction number and/or determination.

(j) Access to the NICS Index for purposes unrelated to NICS background checks required by the Brady Act. Access to the NICS Index for purposes unrelated to NICS background checks pursuant to 18 U.S.C. 922(t) shall be limited to uses for the purpose of:

(1) Providing information to Federal, state, or local criminal justice agencies in connection with the issuance of a firearm-related or explosives-related permit or license, including permits or licenses to possess, acquire, or transfer a firearm, or to carry a concealed firearm, or to import, manufacture, deal in, or purchase explosives; or

(2) Responding to an inquiry from the ATF in connection with a civil or criminal law enforcement activity relating to the Gun Control Act (18 U.S.C. Chapter 44) or the National Firearms Act (26 U.S.C. Chapter 53).

§ 25.7 Querying records in the system.

(a) The following search descriptors will be required in all queries of the system for purposes of a background check:

(1) Name;

(2) Sex;

(3) Race;

(4) Complete date of birth; and

(5) State of residence.

(b) A unique numeric identifier may also be provided to search for additional records based on exact matches by the numeric identifier. Examples of unique numeric identifiers for purposes of this system are: Social Security number (to comply with Privacy Act requirements, a Social Security number will not be required by the NICS to perform any background check) and miscellaneous identifying numbers (e.g., military number or number assigned by Federal, State, or local authorities to an individual's record). Additional identifiers that may be requested by the system after an initial query include height, weight, eye and hair color, and place of birth. At the option of the querying agency, these additional identifiers may also be included in the initial query of the system.

§ 25.8 System safeguards.

(a) Information maintained in the NICS Index is stored electronically for use in an FBI computer environment. The NICS central computer will reside inside a locked room within a secure facility. Access to the facility will be restricted to authorized personnel who have identified themselves and their need for access to a system security officer.

(b) Access to data stored in the NICS is restricted to duly authorized agencies. The security measures listed in paragraphs (c) through (f) of this section are the minimum to be adopted by all POCs and data sources having access to the NICS.

(c) State or local law enforcement agency computer centers designated by a Control Terminal Agency as POCs shall be authorized NCIC users and shall observe all procedures set forth in the NCIC Security Policy of 1992 when processing NICS background checks. The responsibilities of the Control Terminal Agencies and the computer centers include the following:

(1) The criminal justice agency computer site must have adequate physical security to protect against any unauthorized personnel gaining access to the computer equipment or to any of the stored data.

(2) Since personnel at these computer centers can have access to data stored in the NICS, they must be screened thoroughly under the authority and supervision of a State Control Terminal Agency. This authority and supervision may be delegated to responsible criminal justice agency personnel in the case of a satellite computer center being serviced through a State Control Terminal Agency. This screening will also apply to non-criminal justice maintenance or technical personnel.

(3) All visitors to these computer centers must be accompanied by staff personnel at all times.

(4) POCs utilizing a State/NCIC terminal to access the NICS must have the proper computer instructions written and other built-in controls to prevent data from being accessible to any terminals other than authorized terminals.

(5) Each State Control Terminal Agency shall build its data system around a central computer, through which each inquiry must pass for screening and verification.

(d) Authorized State agency remote terminal devices operated by POCs and having access to the NICS must meet the following requirements:

(1) POCs and data sources having terminals with access to the NICS must physically place these terminals in secure locations within the authorized agency;

(2) The agencies having terminals

with access to the NICS must screen terminal operators and must restrict access to the terminals to a minimum number of authorized employees; and

(3) Copies of NICS data obtained from terminal devices must be afforded appropriate security to prevent any unauthorized access or use.

(e) FFL remote terminal devices may be used to transmit queries to the NICS via electronic dial-up access. The following procedures will apply to such queries:

(1) The NICS will incorporate a security authentication mechanism that performs FFL dial-up user authentication before network access takes place;

(2) The proper use of dial-up circuits by FFLs will be included as part of the periodic audits by the FBI; and

(3) All failed authentications will be logged by the NICS and provided to the NICS security administrator.

(f) FFLs may use the telephone to transmit queries to the NICS, in accordance with the following procedures:

(1) FFLs may contact the NICS Operations Center during its regular business hours by a telephone number provided by the FBI;

(2) FFLs will provide the NICS Representative with their FFL Number and code word, the type of sale, and the name, sex, race, date of birth and state of residence of the prospective buyer; and

(3) The NICS will verify the FFL Number and code word before processing the request.

(g) The following precautions will be taken to help ensure the security and privacy of NICS information when FFLs contact the NICS Operations Center:

(1) Access will be restricted to the initiation of a NICS background check in connection with the proposed transfer of a firearm.

(2) The NICS Representative will only provide a response of "Proceed" or "Delayed" (with regard to the prospective firearms transfer), and will not provide the details of any record information about the transferee. In cases where potentially disqualifying information is found in response to an FFL query, the NICS Representative will provide a "Delayed" response to the FFL. Follow-up "Proceed" or "Denied" responses will be provided by the NICS Operations Center during its regular business hours.

(3) The FBI will periodically monitor telephone inquiries to ensure proper use of the system.

(h) All transactions and messages sent and received through electronic access by POCs and FFLs will be automatically logged in the NICS Audit Log described in § 25.9(b). Information in the NICS Audit Log will include initiation and termination messages, failed authentications, and matching records located by each search transaction.

(i) The FBI will monitor and enforce compliance by NICS users with the applicable system security requirements outlined in the NICS POC Guidelines and the NICS FFL Manual (available from the NICS Operations Center, Federal Bureau of Investigation, 1000 Custer Hollow Road, Clarksburg, West Virginia 26306-0147).

§ 25.9 Retention and destruction of records in the system.

(a) The NICS will retain NICS Index records that indicate that receipt of a firearm by the individuals to whom the records pertain would violate Federal or State law. The NICS will retain such records indefinitely, unless they are canceled by the originating agency. In cases where a firearms disability is not permanent, e.g., a disqualifying restraining order, the NICS will automatically purge the pertinent record when it is no longer disqualifying. Unless otherwise removed, records contained in the NCIC and III files that are accessed during a background check will remain in those files in accordance with established policy.

(b) The FBI will maintain an automated NICS Audit Log of all incoming and outgoing transactions that pass through the system.

(1) Contents. The NICS Audit Log will record the following information: Type of transaction (inquiry or response), line number, time, date of inquiry, header, message key, ORI or FFL identifier, and inquiry/response data (including the name and other identifying information about the prospective transferee and the NTN).

(i) NICS Audit Log records relating to denied transactions will be retained for 10 years, after which time they will be transferred to a Federal Records Center for storage;

(ii) NICS Audit Log records relating to transactions in an open status, except the NTN and date, will be destroyed after not more than 90 days from the date of inquiry; and

(iii) In cases of NICS Audit Log records relating to allowed transactions, all identifying information submitted by or on behalf of the transferee will be destroyed within 24 hours after the FFL receives communication of the determination that the transfer may proceed. All other information, except the NTN and date, will be destroyed after not more than 90 days from the date of inquiry.

(2) Use of information in the NICS Audit Log. The NICS Audit Log will be used to analyze system performance, assist users in resolving operational problems, support the appeals process, or support audits of the use and performance of the system. Searches may be conducted on the Audit Log by time frame, i.e., by day or month, or by a particular state or agency. Information in the NICS Audit Log pertaining to allowed transactions may be accessed directly only by the FBI and only for the purpose of conducting audits of the use and performance of the NICS, except that:

(i) Information in the NICS Audit Log, including information not yet destroyed under § 5.9(b)(1)(iii), that indicates, either on its face or in conjunction with other information, a violation or potential violation of law or regulation, may be shared with appropriate authorities responsible for investigating, prosecuting, and/or enforcing such law or regulation; and

(ii) The NTNs and dates for allowed transactions may be shared with ATF in Individual FFL Audit Logs as specified in § 25.9(b)(4).

(3) Limitation on use. The NICS, including the NICS Audit Log, may not be used by any Department, agency, officer, or employee of the United States to establish any system for the registration of firearms, firearm owners, or firearm transactions or dispositions, except with respect to persons prohibited from receiving a firearm by 18 U.S.C. 922(g) or (n) or by State law. The NICS Audit Log will be monitored and reviewed on a regular basis to detect any possible misuse of NICS data.

(4) Creation and Use of Individual FFL Audit Logs. Upon written request from ATF containing the name and license number of the FFL and the proposed date of inspection of the named FFL by ATF, the FBI may extract information from the NICS Audit Log and create an Individual FFL Audit Log for transactions originating at the named FFL for a limited period of time. An Individual FFL Audit Log shall contain all information on denied transactions, and, with respect to all other transactions, only non-identifying information from the transaction. In no instance shall an Individual FFL Audit Log contain more than 60 days worth of allowed or open transaction records originating at the FFL. The FBI will provide POC States the means to provide to the FBI information that will allow the FBI to generate Individual FFL Audit Logs in connection with ATF inspections of FFLs in POC States. POC States that elect not to have the FBI generate Individual FFL Audit Logs for FFLs in their states must develop a means by which the POC will provide such Logs to ATF.

(c) The following records in the FBI-operated terminals of the NICS will be subject to the Brady Act's requirements for destruction:

(1) All inquiry and response messages (regardless of media) relating to a background check that results in an allowed transfer; and

(2) All information (regardless of media) contained in the NICS Audit Log relating to a background check that results in an allowed transfer.

(d) The following records of State and local law enforcement units serving as POCs will be subject to the Brady Act's requirements for destruction:

(1) All inquiry and response messages (regardless of media) relating to the initiation and result of a check of the NICS that allows a transfer that are not part of a record system created and maintained pursuant to independent State law regarding firearms transactions; and

(2) All other records relating to the person or the transfer created as a result of a NICS check that are not part of a record system created and maintained pursuant to independent State law regarding firearms transactions.

§ 25.10 Correction of erroneous system information.

(a) An individual may request the reason for the denial from the agency that conducted the check of the NICS (the "denying agency," which will be either the FBI or the State or local law enforcement agency serving as a POC). The FFL will provide to the denied individual the name and address of the denying agency and the unique transaction number (NTN or STN) associated with the NICS background check. The request for the reason for the denial must be made in writing to the denying agency. (POCs at their discretion may waive the requirement for a written request.)

(b) The denying agency will respond to the individual with the reasons for the denial within five business days of its receipt of the individual's request. The response should indicate whether additional information or documents are required to support an appeal, such as fingerprints in appeals involving questions of identity (i.e., a claim that the record in question does not pertain to the individual who was denied).

(c) If the individual wishes to challenge the accuracy of the record upon which the denial is based, or if the individual wishes to assert that his or her rights to possess a firearm have been restored, he or she may make application first to the denying agency, i.e., either the FBI or the POC. If the denying agency is unable to resolve the appeal, the denying agency will so notify the individual and shall provide the name and address of the agency that originated the document containing the information upon which the denial was based. The individual may then apply for correction of the record directly to the agency from which it originated. If the record is corrected as a result of the appeal to the originating agency, the individual may so notify the denying agency, which will, in turn, verify the record correction with the originating agency (assuming the originating agency has not already notified the denying agency of the correction) and take all necessary steps to correct the record in the NICS.

(d) As an alternative to the above procedure where a POC was the denying agency, the individual may elect to direct his or her challenge to the accuracy of the record, in writing, to the FBI, NICS Operations Center, Criminal Justice Information Services Division, 1000 Custer Hollow Road, Module C–3, Clarksburg, West Virginia 26306–0147. Upon receipt of the information, the FBI will investigate the matter by contacting the POC that denied the transaction or the data source. The FBI will request the POC or the data source to verify that the record in question pertains to the individual who was denied, or to verify or correct the challenged record. The FBI will consider the information it receives from the individual and the response it receives from the POC or the data source. If the record is corrected as a result of the challenge, the FBI shall so notify the individual, correct the erroneous information in the NICS, and give notice of the error to any Federal department or agency or any state that was the source of such erroneous records.

(e) Upon receipt of notice of the correction of a contested record from the originating agency, the FBI or the agency that contributed the record shall correct the data in the NICS and the denying agency shall provide a written confirmation of the correction of the erroneous data to the individual for presentation to the FFL. If the appeal of a contested record is successful and thirty (30) days or less have transpired since the initial check, and there are no other disqualifying records upon which the denial was based, the NICS will communicate a "Proceed" response to the FFL. If the appeal is successful and more than thirty (30) days have transpired since the initial check, the FFL must recheck the NICS before allowing the sale to continue. In cases where multiple disqualifying records are the basis for the denial, the individual must pursue a correction for each record.

(f) An individual may also contest the accuracy or validity of a disqualifying record by bringing an action against the state or political subdivision responsible for providing the contested information, or responsible for denying the transfer, or against the United States, as the case may be, for an order directing that the contested information be corrected or that the firearm transfer be approved.

(g) An individual may provide written consent to the FBI to maintain information about himself or herself in a Voluntary Appeal File to be established by the FBI and checked by the NICS for the purpose of preventing the future erroneous denial or extended delay by the NICS of a firearm transfer. Such file shall be used only by the NICS for this purpose. The FBI shall remove all information in the Voluntary Appeal File pertaining to an individual upon receipt of a written request by that individual. However, the FBI may retain such information contained in the Voluntary Appeal File as long as needed to pursue cases of identified misuse of the system. If the FBI finds a disqualifying record on the individual after his or her entry into the Voluntary Appeal File, the FBI may remove the individual's information from the file.

§ 25.11 Prohibited activities and penalties.

(a) State or local agencies, FFLs, or individuals violating this subpart A shall be subject to a fine not to exceed $10,000 and subject to cancellation of NICS inquiry privileges.

(b) Misuse or unauthorized access includes, but is not limited to, the following:

(1) State or local agencies, FFLs, or individuals purposefully furnishing incorrect information to the system to obtain a "Proceed" response, thereby allowing a firearm transfer;

(2) State or local agencies, FFLs, or individuals purposefully using the system to perform a check for unauthorized purposes; and

(3) Any unauthorized persons accessing the NICS.

Firearms Law Administered by the Postal Service

TITLE 18, UNITED STATES CODE, CHAPTER 83

§ 1715. Firearms as non mailable; regulations.

Pistols, revolvers, and other firearms capable of being concealed on the person are nonmailable and shall not be deposited in or carried by the mails or delivered by any officer or employee of the Postal Service. Such articles may be conveyed in the mails, under such regulations as the Postal Service shall prescribe, for use in connection with their official duty, to officers of the Army, Navy, Air Force, Coast Guard, Marine Corps, or Organized Reserve Corps; to officers of the National Guard or Militia of a State, Territory, Commonwealth, Possession, or District; to officers of the United States or of a State, Territory, Commonwealth, Possession, or District whose official duty is to serve warrants of arrest or commitments; to employees of the Postal Service; to officers and employees of enforcement agencies of the United States; and to watchmen engaged in guarding the property of the United States, a State, Territory, Commonwealth, Possession, or District. Such articles also may be conveyed in the mails to manufacturers of firearms or bona fide dealers therein in customary trade shipments, including such articles for repairs or replacement of parts, from one to the other, under such regulations as the Postal Service shall prescribe.

Whoever knowingly deposits for mailing or delivery, or knowingly causes to be delivered by mail according to the direction thereon, or at any place to which it is directed to be delivered by the person to whom it is addressed, any pistol, revolver, or firearm declared nonmailable by this section, shall be fined under this title or imprisoned not more than two years, or both.

RULINGS, PROCEDURES, AND INDUSTRY CIRCULARS

TABLE OF CONTENTS

RULINGS:

Editor's Note:

The **Rulings**, **Procedures**, and **Industry Circulars** set out herein cannot address all subjects under all possible circumstances. When there is doubt, please contact your nearest ATF office.

These items are often extracts from the original material. Where sections of law have been altered or amended but the purpose or intent of the ruling, procedure, or industry circular remains in effect, a text modification has been made, and the word "amended" appears in brackets at the end in the following manner: [Amended].

"Rev. Rul." or **"RR"** is a revenue ruling published by the Alcohol, Tobacco and Firearms Division of the Internal Revenue Service, the predecessor of the Bureau of Alcohol, Tobacco, Firearms and Explosives. **"ATF Rul."** means ATF Ruling; **"ATFP"** means ATF Procedure; **"I.C."** means Industry Circular; **"ATFB"** means ATF Bulletin; **"ATFQB"** means ATF Quarterly Bulletin; **"ATF C.B."** means ATF Cumulative Bulletin; and **"C.B."** means Internal Revenue Service Cumulative Bulletin.

RULINGS

27 CFR 178.41: General (Also 178.42, 178.50)

Firearms or ammunition may not be sold at gun shows by a licensed dealer, but orders may be taken under specified conditions; Revenue Ruling 66-265 superseded.

Rev. Rul. 69-59

Advice has been requested whether a person who is licensed under 18 U.S.C. Chapter 44 (which superseded the Federal Firearms Act (15 U.S.C. Chapter 18)) or who is continuing operations under a license issued to him under the Federal Firearms Act as a manufacturer, importer or dealer in firearms or ammunition may sell firearms or ammunition at a gun show held on premises other than those covered by his outstanding license.

Under 18 U.S.C. 923(a), "a separate fee" is required to be paid for each place at which business as a licensee is to be conducted. Further, each applicant for a license is required to have in a State "premises from which he conducts business" (18 U.S.C. 923(d)(1)(E)) and to specify such premises in the license application. In addition, records are required to be maintained at the business premises covered by the license (18 U.S.C. 923(g)).

Therefore, a person holding a valid license may engage in the business covered by the license only at the specific business premises for which his license has been obtained. Thus, a licensee may not sell firearms or ammunition at a gun show held on premises other than those covered by his license. He may, however, have a booth or table at such a gun show at which he displays his wares and takes orders for them, provided that the sale and delivery of the firearms or ammunition are to be lawfully effected from his licensed business premises only and his records properly reflect such transactions.

There are no provisions in the law for the issuance of temporary licenses to cover sales at gun shows, and licenses will be issued only for premises where the applicant regularly intends to engage in the business to be covered by the license.

This ruling does not apply to the activities of licensed collectors with respect to the receipt or disposition of curios and relics by such collectors. For provisions relating to transactions by licensed collectors of curios and relics, see 27 CFR 178.50 of the regulations.

Rev. Rul. 66-265, 1966-2, C.B. 559, is hereby superseded.

Editor's Note:

In 1986, the GCA was amended to allow licensees to sell firearms at gunshows in the State in which their licensed premises are located, and was further amended in 1997 to allow licensees to sell curio or relic firearms to other licensees at any location. Moreover, the interstate controls no longer apply to ammunition sales. However, the ruling is still applicable to licensees' off-premises sales not addressed by these amendments.

27 CFR 178.114: IMPORTATION BY MEMBERS OF THE U.S. ARMED FORCES

ATF Rul. 69-309

This ruling was revoked by T.D. ATF-426.

27 CFR 178.41: GENERAL (Also 178.42, 178.45, 178.49)

Licensed firearms dealers operating at multiple locations may establish a common expiration date for all licenses.

ATF Rul. 73-9

Licensed firearms dealers operating more than one location for which a license is required may establish a common expiration date for all licenses issued to their several locations. Dealers wishing to establish such a date for all licenses issued to them may make application in writing to the Regional Director (Compliance) of the region in which the businesses or activities are operated. The application should set out the requested common expiration date and should list all licensed premises in the region covered by the application. The Regional Director (Compliance) will advise the dealer whether the request may be approved and, if approved, will provide the necessary instructions and renewal application. It is pointed out that approval of a request will entail a one-time loss associated with the existing license, as it will be cancelled on and after the date of issuance of the license bearing the requested common expiration date, and the regulations do not provide for prorated refunds.

[73 ATF C.B. 102] [Amended]

27 CFR 178.11: MEANING OF TERMS (Also 178.23, 178.44)

Because of the nature of operations conducted by a gunsmith, he shall not be required to have business premises open to the general public or to have regular business hours.

ATF Rul. 73-13

Because of the nature of operations conducted by a gunsmith, any applicant for a license who intends to engage solely in this type of business and so specifies on his application will not be required to maintain regular business hours. Further, if the business is conducted from a private dwelling, a separate portion should be designated as the business premises, which need not be open to all segments of the public but only accessible to the clientele that the business is set up to serve. However, the licensed premises of the gunsmith are subject to the inspection requirements of 18 U.S.C. 923(g) and 27 CFR 178.23, and the gunsmith must maintain the required records as specified in 27 CFR 178.121 et seq.

Further, since a gunsmith is a licensed firearms dealer, if he engages in the business of buying and selling firearms, he must record his transactions on Form 4473 (Firearms Transaction Record) for each sale, and maintain the firearms acquisition and disposition records required of all licensed dealers. However, if a gunsmith engages in the business of buying and selling firearms during the term of his current license, he may be required to submit a new Form 7 (Firearms) at the time of renewal in accordance with 27 CFR 178.45 and meet the requirements of an applicant engaging in the business of buying and selling firearms, such as having business premises open to the general public and having regular business hours.

(Amplified by ATFR 77-1)

[73 ATF C.B. 92]

27 CFR 178.11: MEANING OF TERMS (Also 178.23, 178.44, 178.99, 178.124)

Because of the nature of operations conducted by a consultant or expert, he shall not be required to have business premises open to the general public or to have regular business hours.

ATF Rul. 73-19

Revenue Ruling 69-248, C.B. 1969-1, 360 (Internal Revenue) permits firearms licensees to ship, transport, or deliver firearms in interstate commerce to their nonlicensed employees, agents, or representatives for business purposes. As was clarified in Industry Circular 72-23 the ruling also permits firearms licensees to similarly transfer firearms to nonlicensed professional writers, consultants, and evaluators for research or evaluation.

Title 18 U.S.C., Section 922(a)(2)(A), permits an individual to ship (and have returned to him) in interstate commerce a firearm to a firearms licensee for repair or customizing. Furthermore, the definition of a firearms dealer in 18 U.S.C. 921 and 27 CFR 178.11 is sufficiently broad that it can be interpreted to include a qualified firearms consultant or expert who is engaged in the business of testing or examining firearms. In view of these provisions, the Bureau has determined that firearms consultants or experts may be licensed as firearms dealers in order that they may receive firearms from nonlicensed individuals for testing and examination.

Because of the nature of operations conducted by a firearms consultant or expert, any licensed dealer who engages solely in this type of business will not be required to maintain regular business hours. If the business is conducted from a private residence, a separate portion of the dwelling should be designated as "business premises." Such premises need not be open to all segments of the public but only accessible to the clientele that the business is set up to serve. However, the licensed premises of the firearms consultant-expert shall be subject to inspection under the authority of 18 U.S.C. 923(g) and 27 CFR 178.23.

A licensed firearms consultant or expert shall maintain records of receipt and delivery of firearms, as is required by 27 CFR 178, Subpart H, except that the licensee need not prepare Forms 4473, Firearms Transaction Record, reflecting the firearms examined.

However, shipments and deliveries of firearms shall not be made in care of persons who are ineligible "to ship or transport in interstate or foreign commerce, or possess in or affecting commerce, any firearm or ammunition; or to receive any firearm or ammunition which has been shipped or transported in interstate or foreign commerce" under 922(g).

A firearms consultant or expert who desires to obtain a license as a dealer in firearms shall file Form 7 (Firearms), Application for License Under 18 U.S.C. Chapter 44, Firearms, in the manner prescribed by 27 CFR 178.44. The application shall include a statement that the applicant is engaged in business as a bona fide firearms consultant or expert and, where the applicant intends to perform testing or examination services for one or more persons on a continuing basis, the statement shall include the name, address, and nature of business of such persons. A license as a dealer in firearms will be issued only after the Regional Director (Compliance) is satisfied that the applicant is a bona fide consultant or expert and is otherwise qualified under the law.

Since a licensed firearms consultant or expert is a firearms dealer, if he engages in the business of buying and selling firearms, he must record his transactions on Form 4473, Firearms Transaction Record, for each sale, and maintain the firearms acquisition and disposition records required of all licensed dealers. If a firearms consultant or expert engages in the business of buying and selling firearms during the term of his current license, he may be required to submit a new Form 7 (Firearms) at the time of renewal in accordance with 27 CFR 178.45 and meet the requirements of an applicant engaging in the business of buying and selling firearms, such as having business premises open to the general public and having regular business hours.

[This ATF ruling does not apply to firearms within the purview of the National Firearms Act (26 U.S.C. Chapter 53)]

[73 ATF C.B. 93] [Amended]

27 CFR 179.104: REGISTRATION OF FIREARMS BY CERTAIN GOVERNMENTAL ENTITIES

When NFA firearms are registered

on Form 10 by governmental entities, subsequent transfers of such firearms shall be made only to other governmental entities.

ATF Rul. 74-8

Advice has been requested whether the Bureau will approve transfer of National Firearms Act weapons by a State or political subdivision (police department) to a special occupational taxpayer where such firearms were registered in the National Firearms Registration and Transfer Record pursuant to 27 CFR 179.104.

27 CFR 179.104 provides that any State, any political subdivision thereof, or any official police organization of such a government entity engaged in criminal investigations, which acquires for official use a firearm not registered to it, such as by abandonment or by forfeiture, will register such firearm with the Director by filing Form 10 (Firearms), Application for Registration of Firearms Acquired by Certain Governmental Entities, and that such registration shall become a part of the National Firearms Registration and Transfer Record.

The purpose of the above regulation was to permit the limited registration of firearms by certain governmental entities for official use only. 27 CFR 179.104 may not be used as a vehicle to register otherwise unregisterable firearms for the purpose of introducing such firearms into ordinary commercial channels. Accordingly, when registration of firearms by governmental entities is approved on Form 10, the form will be marked "official use only." The Bureau will approve subsequent transfers of such firearms only to other governmental entities for official use. Otherwise, such firearms must be destroyed or abandoned to the Bureau.

[74 ATF C.B. 67]

27 CFR 178.114: IMPORTATION BY MEMBERS OF THE U.S. ARMED FORCES

A member of the U.S. Armed Forces who is a resident of any State or territory which requires that a permit or other authorization be issued prior to possessing or owning a handgun shall submit evidence of compliance with State law before an application to import a handgun may be approved.

ATF Rul. 74-13

Handguns have been transported, shipped, received, or imported into the United States by members of the United States Armed Forces to their place of residence without such members having obtained the required permit or other authorization required by their State of residence which would permit them to possess or own (as opposed to a license to purchase) handguns in that State.

18 U.S.C. 925(a)(4) provides that when established to the satisfaction of the Secretary to be consistent with the provisions of 18 U.S.C. Chapter 44 and other applicable Federal and State laws and published ordinances, the Secretary may authorize the transportation, shipment, receipt, or importation into the United States to the place of residence of any member of the United States Armed Forces who is on active duty outside the United States (or has been on active duty outside the United States within the 60-day period immediately preceding the transportation, shipment, receipt, or importation), of any firearm or ammunition which is:

(a) Determined by the Secretary to be generally recognized as particularly suitable for sporting purposes, or determined by the Department of Defense to be a type of firearm normally classified as a war souvenir; and

(b) Intended for the personal use of such member.

27 CFR 178.114(a) provides that an application for a permit to import a firearm or ammunition into the United States to the place of residence of any military member of the United States Armed Forces on active duty outside the United States shall include a certification by the applicant that the transportation, receipt, or possession of the firearm or ammunition to be imported, would not constitute a violation of any State law or local ordinance at the place of the applicant's residence.

In order to assure that the transportation, shipment, receipt, or importation of handguns under 27 CFR 178.114 is not in violation of applicable State laws, it is held that, any member of the United States Armed Forces who is a resident of any State or territory which requires that a permit or authorization be obtained prior to possessing or owning a handgun shall, in addition to making the required certification in the application, submit with his application to the Director a copy of the license, permit, certificate of registration, or firearm identification card, as applicable and as required by his State, in order to obtain a permit to import a handgun into the United States.

[74 ATF C.B. 60]

27 CFR 178.124: FIREARMS TRANSACTION RECORDS (Also 178.123, 178.125, 178.147)

Form 4473 shall not be required to record disposition of a like replacement firearm when such firearm is delivered by a licensee to the person from whom the malfunctioning or damaged firearm was received, provided such disposition is recorded in the licensee's permanent records.

ATF Rul. 74-20

It is held that a firearms transaction record, Form 4473, shall not be required to record the disposition of a replacement firearm of the same kind and type where such a firearm is delivered by a licensee to the person from whom the malfunctioning or damaged firearm was received.

It should be noted, however, that the licensee is required by 27 CFR 178.125 to maintain in his permanent records the disposition of such a replacement firearm. (See also ATFR 76-25)

[74 ATF C.B. 61]

27 CFR 178.11: MEANING OF TERMS (Also 179.11)

A small caliber weapon ostensibly designed to expel only tear gas, similar substances, or pyrotechnic signals, which may readily be converted to expel a projectile by means of an explosive, classified as a firearm.

ATF Rul. 75-7

The term "firearm" as used in 18 U.S.C. 921(a)(3) includes "any weapon (including a starter gun) which will or is designed to or may readily be converted to expel a projectile by the action of an explosive."

A small caliber weapon ostensibly designed to expel only tear gas, similar substances or pyrotechnic signals by the action of an explosive, which may readily be converted to expel a projectile by means of an explosive, constitutes, a "firearm" within the purview of 18 U.S.C. 921(a)(3)(A).

Tests performed on these weapons have established that they may readily be converted to expel a projectile by the action of an explosive, normally by means of a minor alteration of the expended Helix cartridge and/or the simple attachment of a barrel/chamber to the firing mechanism.

Such weapons manufactured within the United States on or after June 1, 1975, will be subject to all of the provisions of Chapter 44 and 27 CFR Part 178. Such weapons manufactured before June 1, 1975, will not be treated as subject to the provisions of Chapter 44 and 27 CFR Part 178 in order to allow persons manufacturing and dealing in such weapons to comply with the provisions of Chapter 44 and 27 CFR Part 178.

Since such weapons are not generally recognized as particularly suitable for or readily adaptable to sporting purposes (18 U.S.C. 925(d)(3)), the importation of such weapons is prohibited unless such importation comes within one of the statutory exceptions provided in 18 U.S.C. 925.

[75 ATF C.B. 55]

27 CFR 178.94: SALES OR DELIVERIES BETWEEN LICENSEES

A firearms licensee may continue operations until his renewal application for a license is finally acted upon.

ATF Rul. 75-27

Under 5 U.S.C. 558, when a licensee has made timely and sufficient application for a renewal in accordance with agency rules, a license with reference to an activity of a continuing nature does not expire until the application has been finally determined by the agency. In accordance with section 558, a firearms licensee who timely applies for renewal of his license is authorized to continue his firearms operations as authorized by his license until his renewal application is finally acted upon. As provided by 27 CFR 178.94, a transferor licensee is authorized to continue to make shipments to a licensee for not more than 45 days following the expiration date of the transferee's license.

Held, a transferor licensee may continue to make firearms and ammunition shipments to a licensee who has timely applied for renewal of his license but has not had his application acted upon within 45 days after the expiration of his license. The transferor licensee shall, however, in cases where the 45-day period has passed, obtain appropriate evidence that the transferee's license renewal application is still pending in the office of the Regional Director (Compliance), Bureau of Alcohol, Tobacco and Firearms. Such evidence should consist of a letter from the Regional Director (Compliance), to the transferee licensee stating that his renewal application has been timely received and that action thereon is currently pending.

[75 ATF C.B. 60]

27 CFR 178.92: IDENTIFICATION OF FIREARMS

Importers may adopt serial numbers placed on certain firearms by foreign manufacturers.

ATF Rul. 75-28

The Bureau has determined that in some cases the serial number placed on a firearm by a foreign manufacturer is adequate to provide the identification number required by section 178.92. See, also, section 178.22(a).

Held, where a serial number has been placed on the frame or receiver of a firearm by a foreign manufacturer in the manner contemplated by 27 CFR 178.92, and such serial number does not duplicate a number previously adopted or assigned by the importer to any other firearm, the importer may adopt the serial number of the foreign manufacturer:

Provided, the importer shall in all cases place his name and address (city and State, or recognized abbreviation thereof), and any other marks necessary to comply with the identification requirements of 27 CFR 178.92, on such imported firearms.

[75 ATF C.B. 59] [Amended]

27 CFR 178.11: MEANING OF TERMS (Also 179.11)

A hand-held device designed to expel by means of an explosive two electrical contacts (barbs) connected by two wires attached to a high voltage source in the device classified as a firearm.

ATF Rul. 76-6

Taser Model TF-1, a hand-held device designed to expel by means of an explosive two electrical contacts (barbs) connected by two wires attached to a high voltage source in the device, is a "firearm" within the purview of 18 U.S.C. 921(a)(3)(A). It is also "any other weapon" under the National Firearms Act (26 U.S.C. 5845(e)).

In order to allow persons manufacturing and dealing in such weapons to comply with the provisions of Chapter 44 and 27 CFR Part 178, this ruling will be applicable to such weapons manufactured within the United States on or after May 1, 1976. Such weapons manufactured before May 1, 1976, will not be treated as subject to the provisions of Chapter 44 and 27 CFR Part 178. With respect to the "any other weapon" classification under the National Firearms Act, pursuant to 26 U.S.C. 7805(b), this ruling will not be applied to such weapons manufactured before May 1, 1976. Accordingly, such weapons manufactured on or after May 1, 1976, will be subject to all the provisions of the National Firearms Act and 27 CFR Part 179.

(Amplified by ATFR 80-20)

[76 ATF C.B. 96]

27 CFR 178.124: FIREARMS TRANSACTION RECORD (Also 178.125, 178.126a)

Certain reporting and recordkeeping requirements of pawnbrokers are explained.

ATF Rul. 76-15

The regulations do not require that a pawnbroker execute Form 4473 when a firearm is pledged for a loan. However, he must record the receipt thereof in his permanent acquisition and disposition record as required by

27 CFR 178.125(e). At the time a firearm is redeemed by a nonlicensee pledgor, Form 4473 must be executed and the appropriate entry made in the permanent acquisition and disposition record. Although a redemption is not considered a sale, it is a disposition for purposes of 27 CFR 178.124(a), 178.125(e), and 178.126a. See *Huddleston v. United States*, 415 U.S. 814 (1974).

However, no report of multiple sales and other dispositions is required to be filed with ATF when the handguns are returned to the person from whom received.

Held, Form 4473, Firearms Transaction Record, need not be executed when a pawnbroker accepts a firearm as a pledge for a loan. However, if a nonlicensee pledgor redeems the firearm or if disposition of the firearm is made to any other nonlicensee, Form 4473 must be executed.

Held further, pawnbrokers must enter into their permanent acquisition and disposition record the receipt of a firearm as a pledge for a loan and any disposition, including redemption, of such firearm.

Held further, pawnbrokers must submit reports of multiple sales and other dispositions of pistols and revolvers as required by 27 CFR 178.126(a) when the person receiving them is not the person who pawned the firearms.

[76 ATF C.B. 100] [Amended]

27 CFR 179.11: MEANING OF TERMS

Mere possession of a license and a special tax stamp as a dealer in firearms does not qualify a person to receive firearms transfer-tax-free.

ATF Rul. 76-22

The mere possession of a license and a special (occupational) tax stamp as a dealer in firearms does not qualify a person to receive firearms transfer-tax-free. Any person holding a license and a special tax stamp as a dealer in firearms and not actually engaged within the United States in the business of selling NFA firearms may not lawfully receive NFA firearms without the transfer tax having been paid by the transferor. Where it is, therefore, determined that the proposed transferee on a Form 3, Application for Tax-Exempt Transfer of Firearm and Registration to Special (Occupational) Taxpayer, is not actually engaged in the business of dealing in NFA firearms, such application will be denied. In addition, if such person receives NFA firearms without the transfer tax having been paid, such firearms may be subject to seizure for forfeiture as having been unlawfully transferred without payment of the transfer tax.

[76 ATF C.B. 103]

27 CFR 178.121: GENERAL (Also 178.147)

Recordkeeping requirements for firearms from which parts are salvaged for use in repairing firearms are clarified.

ATF Rul. 76-25

Section 921(a)(3) of Title 18, United States Code, and the regulations at 27 CFR 178.11 define the term "firearm" to include any weapon which will, or is designed to, or may readily be converted to, expel a projectile by the action of an explosive, and the frame or receiver of any such weapon.

The regulations in 27 CFR 178.122, 178.123 and 178.125 require each licensed importer, licensed manufacturer, and licensed dealer, respectively, to maintain such records of acquisition (including by manufacture) or disposition, whether temporary or permanent, of firearms as therein prescribed.

Held, a licensee who purchases a damaged firearm for the purpose of salvaging parts therefrom shall enter receipt of the firearm in his firearms acquisition and disposition record. If the frame or receiver of the firearm is damaged to the extent that it cannot be repaired, or if the licensee does not desire to repair the frame or receiver, he may destroy it and show the disposition of the firearm in his records as having been destroyed. Before a firearm may be considered destroyed, it must be cut, severed or mangled in such a manner as to render the firearm completely inoperative and such that it cannot be restored to an operative condition.

Where the repair of a customer's firearm results in an exchange of a frame or receiver, an entry shall be made in the licensee's records to show the transfer of such replacement part, as it is a "firearm" as defined in 18 U.S.C. 921(a)(3). Further, as held in ATF Ruling 74-20, 1974 ATF C.B. 61, a Form 4473, Firearms Transaction Record, shall not be required to record the disposition of a replacement firearm of the same kind and type where the firearm is delivered by the licensee to the person from whom the malfunctioning or damaged firearm was received. The frame or receiver received from the customer shall be entered as an acquisition, and if destroyed, it shall be entered in the disposition record as destroyed.

With regard to National Firearms Act firearms as defined in 26 U.S.C. 5845(a), in addition to the above recordkeeping requirements, the registration and transfer procedures of 27 CFR Part 179 must be complied with.

[76 ATF C.B. 99]

27 CFR 178.121: GENERAL (RECORDS)

The recordkeeping requirements for licensed gunsmiths are clarified. ATF Rul. 73-13 amplified.

ATF Rul. 77-1

ATF Ruling 73-13, 1973 ATF C.B. 92, held that a licensed gunsmith must maintain the required records as specified in 27 CFR 178.121 et seq., and if a gunsmith engages in the business of buying and selling firearms, he must record these transactions on a Form 4473 (Firearms Transaction Record) for each sale. However, as provided in Section 178.124(a), a Form 4473 is not required to record the disposition made of a firearm delivered to a gunsmith for repair or customizing when the firearm is returned to the person from whom received.

The Bureau recognizes the necessity for having on-the-spot repairs made to firearms at skeet, trap, target, and similar organized events. It is, therefore, **held** that licensed gunsmiths may take immediate on-the-spot repairs to firearms at skeet, trap, target, and similar organized shooting events.

Held further, a licensed gunsmith must enter into his bound acquisition and disposition record, required to be maintained by 27 CFR 178.125(e), each receipt and disposition of firearms, except that a firearm need not be entered in the bound acquisition and disposition record if the firearm is brought in for adjustment or repair and the owner waits while it is being adjusted or repaired or if the gunsmith returns the firearm to the owner during the same business day it is brought in. If the firearm is retained from one business day to another or longer, it must be recorded in the bound acquisition and disposition record.

Held further, a licensed gunsmith is not required to prepare a Form 4473 (Firearms Transaction Record) where a firearm is delivered to him for the sole purpose of customizing, adjustment, or repair and the firearm is returned to the person from whom received. However, if a licensed gunsmith engages in the business of selling firearms, he must record these transactions on a Form 4473 for each sale in addition to maintaining the bound firearms acquisition and disposition record required by 27 CFR 178.125(e).

ATF Rul. 73-13, 1973 ATF C.B. 92, is hereby amplified.

[77 ATF C.B. 185]

27 CFR 178.124: FIREARMS TRANSACTION RECORD

Means of identification furnished by a nonlicensee purchasing a firearm.

ATF Rul. 79-7

This ruling is superseded by ATF Rul. 2001-5.

27 CFR 178.112: IMPORTATION BY A LICENSED IMPORTER

Applications to import surplus military firearms or non sporting firearms or ammunition for individual law enforcement officers for official use must be accompanied by the agency's purchase order.

ATF Rul. 80-8

The Bureau of Alcohol, Tobacco and Firearms has received several inquiries from firearms importers and dealers, law enforcement agencies, and the public requesting clarification of the statutes, regulations and procedures regarding the importation of firearms for law enforcement agencies.

Importation of surplus military firearms or firearms not particularly suitable for or readily adaptable to sporting purposes is generally prohibited by section 925(d)(3) of Title 18, United States Code. However, section 925(a)(1) provides that this prohibition does not apply to the importation of firearms or ammunition sold or shipped to, or issued for the use of the United States or any department or agency thereof or any State or any department, agency, or political subdivision thereof.

Pursuant to section 925(a)(1), the Bureau has previously allowed the importation of surplus military firearms and nonsporting firearms for individual law enforcement officers for official use. In approving such importation applications, the Bureau required federal firearms licensees to obtain from the agency employing the officer a certificate of the chief law enforcement officer stating that the firearm ammunition is for use in the performance of official duties.

However, once these firearms are imported for the individual officer for "official use," there is no prohibition in the law against the officer's resale or retention of the firearms for personal use. The purpose of section 925(a)(1) is to permit importation of firearms for the exclusive use of government agencies. The statute was not intended and may not be used as a vehicle by which unimportable firearms can be introduced into ordinary commercial channels in the United States.

Held, a licensee's application to import surplus military firearms or nonsporting firearms or ammunition for law enforcement officers will not be approved unless accompanied by a purchase order from a department or agency of the United States or any department, agency or political subdivision of any State. The term "State" includes the District of Columbia, the Commonwealth of Puerto Rico, and the possessions of the United States.

[80 ATF C.B. 20]

27 CFR 178.11: MEANING OF TERMS (Also 27 CFR 179.11)

A hand-held device with a hand grip bent at an angle to the bore and having a rifled bore which is designed to expel, by means of an explosive, two electrical contacts (barbs) connected by two wires to a high voltage source within the device is classified as a firearm. ATF Rul. 76-6 is amplified.

ATF Rul. 80-20

The Bureau has determined in ATF Rul. 76-6 that the Taser Model TF1 was a firearm as that term is defined in Title 18, United States Code (U.S.C.), section 921(a)(3), and that the Model TF1 also met the "any other weapon" definition found in the National Firearms Act (NFA), Title 26, U.S.C., section 5845(e). This ruling was limited in its application to Taser Models TF1 produced on or after May 1, 1976. The Taser Models TF76 and TF76A were subsequently developed and differ from the Taser Model TF1 in that these models each have a hand grip bent at an angle to the bore and the bore of each is rifled.

The changes in the design of the Taser Models TF76 and TF76A bring them within the exclusion found in the "any other weapon" definition of the NFA for pistols and revolvers having a rifled bore or rifled bores.

Held, the Taser Models TF76 and TF76A are not subject to the provisions of the NFA. However, they are firearms as defined in Title 18, U.S.C., section 921(a)(3) and are subject to the provisions of Title 18, U.S.C., Chapter 44 and Title 27, Code of Federal Regulations, Part 178.

ATF Rul. 76-6, 1976, ATF C.B. 96 is hereby amplified.

[ATFB 1980-4 24]

27 CFR 178.11: MEANING OF TERMS

An out-of-State college student may establish residence in a State by residing and maintaining a home in a college dormitory or in a location off-campus during the school term.

ATF Rul. 80-21

"State of residence" is defined by regulation in 27 CFR 178.11 as the State in which an individual regularly resides or maintains a home. The regulation also provides an example of an individual who maintains a home in State X and a home in State

Y. The individual regularly resides in State X except for the summer months and in State Y for the summer months of the year. The regulation states that during the time the individual actually resides in State X he is a resident of State X, and during the time he actually resides in State Y he is a resident of State Y.

Applying the above example to out-of-State college students **it is held**, that during the time the students actually reside in a college dormitory or at an off-campus location they are considered residents of the State where the dormitory or off-campus home is located. During the time out-of-State college students actually reside in their home State they are considered residents of their home State.

[ATFB 1980-4 25]

27 CFR 178.111: GENERAL

Nonresident U.S. citizens returning to the United States and nonresident aliens lawfully immigrating to the United States may obtain a permit to import firearms acquired outside of the United States, provided such firearms may be lawfully imported.

ATF Rul. 81-3

Section 922(a)(3) of Title 18, United States Code, makes it unlawful, with certain exceptions, for a person to bring into his State of residence a firearm which he acquired outside that State. An unlicensed resident of a State must, therefore, arrange for the importation of the firearm through a Federal firearms licensee.

The definition of "State of residence" in 27 CFR 178.11 provides that the State in which an individual regularly resides or maintains a home is the State of residence of that person. U.S. citizens who reside outside of the United States are not residents of a State while so residing. A person lawfully immigrating to the United States is not a resident of a State unless he is residing and has resided in a State for a period of at least 90 days. Therefore, such persons are not precluded by section 922(a)(3) from importing into the United States any firearms acquired outside of the United States that may be lawfully imported. The firearms must accompany such persons since once a person is in the United States and has acquired residence in a State he may import a firearm only by arranging for the importation through a Federal firearms licensee.

As applicable to this ruling, 18 U.S.C. 925(d) provides that firearms are importable if they are generally recognized as particularly suitable for, or readily adaptable to, sporting purposes, excluding National Firearms Act (NFA) firearms and surplus military firearms.

Held: a nonresident U.S. citizen returning to the United States after having resided outside of the United States, or a nonresident alien lawfully immigrating to the United States, may apply for a permit from ATF to import for personal use, not for resale, firearms acquired outside of the United States without having to utilize the services of a Federal firearms licensee. The application on ATF Form 6 Part I (7570.3A), Application and Permit for Importation of Firearms, Ammunition and Implements of War, should include a statement, on the application form or on an attached sheet, that:

(1) the applicant is a nonresident U.S. citizen who is returning to the United States from a residence outside of the United States or, in the case of an alien, is lawfully immigrating to the United States from a residence outside of the United States; and

(2) the firearms are being imported for personal use and not for resale.

[ATFB 1981-3 77]

27 CFR 179.11: MEANING OF TERMS

The AR15 auto sear is a machinegun as defined by 26 U.S.C. 5845(b).

ATF Rul. 81-4

The Bureau of Alcohol, Tobacco and Firearms has examined an auto sear known by various trade names including "AR15 Auto Sear," "Drop In Auto Sear," and "Auto Sear II," which consists of a sear mounting body, sear, return spring, and pivot pin. The Bureau finds that the single addition of this auto sear to certain AR15 type semiautomatic rifles, manufactured with M16 internal components already installed, will convert such rifles into machineguns.

The National Firearms Act, 26 U.S.C. 5845(b), defines "machinegun" to include any combination of parts designed and intended for use in converting a weapon to shoot automatically more than one shot, without manual reloading, by a single function of the trigger.

Held: The auto sear known by various trade names including "AR15 Auto Sear," "Drop In Auto Sear," and "Auto Sear II," is a combination of parts designed and intended for use in converting a weapon to shoot automatically more than one shot, without manual reloading, by a single function of the trigger. Consequently, the auto sear is a machinegun as defined by 26 U.S.C. 5845(b).

With respect to the machinegun classification of the auto sear under the National Firearms Act, pursuant to 26 U.S.C. 7805(b), this ruling will not be applied to auto sears manufactured before November 1, 1981. Accordingly, auto sears manufactured on or after November 1, 1981, will be subject to all the provisions of the National Firearms Act and 27 CFR Part 179.

[ATFQB 1981-3 78]

Editor's Note:

Regardless of the date of manufacture of a drop in auto sear, possession of such a sear and certain M-16 fire control parts is possession of a machinegun as defined by the NFA. Specifically, these parts are a combination of parts designed and intended for use in converting a weapon into a machinegun and are a machinegun as defined in the NFA. (See "Information Concerning AR15-Type Rifles" under "General Information" in this publication.)

27 CFR 179.11: MEANING OF TERMS

The KG-9 pistol is a machinegun as defined in the National Firearms Act.

ATF Rul. 82-2

The Bureau of Alcohol, Tobacco and Firearms has examined a firearm identified as the KG-9 pistol. The KG-9 is a 9 millimeter caliber, semiautomatic firearm which is blowback operated and which fires from the open bolt position with the bolt incorporating a fixed firing pin. In addition, a component part of the weapon is a disconnector which prevents more than one shot being fired with a single function of the trigger.

The disconnector is designed in the KG-9 pistol in such a way that a simple modification to it, such as cutting, filing, or grinding, allows the pistol to operate automatically. Thus, this simple modification to the disconnector together with the configuration of the above design features (blowback operation, firing from the open bolt position, and fixed firing pin) in the KG-9 permits the firearm to shoot automatically more than one shot, without manual reloading, by a single function of the trigger. The above combination of design features as employed in the KG-9 is normally not found in the typical sporting firearm.

The National Firearms Act, 26 U.S.C. 5845(b), defines a machinegun to include any weapon which shoots, is designed to shoot, or can be readily restored to shoot, automatically more than one shot, without manual reloading, by a single function of the trigger.

The "shoots automatically" definition covers weapons that will function automatically. The "readily restorable" definition defines weapons which previously could shoot automatically but will not in their present condition. The "designed" definition includes those weapons which have not previously functioned as machineguns but possess design features which facilitate full automatic fire by simple modification or elimination of existing component parts.

Held: The KG-9 pistol is designed to shoot automatically more than one shot, without function of the trigger. Consequently, the KG-9 pistol is a machinegun as defined in section 5845(b) of the Act.

With respect to the machinegun classification of the KG-9 pistol under the National Firearms Act, pursuant to 26 U.S.C. § 7805(b), this ruling will not be applied to KG-9 pistols manufactured before January 19, 1982. Accordingly, KG-9 pistols manufactured on or after January 19, 1982, will be subject to all the provisions of the National Firearms Act and 27 CFR Part 179.

[ATFB 1982-1 18]

27 CFR 179.11: MEANING OF TERMS

The SM10 and SM11A1 pistols and SAC carbines are machineguns as defined in the National Firearms Act.

ATF Rul. 82-8

The Bureau of Alcohol, Tobacco and Firearms has reexamined firearms identified as SM10 pistols, SM11A1 pistols, and SAC carbines. The SM10 is a 9 millimeter or .45ACP caliber, semiautomatic firearm; the SM11A1 is a .380ACP caliber, semiautomatic firearm; and the SAC carbine is a 9 millimeter or .45ACP caliber, semiautomatic firearm. The weapons are blowback operated, fire from the open bolt position with the bolt incorporating a fixed firing pin, and the barrels of the pistols are threaded to accept a silencer. In addition, component parts of the weapons are a disconnector and a trip which prevent more than one shot being fired with a single function of the trigger.

The disconnector and trip are designed in the SM10 and SM11A1 pistols and in the SAC carbine (firearms) in such a way that a simple modification to them, such as cutting, filing, or grinding, allows the firearms to operate automatically. Thus, this simple modification to the disconnector or trip, together with the configuration of the above design features (blowback operating, firing from the open bolt position, and fixed firing pin) in the SM10 and SM11A1 pistols and in the SAC carbine, permits the firearms to shoot automatically, more than one shot, without manual reloading, by a single function of the trigger. The above combination of design features as employed in the SM10 and SM11A1 pistols and the SAC carbine are normally not found in typical sporting firearms.

The National Firearms Act, 26 U.S.C. § 5845(b), defines a machinegun to include any weapon which shoots, is designed to shoot, or can be readily restored to shoot, automatically more than one shot, without manual reloading, by a single function of the trigger.

The "shoots automatically" definition covers weapons that will function automatically. The "readily restorable" definition defines weapons which previously could shoot automatically but will not in their present condition. The "designed" definition includes those weapons which have not previously functioned as machineguns but possess design features which facilitate full automatic fire by a simple modification or elimination of existing component parts.

Held: The SM10 and SM11A1 pistols and the SAC carbine are designed to shoot automatically more than one shot, without manual reloading, by a single function of the trigger. Consequently, the SM10 and SM11A1 pistols and SAC carbines are machineguns as defined in Section 5845(b) of the Act.

With respect to the machinegun classification of the SM10 and SM11A1 pistols and SAC carbines, under the National Firearms Act, pursuant to 26 U.S.C. 7805(b), this ruling will not be applied to SM10 and SM11A1 pistols and SAC carbines manufactured or assembled before June 21, 1982. Accordingly, SM10 and SM11A1 pistols and SAC carbines, manufactured or assembled on or after June 21, 1982, will be subject to all the provisions of the National Firearms Act and 27 CFR Part 179.

[ATFB 1982-2 49]

27 CFR 179.11: MEANING OF TERMS

The YAC STEN MK II carbine is a machinegun as defined in the National Firearms Act.

ATF Rul. 83-5

The Bureau of Alcohol, Tobacco and Firearms has examined a firearm identified as the YAC STEN MK II carbine. The YAC STEN MK II carbine is a 9 millimeter caliber firearm which has identical design characteristics to the original selective fire STEN submachinegun designed by Reginald Vernon Shepherd and Harold John Turpin. The weapon is blowback operated and fires from the open bolt position with the bolt incorporating a fixed firing pin. In addition, a component part of the weapon is a trip lever (disconnector) which has been modified to prevent more than one shot being fired with a single function of the trigger.

The trip lever (disconnector) is designed in such a way that a simple modification to it, such as bending, breaking or cutting, allows the weapon to operate automatically. Thus, this simple modification to the trip lever (disconnector), together with STEN submachinegun design features and components in the YAC STEN MK II carbine, permits the firearm to shoot automatically, more than one shot, without manual reloading by a single function of the trigger. The above combination of machinegun design features as employed in the YAC STEN MK II carbine are not normally found in the typical sporting firearm.

The National Firearms Act, 26 U.S.C. 5845(b), defines a machinegun to include any weapon which shoots, is designed to shoot, or can be readily restored to shoot, automatically more than one shot, without manual reloading, by a single function of the trigger.

The "shoots automatically" definition covers weapons that will function automatically. The "readily restorable" definition defines weapons which previously could shoot automatically but will not in their present condition. The "designed" definition includes weapons which have not previously functioned as machineguns but possess specific machinegun design features which facilitate automatic fire by simple alteration or elimination of existing component parts.

Held: The YAC STEN MK II carbine is designed to shoot automatically more than one shot, without manual reloading, by a single function of the trigger. Consequently, the STEN MK II semiautomatic carbine is a machinegun as defined in Section 5845(b) of the Act.

[ATFB 1983-3 35]

27 CFR 179.111: IMPORTATION PROCEDURE

A National Firearms Act (NFA) firearm may not be imported for use as a sample for sales to law enforcement agencies if the firearm is a curio or relic unless it is established that the firearm is particularly suitable for use as a law enforcement weapon.

ATF Rul. 85-2

The Bureau of Alcohol, Tobacco and Firearms has approved a number of applications to import National Firearms Act (NFA) firearms for the use of registered importers to generate orders for such firearms from law enforcement agencies.

A review of the characteristics of the NFA firearms approved for importation as sales samples indicates that some of the firearms are not being imported for the purposes contemplated by the statute. Some of the NFA firearms imported are, in fact, curios or relics and are more suitable for use as collector's items than law enforcement weapons.

Importations of NFA firearms are permitted by 26 U.S.C. 5844, which provides in pertinent part:

"No firearms shall be imported or brought into the United States or any territory under its control or jurisdiction unless the importer establishes, under regulations as may be prescribed by the Secretary, that the firearm to be imported or brought in is:

> **(1)** being imported or brought in for the use of the United States or any department, independent establishment, or agency thereof or any State or possession or any political subdivision thereof; or
>
> **(2)** ***
>
> **(3)** being imported or brought in solely for ... use as a sample by a registered importer or registered dealer;

except that, the Secretary may permit the conditional importation or bringing in of a firearm for examination and testing in connection with classifying the firearm."

The sole purpose of the statute permitting the importation of NFA firearms as sales samples is to permit registered importers to generate orders for firearms from government entities, primarily law enforcement agencies, on the basis of the sample.

The implementing regulation, 27 CFR Section 179.111, provides that the person importing or bringing a firearm into the United States or any territory under its control or jurisdiction has the burden of proof to affirmatively establish that the firearm is being imported or brought in for one of the authorized purposes. In addition, a detailed explanation of why the importation falls within one of the authorized purposes must be attached to the application to import. The mere statement that an NFA firearm is being imported as a sales sample for demonstration to law enforcement agencies does not meet the required burden of proof and is not a detailed explanation of why the importation falls within the import standards.

Held, an application to import a National Firearms Act firearm as a sample in connection with sales of such firearms to law enforcement agencies will not be approved if the firearm is determined to be a curio or relic unless it is established by specific information that the firearm is particularly suitable for use as a law enforcement weapon. For example, the importer must provide detailed information as to why a sales sample of a particular weapon is suitable for law enforcement purposes and the expected customers who would require a demonstration of the weapon. Information as to the availability of firearms to fill subsequent orders would help meet the burden of establishing use as a sales sample. Also, letters from law enforcement agencies expressing a need for a particular model or interest in seeing a demonstration of a particular firearm would be relevant.

[ATFB 85-2 62]

Editor's Note:

The importation of machineguns for use as sales samples must also meet the requirements of 27 CFR 179.105(d).

27 CFR 178.118: IMPORTATION OF CERTAIN FIREARMS CLASSIFIED AS CURIOS OR RELICS (Also 178.11 and 178.26)

Surplus military firearms frames or receivers alone not specifically classified as curios or relics by ATF will be denied importation.

ATF Rul. 85-10

Section 233 of the Trade and Tariff Act of 1984, 98 Stat. 2991, amended Title 18, United States Code, section 925 to allow licensed importers to import firearms listed by the Secretary as curios or relics, excluding handguns not generally recognized as particularly suitable for or readily adaptable to sporting purposes. The amendment had the effect of allowing the importation of surplus military curio or relic firearms that were previously prohibited from importation by 18 U.S.C. section 925(d)(3).

Congressional intent was expressed by Senator Robert Dole in 130 CONG. REC. S2234 (daily ed., Mar. 2, 1984), as follows:

First. This provision is aimed at allowing collectors to import fine works of art and other valuable weapons.

Second. This provision would allow the importation of certain military surplus firearms that are classified as curios and relics by regulations of the Secretary of the Treasury.

Third. In order for an individual or firm to import a curio or relic it must first be put on a list by petitioning the Secretary of the Treasury. The Secretary must find the firearm's primary value is that of being a collector's item.

Fourth. The only reason a person would purchase these firearms is because of their peculiar collector's status. And, in fact, they must be special firearms and classified as such in order to import.

This language clearly shows that Congress intended to permit the importation of surplus military firearms of special interest and value to collectors and recognized by ATF as meeting the curio or relic definition in 27 CFR 178.11. The regulation defines "curios or relics" as firearms of "special interest to collectors by reason of some quality other than is ordinarily associated with firearms intended for sporting use or as offensive or defensive weapons." The regulation further defines curios or relics to include "firearms which derive a substantial part of their monetary value from the fact that they are novel, rare, bizarre, or because of their association with some historical figure, period or event."

In classifying firearms as curios or relics under this regulation, ATF has recognized only assembled firearms as curios or relics.

Moreover, ATF's classification of surplus military firearms as curios or relics has extended only to those firearms in their original military configuration. Frames or receivers of curios or relics and surplus military firearms not in their original military configuration were not generally recognized as curios or relics by ATF since they were not of special interest or value as collector's items.

Specifically, they did not meet the definition of curio or relic in section 178.11 as firearms of special interest to collectors by reason of a quality other than is ordinarily associated with sporting firearms or offensive or defensive weapons.

Furthermore, they did not ordinarily have monetary value as novel, rare, or bizarre firearms; nor were they generally considered curios or relics because of their association with some historical figure, period or event.

It is clear from the legislative history that Congress did not intend for the frames or receivers alone of surplus military firearms, or any other surplus military firearms not in their original military configuration, to be importable under section 925(e). It is also clear that only those firearms classified by ATF as curios or relics were intended to be approved by ATF for importation.

Held: to be importable under 18 U.S.C. section 925(e), surplus military firearms must be classified as curios or relics by ATF. Applications by licensed importers to import frames or receivers alone of surplus military curio or relic firearms will not be approved under section 925(e). Surplus military firearms will not be classified as curios or relics unless they are assembled in their original military configuration, and applications for permits to import such firearms will not be approved.

[ATFB 85-3 46]

26 U.S.C. § 5845(f)(2): DESTRUCTIVE DEVICE

(Nonsporting shotgun having a bore of more than one-half inch in diameter)

The USAS-12 shotgun has a bore of more than one-half inch in diameter and is not generally recognized as particularly suitable for sporting purposes. Therefore, it is classified as a destructive device for purposes of the National Firearms Act, 26 U.S.C. Chapter 53.

ATF Rul. 94-1

The Bureau of Alcohol, Tobacco and Firearms (ATF) has examined a firearm identified as the USAS-12 shotgun to determine whether it is a destructive device as that term is used in the National Firearms Act (NFA), 26 U.S.C. Chapter 53.

The USAS-12 is a 12-gauge, gas-operated, autoloading semiautomatic shotgun which is chambered for 12-gauge 2 3/4-inch ammunition. It has an 18 1/4-inch barrel, is approximately 38 inches long, and weighs 12.4 pounds unloaded and approximately 15 pounds with a loaded magazine, depending on the capacity of the magazine. The USAS-12 is equipped with a 12-round detachable box magazine, but a 28-round detachable drum magazine is also available. The shotgun is approximately 11 inches deep with a box magazine. There is an integral carrying handle on top of the receiver, which houses a rifle-type aperture rear and adjustable post-type front sight. The USAS-12 has a separate combat-style pistol grip located on the bottom of the receiver, forward of the buttstock. An optional telescopic sight may be attached to the carrying handle. The barrel is located below the operating mechanism in such fashion that the barrel is in a straight line with the center of the buttstock.

Section 5845(f), Title 26, U.S.C., classifies certain weapons as "destructive devices" which are subject to the registration and tax provisions of the NFA. Section 5845(f)(2) provides as follows:

(f) Destructive device. – The term "destructive device" means * * *

(2) any type of weapon by whatever name known which will, or which may be readily converted to, expel a projectile by the action of an explosive or other propellant, the barrel or barrels of which have a bore of more than one-half inch in diameter, except a shotgun or shotgun shell which the Secretary or his delegate finds is generally recognized as particularly suitable for sporting purposes;

A "sporting purposes" test which is almost identical to that in section 5845(f)(2) appears in 18 U.S.C. § 925(d)(3). This provision of the Gun Control Act of 1968 (GCA) provides that the Secretary shall authorize a firearm to be imported into the United States if the firearm is "generally recognized as particularly suitable for or readily adaptable to sporting purposes." With the exception of the "readily adaptable" language, this provision is identical to the sporting shotgun exception to the destructive device definition. The definition of "destructive device" in the GCA (18 U.S.C. § 921(a)(4)) is identical to that in the NFA.

In determining whether shotguns with a bore of more than one-half inch in diameter are "generally recognized as particularly suitable for sporting purposes" and thus are not destructive devices under the NFA, we believe it is appropriate to use the same criteria used for evaluating shotguns under the "sporting purposes" test of section 925(d)(3). Congress used virtually identical language in describing the weapons subject to the two statutory schemes, and the language was added to the GCA and NFA at the same time.

In connection with the determination of importability, ATF determined that the USAS-12 shotgun was not eligible for importation under the sporting purposes test in section 925(d)(3). In reaching this determination, ATF evaluated the weight, size, bulk, designed magazine capacity, configuration, and other characteristics of the USAS-12. It was determined that the weight of the USAS-12, 12.4 pounds, made it much heavier than traditional 12-gauge sporting shotguns, which made it awkward to carry for extended periods, as in hunting, and cumbersome to fire at multiple small moving targets, as in skeet and trap shooting. The width of the USAS-12 with drum magazine, approximately 6 inches, and the depth with box magazine, in excess of 11 inches, far exceeded that of traditional sporting shotguns, which do not exceed 3 inches in width or 4 inches in depth. The large size and bulk of the USAS-12 made it extremely difficult to maneuver quickly enough to engage moving targets as is necessary in hunting, skeet, and trap shooting. The detachable box magazine with 12-cartridge capacity and the detachable drum magazine with 28-cartridge capacity were of a larger capacity than traditional repeating sporting shotguns, which generally contain tubular magazines with a capacity of 3-5 cartridges. Additionally, detachable magazines permit more rapid reloading than do tubular magazines. Finally, the combat-style pistol grip, the barrel-to-buttstock configuration, the bayonet lug, and the overall appearance and general shape of the weapon were radically different from traditional sporting shotguns and strikingly similar to shotguns designed specifically for or modified for combat and law enforcement use.

Section 7805(b), Title 26, U.S.C., provides that the Secretary may prescribe the extent, if any, to which any ruling relating to the internal revenue laws shall be applied without retroactive effect. Accordingly, all rulings issued under the Internal Revenue Code are applied retroactively unless they specifically provide otherwise. Pursuant to section 7805(b), the Director, as the delegate of the Secretary, may prescribe the extent to which any ruling will apply without retroactive effect.

Held*:* The USAS-12 is a shotgun with a bore of more than one-half inch in diameter which is not particularly suitable for sporting purposes. The weight, size, bulk, designed magazine capacity, configuration, and other factors indicate that the USAS-12 is a semiautomatic version of a military-type assault shotgun. Accordingly, the USAS-12 is a destructive device as that term is used in 26 U.S.C. § 5845(f)(2). Pursuant to section 7805(b), this ruling is applied prospectively effective March 1, 1994, with respect to the making, transfer, and special (occupational) taxes imposed by the NFA. All other provisions of the NFA apply retroactively effective March 1, 1994.

[ATFB 1993-94-1 21]

26 U.S.C. § 5845(f)(2): DESTRUCTIVE DEVICE

(Nonsporting shotgun having a bore of more than one-half inch in diameter)

The Striker-12/ Streetsweeper shotgun has a bore of more than one-half inch in diameter and is not generally recognized as particularly suitable for sporting purposes. Therefore, it is classified as a destructive for purposes of the National Firearms Act, 26 U.S.C. Chapter 53.

ATF Rul. 94-2

The Bureau of Alcohol, Tobacco and Firearms (ATF) has examined a firearm identified as the Striker-12/Streetsweeper shotgun to determine whether it is a destructive device as that term is used in the National Firearms Act (NFA), 26 U.S.C. Chapter 53.

The Striker-12 and Streetsweeper shotguns are virtually identical 12-gauge shotguns with a spring-driven revolving magazine. The magazine has a 12-round capacity. The shotgun has a fixed stock or folding shoulder stock and may be fired with the folding stock collapsed. The shotgun with an 18-inch barrel is 37 inches in length with the stock extended, and 26.5 inches in length with the stock folded. The shotgun is 5.7 inches in width and weighs 9.24 pounds unloaded. The Striker/Streetsweeper has two pistol grips, one in the center of the firearm below the buttstock, and one on the forearm. The Striker/Streetsweeper was designed and developed in South Africa as a military, security, and anti-terrorist weapon. Various types of 12-gauge cartridges can be fired from the shotgun, and a rapid indexing procedure allows various types of ammunition to be loaded into the cylinder and selected for firing. All 12 rounds can be fired from the shotgun in 3 seconds or less.

Section 5845(f), Title 26, U.S.C., classifies certain weapons as "destructive devices" which are subject to the registration and tax provisions of the NFA. Section 5845(f)(2) provides as follows:

> **(f) Destructive device** – The term "destructive device" means * * *
>
> **(2)** any type of weapon by whatever name known which will, or which may be readily converted to, expel a projectile by the action of an explosive or other propellant, the barrel or barrels of which have a bore of more than one-half inch in diameter, except a shotgun or shotgun shell which the Secretary or his delegate finds is generally recognized as particularly suitable for sporting purposes; ..."

A "sporting purposes" test which is almost identical to that in section 5845(f)(2) appears in 18 U.S.C. § 925(d)(3). This provision of the Gun Control Act of 1968 (GCA) provides that the Secretary shall authorize a firearm to be imported into the United States if the firearm is "generally recognized as particularly suitable for or readily adaptable to sporting purposes." With the exception of the readily adaptable language, this provision is identical to the sporting shotgun exception to the destructive devices definition. The definition of "destructive device" in the GCA (18

U.S.C. § 921(a)(4)) is identical to that in the NFA.

In determining whether shotguns with a bore of more than one-half inch in diameter are "generally recognized as particularly suitable for sporting purposes" and thus are not destructive devices under the NFA, we believe it is appropriate to use the same criteria used for evaluating shotguns under the "sporting purposes" test of section 925(d)(3). Congress used virtually identical language in describing the weapons subject to the two statutory schemes, and the language was added to the GCA and NFA at the same time.

In 1984, ATF ruled that the Striker-12 was not eligible for importation under section 925(d)(3) since it is not particularly suitable for sporting purposes. In making this determination, the 1984 letter-ruling notes that the Striker was being used in a number of "combat" shooting events. In a letter dated June 30, 1986, ATF again denied importation to the Striker-12, on the basis that it did not meet the "sporting purposes" test of section 925(d)(3). This letter states that, "We believe the weapon to have been specifically designed for military and law enforcement uses."

In evaluating the physical characteristics of the Striker 12/Streetsweeper, ATF concludes that the weight, bulk, designed magazine capacity, configuration, and other features indicate that it was designed primarily for military and law enforcement use and is not particularly suitable for sporting purposes.

The weight of the Striker-12/Streetsweeper, 9.24 pounds unloaded, is on the high end for traditional 12-gauge sporting shotguns, which generally weigh between 7 and 10 pounds. Thus, the weight of the Striker-12/Streetsweeper makes it awkward to carry for extended periods, as in hunting, and cumbersome to fire at multiple small moving targets, as in skeet and trap shooting. The width of the Striker-12/Streetsweeper, 5.7 inches, far exceeds that of traditional sporting shotguns, which do not exceed three inches in width or four inches in depth. The large size and bulk of the Striker-12/Streetsweeper make it extremely difficult to maneuver quickly enough to engage moving targets as is necessary in hunting, skeet, and trap shooting. The spring driven revolving magazine with 12-cartridge capacity is a much larger capacity than traditional repeating sporting shotguns, which generally contain tubular magazines with a capacity of 3-5 cartridges. The folding shoulder stock and the two pistol grips are not typical of sporting-type shotguns. Finally, the overall appearance and general shape of the weapon are radically different from traditional sporting shotguns and strikingly similar to shotguns designed specifically for or modified for combat and law enforcement use.

Section 7805(b), Title 26, U.S.C., provides that the Secretary may prescribe the extent, if any, to which any ruling relating to the internal revenue laws shall be applied without retroactive effect. Accordingly, all rulings issued under the Internal Revenue Code are applied retroactively unless they specifically provide otherwise. Pursuant to section 7805(b), the Director, as the delegate of the Secretary, may prescribe the extent to which any ruling will apply without retroactive effect.

Held: The Striker-12/Streetsweeper is a shotgun with a bore of more than one-half inch in diameter which is not particularly suitable for sporting purposes. The weight, size, bulk, designed magazine capacity, configuration, and other factors indicate that the Striker-12/Streetsweeper is a military-type shotgun, as opposed to a shotgun particularly suitable for sporting purposes. Accordingly, the Striker-12/Streetsweeper is a destructive device as that term is used in 26 U.S.C. § 5845(f)(2). Pursuant to section 7805(b), this ruling is applied prospectively effective March 1, 1994, with respect to the making, transfer, and special (occupational) taxes imposed by the NFA. All other provisions of the NFA apply retroactively effective March 1, 1994.

[ATFB 1993-1994-1 23]

18 U.S.C. § 921(a)(4): DESTRUCTIVE DEVICE

26 U.S.C. § 5845(f)(2): DESTRUCTIVE DEVICE

(Firearm having a bore of more than one-half inch in diameter)

37/38 mm gas/flare guns possessed with cartridges containing wood pellets, rubber pellets or balls, or bean bags are classified as destructive devices for purposes of the Gun Control Act, 18 U.S.C. Chapter 44, and the National Firearms Act, 26 U.S.C. Chapter 53.

ATF Rul. 95-3

The Bureau of Alcohol, Tobacco and Firearms (ATF) has examined various 37/38 mm gas/flare guns in combination with certain types of ammunition to determine whether these are destructive devices as defined in the Gun Control Act (GCA), 18 U.S.C. Chapter 44, and the National Firearms Act (NFA), 26 U.S.C. Chapter 53.

Section 5845(f), Title 26, United States Code, classifies certain weapons as "destructive devices" which are subject to the registration and tax provisions of the National Firearms Act (NFA). Section 5845(f)(2) provides as follows:

(f) Destructive device. — The term "destructive device" means * * *

> **(2)** any type of weapon by whatever name known which will, or which may be readily converted to, expel a projectile by the action of an explosive or other propellant, the barrel or barrels of which have a bore of more than one-half inch in diameter, except a shotgun or shotgun shell which the Secretary or his delegate finds is generally recognized as particularly suitable for sporting purposes. "

Section 5845(f)(3) excludes from the term "destructive device" any device which is neither designed or redesigned for use as a weapon and any device, although originally designed for use as a weapon, which is redesigned for use as a signaling, pyrotechnic, line throwing, safety, or similar device.

The definition of "destructive device" in the GCA (18 U.S.C. § 921(a)(4)) is identical to that in the NFA.

ATF has previously held that devices designed for expelling tear gas or pyrotechnic signals are not weapons and are exempt from the destructive device definition. However, ammunition designed to be used against individuals is available for these 37/38 mm devices. This "anti-personnel" ammunition consists of cartridges containing wood pellets, rubber pellets or balls, and bean bags.

When a gas/flare gun is possessed with "anti-personnel" type ammunition, it clearly becomes an instrument of offensive or defensive combat and is capable of use as a weapon. Since these gas/flare guns have a bore diameter of greater than one-half inch, fire a projectile by the means of an explosive, and, when possessed with "anti-personnel" ammunition, are capable of use as weapons, the combination of the gas/flare gun and "anti-personnel" ammunition is a destructive device as defined in the GCA and NFA. As a result, registration as a destructive device is required. Any person possessing a gas/flare gun with which "anti-personnel" ammunition will be used must register the making of a destructive device prior to the acquisition of any "anti-personnel" ammunition. In addition, the gas/flare guns are classified as firearms as defined by the GCA when possessed with "anti-personnel" type ammunition.

Each gas/flare gun possessed with anti-personnel ammunition will be required to be identified as required by law and regulations (27 CFR §§ 178.92 and 179.102), including a serial number. Any person manufacturing the gas/flare gun and the "anti-personnel" ammunition must, if selling them in combination, have the appropriate Federal firearms license as a manufacturer of destructive devices and must have paid the special (occupational) tax as a manufacturer of National Firearms Act firearms. Any person importing the gas/flare gun and the "anti-personnel" ammunition must, if importing them in combination, have the appropriate Federal firearms license as an importer of destructive devices and must have paid the special (occupational) tax as an importer of National Firearms Act firearms.

Further, the "anti-personnel" ammunition to be used in the gas/flare launchers is ammunition for destructive devices for purposes of the GCA. Any person manufacturing the "anti-personnel" ammunition must have the appropriate Federal firearms license as a manufacturer of ammunition for destructive devices. Any person importing the "anti-personnel" ammunition must have the appropriate Federal firearms license as an importer of ammunition for destructive devices.

Held: 37/38 mm gas/flare guns possessed with "anti-personnel" ammunition, consisting of cartridges containing wood pellets, rubber pellets or balls, or bean bags, are destructive devices as that term is used in 18 U.S.C. § 921(a)(4) and 26 U.S.C. 5845(f)(2).

[ATFB 95-3 28]

18 U.S.C. § 923 (a): ENGAGING IN THE BUSINESS OF DEALING IN FIREARMS (Auctioneers)

Auctioneers who regularly conduct consignment-type auctions of firearms, for example, held every 1-2 months, on behalf of firearms owners where the auctioneer takes possession of the firearms pursuant to a consignment contract with the owner of the firearms giving the auctioneer authority to sell the firearms and providing for a commission to be paid by the owner upon sale of the firearms are required to obtain a license as a dealer in firearms.

ATF Rul. 96-2

An association of auctioneers has asked the Bureau of Alcohol, Tobacco and Firearms (ATF) for a ruling concerning the auctions conducted by their members and whether the sale of firearms at such auctions requires a Federal firearms license as a dealer in firearms.

The auctioneers' association stated that their members generally conduct two types of auctions: estate-type auctions and consignment auctions. In estate-type auctions, articles to be auctioned, including firearms, are sold by the executor of the estate of an individual. In these cases the firearms belong to and are possessed by the executor. The auctioneer acts as an agent of the executor and assists the executor in finding buyers for the firearms. The firearms are possessed by the estate and their sale to third parties is controlled by the estate. The auctioneer is paid a commission on the sale of each firearm by the estate at the conclusion of the auction.

The association states that, in consignment-type auctions, an auctioneer may take possession of firearms in advance of the auction. The firearms are inventoried, evaluated, and tagged for identification. The firearms belong to individuals or businesses who have entered into a consignment agreement with the auctioneer giving the auctioneer authority to sell the firearms. The agreement states that the auctioneer has the exclusive right to sell the items listed on the contract at a location, time, and date to be selected by the auctioneer. The consignment-type auctions generally involve accepting firearms for auction from more than one owner. Also, these auctions are held on a regular basis, for example, every 1-2 months.

Section 923(a), Title 18, U.S.C., provides that no person shall engage in the business of dealing in firearms until he has filed an application and received a license to do so. Section 922(a)(1), Title 18, U.S.C., provides that it is unlawful for any person, other than a licensee, to engage in the business of dealing in firearms. Licensees generally may not conduct business away from their licensed premises.

The term "dealer" is defined at 18 U.S.C. § 921(a)(11)(A) to include any person engaged in the business of selling firearms at wholesale or retail. The term "engaged in the business" as applied to a dealer in firearms means a person who devotes time, attention, and labor to dealing in firearms as a regular course of trade or business with the principal objective of livelihood and profit through the repetitive purchase and resale of firearms. A dealer can be "engaged in the business" without taking title to the firearms that are sold. However, the term does not include a person who makes occasional sales, exchanges, or purchases of firearms for the enhancement of a personal collection or for a hobby, or who sells all or part of his personal collection of firearms. 18 U.S.C. § 921(a)(21)(C).

In the case of estate-type auctions, the auctioneer acts as an agent of the executor and assists the executor in finding buyers for the estate's firearms. The firearms are possessed by the estate, and the sales of firearms are made by the estate. In these cases, the auctioneer does not meet the definition of "engaging in the business" as a dealer in firearms and would not require a license. An auctioneer engaged in estate-type auctions, whether licensed or not, may perform this function, including delivery of the firearms, away from the business premises.

In the case of consignment-type auctions held on a regular basis, for example, every 1-2 months, where persons consign their firearms to the auctioneer for sale pursuant to an agreement as described above, the auctioneer would be "engaging in the business" and would require a license. The auctioneer would be disposing of firearms as a regular course of trade or business within the definition of a "dealer" under § 921(a)(11)(A) and must comply with the licensing requirements of the law.

As previously stated, licensed auctioneers generally must engage in the business from their licensed premises. However, an auctioneer may conduct an auction at a location other than his licensed premises by displaying the firearms at the auction site, agreeing to the terms of sale of the firearms, then returning the firearms to the licensed premises for delivery to the purchaser.

Held: Persons who conduct estate-type auctions at which the auctioneer assists the estate in selling the estate's firearms, and the firearms are possessed and transferred by the estate, do not require a Federal firearms license.

Held further: Persons who regularly conduct consignment-type auctions, for example, held every 1-2 months, where the auctioneer takes possession of the firearms pursuant to a consignment contract giving the auctioneer the exclusive right and authority to sell the firearms at a location, time and date to be selected by the auctioneer and providing for a commission to be paid upon sale are required to obtain a license as a dealer in firearms pursuant to 18 U.S.C. § 923(a).

[ATFB 96-2 101]

18 U.S.C. 921(a)(4): DESTRUCTIVE DEVICE

26 U.S.C. 5845(f)(2): DESTRUCTIVE DEVICE

(Nonsporting shotgun having a bore of more than one-half inch in diameter)

The registration period for the USAS-12, Striker-12, and Streetsweeper shotguns closed on May 1, 2001, pursuant to ATF Rul. 2001-1.

ATF Rul. 2001-1

Pursuant to ATF Rulings 94-1 (ATF Q.B. 1994-1, 22) and 94-2 (ATF Q.B. 1994-1, 24), the Bureau of Alcohol, Tobacco and Firearms (ATF) classified the USAS-12, Striker 12, and Streetsweeper shotguns as destructive devices under the National Firearms Act (NFA), 26 U.S.C. Chapter 53. The NFA requires that certain "firearms" be registered and imposes taxes on their making and transfer. The term "firearm" is defined in section 5845 to include "destructive devices." The term "destructive device" is defined in section 5845(f)(2) as follows:

> [T]he term **'destructive device'** means . . . (2) any type of weapon by whatever name known which will, or which may be readily converted to, expel a projectile by the action of an explosive or other propellant, the barrel or barrels of which have a bore of more than one-half inch in diameter, except a shotgun or shotgun shell which the Secretary finds is generally recognized as particularly suitable for sporting purposes;
>
> The USAS-12, Striker 12, and Streetsweeper shotguns were classified as destructive devices pursuant to section 5845(f) because they are shotguns with a bore of more than one-half inch in diameter which are not generally recognized as particularly suitable for sporting purposes.

Pursuant to 26 U.S.C. 7805(b), ATF. Ruls. 94-1 and 94-2 were issued prospectively with respect to the making, transfer, and special (occupational) taxes imposed by the NFA. Thus, although the classification of the three shotguns as NFA weapons was retroactive, the prospective application of the tax provisions allowed registration without payment of tax. ATF has contacted all purchasers of record of the shotguns to advise them of the classification of the weapons as destructive devices and that the weapons must be registered. ATF has registered approximately 8,200 of these weapons to date.

Held, the registration period for the USAS-12, Striker-12, and Streetsweeper shotguns will close on May 1, 2001. No further registrations will be accepted after that date. Persons in possession of unregistered NFA firearms are subject to all applicable penalties under 26 U.S.C. Chapter 53.

Date signed: February 2, 2001.

27 CFR 47.45: IMPORTATION OF SURPLUS MILITARY CURIO OR RELIC FIREARMS

Importers of surplus military curio or relic firearms must submit originals of all appropriate statements supporting the Form 6 application.

ATF Rul. 2001-3

The Bureau of Alcohol, Tobacco and Firearms (ATF) has received inquiries from firearms importers concerning supporting documents required to be submitted with applications to import surplus military curio or relic firearms. Importers often rely upon documents obtained by the foreign shipper or seller and have asked whether copies of the documents, rather than originals, may be submitted with the application. For the reasons stated below, ATF has found that importers of surplus military curio or relic firearms must submit *originals* of all appropriate statements supporting the application.

ATF has the authority pursuant to section 38 of the Arms Export Control Act (AECA), 22 U.S.C. 2778, and implementing regulations, to approve import permits, as well as to deny, revoke, suspend, or revise import permits without prior notice whenever the proposed importation is found to be inconsistent with the purpose or in violation of section 38 or its implementing regulations. *See* 27 CFR 47.41, 47.44(a).

Under the AECA and implementing regulations, it is the policy of the United States to deny licenses and other approvals with respect to defense articles and defense services originating in certain countries or areas as determined by the Department of State. This policy applies to countries or areas with respect to which the United States maintains an arms embargo. *See* 27 CFR 47.52(a). Nonetheless, applications for permits to import articles that were manufactured in, or have been in, a proscribed country or proscribed area may be approved where the articles:

> Are covered by Category I(a) of the Import List (other than those subject to the provisions of 27 CFR Part 179);
>
> Are importable as curios or relics under the provisions of 27 CFR 178.118;

Were manufactured in a proscribed country or area prior to the date the country or area became proscribed, or, were manufactured in a non-proscribed country or area; *and*,

The articles have been stored for the five year period immediately prior to importation in a non-proscribed country or area.

22 U.S.C. 2778(b)(1)(B);
27 CFR 47.52(e).

Any persons seeking to import articles under these provisions must explain and certify how the firearms meet the applicable criteria. The certification statement must be executed under the penalties of perjury. In addition, the statement must be accompanied by documentary information both on the country or area of original manufacture, and on the country or area of storage for the five year period immediately prior to importation. Such information may, for example, include a verifiable statement in the English language of a government official or any other person having knowledge of the date and place of manufacture and/or the place of storage. ATF reserves the right to determine whether documentation provided is acceptable, and to require the submission of additional documentation as may be necessary.

27 CFR 47.52(f).

To ensure the lawfulness of the importation of surplus military defense articles, ATF must be able to rely upon the validity of import permit applications and all supporting documentation. This documentation includes but is not limited to appropriate and verifiable documentation of the above-referenced:

(1) Importer certification statement;

(2) Statement on the country or area of original manufacture; and,

(3) Statement on the country or area of storage for the five year period immediately prior to importation.

In the past, import permit applicants have submitted photocopies of the required statements. ATF has become aware that, in some cases, the photocopies are fraudulent. To assist ATF in confirming the validity and authenticity of these statements, importers must submit *original* statements in support of all import permit applications. Consistent with the purpose of section 38 of the AECA and implementing regulations, ATF will deny all permit applications that fail to include the above-described original statements.

Held, all importers submitting permit applications to import surplus military defense articles, importable as curio or relics, must provide with the permit applications originals of all necessary supporting statements. ATF will deny permit applications when applicants fail to provide appropriate original statements, as ATF finds that copies are not acceptable documentation within the meaning of 27 CFR 47.52(f).

Date signed: October 31, 2001

18 U.S.C. 922(t)(1)(C): IDENTIFICATION OF TRANSFEREE

27 CFR 178.124: FIREARMS TRANSACTION RECORD

Licensees may accept a combination of valid government-issued documents to satisfy the identification document requirements of the Brady Act. The required valid government-issued photo identification document bearing the name, photograph, and date of birth of the transferee may be supplemented by another valid, government-issued document showing the transferee's residence address. A member of the Armed Forces on active duty is a resident of the State in which his or her permanent duty station is located, and may satisfy the identification document requirement by presenting his or her military identification card along with official orders showing that his or her permanent duty station is within the State where the licensed premises are located.

ATF Rul. 2001-5

The Bureau of Alcohol, Tobacco and Firearms (ATF) has received numerous inquiries from Federal firearms licensees (FFLs) regarding the acceptance of identification documents that do not show the purchaser's current residence address. FFLs have asked whether they may accept other documents, such as tax bills or vehicle registration documents, to establish the current residence address of the purchaser.

It has been ATF's longstanding position that licensees may accept a combination of documents to establish the identity of a firearm purchaser. ATF Rul. 79-7, ATFQB 79-1, 26, interpreted a licensee's obligation to obtain satisfactory identification from a purchaser in the manner customarily used in commercial transactions, pursuant to the existing regulations under the Gun Control Act of 1968 (GCA). The ruling held that satisfactory identification of a firearms purchaser must include the purchaser's name, age or date of birth, place of residence, and signature. The ruling also held that while a particular document may not be sufficient to meet the statutory requirement for identifying the purchaser, any combination of documents that together disclosed the required information would be acceptable.

ATF Rul. 79-7 has been superseded by an amendment to the GCA. The Brady Handgun Violence Prevention Act (Brady Act), which took effect in 1994, mandated the use of photo identification documents for transfers subject to the Act. Under the permanent provisions of the Brady Act, which went into effect on November 30, 1998, a licensed importer, manufacturer, or dealer is generally required to initiate a background check through the National Instant Criminal Background Check System (NICS) prior to transferring a firearm to an unlicensed individual.

The Brady Act requires a licensee to identify the nonlicensed transferee by examining a valid government-issued identification document that contains the photograph of the holder. *See* 18 U.S.C. 922(t)(1)(C). This requirement applies to all over-the-counter transfers, even where the transferee holds a permit that qualifies as an exception to the requirement for a NICS check at the time of transfer. 27 CFR 178.124(c)(3)(i).

The Brady Act incorporates the definition of an "identification document" provided by 18 U.S.C. 1028(d)(2), which is set forth in relevant part as follows:

[A] document made or issued by or under the authority of the United States Government, a State, political subdivision of a State, a foreign government, political subdivision of a foreign government, an international governmental or an international quasi-governmental organization which, when completed with information concerning a particular individual, is of a type intended or commonly accepted for the purpose of identification of individuals.

ATF regulations further require that the identification document must contain the name, residence address, date of birth, and photograph of the holder. 27 CFR 178.11.

ATF has received questions from licensees regarding purchasers who present a State-issued driver's license or other identification document that shows either an out-of-date residence address or a mailing address (such as a post office box) in lieu of a residence address. ATF has advised that these identification documents, standing alone, would not satisfy the requirements of the regulations implementing the Brady Act.

It is ATF's position that a combination of documents may be used to satisfy the Brady Act's requirement for an identification document. The prospective transferee must present at least one valid document that meets the statutory definition of an identification document; i.e., it must bear the transferee's name and photograph, it must have been issued by a governmental entity, and it must be of a type intended or commonly accepted for identification purposes. ATF recognizes, however, that some valid government-issued identification documents do not include the bearer's current residence address. Such an identification document may be supplemented with another valid government-issued document that contains the necessary information.

Thus, for example, a licensee may accept a valid driver's license that accurately reflects the purchaser's name, date of birth, and photograph, along with a vehicle registration issued by the State indicating the transferee's current address. Licensees should note that if the law of the State that issued the driver's license provides that the driver's license is invalid due to any reason (*i.e.*, the license is expired or is no longer valid due to an unreported change of address), then the driver's license may not be used for identification purposes under the Brady Act. If a licensee has reasonable cause to question the validity of an identification document, he or she should not proceed with the transfer until those questions can be resolved.

The licensee must record on the Form 4473 the type of identification document(s) presented by the transferee, including any document number. Examples of documents that may be accepted to supplement information on a driver's license or other identification document include a vehicle registration, a recreation identification card, a fishing or hunting license, a voter identification card, or a tax bill. However, the document in question must be valid and must have been issued by a government agency.

ATF has also received questions from licensees as to how to comply with the identification document requirement in the case of purchasers who are in the military. Some active duty military personnel may not have driver's licenses from the State in which they are stationed. The only identification document carried by some active duty military personnel is a military identification card that bears the holder's name, date of birth, and photograph, but does not reflect the holder's residence address.

Section 921(b) of the GCA provides that a member of the Armed Forces on active duty is a resident of the State in which his permanent duty station is located. The purchaser's official orders showing that his or her permanent duty station is within the State where the licensed premises are located suffice to establish the purchaser's residence for GCA purposes. In combination with a military identification card, such orders will satisfy the Brady Act's requirement for an identification document, even though the purchaser may actually reside in a home that is not located on the military base.

Licensees should note that for purposes of the GCA, military personnel may in some cases have two States of residence. For example, a member of the Armed Forces whose permanent duty station is Fort Benning, Georgia, may actually reside in a home in Alabama. For GCA purposes, that individual is a resident of Georgia when he or she is in Georgia and a resident of Alabama when he or she is in Alabama. If such an individual wishes to purchase a firearm in Alabama, he or she must of course comply with the identification document requirement in the same way as any other Alabama resident.

Held: the Brady Act and the implementing ATF regulations require licensed importers, manufacturers, and dealers to examine a valid government-issued identification document that bears the name, residence address, date of birth, and photograph of the holder prior to making an over-the-counter transfer to any unlicensed transferee. Licensees may accept a combination of valid, government-issued documents to satisfy the identification document requirements of the Brady Act. A government-issued photo identification document bearing the name, photograph, and date of birth of the transferee may be supplemented by another valid, government-issued document showing the transferee's current residence address.

Held further, a purchaser who is a member of the Armed Forces on active duty is a resident of the State in which his or her permanent duty station is located, and may satisfy the identification document requirement by presenting his or her military identification card along with official orders showing that his or her permanent duty station is located within the State where the licensed premises are located.

ATF Ruling 79-7, ATFQB 79-1, 26, is hereby superseded.

Date signed: December 31, 2001

Editor's Note:

"Identification document" currently is defined in 18 U.S.C. 1028(d)(3).

27 CFR 179.105: TRANSFER AND POSSESSION OF MACHINEGUNS

Applications to transfer two (2) machineguns of a particular model to a Federal firearms licensee as sales samples will be approved if documentation shows necessity for demonstration to government agencies.

ATF Rul. 2002-5

The Bureau of Alcohol, Tobacco and Firearms (ATF) has received inquiries from dealers in machineguns concerning the justification necessary to obtain more than one machinegun of a particular model as dealer sales samples. Specifically, the inquiries are from machinegun dealers who demonstrate machineguns to large police departments and Special Weapons and Tactics (SWAT) teams, which requires the firing of thousands of rounds of ammunition during a single demonstration. Section 922(o) of Title 18, United States Code, makes it unlawful for any person to transfer or possess a machinegun, except a transfer to or by or under the authority of the United States or any department or agency thereof or a State or a department, agency, or political subdivision of; or any lawful transfer or lawful possession of a machinegun lawfully possessed before May 19, 1986.

The regulations in 27 CFR 179.105(d) provide that applications to register and transfer a machinegun manufactured or imported on or after May 19, 1986, to dealers registered under the National Firearms Act (NFA), 26 U.S.C. Chapter 53, will be approved if three conditions are met. The conditions required to be established include **(1)** a showing of the expected government customers who would require a demonstration of the weapon; **(2)** information as to the availability of the machinegun to fill subsequent orders; and **(3)** letters from government entities expressing a need for a particular model or interest in seeing a demonstration of a particular weapon. The regulation further provides that applications to transfer more than one machinegun of a particular model must also establish the dealer's need for the quantity of samples sought to be transferred.

The dealer sales sample regulation in section 179.105(d) is a narrow exception to the general prohibition on possession of post-1986 machineguns imposed by section 922(o). It requires that dealers submit letters of interest from law enforcement agencies to ensure that dealers possess post-1986 machineguns only for the purposes permitted by law, i.e., for sale or potential sale to government agencies.

Qualified dealers in machineguns often demonstrate weapons to all officers of the department, requiring the machinegun to fire thousands of rounds of ammunition during a single demonstration. In the case of new model machineguns, a department may wish to have thousands of rounds fired from the weapon before they are fully satisfied of its reliability. ATF is aware that after firing hundreds of rounds a machinegun often gets too hot to safely handle, resulting in the dealer's inability to demonstrate the weapon until it cools. In addition, it is not uncommon for machineguns to jam or misfeed ammunition after a large quantity of ammunition has been fired. Accordingly, dealers who demonstrate machineguns to departments with a large number of officers have asked that ATF approve the transfer of two (2) machineguns of each model as dealer sales samples.

The purpose of the dealer sales sample provision is to permit properly qualified dealers to demonstrate and sell machineguns to law enforcement agencies. Neither the law nor the implementing regulations were intended to impose unnecessary obstacles to police departments and other law enforcement agencies in obtaining the weapons they need to carry out their duties. Accordingly, if a dealer can provide documentation that the dealer needs to demonstrate a particular model of machinegun to an entire police department or SWAT team, ATF will approve the transfer of two (2) machineguns of that model to the dealer as sales samples.

This ruling should not be interpreted to imply that under no circumstances may a Federal firearms licensee (FFL) receive more than two (2) machineguns as sales samples. Consistent with past practice, an FFL who can show a bona fide reason as to why they need more than two (2) machineguns, may be able to receive more than two (2) if the request is accompanied by specific documentation.

Held: applications to transfer two (2) machineguns of a particular model to a Federal firearms licensee as sales samples will be approved if the dealer provides documentation that the dealer needs to demonstrate the machinegun to all the officers of a police department or the department's SWAT team or special operations team. An FFL who offers other bona fide reasons for their need for two (2) or more machineguns may get more than two (2) with specific documentation.

Date signed: September 6, 2002

18 U.S.C. 923(i): LICENSING

26 U.S.C. 5842: IDENTIFICATION OF FIREARMS

27 CFR 179.102: IDENTIFICATION OF FIREARMS

27 CFR 178.92: IDENTIFICATION OF FIREARMS, ARMOR PIERCING AMMUNITION, AND LARGE CAPACITY AMMUNITION FEEDING DEVICES

In accordance with 27 CFR 178.92 and 27 CFR 179.102, identification of firearms, armor piercing ammunition, and large capacity ammunition feeding devices, the terms "conspicuously" and "legibly" as used therein mean, respectively, that the markings are wholly unobstructed from plain view and that the markings contain exclusively Roman letters and Arabic numerals.

ATF Rul. 2002-6

The Bureau of Alcohol, Tobacco and Firearms (ATF) has been asked by State and local law enforcement officials to trace firearms that are marked, in part, with non-Roman letters, and/or non-Arabic numbers. Specifically, ATF received a request to trace a Makarov type pistol made in Bulgaria. The original manufacturer marking was ИМ 18 355. Because the importer did not stamp the firearm with a unique identifier that could be recognized by either ATF or a State or local law enforcement official, and because the marking contained a Cyrillic character, the firearm was not properly recorded, resulting in a failed trace of the weapon.

Because markings with non-Roman characters or non-Arabic numbers are not easily recorded or transmitted through ordinary means by importers, dealers or distributors, many firearm traces have proved unsuccessful. In some cases, an importer attempts to translate portions of the markings into Roman letters and Arabic numbers and re-marks the weapon with "translated" symbols. For example, an imported SKS rifle was marked with the serial number ДМ7639И. The importer translated the marking as LM7639i, but rather than restamp the entire number merely added the letters "L" and "i" below the original markings. This practice often results in failed traces because those required to record the markings (importers, dealers, or distributors) may record only the translated portions or both sets of markings. Moreover, law enforcement recovering a firearm with such markings may submit a trace uest lacking some portion of the rkings, further impeding efforts to ccessfully trace the firearm.

n addition, ATF has found that ne traces have failed because the uired markings on the firearms rel were wholly or partially ob- ucted from plain view by a flash opressor or bayonet mount, result- in the Federal Firearms Licensee ating an inaccurate record. ATF s been unable to trace hundreds of arms as a result of nonstandard or scured markings.

As a result of these practices, ne licensed importers may not be compliance with the marking re- rements set forth in 27 CFR 8.92 and 27 CFR 179.102 because y have marked using non-Roman ers (such as Greek or Russian ers, Δ or Д) or non-Arabic numbers g., XXV).

The above regulations require rkings that legibly identify each n or package and require that such rkings are conspicuous. ATF has nsistently taken the position that gibly" marked means using exclu- ely Roman letters (A, a, B, b, C, c, d so forth) and Arabic numerals (1, 3, 4, 5, 6, and so forth), and "con- cuous" means that all required rkings must be placed in such a nner as to be wholly unobstructed m plain view. These regulations ply to licensed manufacturers and ensed importers relative to firearms, nor piercing ammunition, and large pacity ammunition feeding devices, d to makers of National Firearms firearms.

Firearms, armor piercing ammuni- n, and large capacity ammunition eding devices which contain re- red markings or labels using non- man letters (such as Greek or ssian letters, Δ or Д) or non-Arabic nbers (e.g., XXV), must be com- tely remarked or relabeled with a w serial number or other required rkings that satisfy the legibility uirements described above. It is sufficient to simply add an addi- nal Roman letter or Arabic numeral a nonconforming marking; a new d unique marking using Roman ers and Arabic numerals is re- red. Where feasible, the new rkings should be placed directly ove the non-compliant markings.

Similarly, firearms and large capac- ammunition feeding devices which ntain required markings obstructed in whole or in part from plain view must be remarked with required markings that satisfy the conspicuousness requirements described above. For example, required markings may not be placed on a portion of the barrel where the markings would be wholly or partially obstructed from view by another part of the firearm, such as a flash suppressor or bayonet mount.

In certain unavoidable circumstances owing mainly to firearms of unusual design or other limiting factor(s) which would limit the ability of the manufacturer or importer to comply with the above legibility and conspicuousness requirements, alternate means of identification may be authorized as described in 27 CFR 178.92(a)(3)(i), (ii), or (iii) and 27 CFR 178.92(c)(3)(ii).

Held, a Makarov type pistol imported from Bulgaria utilizing Cyrillic letters or non-Arabic numbers is not marked in accordance with 27 CFR 178.92 and 27 CFR 179.102.

Held further, an imported firearm with any part of the required marking partially or wholly obstructed from plain view is not marked in accordance with section 27 CFR 178.92 and 27 CFR 179.102.

Date signed: November 5, 2002

26 USC 5844, 18 USC 922(o), 22 USC 2778

IMPORTATION OF BROWNING M1919 TYPE RECEIVERS FOR UNRESTRICTED COMMERCIAL SALE

An ATF-approved method of destruction for the Browning M1919 type machinegun will result in the severed portions of the receiver being importable for unrestricted commercial sale.

ATF Rul. 2003–1

The Bureau of Alcohol, Tobacco and Firearms (ATF) has received inquiries about modifications necessary to the receiver of a Browning M1919 type machinegun to make it importable under 26 U.S.C. 5844 and 18 U.S.C. 922(o) for unrestricted commercial sale.

The Browning M1919 is a machinegun as defined in 26 U.S.C. 5845(b). The receiver of a Browning M1919 is also a machinegun as defined. Various manufacturers made Browning M1919 style machineguns in caliber .30-06 and 7.62x51mm (.308). The M1919 is a recoil-operated, belt-fed machinegun designed to be fired from a mount.

Section 5844 of Title 26, United States Code, makes it unlawful to import any firearm into the United States, unless the firearm to be imported or brought in is: (1) being imported for use by the United States or any department, independent establishment, or agency thereof or any State or possession or any political subdivision thereof; or (2) the firearm is being imported for scientific or research purposes; or (3) it is being imported solely for testing or use as a model by a registered manufacturer or solely for use as a sample by a registered importer or dealer. Additionally, the Secretary may permit the conditional importation of a firearm for examination and testing in connection with classifying the firearm.

Section 922(o) of Title 18, United States Code, makes it unlawful for any person to transfer or possess a machinegun, except a transfer to or by the United States or any department or agency thereof or a State or a department, agency, or political subdivision thereof; or any lawful transfer or lawful possession of a machinegun

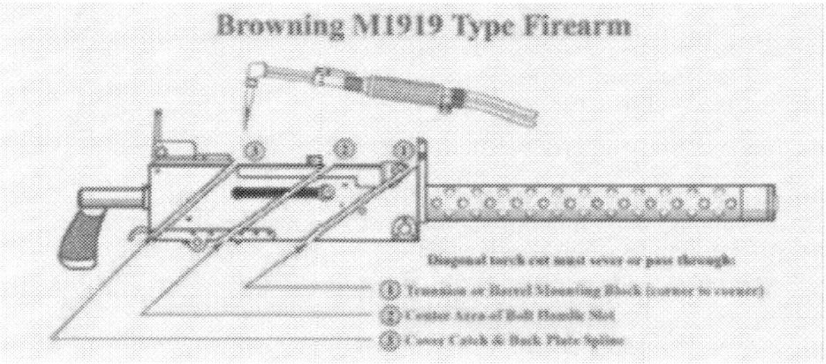

lawfully possessed before May 19, 1986.

A review of the statutes above indicates that machineguns and machinegun receivers cannot be lawfully imported for unrestricted commercial sale. Accordingly, machinegun receivers may be imported for commercial sale only if they are destroyed in a manner that will prevent their function and future use as a firearm. The resulting severed receiver portions would not be subject to the provisions of 26 U.S.C. 5844 or 18 U.S.C. 922(o); however, these articles would be subject to the provisions of the Arms Export Control Act, 18 U.S.C. 925, 22 U.S.C. 2778, and implementing regulation at 27 CFR Part 47. It is important to note that these machinegun receivers must be destroyed and cannot be imported whether they are serviceable or unserviceable.

An ATF-approved method of destruction for a Browning M1919 type machinegun receiver requires three diagonal torch cuts that sever or pass through the following areas: (1) the trunnion or barrel mounting block (corner to corner), (2) the center area of the bolt handle slot, and (3) the cover catch and back plate spline. All cutting must be done with a cutting torch having a tip of sufficient size to displace at least ¼ inch of material at each location. Each cut must completely sever the receiver in the designated areas and must be done with a diagonal torch cut. Using a bandsaw or a cut-off wheel to destroy the receiver does not ensure destruction of the weapon.

This method of destruction is illustrated in the diagram below.

Alternative methods of destruction may also be acceptable. These alternative methods must be equivalent in degree to the approved method of destruction. Receivers that are not sufficiently modified cannot be approved for importation. To ensure compliance with the law, it is recommended that the importer submit in writing the alternative method of destruction to the ATF Firearms Technology Branch (FTB) for review and approval prior to importation.

Held, an ATF-approved method of destruction for a Browning M1919 type machinegun receiver will result in the severed portions of the receiver being importable for unrestricted commercial sale. The severed articles would not be subject to the provisions of 26 U.S.C. 5844 or 18 U.S.C. 922(o), but would continue to be subject to the provisions of the Arms Export Control Act, 22 U.S.C. 2778. Alternative methods of destruction may also be acceptable. It is recommended that such methods be reviewed and approved by the ATF Firearms Technology Branch prior to the weapon's importation.

Date signed: January 24, 2003

26 USC 5844, 18 USC 922(o), 22 USC 2778

IMPORTATION OF FN FAL TYPE RECEIVERS FOR UNRESTRICTED COMMERCIAL SALE

An ATF-approved method of destruction for the FN FAL type machinegun will result in the severed portions of the receiver being importable for unrestricted commercial sale.

ATF Rul. 2003–2

The Bureau of Alcohol, Tobacco and Firearms (ATF) has received inquiries about modifications necessary to the receiver of an FN FAL type machinegun to make it importable under 26 U.S.C. 5844 and 18 U.S.C. 922(o) for unrestricted commercial sale.

The FN FAL is a machinegun as defined in 26 U.S.C. 5845(b). The receiver of an FAL is also a machinegun as defined. Various manufacturers made FAL style machineguns in caliber 7.62x51mm (.308). The FAL is a gas-operated, shoulder-fired, magazine-fed, selective-fire machinegun.

Section 5844 of Title 26, United States Code, makes it unlawful to import any firearm into the United States, unless the firearm to be imported or brought in is: (1) being imported for use by the United States or any department, independent establishment, or agency thereof or any State or possession or any political subdivision thereof; or (2) the firearm is being imported for scientific or research purposes; or (3) it is being imported solely for testing or use as a model by a registered manufacturer or solely for use as a sample by a registered importer or dealer. Additionally, the Secretary may permit the conditional importation of a firearm for examination and testing in connection with classifying the firearm.

Section 922(o) of Title 18, United States Code, makes it unlawful for any person to transfer or possess a machinegun, except a transfer to or by the United States or any department or agency thereof or a State or a department, agency, or political subdivision thereof; or any lawful transfer or lawful possession of a machinegun lawfully possessed before May 19, 1986.

A review of the statutes above indicates that machineguns and machinegun receivers cannot be lawfully imported for unrestricted commercial sale. Accordingly, machinegun receivers may be imported for commercial sale only if they are destroyed in a manner that will prevent their function and future use as a firearm. The resulting severed receiver portions would not be subject to the provisions of 26 U.S.C. 5844 or 18 U.S.C. 922(o); however, these articles would be subject to the provisions of the Arms Export Control Act, 18 U.S.C. 925, 22 U.S.C. 2778, and implementing regulation at 27 CFR Part 47. It is important to note that these machinegun receivers must be destroyed and cannot be imported whether they are serviceable or unserviceable.

An ATF-approved method of destruction for an FN FAL type machinegun receiver requires three

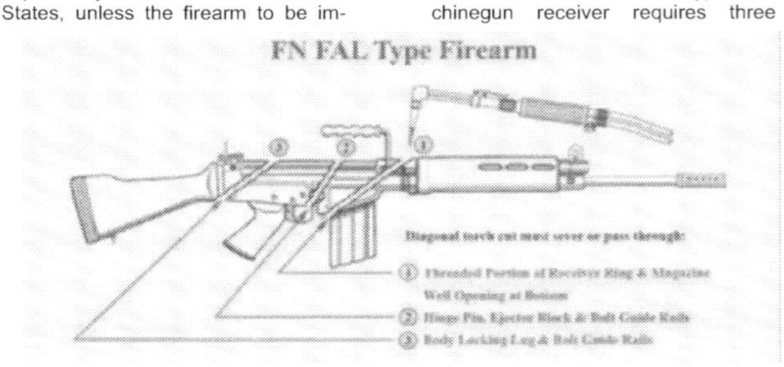

diagonal torch cuts that sever or pass through the following areas: (1) the threaded portion of the receiver ring and magazine well opening at bottom, (2) the hinge pin, ejector block and bolt guide rails, and (3) the body locking lug and bolt guide rails. All cutting must be done with a cutting torch having a tip of sufficient size to displace at least ¼ inch of material at each location. Each cut must completely sever the receiver in the designated areas and must be done with a diagonal torch cut. Using a bandsaw or a cut-off wheel to destroy the receiver does not ensure destruction of the weapon.

This method of destruction is illustrated in the diagram below.

Alternative methods of destruction may also be acceptable. These alternative methods must be equivalent in degree to the approved method of destruction. Receivers that are not sufficiently modified cannot be approved for importation. To ensure compliance with the law, it is recommended that the importer submit in writing the alternative method of destruction to the ATF Firearms Technology Branch (FTB) for review and approval prior to importation.

Held, an ATF-approved method of destruction for an FN FAL type machinegun receiver will result in the severed portions of the receiver being importable for unrestricted commercial sale. The severed articles would not be subject to the provisions of 26 U.S.C. 5844 or 18 U.S.C. 922(o), but would continue to be subject to the provisions of the Arms Export Control Act, 22 U.S.C. 2778. Alternative methods of destruction may also be acceptable. It is recommended that such methods be reviewed and approved by the ATF Firearms Technology Branch prior to the weapon's importation.

Date signed: January 24, 2003.

26 USC 5844, 18 USC 922(o), 22 USC 2778

IMPORTATION OF HECKLER & KOCH G3 TYPE RECEIVERS FOR UNRESTRICTED COMMERCIAL SALE

An ATF-approved method of destruction for the Heckler & Koch G3 type machinegun will result in the severed portions of the receiver being importable for unrestricted commercial sale.

ATF Rul. 2003–3

The Bureau of Alcohol, Tobacco and Firearms (ATF) has received inquiries about modifications necessary to the receiver of a Heckler and Koch G3 type machinegun to make it importable under 26 U.S.C. 5844 and 18 U.S.C. 922(o) for unrestricted commercial sale.

The G3 is a machinegun as defined in 26 U.S.C. 5845(b). The receiver of a G3 is also a machinegun as defined. Various manufacturers made G3 style machineguns in caliber 7.62x51mm (.308). The G3 is a delayed blowback, shoulder-fired, magazine-fed, selective-fire machinegun.

Section 5844 of Title 26, United States Code, makes it unlawful to import any firearm into the United States, unless the firearm to be imported or brought in is: (1) being imported for use by the United States or any department, independent establishment, or agency thereof or any State or possession or any political subdivision thereof; or (2) the firearm is being imported for scientific or research purposes; or (3) it is being imported solely for testing or use as a model by a registered manufacturer or solely for use as a sample by a registered importer or dealer. Additionally, the Secretary may permit the conditional importation of a firearm for examination and testing in connection with classifying the firearm.

Section 922(o) of Title 18, United States Code, makes it unlawful for any person to transfer or possess a machinegun, except a transfer to or by the United States or any department or agency thereof or a State or a department, agency, or political subdivision thereof; or any lawful transfer or lawful possession of a machinegun lawfully possessed before May 19, 1986.

A review of the statutes above indicates that machineguns and machinegun receivers cannot be lawfully imported for unrestricted commercial sale. Accordingly, machinegun receivers may be imported for commercial sale only if they are destroyed in a manner that will prevent their function and future use as a firearm. The resulting severed receiver portions would not be subject to the provisions of 26 U.S.C. 5844 or 18 U.S.C. 922(o); however, these articles would be subject to the provisions of the Arms Export Control Act, 18 U.S.C. 925, 22 U.S.C. 2778, and implementing regulation at 27 CFR Part 47. It is important to note that these machinegun receivers must be destroyed and cannot be imported whether they are serviceable or unserviceable.

An ATF-approved method of destruction for a Heckler and Koch G3 type machinegun receiver requires four diagonal torch cuts that sever or pass through the following areas: (1) the chamber area, (2) the grip assembly locking pin hole, (3) the ejection port, and (4) the buttstock locking pin hole. All cutting must be done with a cutting torch having a tip of sufficient size to displace at least ¼ inch of material at each location. Each cut must completely sever the receiver in the designated areas and must be done with a diagonal torch cut. Using a bandsaw or a cut-off wheel to destroy the receiver does not ensure destruction of the weapon.

This method of destruction is illustrated in the diagram below.

Alternative methods of destruction may also be acceptable. These alternative methods must be equivalent in degree to the approved method of destruction. Receivers that are not sufficiently modified cannot be ap-

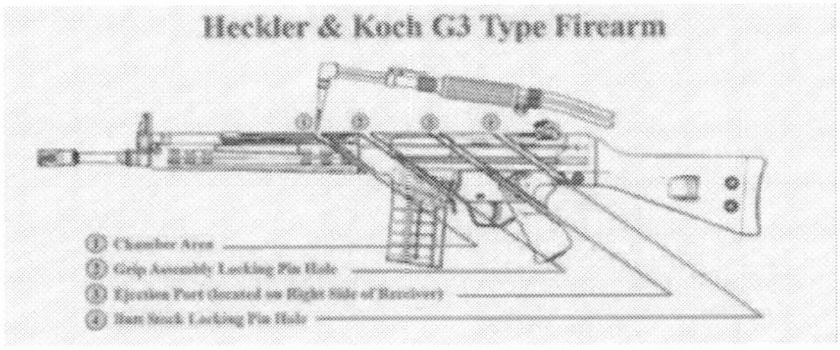

proved for importation. To ensure compliance with the law, it is recommended that the importer submit in writing the alternative method of destruction to the ATF Firearms Technology Branch (FTB) for review and approval prior to importation.

Held, an ATF-approved method of destruction for a Heckler and Koch G3 type machinegun receiver will result in the severed portions of the receiver being importable for unrestricted commercial sale. The severed articles would not be subject to the provisions of 26 U.S.C. 5844 or 18 U.S.C. 922(o), but would continue to be subject to the provisions of the Arms Export Control Act, 22 U.S.C. 2778. Alternative methods of destruction may also be acceptable. It is recommended that such methods be reviewed and approved by the ATF Firearms Technology Branch prior to the weapon's importation.

Date signed: January 24, 2003.

26 USC 5844, 18 USC 922(o), 22 USC 2778

IMPORTATION OF STEN TYPE RECEIVERS FOR UNRESTRICTED COMMERCIAL SALE

An ATF-approved method of destruction for the Sten ty pe machinegun will result in the se vered portions of t he receiv er being importable for u nrestricted co mmercial sale.

ATF Rul. 2003–4

The Bureau of Alcohol, Tobacco and Firearms (ATF) has received inquiries about modifications necessary to the receiver of a Sten type machinegun to make it importable under 26 U.S.C. 5844 and 18 U.S.C. 922(o) for unrestricted commercial sale.

The Sten is a machinegun as defined in 26 U.S.C. 5845(b). The receiver of a Sten is also a machinegun as defined. Various manufacturers made Sten style machineguns in caliber 9x19mm (9mm Luger). The Sten is a blowback-operated, shoulder-fired, magazine-fed, selective-fire submachinegun.

Section 5844 of Title 26, United States Code, makes it unlawful to import any firearm into the United States, unless the firearm to be imported or brought in is: (1) being imported for use by the United States or any department, independent establishment, or agency thereof or any State or possession or any political subdivision thereof; or (2) the firearm is being imported for scientific or research purposes; or (3) it is being imported solely for testing or use as a model by a registered manufacturer or solely for use as a sample by a registered importer or dealer. Additionally, the Secretary may permit the conditional importation of a firearm for examination and testing in connection with classifying the firearm.

Section 922(o) of Title 18, United States Code, makes it unlawful for any person to transfer or possess a machinegun, except a transfer to or by the United States or any department or agency thereof or a State or a department, agency, or political subdivision thereof; or any lawful transfer or lawful possession of a machinegun lawfully possessed before May 19, 1986.

A review of the statutes above indicates that machineguns and machinegun receivers cannot be lawfully imported for unrestricted commercial sale. Accordingly, machinegun receivers may be imported for commercial sale only if they are destroyed in a manner that will prevent their function and future use as a firearm. The resulting severed receiver portions would not be subject to the provisions of 26 U.S.C. 5844 or 18 U.S.C. 922(o); however, these articles would be subject to the provisions of the Arms Export Control Act, 18 U.S.C. 925, 22 U.S.C. 2778, and implementing regulation at 27 CFR Part 47. It is important to note that these machinegun receivers must be destroyed and cannot be imported whether they are serviceable or unserviceable.

An ATF-approved method of destruction for a Sten type machinegun receiver requires three diagonal torch cuts that sever or pass through the following areas: (1) the threaded portion of the receiver/chamber area, (2) the return spring cap socket, and (3) the sear slot in the lower side of the receiver. All cutting must be done with a cutting torch having a tip of sufficient size to displace at least ¼ inch of material at each location. Each cut must completely sever the receiver in the designated areas and must be done with a diagonal torch cut. Using a bandsaw or a cut-off wheel to destroy the receiver does not ensure destruction of the weapon.

This method of destruction is illustrated in the diagram below.

Alternative methods of destruction may also be acceptable. These alternative methods must be equivalent in degree to the approved method of destruction. Receivers that are not sufficiently modified cannot be approved for importation. To ensure compliance with the law, it is recommended that the importer submit in writing the alternative method of destruction to the ATF Firearms Technology Branch (FTB) for review and approval prior to importation.

Held, an ATF-approved method of destruction for a Sten type machinegun receiver will result in the severed portions of the receiver being importable for unrestricted commercial sale. The severed articles would not be subject to the provisions of 26 U.S.C. 5844 or 18 U.S.C. 922(o), but would continue to be subject to the provisions of the Arms Export Control Act, 22 U.S.C. 2778. Alternative methods of destruction may also be acceptable. It is recommended that such methods be reviewed and approved by the ATF Firearms Technology Branch prior to the weapon's importation.

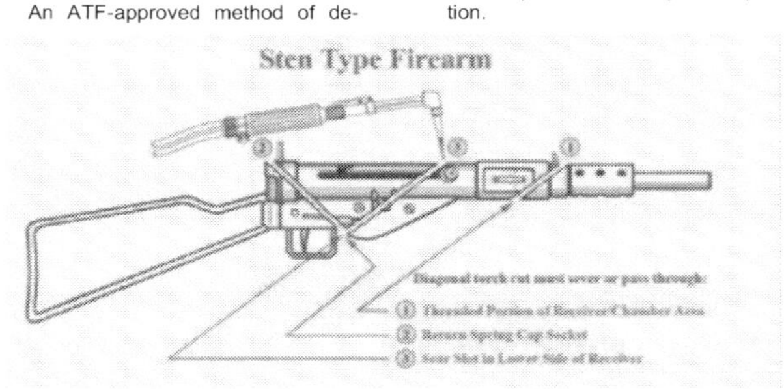

Date signed: January 24, 2003.

18 U.S.C. 925(d): EXCEPTIONS

22 U.S.C. 2778: IMPORTATION

26 U.S.C. 5844: IMPORTATION

27 CFR 478.111, 478.112, 478.113: IMPORTATION

27 CFR 479.111, 479.112, 479.113: IMPORTATION

27 CFR 447.42: APPLICATION FOR PERMIT

Persons with a valid Federal Firearms license and/or registered as an importer of articles enumerated on the U.S. Munitions Import List seeking to import firearms, ammunition and implements of war may submit the ATF Form 6, Application and Permit for Importation of Firearms, Ammunition and Implements of War, electronically using the eForm 6 online electronic filing system, provided such persons have met certain registration requirements.

ATF Rul. 2003-6

The Gun Control Act of 1968 (GCA), 18 U.S.C. Chapter 44, and the National Firearms Act (NFA), 26 U.S.C. Chapter 53, provide that, with certain exceptions, no firearm, firearm barrel, or ammunition shall be imported or brought into the United States unless the Director of the Bureau of Alcohol, Tobacco, Firearms and Explosives (ATF) has authorized its importation. See 18 U.S.C. 925(d); 26 U.S.C. 5844. The Arms Export Control Act (AECA), 22 U.S.C. 2778, gives the President the authority to control the export and import of defense articles and defense services in furtherance of world peace and the security and foreign policy of the United States. Authority to administer the permanent import provisions of the AECA was delegated to the Attorney General, while the authority to administer the export and temporary import provisions of the AECA was delegated to the Secretary of State. Executive Order 11958 of January 18, 1977, as amended by Executive Order 13284 of January 23, 2003, 3 CFR Executive Order 13284.

Persons who wish to import firearms or ammunition must file with the Director an ATF Form 6 (Firearms), Application and Permit for Importation of Firearms, Ammunition and Implements of War, in triplicate, executed under the penalties of perjury. See 27 CFR 478.112, 478.113, 479.111, 479.112, 479.113, and 447.42. The Form 6 must contain the information specified in 27 CFR Subpart G. If the Director approves the application, the approved application will serve as the import permit. See 27 CFR 478.112(b).

The Government Paperwork Elimination Act (GPEA), enacted in 1998, requires executive agencies to provide for the option of the electronic maintenance, submission, or disclosure of information, as a substitute for paper, and for the use and acceptance of electronic signatures, when practicable, by October 2003. See Government Paperwork Elimination Act, Pub. L. No. 105-277, § 1704, 112 Stat. 2681-749, 2681-750 (1998). In accordance with the GPEA's mandate, ATF developed the eForm 6 online electronic filing system for persons with a valid Federal Firearms license and persons registered as an importer of articles enumerated on the U.S. Munitions Import List.

The eForm 6 online electronic filing system enables licensees and registered importers to file the ATF Form 6 and obtain an approved import permit from ATF electronically via the Internet. eForm 6 online applications will be approved, returned for correction, or denied and a paper copy returned to the applicant. If approved, the paper copy will serve as the import permit and may be submitted to United States Bureau of Customs and Border Protection. The system also enables licensees and registered importers to obtain status updates regarding both electronic and paper import permit applications they have filed. The applicable laws, regulations, policies, and procedures pertaining to import applications also apply to the eForm 6.

To register for the eForm 6 online electronic filing system, licensees and registered importers (or employees of licensees and registered importers) must complete a registration form, the ATF Form 5013.3, eForm 6 Access Request. Each individual requesting access to the system must sign the ATF Form 5013.3 certifying that they intend the electronic credentials assigned to them to substitute for their original signature and that any eForm 6 submissions will be treated as bearing an original signature. The user also agrees to be bound by the Notices and Agreement governing the use of the eForm 6 system.

Each ATF Form 5013.3 must also include the name, title and signature of a responsible person for the Federal firearms licensee or AECA registrant. The responsible person authorizes the user to complete and execute import applications on behalf of the Federal firearms licensee or AECA registrant. The responsible person also agrees that the licensee or registered importer will be bound by the entries on applications filed via the eForm 6 system and intends that such applications be treated as bearing an original signature, and agrees to be bound by the Notices and Agreement governing the use of the eForm 6 system.

Upon proper registration, ATF will issue each registrant a user ID and password allowing access to the eForm 6 system. Each individual user will be issued a separate user ID and password.

The eForm 6 system will require users to attest that the information submitted via the eForm 6 system are statements made subject to penalty of perjury and confirm their ATF-issued electronic credentials in order to complete the application process. Specifically, in order to complete the application process, a user will be required to declare first that all the statements contained in the application are true and correct and that the user has read, understood, and complied with the conditions and instructions for the import application. Second, the user will be required to declare that the user authorizes the transmittal via the eForm 6 system of what may constitute tax return information, as defined in section 6103 of the Internal Revenue Code, Title 26, United States Code.

The eForm 6 online electronic filing system is accessible on ATF's Firearms and Explosives Imports Branch website at :

http://www.atfonline.gov/eforms6

This site contains the instructions and forms necessary to register as an eForm 6 user.

Licensees and registered importers are not required to use the eForm 6 electronic filing system, and in certain circumstances may not be able to participate. Licensees and registered importers may continue to submit ATF Form 6 on paper to ATF.

Sections 478.22 and 479.26, Title 27, CFR, provide that the Director may approve an alternate method or procedure in lieu of a method or procedure specifically prescribed in the regulations when he finds that:

(1) Good cause is shown for the use of the alternate method or procedure;

(2) The alternate method or procedure is within the purpose of, and consistent with the effect intended by, the specifically prescribed method or procedure and that the alternate method or procedure is substantially equivalent to that specifically prescribed method or procedure; and

(3) The alternate method or procedure will not be contrary to any provision of law and will not result in an increase in cost to the Government or hinder the effective administration of Parts 478 and 479.

ATF finds that there is good cause to authorize a variance to the provisions of 27 CFR 478.111, 478.112, 478.113, 479.111, 479.112, 479.113 and 447.42 requiring the filing of ATF Form 6, Application and Permit for Importation of Firearms, Ammunition and Implements of War, in paper form due to the mandate of the GPEA that executive agencies provide the option of electronic submission of information as a substitute for paper, and for the use and acceptance of electronic signatures. Accordingly, ATF authorizes the following alternate method or procedure to the ATF Form 6 filing requirements of 27 CFR 478.111, 478.112, 478.113, 479.111, 479.112, 479.113, and 447.42:

The ATF Form 6 may be filed in electronic form on ATF eForm 6, provided that:

(1) The applicant has registered with ATF by submitting the registration form, ATF Form 5013.3, eForm 6 Access Request;

(2) The applicant has received a unique user ID and password, and has agreed that the electronic signature assigned to them is intended as their original signature on eForm 6 submissions; and

(3) The applicant has agreed to be bound by the Notices and Agreement governing the use of the eForm 6 system.

Licensees and registered importers who fail to abide by the conditions outlined above may be advised by ATF that their privilege of utilizing the eForm 6 electronic filing system has been terminated.

ATF finds that the above alternate method is consistent with the provisions of 27 CFR 478.111, 478.112, 478.113, 479.111, 479.112, 479.113 and 447.42 because it will ensure that the required information is captured on the eForm 6 and that the eForm 6 is signed under penalties of perjury. The alternate method is not contrary to any provision of law, will not increase costs to ATF, and will not hinder the effective administration of the regulations in 27 CFR Parts 478, 479, and 447.

Held, pursuant to 27 CFR 478.22 and 479.26, ATF authorizes a variance from the requirements of 27 CFR 478.111, 478.112, 478.113, 479.111, 479.112, 479.113 and 447.42 for Federal Firearms licensees and registered importers of articles enumerated on the U.S. Munitions Import List filing ATF Form 6, Application and Permit for Importation of Firearms, Ammunition and Implements of War. As an alternate method or procedure, the ATF Form 6 may be filed in electronic form on ATF eForm 6, provided that:

(1) The applicant has registered with ATF by submitting the registration form, ATF Form 5013.3, eForm 6 Access Request;

(2) The applicant has received a unique user ID and password, and has agreed that the electronic signature assigned to them is intended as their original signature on eForm 6 submissions; and

(3) The applicant has agreed to be bound by the Notices and Agreement governing the use of the eForm 6 system.

Date signed: July 11, 2003

18 U.S.C. 922(b)(3): TRANSFER OF FIREARMS

27 CFR 478.11, 478.124(c)(3)(ii): DEFINITIONS, TRANSFER OF FIREARMS

A Federal firearms licensee (FFL) may not lawfully transfer a firearm to a nonimmigrant alien who has not resided in a State continuously for at least 90 days immediately prior to the FFL conducting a National Instant Criminal Background Check System (NICS) check.

ATF Rul. 2004-1

The Bureau of Alcohol, Tobacco, Firearms and Explosives (ATF) has received questions from nonimmigrant aliens concerning how aliens satisfy the Gun Control Act's (GCA) State of residence requirement. Several nonimmigrant aliens have asked why they have been prohibited from purchasing a firearm from a Federal firearms licensee (FFL), even though the aliens believe they have lived in the State where the FFL is licensed for more than 90 days.

The GCA provides that an FFL generally may not transfer a handgun to an unlicensed person who does not reside in the State where the licensee's premises are located. 18 U.S.C. 922(b)(3). FFLs may transfer long guns to residents of other States in over-the-counter transactions, if the sale, delivery, and receipt of the firearm comply with the laws of the FFL's State and the buyer's State. *Id.* In order to satisfy these long gun sale requirements, the buyer must reside in a State within the United States.

The regulations implementing the GCA define "State of residence" as: "[t]he State in which an individual resides. An individual resides in a State if he or she is present in a State with the intention of making a home in that State." For aliens, the definition also provides that a legal alien "shall be considered to be a resident of a State only if the alien is residing in the State and has resided in the State for a period of at least 90 days prior to the date of sale or delivery of a firearm." 27 CFR 478.11.

Moreover, the GCA regulations require that after an alien completes ATF Form 4473 (Firearms Transaction Record), the FFL shall have the alien present documentation establishing that the alien is a resident of the State (as defined in section 478.11) in which the FFL's business premises are located. The regulation states that "[e]xamples of acceptable documentation include utility bills or a lease agreement which show that the transferee has resided in the State continuously for at least 90 days prior to the transfer of the firearm." 27 CFR 478.124(c)(3)(ii).

ATF interprets these provisions to mean a nonimmigrant alien must reside in the State continuously for 90 days *immediately* preceding the NICS check. If this temporal requirement is not imposed, the purpose of the 90 continuous days requirement will be defeated. Documentation that a nonimmigrant alien resided in a State continuously for 90 days at some point in the past does not establish the alien's State of residence at the time of the NICS check.

Moreover, for all non-U.S. citizens, NICS checks include a check of Bureau of Immigration and Customs Enforcement (ICE) databases. These databases generally contain records of when nonimmigrant aliens enter and exit the United States. Accordingly, when a nonimmigrant alien attempts to receive a firearm from an FFL, the background check generally will show if the nonimmigrant alien has left the United States in the preceding 90 days.

ATF recognizes that some nonimmigrant aliens who temporarily leave the United States may have an intent to reside in a State within the United States, and are simply going on a trip abroad. However, ATF has determined that evidence of a nonimmigrant alien leaving the country represents a break in residency that requires a subsequent 90-day residence in a State before the alien can lawfully purchase a firearm from an FFL.

Accordingly, if the ICE records show the nonimmigrant alien has left the United States in the preceding 90 days, NICS will tell the FFL to cancel the transaction because the 90 day residency requirement has not been met. The transaction is canceled (rather than denied) because, since the nonimmigrant alien did not meet the residency requirement, a NICS check was not appropriate. Further, a denial is not appropriate because the 90 day residency requirement is not a prohibited category under 18 U.S.C. 922(g) or (n), and so is not grounds for NICS denying a transaction.

Held, pursuant to 18 U.S.C. 922(b)(3), 27 CFR 478.11, and 478.124(c)(3)(ii), Federal firearms licensees may not lawfully transfer firearms to nonimmigrant aliens who have not resided in the State where their business premises are located (or in the case of a long gun, in any State) for at least 90 continuous days immediately preceding the National Instant Criminal Background Check System (NICS) check.

Held further, that if a National Instant Criminal Background Check System check demonstrates a nonimmigrant alien has left the United States during the 90 days immediately preceding the NICS check, the nonimmigrant alien does not satisfy the 90 day State of residency requirement. This is the case even if the nonimmigrant alien has provided other documentation, such as utility bills or a lease, to demonstrate 90 days of continuous residency immediately preceding the NICS check.

Date signed: March 22, 2004

26 U.S.C. 5812, 5841, 5844, 5861, 5872
27 CFR 479.11, 479.26, 479.105, 479.111, 479.112, 479.114 – 479.119:
IMPORTATION OF FIREARMS SUBJECT TO THE NATIONAL FIREARMS ACT

18 U.S.C. 921(a)(3), 922(l), 922(o), 923(e), 924(d), 925(d)(3)
27 CFR 478.11, 478.22, 478.111 – 478.113: IMPORTATION OF MACHINEGUNS, DESTRUCTIVE DEVICES, SHORT-BARREL SHOTGUNS, SHORT-BARREL RIFLES, FIREARMS SILENCERS, AND OTHER FIREARMS SUBJECT TO THE NATIONAL FIREARMS ACT

22 U.S.C. 2778
27 CFR 447.11, 447.21:
TEMPORARY IMPORTATION OF DEFENSE ARTICLES

The Bureau of Alcohol, Tobacco, Firearms and Explosives has approved an alternate method or procedure for importers to use when temporarily importing firearms subject to the National Firearms Act, the Gun Control Act and the Arms Export Control Act for inspection, testing, calibration, repair, or incorporation into another defense article.

ATF Rul. 2004-2

The Bureau of Alcohol, Tobacco, Firearms and Explosives (ATF) has received numerous inquiries from importers who wish to temporarily import firearms subject to the Gun Control Act of 1968 (GCA), 18 U.S.C. Chapter 44, and the National Firearms Act (NFA), 26 U.S.C. Chapter 53, for inspection, testing, calibration, repair, or incorporation into another defense article. Importers advise ATF that they generally obtain a temporary import license, DSP-61, from the Department of State authorizing the importation or comply with one of the regulatory exemptions from licensing in 22 CFR 123.4. They ask whether such a license or exemption is sufficient to satisfy the requirements of the GCA and NFA.

Statutory Background

1. *The National Firearms Act*

The NFA imposes restrictions on certain firearms, including registration requirements, transfer approval requirements, and import restrictions. 26 U.S.C. 5812, 5841, 5844. The term "firearm" is defined in 26 U.S.C. 5845(a) to include machineguns, short-barrel shotguns, short-barrel rifles, silencers, destructive devices, and "any other weapons." Section 5844 of the NFA provides that no firearm may be imported into the United States unless the importer establishes that the firearm to be imported is—

(1) Being imported or brought in for the use of the United States or any department, independent establishment, or agency thereof or any State or possession or any political subdivision thereof; or

(2) Being imported or brought in for scientific or research purposes; or

(3) Being imported or brought in solely for testing or use as a model by a registered manufacturer or solely for use as a sample by a registered importer or registered dealer.

Regulations implementing the NFA in 27 CFR Part 479 require importers to obtain an ATF Form 6, Application and Permit for Importation of Firearms, Ammunition and Implements of War, prior to importing NFA firearms into the United States. 27 CFR 479.111. In addition, the regulations require importers to register the firearms they import by filing with the

Director an accurate notice on Form 2, Notice of Firearms Manufactured or Imported, executed under the penalties of perjury, showing the importation of a firearm. 27 CFR 479.112. When an NFA firearm is to be exported from the United States, the exporter must file with the Director an application on Form 9, Application and Permit for Exportation of Firearms, to obtain authorization to export the firearm. 27 CFR 479.114-119.

Regulations in 27 CFR Part 479 indicate that NFA firearms may be imported for scientific or research purposes or for testing or use as a model by a registered manufacturer or as a sample by a registered importer or registered dealer. 27 CFR 479.111(a). However, section 479.105(c), implementing section 922(o) of the GCA, clarifies that machineguns manufactured on or after May 19, 1986, may be imported only with a purchase order for transfer to a governmental entity, or as a dealer's sales sample pursuant to section 479.105(d).

The regulations in Part 479 give the Director the authority to approve an alternate method or procedure in lieu of a method or procedure specifically prescribed in the regulations when it is found that:

> **(1)** Good cause is shown for the use of the alternate method or procedure;
>
> **(2)** The alternate method or procedure is within the purpose of, and consistent with the effect intended by, the specifically prescribed method or procedure and that the alternate method or procedure is substantially equivalent to that specifically prescribed method or procedure; and
>
> **(3)** The alternate method or procedure will not be contrary to any provision of law and will not result in an increase in cost to the Government or hinder the effective administration of the GCA or regulations issued thereunder.

27 CFR 479.26.

2. *The Gun Control Act*

Import provisions of the GCA, 18 U.S.C. 922(l) and 925(d)(3), generally prohibit the importation of firearms subject to the NFA, except for the use of governmental entities. 18 U.S.C. 925(a)(1). The term "firearm" is defined in section 921(a)(3) to include any weapon which will or is designed to or may be readily converted to expel a projectile by the action of an explosive; the frame or receiver of such weapon; any firearm silencer; and any destructive device. In addition, section 922(o) of the GCA prohibits the transfer or possession of a machinegun manufactured on or after May 19, 1986, except for the official use of governmental entities.

Regulations implementing the GCA in 27 CFR Part 478 require that persons importing firearms into the United States obtain an approved ATF Form 6, Application and Permit for Importation of Firearms, Ammunition and Implements of War, prior to bringing the firearms into the United States. 27 CFR 478.111-114. Regulations in Part 478 provide that the Director may approve an alternate method or procedure in lieu of a method or procedure specifically prescribed by the GCA and regulations when it is found that:

> **(1)** Good cause is shown for the use of the alternate method or procedure;
>
> **(2)** The alternate method or procedure is within the purpose of, and consistent with the effect intended by, the specifically prescribed method or procedure and that the alternate method or procedure is substantially equivalent to that specifically prescribed method or procedure; and
>
> **(3)** The alternate method or procedure will not be contrary to any provision of law and will not result in an increase in cost to the Government or hinder the effective administration of the GCA or regulations issued thereunder.

27 CFR 478.22.

3. *The Arms Export Control Act*

The Arms Export Control Act (AECA), 22 U.S.C. 2778, gives the President the authority to control the export and import of defense articles and defense services in furtherance of world peace and the security and foreign policy of the United States. Authority to administer the permanent import provisions of the AECA was delegated to the Attorney General, while the authority to administer the export and temporary import provisions of the AECA was delegated to the Secretary of State. Executive Order 11958 of January 18, 1977, as amended by Executive Order 13333 of January 23, 2003, 3 CFR Executive Order 13284.

The term "defense article" is defined in 27 CFR 447.11 as any item designated in sections 447.21 or 447.22. Section 447.21, the U.S. Munitions Import List, includes a number of defense articles that are also subject to the GCA and NFA. Category I, "Firearms," includes nonautomatic and semiautomatic firearms to caliber .50 inclusive, combat shotguns, shotguns with barrels less than 18 inches in length, and firearms silencers and suppressors. All Category I firearms are subject to the GCA. "Combat shotguns" include the USAS-12 shotgun and the Striker-12/Streetsweeper shotgun, which have been classified as destructive devices under the GCA and NFA. In addition, all shotguns with barrels of less than 18 inches in length are subject to both the GCA and NFA. All rifles with barrels of less than 16 inches in length are subject to both the GCA and NFA, and silencers are subject to both the GCA and NFA.

Category II, "Artillery Projectors," includes guns over caliber .50, howitzers, mortars, and recoilless rifles. Firearms over .50 caliber have a bore of more than one-half inch in diameter and are "destructive devices" as defined in the GCA and NFA.

Category IV, "Launch Vehicles, Guided Missiles, Ballistic Missiles, Rockets, Torpedoes, Bombs and Mines," includes rockets, bombs, grenades, torpedoes, and land and naval mines. All these articles are "destructive devices" as defined in the GCA and NFA.

Regulations of the Department of State implementing the AECA generally require a temporary import license, DSP-61, for the temporary import and subsequent export of unclassified defense articles, unless otherwise exempted. 22 CFR 123.3. Regulations in 22 CFR 123.4 provide an exemption from licensing if the item temporarily imported:

> **(1)** Is serviced (e.g., inspection, testing, calibration or repair, including overhaul, reconditioning and one-to-one replacement of any defective items, parts or components, but excluding any modification, enhancement, upgrade or other form

of alteration or improvement that changes the basic performance of the item), and is subsequently returned to the country from which it was imported. Shipment may be made by the U.S. importer or a foreign government representative of the country from which the goods were imported; or

(2) Is to be enhanced, upgraded or incorporated into another item which has already been authorized by the Office of Defense Trade Controls for permanent export; or

(3) Is imported for the purpose of exhibition, demonstration or marketing in the United States and is subsequently returned to the country from which it was imported; or

(4) Has been rejected for permanent import by the Department of the Treasury [after January 24, 2003, the Department of Justice] and is being returned to the country from which it was shipped; or

(5) Is approved for such import under the U.S. Foreign Military Sales (FMS) program pursuant to an executed U.S. Department of Defense Letter of Offer and Acceptance (LOA).

Willful violations of the AECA are punishable by imprisonment for not more than 10 years, a fine of not more than $1,000,000, or both. 22 U.S.C. 2778(c). Articles imported in violation of the AECA are also subject to seizure and forfeiture. 18 U.S.C. 545.

Discussion

A temporary import license authorizing the temporary importation and subsequent export of a defense article by the Department of State satisfies all legal requirements under the AECA. Importers may also comply with AECA requirements if the importation meets one of the exemptions in 22 CFR 123.4. However, if the defense article is subject to the GCA and NFA, the importer must also comply with the requirements of those statutes. Neither the GCA nor NFA make a distinction between temporary importation and permanent importation, as is the case under the AECA. Regulations implementing the GCA and NFA make it clear that an "importation" occurs when firearms are brought within the territory of the United States. 27 CFR 478.11 and 479.11. Accordingly, any bringing of firearms into the territory of the United States is subject to the import provisions of the GCA and NFA. Issuance of a temporary import license by the Department of State, or exemption from licensing under regulations in 22 CFR Part 123, will not excuse compliance with the GCA and NFA.

The statutes and regulations outlined above do not address the importation of machineguns manufactured after May 19, 1986, for scientific or research purposes or for testing, repair, or use as a model by a manufacturer or importer. Nor do the regulations address the importation of post-86 machineguns for repair, inspection, calibration, or incorporation into another defense article.

For other "defense articles" that are subject to the requirements of the GCA and NFA, such as silencers, destructive devices, and short-barrel weapons, ATF has the authority to approve the importation of such firearms for scientific or research purposes or for testing or use as a model or sample by a registered importer or registered dealer. However, such importations must comply with all applicable provisions of the NFA, including filing of a Form 2, Notice of Firearms Manufactured or Imported, to effect registration. If such articles are subsequently exported, a Form 9, Application and Permit for Permanent Exportation of Firearms, must also be approved prior to exportation.

As with post-86 machineguns, neither the law nor regulations specifically address the importation of firearms subject to the NFA for purposes of repair, inspection, calibration, or for incorporation into another defense article.

ATF recognizes that inspection, repair, calibration, incorporation into another defense article, and reconditioning of machineguns, destructive devices, and other NFA firearms is often necessary for National defense. These defense articles are frequently sold to allies of the United States for their legitimate defense needs. Accordingly, ATF believes it is appropriate to recognize an alternate method that allows importers to temporarily import these firearms, subject to requirements to ensure the security of these articles while they are in the United States and accountability of the persons who import them.

Pursuant to 27 CFR 478.22 and 479.26, ATF hereby authorizes an alternate method or procedure for importers of defense articles to use for temporary importation of such articles for inspection, calibration, repair, or incorporation into another defense article when such articles are subject to the requirements of the NFA and GCA. The procedure requires that importers--

(1) Be qualified under the GCA and NFA to import the type of firearms sought for importation;

(2) Obtain a temporary import license, DSP-61, from the Department of State in accordance with 22 CFR 123.3 OR qualify for a temporary import license exemption pursuant to 22 CFR 123.4;

(3) Within 15 days of the release of the firearms from Customs custody, file an ATF Form 2, Notice of Firearms Manufactured or Imported, showing the importation of the firearms. The DSP-61 must be attached to the Form 2. If the importation is subject to a licensing exemption under 22 CFR 123.4, the importer must submit with the ATF Form 2 a statement, under penalty of perjury, attesting to the exemption and stating that the article will be exported within four years of its importation into the United States;

(4) Maintain the defense articles in a secure place and manner to ensure that the articles are not diverted to criminal or terrorist use; and

(5) Export the articles within 4 years of importation into the United States.

Importers who follow the procedures outlined above will be in compliance with all the provisions of the GCA, NFA, and AECA administered and enforced by ATF. All other provisions of the law must be followed.

ATF finds that the procedure outlined above meets the legal requirements for an alternate method or procedure because there is good cause to authorize the importation of defense articles for repair, inspection, calibration, or incorporation into another defense article. Because such defense articles are often provided to allies of the United States, it is imperative that the original manufacturers have a lawful method of importing such articles for repair and routine maintenance. The alternate method or procedure is consistent with the effect intended by the procedure set forth in the GCA and NFA, because the firearms must be registered and stored securely. Finally, the alternate method is consistent with the requirements of the GCA and NFA and will not result in any additional costs to ATF or the Department of State.

"Transfers" of NFA Weapons After Importation

ATF recognizes that temporarily imported NFA firearms are sometimes "transferred" from the importer to a contractor within the United States for inspection, testing, calibration, repair, or incorporation into another defense article. ATF has approved a procedure for authorizing the transportation or delivery of temporarily imported NFA firearms to licensed contractors for repair or manipulation, as noted above.

Conveyance of an NFA weapon to a licensee for purposes of inspection, testing, calibration, repair, or incorporation into another defense article is generally not considered to be a "transfer" under 26 U.S.C. 5845(j). ATF has taken the position that temporary custody by a licensee is not a transfer for purposes of the NFA since no sale, lease, or other disposal is intended by the owner. However, in order to document the transaction as a temporary conveyance and make clear that an actual "transfer" of a firearm has not taken place, ATF strongly recommends that the importer submit a Form 5, Application for Tax Exempt Transfer and Registration of Firearm, for approval prior to conveying a firearm for repair or manipulation. In the alternative, the importer should convey the weapon with a letter to the contractor, stating: (1) the weapon is being temporarily conveyed for inspection, testing, calibration, repair, or incorporation into another defense article; and (2) the approximate time period the weapon is to be in the contractor's possession. The transferee must be properly licensed to engage in an NFA firearms business.

Held, pursuant to 27 CFR 478.22 and 479.26, the Bureau of Alcohol, Tobacco, Firearms and Explosives has approved an alternate method or procedure for importers to use when temporarily importing firearms subject to the Gun Control Act, National Firearms Act, and the Arms Export Control Act for inspection, testing, calibration, repair, or incorporation into another defense article. This procedure applies to all defense articles that are also subject to the NFA and GCA. The procedure requires that importers--

(1) Be qualified under the GCA and NFA to import the type of firearms sought for importation;

(2) Obtain a temporary import license, DSP-61, from the Department of State in accordance with 22 CFR 123.3 or qualify for a temporary import license exemption pursuant to 22 CFR 123.4;

(3) Within 15 days of the release of the firearms from Customs custody, file an ATF Form 2, Notice of Firearms Manufactured or Imported, showing the importation of the firearms. The DSP-61 must be attached to the Form 2. If the importation is subject to a licensing exemption under 22 CFR 123.4, the importer must submit with the ATF Form 2 a statement, under penalty of perjury, attesting to the exemption and stating that the article will be exported within four years of its importation into the United States;

(4) Maintain the defense articles in a secure place and manner to ensure that the articles are not diverted to criminal or terrorist use; and

(5) Export the articles within 4 years of importation into the United States.

Held further, temporary conveyance of NFA weapons from the importer to a contractor within the United States for purposes of inspection, testing, calibration, repair, or incorporation into another defense article may be accomplished through advance approval of ATF Form 5, Application for Tax Exempt Transfer and Registration of Firearm, or with a letter from the importer to the contractor stating: (1) the weapon is being temporarily conveyed for inspection, testing, calibration, repair, or incorporation into another defense article; and (2) the approximate time period the weapon is to be in the contractor's possession. The transferee must be properly licensed to engage in an NFA firearms business.

Date signed: April 7, 2004

26 U.S.C. 5845(b): DEFINITIONS (MACHINEGUN)

27 CFR 179.11: MEANING OF TERMS

The 7.62mm Aircraft Machine Gun, identified in the U.S. military inventory as the "M-134" (Army), "GAU-2B/A" (Air Force), and "GAU-17/A" (Navy), is a machinegun as defined by 26 U.S.C. 5845(b). Rev. Rul. 55-528 modified.

ATF Rul. 2004-5

The Bureau of Alcohol, Tobacco, Firearms and Explosives (ATF) has examined the 7.62mm Aircraft Machine Gun, commonly referred to as a "Minigun." The Minigun is a 36-pound, six-barrel, electrically powered machinegun. It is in the U.S. military inventory and identified as the "M-134" (Army), "GAU-2B/A" (Air Force), and "GAU-17/A" (Navy). It is a lightweight and extremely reliable weapon, capable of discharging up to 6,000 rounds per minute. It has been used on helicopters, fixed-wing aircraft, and wheeled vehicles. It is highly adaptable, being used with pintle mounts, turrets, pods, and internal installations.

The Minigun has six barrels and bolts which are mounted on a rotor. The firing sequence begins with the manual operation of a trigger. On an aircraft, the trigger is commonly found on the control column, or joystick. Operation of the trigger causes an electric motor to turn the rotor. As the rotor turns, a stud on each bolt travels along an elliptical groove on the inside of the housing, which causes the bolts to move forward and rearward on tracks on the rotor. A triggering cam, or sear shoulder, trips the firing pin when the bolt has traveled forward through the full length of the bolt track. One complete revolution of the rotor discharges cartridges in all six barrels. The housing that surrounds the rotor, bolts and firing mechanism constitutes the frame or receiver of the firearm.

The National Firearms Act defines "machinegun" as "any weapon which shoots, is designed to shoot, or can be readily restored to shoot, automatically more than one shot, without manual reloading, by a single function of the trigger." 26 U.S.C. 5845(b). The term also includes "the frame or receiver of any such weapon, any part designed and intended solely and exclusively, or combination of parts designed and intended, for use in converting a weapon into a machinegun, and any combination of parts from which a machinegun can be assembled if such parts are in the possession or under the control of the person." Id.; see 18 U.S.C. 921(a)(23); 27 CFR 478.11, 479.11.

ATF and its predecessor agency, the Internal Revenue Service (IRS), have historically held that the original, crank-operated Gatling Gun, and replicas thereof, are not automatic firearms or machineguns as defined. See Rev. Rul. 55-528, 1955-2 C.B. 482. The original Gatling Gun is a rapid-firing, hand-operated weapon. The rate of fire is regulated by the rapidity of the hand-cranking movement, manually controlled by the operator. It is not a "machinegun" as that term is defined in 26 U.S.C. 5845(b) because it is not a weapon that fires automatically.

The Minigun is not a Gatling Gun. It was not produced under the 1862 - 1893 patents of the original Gatling Gun. While using a basic design concept of the Gatling Gun, the Minigun does not incorporate any of Gatling's original components and its feed mechanisms are entirely different. Critically, the Minigun shoots more than one shot, without manual reloading, by a single function of the trigger, as prescribed by 26 U.S.C. 5845(b). See United States v. Fleischli, 305 F.3d 643, 655-656 (7th Cir. 2002). See also Staples v. United States, 511 U.S. 600, 603 (1994) (automatic refers to a weapon that "once its trigger is depressed, the weapon will automatically continue to fire until its trigger is released or the ammunition is exhausted"); GEORGE C. NONTE, JR., FIREARMS ENCYCLOPEDIA 13 (Harper & Rowe 1973) (the term "automatic" is defined to include "any firearm in which a single pull and continuous pressure upon the trigger (or other firing device) will produce rapid discharge of successive shots so long as ammunition remains in the magazine or feed device - in other words, a machinegun"); WEBSTER'S II NEW RIVERSIDE - UNIVERSITY DICTIONARY (1988) (defining automatically as "acting or operating in a manner essentially independent of external influence or control"); JOHN QUICK, PH.D., DICTIONARY OF WEAPONS AND MILITARY TERMS 40 (McGraw-Hill 1973) (defining automatic fire as "continuous fire from an automatic gun, lasting until pressure on the trigger is released").

The term "trigger" is generally held to be the part of a firearm that is used to initiate the firing sequence. See United States v. Fleischli, 305 F.3d at 655-56 (and cases cited therein); see also ASSOCIATION OF FIREARMS AND TOOLMARK EXAMINERS (AFTE) GLOSSARY 185 (1st ed. 1980) ("that part of a firearm mechanism which is moved manually to cause the firearm to discharge"); WEBSTER'S II NEW RIVERSIDE- UNIVERSITY DICTIONARY (1988) ("lever pressed by the finger in discharging a firearm").

Held, the 7.62mm Minigun is designed to shoot automatically more than one shot, without manual reloading, by a single function of the trigger. Consequently, the 7.62mm Minigun is a machinegun as defined in section 5845(b) of the National Firearms Act. See United States v. Fleischli, 305 F.3d at 655-56. Similarly, the housing that surrounds the rotor is the frame or receiver of the Minigun, and thus is also a machinegun. Id.; see 18 U.S.C. 921(a)(23); 27 CFR 478.11, 479.11.

To the extent this ruling is inconsistent with Revenue Ruling 55-528 issued by the IRS, Revenue Ruling 55-528, 1955-2 C.B. 482, is hereby modified.

Date signed: August 18, 2004

PART 178: COMMERCE IN FIREARMS (Also 27 CFR 178.94, 178.12)

Recordkeeping procedures for "drop shipments" of firearms are prescribed.

ATF Proc. 75-3

Purpose: This ATF procedure sets forth the recordkeeping procedures for "drop shipments" of firearms (other than National Firearms Act firearms as defined in section 5845(a) of Chapter 53, Title 26, U.S.C.) and ammunition between federally licensed firearms dealers, importers, and manufacturers.

Background: The Bureau has experienced difficulty in tracing firearms in instances where drop shipments have been made to third parties and where the recordkeeping procedures employed by the three parties do not lend themselves to easy and fast tracing of firearms and ammunition. For this reason, the Bureau has prescribed recordkeeping procedures for "drop shipments" as set forth below.

Procedures: (1) Where licensee "A" places an order for firearms or ammunition with licensee "B" and "B" transmits the order to licensee "C" for direct shipment (drop shipment) to "A," a certified copy of the license of "A" must be forwarded to "C" prior to shipment of the order. On shipment of the order to "A," "C" shall enter in his bound record the disposition of the firearms or ammunition to "A." On receipt of the shipment by "A," he shall enter the acquisition of the firearms or ammunition in his bound record. Both licensees shall make such entries in the manner prescribed by regulations. Since the actual movement of the firearms or ammunition is between "C" and "A" and since "B" does not take physical possession of them, "B" will make no entry in his bound record. However, "B" should make appropriate entries or notations in his commercial records to reflect the transaction.

(2) For example, where a licensed dealer orders firearms from a wholesaler and the wholesaler requests drop shipment from the manufacturer to the dealer, a certified copy of the dealer's license shall accompany the wholesaler's order to the manufacturer. The manufacturer shall enter in his bound record the disposition of the firearms to the dealer, and the dealer shall enter the acquisition of the firearms in his bound record reflecting receipt from the manufacturer. The wholesaler, although a part of the business transaction, neither acquires nor disposes of the firearms and would, therefore, enter nothing of the transaction in his bound record.

NFA Firearms: Transfer of National Firearms Act firearms may be accomplished only pursuant to the manner outlined in Subpart F, Part 179, Title 27, Code of Federal Regulations.

Inquiries: Inquiries concerning this procedure should refer to its number and be addressed to the office of the appropriate Regional Director.

[75 ATF C.B. 78]

27 CFR 179.35: EMPLOYER IDENTIFICATION NUMBER (Also see 179.34, 179.84 179.88, 179.90, 179.103 and 179.112)

Identification Number for Special (Occupational) Taxpayer

ATF Proc. 90-1

Purpose: The purpose of this ATF procedure is to inform Federal Firearms licensees who have paid the special (occupational) tax to import, manufacture, or deal in National Firearms Act (NFA) firearms of the discontinuance of the use of the ATF Identification Number and the replacement with the use of the Employer Identification Number (EIN) on all NFA transaction forms.

Background: Section 5801 of Title 26, U.S.C. provides that on first engaging in business, and thereafter on or before the first day of July of each year, every importer, manufacturer, and dealer in NFA firearms shall pay the appropriate special (occupational) tax. In addition, section 5802 requires each importer, manufacturer, and dealer to register with the Secretary his name and the address of each location where he will conduct business. The filing of ATF Form 5630.5, with payment of the appropriate tax required by section 5801, also accomplishes registration requirements under section 5802.

The regulations at 27 CFR 179.34 require that the special tax be paid by return (ATF Form 5630.5, Special Tax Registration and Return) and require that all information called for on the return be provided, including the Employee Identification Number. 27 CFR 179.35 provides the instruction for applying for an EIN.

The regulations in 27 CFR 179.84, 179.88, and 179.90 require that the application to transfer an NFA firearm identify the special tax stamp, if any, of the transferor and transferee. The regulations in 27 CFR 179.103 and 179.112 require that the notice submitted to register NFA firearms identify the special tax stamp of the manufacturer or importer respectively. Identification of the tax stamp is necessary to ensure the tax liability has been satisfied, that the parties are qualified to import, manufacture, or deal in NFA firearms, and, in certain instances, is necessary to ensure that both parties in a transfer application are entitled to an exemption from the transfer tax.

In 1980, because of delays in the issuance of special tax stamps resulting in the inability of special taxpayers to conduct business operations, ATF Procedure 80-6 was implemented. This procedure notified taxpayers that they could obtain an ATF identification number which should be used in lieu of the IRS special tax stamp number on all NFA transaction forms. This procedure was established to facilitate the processing of NFA forms and to eliminate the delay caused by the time period required for IRS processing of the special tax stamp.

ATF has recently taken over the collection of special tax from the Internal Revenue Service, and is now issuing the special tax stamps. The number used to identify the special tax stamp is the EIN.

Because the number used to identify the special tax stamp is the EIN, this number must appear on all forms (applications, notices, and returns) involving NFA firearms. The problems that caused the implementation of the procedure in 1980 have been resolved. In fact, the assignment of an ATF identification number is now duplicative and requires more paperwork of the taxpayer. Accordingly, the use of the ATF identification number is no longer necessary and is discontinued.

ATF Procedure 80-6 is cancelled.

Inquiries: Inquiries regarding this ATF procedure should refer to its number and be addressed to the Bureau of Alcohol, Tobacco and Firearms, Chief, National Firearms Act Branch, Washington, DC 20226.

[ATFB 1990-1 55]

INDUSTRY CIRCULARS

Industry Circular 72-23

SHIPMENT OR DELIVERY OF FIREARMS BY LICENSEES TO EMPLOYEES, AGENTS, REPRESENTATIVES, WRITERS, AND EVALUATORS

Purpose. The purpose of this circular is to clarify the provisions of 18 U.S.C. Chapter 44, and Subpart F of the regulations thereunder (27 CFR 178) pertaining to the shipment of firearms in interstate commerce by a firearms licensee to its own nonlicensed employees, agents, and representatives, for the use and benefit of the licensee's business. The position of the Bureau is set out in Revenue Ruling 69-248.

Background. Revenue Ruling 69-248 provides as follows: [See RR 69-248].

Scope. Included within the category of agents and representatives discussed in the Revenue Ruling are professional writers, consultants and evaluators who in the course of their professions acquire firearms from a licensee for research or evaluation. The Revenue Ruling applies only to firearms acquired from a licensee for limited lengths of time and where the title to and ultimate control of the firearm remains in the licensee. Should the writer or evaluator desire to permanently keep the examined firearm, prior arrangements must be made to acquire the firearm through a licensee in such writer's or evaluator's State of residence and the Revenue Ruling would have no application.

Restriction. This Revenue Ruling also does not apply to firearms and ammunition within the purview of the National Firearms Act (26 U.S.C. Chapter 53).

Records. The licensee should enter in his firearms records the shipment or delivery of firearms to the employee, agent, representative, writer, consultant, or evaluator in accordance with Subpart H of the regulations. Upon the completion of the business purpose for which the firearms were received the firearms must be returned to the licensee who should enter their receipt in his records.

Industry Circular 72-30

IDENTIFICATION OF PERSONAL FIREARMS ON LICENSED PREMISES NOT OFFERED FOR SALE

Purpose. The purpose of this circular is to urge licensed firearms dealers to identify their personal collection of firearms kept at the business premises.

Scope. The provisions of Section 923(g), 18 U.S.C. Chapter 44, and Subpart H of the regulations (27 CFR 178) require all licensed firearms dealers to maintain records of their receipt and disposition of all firearms at the licensed premises. Section 178.121(b) of the regulations and the law further provide for the examination and inspection during regular business hours or other reasonable times of firearms kept or stored on business premises by licensees and any firearms record or document required to be maintained.

Guidelines for Identifying Personal Firearms on the Business Premises of Licensed Dealers. A presumption exists that all firearms on a business premises are for sale and accordingly must be entered in the records required to be maintained under the law and regulations. However, it is recognized that some dealers may have personal firearms on their business premises for purposes of display or decoration and not for sale. Firearms dealers who have such personal firearms on licensed premises should not intermingle such firearms with firearms held for sale. Such firearms should be segregated from firearms held for sale and appropriately identified (for example, by attaching a tag) as being "not for sale". Personal firearms on licensed premises which are segregated from firearms held for sale and which are appropriately identified as not being for sale need not be entered in the dealer's records.

There may be occasions where a firearms dealer utilizes his license to acquire firearms for his personal collection. Such firearms must be entered in his permanent acquisition records and subsequently be recorded as a disposition to himself in his private capacity. If such personal firearms remain on the licensed premises, the procedures described above with respect to segregation and identification must be followed.

The above procedures will facilitate the examination and inspection of the records of firearms dealers and result in less inconvenience to licensees.

Industry Circular 74-13

GUIDELINES FOR VERIFYING IDENTITY AND LICENSED STATUS OF TRANSFEREE

General: The Bureau urges all firearms licensees to require whatever information they deem necessary and within reason in order to verify the identity and licensed status of transferee licensees with whom they do business.

Personal Appearance: A licensee who appears in person at another licensee's business premises for the purpose of acquiring firearms should be required to furnish, to the transferor, positive identification in addition to a certified copy of his license [or in addition to a copy of his certified list, if a multi-licensed entity]. Such identification should prove to the satisfaction of the transferor that the person receiving firearms is, in fact, the same person to whom the license was issued.

Mail Order Sales: When the shipment is to be made to an address other than the transferee's premises as listed on his license or on his certified list, it is suggested that the transferor verify the address as being that of the transferee.

Industry Circular 77-20

DUPLICATION OF SERIAL NUMBERS BY LICENSED IMPORTERS

ATF has noted cases where some licensed importers have adopted the same serial number for more than one firearm. These instances of duplication have generally occurred when firearms are received from more than one source.

Title 27, CFR section 178.92 requires that the serial number affixed to a firearm must not duplicate the number affixed to any other firearm that you import into the United States. Those of you who import destructive devices are under the same requirement due to the inclusion of destructive devices in the definition of firearm as used in 27 CFR 178.11. ATF Ruling 75-28 stated that a serial number affixed by the foreign manufacturer may be adopted to fulfill this unique serial number requirement. However, the manufacturer's serial number must be affixed in the manner set forth in 27 CFR 178.92 and must not duplicate a number previously adopted by you for another firearm.

If you receive two or more firearms with the same serial number, it is your responsibility to affix additional markings to make each serial number unique.

ATF Ruling 75-28 also reminds you of the other identifying marks required by 27 CFR 178.92. In addition to a unique serial number, each firearm must be marked to show the model (if any); the caliber or gauge; the name of the manufacturer and importer, or recognizable abbreviations; the country of manufacture; and the city and State (or recognized abbreviations) in which your licensed premises are located.

GENERAL INFORMATION

TABLE OF CONTENTS

GENERAL INFORMATION

1. INFORMATION ABOUT ATF

For assistance with applications, records, transactions, and other regulatory matters, contact your local ATF office. Information about criminal violations of Federal firearms laws also should be referred to that office.

A list of ATF field offices is on page 198 of this publication.

Requests for ATF forms and publications should be directed to the ATF Distribution Center, P.O. Box 5950, Springfield, Virginia 22150-5950. Most forms and publications also are available at www.ATF.gov.

ATF publishes the ATF Quarterly Bulletin, by which the agency informs interested persons about current alcohol, tobacco, firearms, and explosives matters, including regulatory, procedural, and administrative information; items of general interest; and, excerpts from public laws and congressional committee reports. The ATF Quarterly Bulletin is available at www.ATF.gov.

2. INFORMATION CONCERNING AR-15 TYPE RIFLES

ATF has encountered various AR-15 type assault rifles such as those manufactured by Colt, E.A. Company, SGW, Sendra and others, which have been assembled with fire control components designed for use in M16 machineguns. The vast majority of these rifles which have been assembled with an M16 bolt carrier, hammer, trigger, disconnector and selector will fire automatically merely by manipulation of the selector or removal of the disconnector. Many of these rifles using less than the 5 M16 parts listed above also will shoot automatically by manipulation of the selector or removal of the disconnector.

Any weapon which shoots automatically more than 1 shot without manual reloading, by a single function of the trigger, is a machinegun as defined in 26 U.S.C. 5845(b), the National Firearms Act (NFA). The definition of a machinegun also includes any combination of parts from which a machinegun may be assembled, if such parts are in possession or under the control of a person. An AR-15 type assault rifle which fires more than 1 shot by a single function of the trigger is a machinegun under the NFA. Any machinegun is subject to the NFA and the possession of an unregistered machinegun could subject the possessor to criminal prosecution.

Additionally, these rifles could pose a safety hazard in that they may fire automatically without the user being aware that the weapon will fire more than 1 shot with a single pull of the trigger.

In order to avoid violations of the NFA, M16 hammers, triggers, disconnectors, selectors and bolt carriers must not be used in assembly of AR-15 type semiautomatic rifles, unless the M16 parts have been modified to AR-15 Model SP1 configuration. Any AR-15 type rifles which have been assembled with M16 internal components should have those parts removed and replaced with AR-15 Model SP1 type parts which are available commercially. The M16 components also may be modified to AR-15 Model SP1 configuration.

It is important to note that any modification of the M16 parts should be attempted by fully qualified personnel only.

Should you have any questions concerning AR-15 type rifles with M16 parts, please contact your nearest ATF office. A list of ATF field offices is on page 198 of this publication.

3. FEDERAL AGE RESTRICTIONS

Federal law prohibits Federal firearms licensees from selling or delivering any firearm or ammunition to any individual who the licensee knows or has reasonable cause to believe is less than 18 years of age, and, if the firearm is other than a shotgun or rifle, or ammunition for a shotgun or rifle, to any individual who the licensee knows or has reasonable cause to believe is less than 21 years of age. (18 U.S.C. 922(b)(1), 27 CFR 478.99(b)(1).)

Ammunition interchangeable between rifles and handguns (such as .22 caliber rimfire) may be sold to an individual 18 years of age, but less than 21, if the licensee is satisfied that the ammunition is being acquired for use in a rifle.

Additionally, it generally is unlawful for a person to transfer a handgun to a juvenile (a person less than 18 years of age) and it generally is unlawful for a juvenile to possess a handgun. Exceptions are provided for the transfer of a handgun to and possession by a juvenile for the purposes of employment, ranching, farming, target practice or hunting as provided for in 18 U.S.C. 922(x).

Pursuant to regulations at 27 CFR 478.103, ATF distributes posters to licensed importers, manufacturers, and dealers that caution against the transfer of handguns to juveniles, as well as the possession of handguns by juveniles, in violation of 18 U.S.C. 922(x). The regulations require licensees to display the posters on their premises. Some licensees erroneously interpreted the poster to mean that that they may now lawfully transfer handguns to any person over 17 years of age. Enactment of section 922(x) making it unlawful to sell handguns to juveniles (persons under 18 years of age) and for juveniles to possess handguns did not alter section 922(b)(1) of the GCA that continues to prohibit licensees from transferring handguns to persons under 21 years of age.

4. SALES TO LAW ENFORCEMENT OFFICERS

Section 925(a)(1) of the GCA exempts law enforcement agencies from the transportation, shipment, receipt, or importation controls of the GCA when firearms are to be used for the official business of the agency. An individual law enforcement officer's receipt and possession of firearms for use in carrying out official duties on behalf of an agency is also exempt, unless the officer has been convicted of a misdemeanor crime of domestic violence.

If a law enforcement officer is issued a certification letter on the agency's letterhead, signed by a person in authority within the agency stating the officer will use the firearm in performance of official duties and that a records check reveals that the purchasing officer has not been convicted of a misdemeanor crime of domestic violence, the officer specified in the certification may purchase a firearm from a licensee regardless of the State in which the officer resides, or in which the agency is located. A licensee is not required to prepare a Form 4473 covering this particular sale, as the certification letter is evidence of the transaction. Moreover, a licensee is not required to comply with the Brady law (e.g., conduct a NICS check) where the purchase is made pursuant to such letter. However, disposition to the officer is to be entered in the licensee's permanent records and the certification letter kept in his files. The permanent records should show the residence address of the purchasing officer, not the address of the officer's employing agency.

If a law enforcement officer desiring to purchase a firearm does not have a certification letter, a licensee may still make the sale if the requirements of the Brady law (18 U.S.C. 922(t)), including a NICS background check of the purchaser, are met. A Form 4473 covering such a sale must be prepared, the transaction must be entered in the licensee's permanent records, and all other applicable requirements of the law and regulations must be met.

ATF considers the following as persons having authority to certify that law enforcement officers purchasing firearms will use the firearms in performance of official duties:

a. In a city or county police department, the director of public safety or the chief or commissioner of police.

b. In a sheriff's office, the sheriff.

c. In a State police or highway patrol department, the superintendent or the supervisor in charge of the office to which the State officer or employee is assigned.

d. In Federal law enforcement offices, the supervisor in charge of the office to which the Federal officer or employee is assigned.

The Bureau would also recognize someone signing on behalf of a person in authority, provided there is a proper delegation of authority and overall responsibility has not changed in any way. If the purchasing officer is a supervisory officer, the certification must be made by that officer's supervisor. In other words, the purchasing officer and the certifying officer may not be the same person.

5. SALES TO ALIENS IN THE UNITED STATES

a. Domestic Sales

In order to purchase firearms in the United States from a Federal firearms licensee, an alien must:

(1) Be 18 years of age (21 for handguns);

(2) Provide the licensee with a government-issued photo identification document;

(3) Complete ATF Form 4473, Firearms Transaction Record;

(4) Comply with the Brady law, 18 U.S.C. 922(t);

(5) Be a resident of the State in which the firearm purchase is made for a period of 90 continuous days before the transfer (if purchasing a long gun, the alien must establish 90-day continuous residency in any State) and substantiate residency by documentation (for example, utility bills or a lease agreement);

(6) Not be an illegal or nonimmigrant alien (for exceptions to the nonimmigrant alien prohibition, see 18 U.S.C. 922(y)(2) and (3)) ; and

(7) Not be a felon or within any other category of prohibited person.

An alien who is legally in the United States will be considered to be a resident of a State for the purpose of complying with the GCA if he or she is residing in that State and has resided in that State continuously for at least 90 days before purchasing a firearm. Note, however, that even a legal resident alien who has lived in the United States for many years will have to wait 90 days before purchasing a firearm if the alien changes his or her State of residence.

b. Export Sales

Removal of a firearm or ammunition from the U.S. by anyone is an exportation. With few exceptions, the firearms licensee must obtain an export license (Form DSP-5) from the State Department's Directorate of Defense Trade Controls (DDTC) or the Commerce Department's Bureau of Industry and Security (BIS) prior to exportation. When a licensee exports firearms directly to an alien's residence outside the United States, the licensee need only record the name and address of the foreign customer in his or her bound book. ATF Form 4473 need not be completed.

DDTC takes the position that when a dealer knows or believes that a foreign customer intends to take a rifle or handgun out of the U.S., the dealer is legally obligated to notify DDTC that the firearm was sold for the purpose of exportation.

Exportation Guidelines

Exportation of firearms other than sporting shotguns is regulated by the Department of State, DDTC.

Any person who intends to export, temporarily export or to import temporarily firearms, ammunition or components as defined under the United States Munitions List (22 CFR 121) must obtain the approval of the DDTC prior to export or temporary import, unless the export or temporary import qualifies for an exemption under the provisions of the International Traffic in Arm Regulations (22 CFR 120-130).

For further information about the exportation or temporary importation of firearms and ammunition, contact:

DIRECTORATE OF DEFENSE
TRADE CONTROLS
U.S. DEPARTMENT OF STATE
PM/DDTC, SA-1, ROOM 1200
WASHINGTON, DC 20037

TELEPHONE: 202-663-1282
WEBSITE: WWW.PMDTC.ORG

Exportation Guidelines (Shotguns and Related Parts, Components, Shotgun Shells, and C ertain Muzzle Loading Firearms)

The Department of Commerce regulates the exportation of shotguns with a barrel length of 18 inches and over, as well as related parts, components, shotgun shells, and certain muzzle loading (black powder) firearms. The Department of Commerce requires a specific license to export or re-export these items to most destinations. For further information, contact:

OUTREACH AND EDUCATION
SERVICES DIVISION
OFFICE OF EXPORTER SERVICES
BUREAU OF INDUSTRY AND
SECURITY
DEPARTMENT OF COMMERCE
1401 CONSTITUTION AVENUE, N.W.,
ROOM 1099-C
WASHINGTON, DC 20230

TELEPHONE: 202-482-4811
WEBSITE: WWW.BIS.DOC.GOV

c. Sales to Diplomats

Diplomats, as individuals, are not exempt from Federal, State or local firearms laws. Sales to individuals, including diplomats and embassy personnel, must comply with all requirements of the GCA and the firearms regulations (27 CFR Part 478), including the general prohibition on nonimmigrant aliens unless the diplomat is purchasing the firearm/ammunition for official purposes.

d. Sales to Embassies or Consulates

Special provisions have been made to allow for the sale of small quantities of firearms to foreign missions for the purpose of the physical security of embassy grounds. The arms become the property of the government whose embassy made the purchase, not the private property of an individual.

The dealer should obtain documentation which will show that the sale was a bona fide sale to a foreign mission and not a sale to an individual diplomat. Documentation should contain one of the following:

(1) A purchase order or invoice from the foreign mission;

(2) Payment out of government funds rather than from private funds; or

(3) A written statement by the principal officer of the embassy or consulate that the weapons are being purchased by, and will be the property of, the mission.

Once the dealer has documented that a sale is to a foreign mission, he or she may complete the transaction by shipping or delivering the firearms directly to the foreign mission. Form 4473 need not be completed because ATF considers the sale to be an exportation. ATF views the transaction as an exportation because embassy grounds are regarded as foreign territory.

However, DDTC does not view the sale of "reasonable quantities" of firearms to a foreign embassy to be an exportation. Consequently, the dealer need not obtain an export license from DDTC to deliver firearms to the embassy. DDTC should be contacted for further information.

6. CANADIAN FIREARMS INFORMATION

a. General

Implementation of the Firearms Act on December 1, 1998, brought about extensive changes to Canadian firearm regulations. Most changes affecting visitors bringing firearms into Canada came into effect on January 1, 2001.

b. Individuals Bringing Firearms Into Canada

An individual must be at least 18 years of age to bring a firearm into Canada. Prohibited firearms (see discussion below on prohibited firearms) may not be brought into Canada or transported through Canada.

Restricted firearms (see discussion below on restricted firearms) may only be imported with prior authorization from the Chief Firearms Officer of the province or territory to which you are traveling.

Anyone entering Canada must declare all firearms to Canadian Customs.

For more information on bringing firearms into Canada, contact the Canada Firearms Centre at 1-800-731-4000.

1. Prohibited firearms

The following firearms are classified as prohibited firearms and cannot be brought into Canada:

- short-barreled handguns (handguns with a barrel length equal to or less than 105 mm)
- .25 caliber handguns
- .32 caliber handguns

No handgun listed above is prohibited if it is prescribed by regulation for use in competitions governed by the rules of the International Shooting Union.

- sawed-off rifles or sawed-off shotguns less than 660 mm in overall length
- sawed-off rifles or sawed-off shotguns which have a barrel length of less than 457 mm and are equal to or more than 660 mm in overall length
- all automatic firearms
- automatic firearms that have been converted to semiautomatic or single shot
- all firearms prescribed by regulation (military and paramilitary firearms and firearms deemed to have no legitimate sporting or recreational use).

Cartridge magazines are also regulated. Generally, magazines used in semiautomatic, centerfire rifles and shotguns, with a capacity to contain more than 5 cartridges are prohibited. Magazines for semi-automatic handguns that can contain more than 10 cartridges are also prohibited.

When a prohibited firearm is declared at Canadian Customs, a customs officer may allow the firearm to be exported back to its country of origin. Firearms that are not immediately exported are forfeited.

For information on firearms prohibited by regulations or on firearms prescribed as International Shooting Union handguns, contact the Canada Firearms Centre at 1-800-731-4000.

2. Restricted firearms

The following firearms are classified as restricted firearms requiring an Authorization to Transport from a Chief Firearms Officer to bring into Canada:

- all handguns which are not prohibited firearms
- semiautomatic centerfire rifles and shotguns that have a barrel length less than 470 mm and are not prohibited
- rifles and shotguns that can fire after being reduced to an overall length of less than 660 mm, by any temporary means such as folding or telescoping
- all firearms prescribed by regulation as restricted (for information on these regulations, contact the Canada Firearms Centre).

Anyone bringing a restricted firearm into Canada must have an Authorization to Transport for the restricted firearm. This authorization will permit transport of the restricted firearm between specified places within Canada. This authorization must be obtained in advance from the Chief Firearms Officer of the Canadian province or territory to be visited.

An applicant for an Authorization to Transport must have a valid purpose for bringing restricted firearms to Canada, such as for use in target practice, or a target shooting competition at an approved shooting club or range. Restricted firearms cannot be used for hunting.

For more information on Authorizations to Transport, call the Chief Firearms Officer of the Canadian province or territory that you will be visiting. You can obtain the address, telephone and fax numbers from the Canada Firearms Centre.

3. January 1, 2001: Confirmed Firearms Declarations

All persons who possess or use firearms in Canada must have a license authorizing possession of firearms as of January 1, 2001. Visitors to Canada will require authorization to possess firearms in the form of a confirmed Firearms Declaration. When firearms are declared on entry into Canada, the declaration will be confirmed by a customs officer.

Firearms Declarations must normally be made in writing. In certain cases, a customs officer can accept an oral declaration. In some cases, it will be possible to complete and mail in a predeclaration form, which can be confirmed when you arrive at the border. Firearms Declaration forms are available through outfitters, hunting and shooting clubs, Canadian tourism offices, and the Canada Firearms Centre.

The declaration requires basic information about the visiting firearm user, the destination in Canada, and the reason for bringing a firearm into Canada. Background checks, including a criminal history search, may be conducted. Descriptive information about each firearm being brought into the country will also be required (i.e., make, model, serial number, caliber/gauge).

4. Confirmation

On arrival at the border, a customs officer may review the Firearms Declaration and examine the firearms. If everything checks out, the customs officer will confirm the Firearms Declaration and, where restricted firearms are being imported, the Authorization to Transport. The visitor will receive a confirmation number either in writing or orally.

A confirmed Firearms Declaration will serve two purposes:

- it will act as a temporary license authorizing possession of the firearm(s) listed thereon; and
- it will serve as a temporary registration certificate for the firearms imported.

A Firearms Declaration will expire after 60 days. However, a Firearms Declaration for restricted firearms will expire on the earlier of 60 days or the expiration date of the Authorization to Transport. A new Firearms Declaration will be required for every reentry into Canada. A Firearms Declaration may be extended from within Canada by contacting the Chief Firearms Officer of the province or territory visited.

5. Refusal to confirm

A customs officer can refuse to confirm a Firearms Declaration if:

- the declaration form is not completed truthfully, or required information is not provided;
- the requirements set out in the Firearms Act and regulations are not met; or
- he or she concludes that it is desirable, in the interests of the visitor's safety and/or the safety of others, that the firearm not be allowed to enter Canada.

If the customs officer refuses to confirm the Firearms Declaration, then, depending on the situation, the officer may:

- require that the firearm be exported from Canada; or
- detain the firearm and give the visitor a reasonable amount of time to comply with requirements.

6. Fee

The fee for a 60-day Firearms Declaration is $50 in Canadian funds. This amount will be payable only once every 12 months. If, within the 12-month period, the Firearms Declaration is extended, or a new declaration is issued on reentry to Canada, the fee will be waived. The $50 fee covers all firearms imported by the same individual.

There is no fee for the Authorization to Transport restricted firearms.

c. Borrowing Firearms while in Canada

On April 1, 1999, a temporary license authorizing non-residents to borrow firearms was introduced. The Non-Resident's Sixty-Day Possession License allows a holder to borrow **nonrestricted** firearms (ordinary rifles and shotguns) in Canada. Persons who are **18 years of age or over** may get a license to borrow firearms for the following purposes:

- hunting under the supervision of an outfitter or other person authorized to provide organized hunting services in Canada;
- hunting with a Canadian resident who has the proper firearms license and hunting license;
- competing in a shooting competition;
- target practice at an approved shooting club or range;
- participating in a historical reenactment or display;
- engaging in a business or scientific activity being carried on in a remote area where firearms are necessary for the control of predators;
- participating in a parade, pageant or other similar event; or
- using firearms for movie, television, video or theatrical productions or publishing activities.

Visitors are advised to apply for this temporary license well in advance of arrival at the Canadian border. License application forms are available from the Canada Firearms Centre.

Fee: The fee for a Non-Resident's Sixty-Day Possession License for borrowed firearms will be $30 in Canadian funds. This license may be renewed once in a 12-month period at no extra charge. Subsequent renewals cost $30.

d. Storing, Transporting and Handling Firearms in Canada

Canadian federal law requires that firearms be transported unloaded. Restricted firearms must also be rendered inoperable with a secure locking device and locked inside an opaque container that cannot be readily broken open or into or accidentally opened during transportation. When left in an unattended vehicle, the firearms must be in a locked trunk or a similar compartment. Where the vehicle does not have a trunk or similar compartment, the firearms must be locked inside the vehicle and out of sight. Provincial and municipal regulations may require a nonrestricted firearm to be transported in a case in certain areas and at certain times.

When stored, firearms must be unloaded and rendered incapable of being fired, by using a secure locking device, by removing the bolt or bolt carrier or by storing the firearm in a sturdy, secure, locked container or room. Restricted firearms must be stored in a sturdy, secure, locked container or room and rendered inoperable with a secure locking device. Ammunition must be stored separately, unless it is in a securely locked container (ammunition and firearms may be stored in the same secure, locked container).

More information on requirements for the safe storage, transportation and handling of firearms in Canada is available from the Canada Firearms Centre.

e. Canada Firearms Centre

The Canada Firearms Centre Administration is an agency of the Government of Canada responsible for the implementation of the Firearms Act. For more information or to order a copy of the Firearms Act and regulations, or other Canada Firearms Centre publications, contact the CFC at:

CANADA FIREARMS CENTRE
OTTAWA, ONTARIO, CANADA
K1A 1M6

TELEPHONE: 1-800-731-4000
WEBSITE: http://www.cfc.gc.ca
E-MAIL: cfc-cafc@cfc-cafc.gc.ca

This article is intended to provide general information only. For legal references, refer to the Firearms Act and its regulations. Provincial, territorial and municipal laws, regulations, and policies may also apply.

7. OPERATIONS BY LICENSED COLLECTORS

a. Licensing

A collector of curios or relics may obtain a collector's license under the GCA. The privileges conferred by this license extend only to curio or relic transactions, as discussed in detail below. In transactions involving firearms not classified as curios or relics, the licensed collector has the same status as a nonlicensee. A person need not be federally licensed to collect curios or relics. However, the individual must be licensed in order to lawfully receive curios or relics from outside his or her State of residence. Federal law, regulations, and general information pertaining to licensed collectors and curios or relics can be found in this publication.

Recordkeeping requirements for licensed collectors are discussed in detail in 27 CFR Part 478.

b. What are Curios or Relics?

As set out in the regulations (27 CFR 478.11), curios or relics include firearms which have special value to collectors because they possess some qualities not ordinarily associated with firearms intended for sporting use or as offensive or defensive weapons.

Please note that ammunition is no longer classified as curios or relics since the Congress in 1986 removed the interstate controls over ammunition under the GCA.

To be recognized as curios or relics, firearms must:

(1) Have been manufactured at least 50 years prior to the current date, but does not include replicas thereof; **or**

(2) Be certified by the curator of a municipal, State or Federal museum which exhibits firearms to be curios or relics of museum interest; **or**

(3) Derive a substantial part of their monetary value from the fact that they are novel, rare, or bizarre, or from the fact of their association with some historical figure, period, or event.

ATF has recognized only complete, assembled firearms as curios or relics. ATF's classification of surplus military firearms as curios or relics has extended only to those firearms in their original military configuration.

Frames or receivers of curios or relics are not generally recognized as curios or relics by ATF since they are not of special interest or value as collectors' items. Specifically, they do not meet the definition of curio or relic in 27 CFR 478.11 as firearms of special interest to collectors by reason of a quality other than is ordinarily associated with sporting firearms or offensive or defensive weapons.

Collectors wishing to obtain a determination whether a particular firearm qualifies for classification as a curio or relic in accordance with 27 CFR 478.26 should submit a written request for a ruling. The letter should include:

(1) A complete physical description of the item;

(2) Reasons the collector believes the item merits the classification;

(3) Data concerning the history of the item, including production figures, if available, and market value.

In some cases, actual submission of the firearm may be required prior to a determination being made. Requests should be sent to the Bureau of ATF, Firearms Technology Branch, 244 Needy Road, Martinsburg, West Virginia 24501.

ATF's classifications of curios or relics are published in ATF P 5300.11, Firearms Curios or Relics List. Curios or relics are listed in the publication under the following headings:

Section I. Ammunition Classified as Curios or Relics: As noted above, Congress ended the recognition of ammunition curios or relics. Thus, no ammunition has received curio or relic classification since August 1986.

Section II. Firearms Classified as Curios or Relics Under the GCA: Licensed collectors may acquire, hold or dispose of these firearms as curios or relics. However, they are still firearms as defined in 18 U.S.C. 921(a)(3) and are, therefore, subject to all GCA controls. Generally, this category includes commemorative handguns, semiautomatic pistols, revolvers and rifles.

Section III. NFA Firearms Removed From the NFA as Collectors' Items and Classified as Curios or Relics Under the GCA: Weapons in this section are excluded entirely from the provisions of the NFA. Thus, approval from ATF to transfer these weapons is not required. They need not be registered in the National Firearms Registration and Transfer Record and they are not subject to the transfer tax. These weapons are still firearms under the GCA and remain subject to regulation under 27 CFR Part 478.

Section IIIA. Weapons Removed from the NFA as Collectors' Items Which are Antiques not Subject to the Provisions of the GCA: Weapons in this section are not subject to the provisions of either the NFA or the GCA.

Section IV. NFA Firearms Classified as Curios or Relics Under the GCA: These weapons (e.g., machineguns) are firearms within the scope of the NFA and are subject to all the Act's provisions. Accordingly, these weapons cannot be lawfully transferred or received unless they are registered with ATF in the National Firearms Registration and Transfer Record.

c. Licensed Collector's Activities

Subject to other applicable provisions of the law and regulations, a collector's license entitles its holder to transport, ship, receive and acquire curios or relics in interstate or foreign commerce and to dispose of curios or relics in interstate or foreign commerce to any other Federal firearms licensee. Dispositions of curios or relics by licensed collectors are not subject to the requirements of the Brady law; however, dispositions should not be made to any person whom the collector knows or has reasonable cause to believe is a felon or is within any other category of persons to whom sales are prohibited by 18 U.S.C. 922(d).

Those collectors having questions concerning the importability of specific curio or relic firearms should contact the Bureau of ATF, Firearms and Explosives Imports Branch.

The principal advantage of a collector's license is that the collector can lawfully acquire curios or relics from both licensees and nonlicensees without regard to his/her State of residence. A licensed collector may acquire and dispose of curios or relics at any location, the only limitation being that a disposition made to a nonlicensee is to be made to a resident of the same State in which the collector is licensed.

d. Restrictions on Licensed Collector's Activities

As stated earlier, the collector's license covers only transactions in curios or relics. A licensed collector has the same status as a nonlicensee with respect to transactions in firearms that are not curios or relics.

While a licensed collector may acquire and dispose of curios or relics from a personal collection, the collector is not authorized to engage in a firearms dealing business in curios or relics pursuant to a collector's license. As stated in 27 CFR 478.41(d), "...if the acquisitions and dispositions of curios or relics by a collector bring the collector within the definition of a manufacturer, importer or dealer under this part, he shall qualify as such." For example, if a collector acquires curios or relics for the purpose of sale rather than to enhance a collection, the collector would have to be licensed as a dealer in firearms under the GCA. Additionally, if the collector is dealing in NFA firearms, the collector would be liable for the special (occupational) tax prescribed by the NFA. The sole intent and purpose of the collector's license is to enable a firearms collector to obtain a curio or relic from outside his or her State of residence.

8. ANTIQUE FIREARMS

Under section 921(a)(16) of the GCA, the term antique firearm means:

(A) any firearm (including any firearm with a matchlock, flintlock, percussion cap, or similar type of ignition system) manufactured in or before 1898; **or**

(B) any replica of any firearm described in subparagraph (A) if such replica—

(i) is not designed or redesigned for using rimfire or conventional centerfire fixed ammunition, or

(ii) uses rimfire or conventional centerfire fixed ammunition which is no longer manufactured in the United States and which is not readily available in the ordinary channels of commercial trade; or

(C) any muzzle loading rifle, muzzle loading shotgun, or muzzle loading pistol, which is designed to use black powder, or a black powder substitute, and which cannot use fixed ammunition.

For purposes of subparagraph (C), a muzzle loading rifle, shotgun, or pistol is not an "antique firearm" for purposes of the GCA if it incorporates a firearm frame or receiver, if it is a firearm which is converted into a muzzle loading weapon, or if it can be readily converted to fire fixed ammunition by replacing the barrel, bolt, breechblock, or any combination of such parts.

Under section 5845(g) of the NFA, **antique firearm** means:

"...Any firearm not designed or redesigned for using rim fire or conventional center fire ignition with fixed ammunition and manufactured in or before 1898 (including any matchlock, flintlock, percussion cap, or similar type of ignition system or replica thereof, whether actually manufactured before or after 1898) and also any firearm using fixed ammunition manufactured in or before 1898, for which ammunition is no longer manufactured in the United States and is not readily available in the ordinary channels of commercial trade."

To illustrate the distinction between the two definitions of antique firearm under the GCA and NFA, a rifle manufactured in or before 1898 would be an antique firearm under the provisions of the GCA, even though it uses conventional ammunition. However, if such rifle has a barrel of less than 16 inches in length AND uses conventional fixed ammunition which is available in the ordinary channels of commercial trade, it would not be an antique firearm under the NFA.

An antique firearm as defined in both the GCA and NFA is exempt from all of the provisions and restrictions contained in both laws. Consequently, such an antique firearm may be bought, sold, transported, shipped, etc., without regard to the requirements of these laws.

Under the Arms Export Control Act certain "antique firearms" are not subject to the import controls under that Act. These "antique firearms" are defined as "muzzle loading (black powder) firearms (including any firearm with a matchlock, flintlock, percussion cap, or similar type of ignition system) or firearms covered by Category I(a) established to have been manufactured in or before 1898." No all-inclusive list of antique firearms is published by ATF.

Under the Export Administration Act, the Bureau of Industry and Security (BIS) maintains export control requirements on a wide range of dual-use commodities, software, and technology. BIS requires an export license for muzzle loading (black powder) firearms with a caliber less than 20 mm that were manufactured later than 1937 and that are not reproductions of firearms manufactured earlier than 1890. This BIS control does not apply to weapons used for hunting or sporting purposes that were not specially designed for military use and are not of the fully automatic type. This exemption does not apply if the weapon meets certain control requirements. Contact BIS at 202-482-4811 for additional information.

9. IMPORTATION BY NONLICENSEES

a. Nonlicensed U.S. Residents

A permit must be obtained to import or bring into the United States any firearm or ammunition. The firearm or ammunition must be generally recognized as particularly suitable for, or readily adaptable to, sporting purposes.

Surplus military firearms are generally excluded from importation into the United States except for certain curio or relic surplus military firearms imported by licensed importers only.

A federally licensed firearms dealer located in the nonlicensee's State of residence may act as an agent to import the nonlicensee's personal firearm, provided that the firearm is lawfully importable. The form to be used by the licensee is ATF Form 6, Part I, Application and Permit For Importation of Firearms, Ammunition and Implements of War, and may be obtained from the Bureau of ATF, Firearms and Explosives Imports Branch, or from the ATF website at www.ATF.gov.

A nonlicensee may obtain a permit to import sporting ammunition for personal use (excluding armor piercing handgun ammunition, or tracer or incendiary ammunition) or firearm parts (other than frames or receivers) without engaging the services of a Federal firearms licensee. Silencer parts and certain machinegun parts are subject to the NFA and may not be imported. If the nonlicensee chooses to have a licensee handle the importation, the licensee should complete and send to ATF an ATF Form 6, Part I, in accordance with the instructions on the form. The nonlicensee's name, address, and telephone number should appear in Item 9, "Specific purpose of importation."

No permit or authorization from ATF is required to bring into the United States a firearm or ammunition that was ***previously taken out*** of the U.S. by the person bringing it in. U.S. Customs and Border Protection (CBP) is authorized to release a firearm or ammunition without a permit from ATF upon a proper showing of proof that the firearm or ammunition was taken out of the country by the person bringing it in. This proof is best established by having registered the item or items on CBP Form 4457, Certificate of Registration, at the point and time of departure.

For further information, see ATF Rul. 81-3 and ATF Rul. 85-10, set out within the Rulings, Procedures, and Industry Circulars portion of this publication.

b. Non-resident U.S. Citizens Returning to the United States and Non-resident Aliens Immigrating to the United States

A non-resident U.S. citizen returning to the United States, or a non-resident alien lawfully immigrating to the United States, may apply for a permit from ATF to import for personal use, and not for resale, firearms and ammunition without having to utilize the services of a federally licensed firearms dealer. ATF Form 6, Part I application should include a statement, on the application form or on an attached sheet, that:

(1) the applicant is a non-resident U.S. citizen who is returning to the United States from a residence outside of the United States or, in the case of an alien, is lawfully immigrating to the United States from a residence outside of the United States, and

(2) the firearms and ammunition are being imported for personal use and not for resale.

No permit will be issued to import firearms or ammunition which are not generally recognized as particularly suitable for, or readily adaptable to, sporting purposes, surplus military firearms, or National Firearms Act (NFA) firearms (e.g., machineguns, silencers, destructive devices, short-barreled rifles, short-barreled shotguns, etc.).

The firearms must accompany non-resident U.S. citizens, since once a person is in the United States and has acquired residence in a State, he or she may import a firearm only by arranging for the importation through a federally-licensed firearms dealer. Non-resident aliens must bring in the firearms within 90 days of their arrival, which is when they obtain State residency.

Section 922(a)(3) of Title 18, U.S.C. makes it unlawful, with certain exceptions, for a person to bring into his State of residence a firearm which he or she acquired outside that State. An unlicensed resident of a State must, therefore, arrange for the importation of the firearm through a federally licensed firearms dealer.

c. Members of the Armed Forces

(1) Import Permit Requirements

Section 925(a)(4) of the GCA provides that:

When established to the satisfaction of the Attorney General to be consistent with the provisions of this chapter [the GCA] and other applicable Federal and State laws and published ordinances, the Attorney General may authorize the transportation, shipment, receipt or importation into the United States to the place of residence of any member of the United States Armed Forces who is on active duty outside the United States (or who has been on active duty outside the United States within the 60 day period immediately preceding the transportation, shipment, receipt, or importation), of any firearm or ammunition which is:

(A) determined by the Attorney General to be generally recognized as particularly suitable for sporting purposes, or determined by the Department of Defense to be a type of firearm normally classified as a war souvenir, and

(B) intended for the personal use of such member.

Applications to import such firearms are filed on ATF Form 6, Part II and should include a detailed description of each firearm to be imported. Incomplete information will cause return of your application. Applications should be completed in triplicate and mailed to the Bureau of ATF, Firearms and Explosives Imports Branch.

A member of the Armed Forces who does not meet the above criteria must obtain the services of a Federal firearms licensee located in his or her State of residence to import a firearm on behalf of the member. The licensee would submit an application on ATF Form 6, Part I.

If your application is approved, the original will be returned to you. This will be your authorization to import the firearm(s) described on the form. The permit is valid for 1 year from the date of approval. If disapproved, your application will be stamped **disapproved** and returned to you with the reason for disapproval stated.

A permit must be obtained for all firearms to be imported, regardless of the date purchased. However, this does not apply to a firearm previously taken out of the United States by the person bringing it in (if they can prove they previously took the firearm out of the United States), nor to a firearm shipped by a licensee in the United States to a serviceperson on active duty outside the United States or to an authorized rod and gun club abroad specifically for the serviceperson importing the firearm.

Authorization will not be given to import a machinegun, or any other firearm as defined in the NFA, regardless of the degree of serviceability.

Authorization will not be given to import any surplus military firearm (unless it has been listed as a curio or relic in accordance with 18 U.S.C. 921(a)(13) and 925(e)) and is being imported by a licensed importer.

To determine whether or not a handgun may be authorized for importation ***as particularly suitable for sporting purposes***, the "factoring" criteria for pistols and revolvers (ATF Form 4590) is used.

(2) Importation of War Souvenirs or War Trophy Firearms

The regulations (27 CFR 478.114(c)) provide that firearms determined by the Department of Defense to be war souvenirs may be imported into the United States by members of the U.S. Armed Forces under such provisions and procedures as the Department of Defense may issue.

For information regarding the classification of war souvenirs or trophies by the Department of Defense, see DOD regulations AR 608-4, OPNAVINST 3460.7A, AFR 125-13, and MCO 5800.6A, describing articles and material that are not considered war trophies and may not be kept or imported into the United States by members of the U.S. Armed Forces.

The aforementioned Department of Defense regulations list machineguns and other firearms coming within the purview of the NFA, regardless of the degree of serviceability, among the items which are prohibited from being retained and introduced into the United States by Armed Forces personnel.

CBP is authorized to release a firearm without an import permit from ATF where a properly executed DD Form 603, **Registration of War Trophy Firearms**, is presented certifying that the firearm to be brought in has been classified as a war souvenir under DOD regulations. To be valid, the DOD Form 603 must have been issued during a period authorized by DOD.

(3) ATF Ruling 74-13

ATF was informed by State and local authorities that handguns were being transported, shipped, received, or imported into the United States by members of the U.S. Armed Forces to their State of residence without such members having obtained the required permits or other authorizations required by the State for lawful possession or ownership of handguns in that State.

Ruling 74-13 holds that a member of the U.S. Armed Forces who is a resident of any State or territory which requires that a permit or other authorization be issued prior to possessing or owning a handgun shall submit evidence of compliance with State law before an application to import a handgun may be approved.

ATF Rul. 74-13 is set out within the Rulings, Procedures, and Industry Circulars portion of this publication.

10. SPECIAL TAXPAYERS AND NFA FIREARMS

a. General

Anyone wishing to manufacture, import, or deal in firearms as defined in the NFA must:

1. Be properly licensed as a Federal firearms licensee;

2. Have an employer identification number (even if you have no employees); and

3. Pay the Special (Occupational) Tax required of those manufacturing, importing, or dealing in NFA firearms.

Those weapons defined as NFA firearms can be found in sections 5845(a)-(f) of the NFA.

After payment of the tax, you will receive a Special (Occupational) Tax Stamp as evidence you have paid the required occupational tax as a NFA manufacturer, importer, or dealer.

b. What You Need To Proceed

If you do not already have an employer identification number (EIN), you must obtain and complete a Form SS-4 application to obtain such a number. This number must appear on all registration documents when you apply to receive or transfer any NFA firearm. You may obtain the Form SS4 from any Social Security Administration Office, any IRS Service Center, or IRS District Office.

Federal firearms licensees who wish to engage in the business of importing, manufacturing, or dealing in NFA firearms are required to pay Special (Occupational) Tax for each business location. The tax year begins July 1st and ends June 30th of the following year. If you begin business any time during the tax year, you are responsible for the full amount of tax for the entire year, i.e., the taxes are not prorated.

Class of Special Tax	Activity Covered	Annual Tax Amount	Type of Firearms License
Class 1	Importer	$1000*	08 or 11
Class 2	Manufacturer	$1000*	07 or 10
Class 3	Dealer	$ 500	01

*** If your gross receipts for the prior Fiscal Year were less than $500,000, the tax is $500.**

If you want to be a Class 3 dealer, you could have a Type 01, Type 02, Type 07, or Type 08 Federal firearms license. The tax you would pay (Class 3) only allows you to deal in NFA firearms. Being a Class 3 dealer will not, however, have any effect on your business activity involving non-NFA firearms.

Submit ATF Form 5630.7, Annual Special Tax Registration and Return, along with your check or money order [not cash] to:

BUREAU OF ALCOHOL, TOBACCO, FIREARMS AND EXPLOSIVES
P.O. BOX 371970
PITTSBURGH, PENNSYLVANIA
15250-7970

Upon receipt of your properly completed ATF Form 5630.7, together with your remittance, a Special Tax Stamp will be mailed directly to you.

c. Permanent Changes

If you change your address, location, or trade name, you must file a new ATF Form 5630.7 advising us of that change. You may accomplish this easily by attaching ATF F 5630.7 to your Special Tax Stamp and mailing them to the Pittsburgh address shown above.

To change your trade name, you must also obtain an amended Federal firearms license. This is done by sending a copy of your license (with the changes noted thereon) to the Federal Firearms Licensing Center.

To change your location, you must file an Application for an Amended Federal Firearms License, ATF Form 5300.38, with the Chief, Federal Firearms Licensing Center, not less than 30 days prior to the move. You must obtain the amended license before commencing business at the new location.

We suggest that you contact the NFA Branch if there is a change in who controls the business or in business structure.

d. Applications to Make or Transfer NFA Firearms

All applications to transfer or make NFA firearms must be submitted in duplicate, with both copies bearing original signatures. Extra care in ensuring that the applications are completed accurately will expedite the flow of your paperwork. Particular attention should be given to the serial number of the weapon to ensure that it does not have suffixes or prefixes.

With regard to transfers going to individuals, please ensure that the law enforcement certification is signed by someone acceptable to sign, and that the certifying officer does, in fact, have jurisdiction where the transferee resides. See Question **M18** on acceptable certifying officials.

All ATF Form 4 applications must be accompanied by 2 properly completed sets of fingerprint cards (FBI Form FD-258). In some cases, particularly when fingerprints have not been properly taken, fingerprint classification can, and does, take several months. For your Form 4 applications to be expeditiously acted upon, it is imperative that the fingerprint cards you submit be complete.

All applications for taxpaid making or transfer (ATF Forms 1 and ATF Forms 4) should be forwarded, together with proper remittance, to the following address:

BUREAU OF ATF
P.O. BOX 73201
CHICAGO, ILLINOIS 60673

All other applications and correspondence should be forwarded to the National Firearms Act Branch at the address appearing at the end of this item.

e. Machineguns

Machineguns produced, imported, or registered after May 19, 1986, the effective date of 18 U.S.C. 922(o), generally are unlawful except for use by a government agency or for exportation. We will allow Class 3 dealers to receive necessary "sales samples" of these firearms if they obtain a letter from a local law enforcement agency, on the agency's letterhead, indicating a ***bona fide*** need to see the weapon.

If we, through an error in processing, fail to note on your transfer document(s) that certain weapons are restricted, such error will not exempt you from complying with the restrictions of 18 U.S.C. 922(o).

f. Forms

Forms you may need to conduct your business (but not bound books which are privately sold) are available from:

ATF DISTRIBUTION CENTER
PO BOX 5950
SPRINGFIELD, VIRGINIA 22150-5950

g. Going Out Of Business

(1) NFA Activities Only

Machineguns. If you, as a Special (Occupational) Taxpayer, decide not to renew your payment of the special tax, all machineguns which you possess that are restricted under 18 U.S.C. 922(o) must be transferred to a Special (Occupational) Taxpayer having a legitimate need for the weapon(s) or be exported. Such transfer must occur before you allow your Federal firearms license and special tax status to expire. Otherwise, these firearms must be abandoned to ATF or be subject to seizure and forfeiture. Additional information relating to possession of registered machineguns, other NFA firearms, and record keeping requirements following discontinuance of business is provided below.

Pre-86 Machineguns. When a Special (Occupational) Taxpayer goes out of business as a dealer in NFA firearms, if the business is a sole proprietor, the proprietor may continue to possess machineguns lawfully imported or manufactured prior to May 19, 1986 (the effective date of 18 U.S.C. 922(o)) and which are registered to the sole proprietor. The proprietor should record the disposition of the firearms by the business to the individual proprietor in the acquisition and disposition book. If all firearms business is being discontinued and there is no successor, all records required by the GCA must be delivered within 30 days to the Out-of-Business Records Center (see Questions **C5-C7**, later in this publication). As these firearms are "grandfathered" weapons not subject to the restrictions of 18 U.S.C. 922(o), they may be transferred to any person who is not otherwise prohibited by law from possessing them in accordance with the NFA.

When a Special (Occupational) Taxpayer goes out of business as a dealer in NFA firearms and the business is a corporation, partnership, or other type of artificial business entity, the business may continue to possess pre-86 machineguns registered to the business only if the corporation or partnership continues to exist under State law and only if the title to the machineguns remains in the business after the Special (Occupational) tax stamp expires. If the corporation or partnership is dissolved or otherwise ceases to exist under State law, then the machineguns will have been transferred to whomever possesses them after dissolution. If all firearms business is being discontinued and there is no successor, all records required by the GCA must be delivered within 30 days to the Out-of-Business Records Center (see Questions **C5-C7**, later in this publication). Transfer applications must be submitted and approved before dissolution occurs to avoid placing the possessors in violation of the NFA. If the registered machineguns are transferred to officers or directors of a corporate registrant or individual partners of a partnership, the transaction is a transfer subject to all applicable provisions of the NFA and GCA, including completion of a Form 4473 and compliance with the Brady law.

Post-86 Machineguns. When a Special (Occupational) Taxpayer goes out of business as a dealer in NFA firearms, machineguns lawfully imported or manufactured on or after May 19, 1986, may not be retained by the business. Such machineguns are subject to the restrictions of 18 U.S.C. 922(o) and must be transferred to a law enforcement agency for official use or to another qualified NFA dealer as dealer sales samples in accordance with 27 C.F.R. 479.105(d). Post-86 machineguns that are not transferred prior to expiration of the special tax stamp are subject to seizure and forfeiture.

NFA Firearms Other Than Machineguns. When a Special (Occupational) Taxpayer goes out of business as a dealer in NFA firearms and the business is a sole proprietor, the proprietor may continue to possess firearms registered to the business. The proprietor should record the disposition of the firearms by the business to the individual proprietor in the acquisition and disposition book. If all firearms business is being discontinued and there is no successor, all records required by the GCA must be delivered within 30 days to the Out-of-Business Records Center (see Questions **C5-C7**, later in this publication). Subsequent transfers of the registered weapons are subject to all the provisions of the NFA and GCA.

When a Special (Occupational) Taxpayer goes out of business as a dealer in NFA firearms and the business is a corporation, partnership, or other type of artificial business entity, the business may continue to possess firearms other than machineguns registered to the business only if the corporation or part nership continues to exist under State law and only if title to the firearms remains in the business after the Special (Occupational) tax stamp expires. If the corporation or partnership ceases to exist under State law, then the firearms will have to be transferred to whoever possesses them after dissolution. Transfer applications must be submitted and approved before this occurs to avoid placing the possessors in violation of the NFA. If the registered firearms are transferred to officers or directors of a corporate registrant or individual partners of a partnership, the transaction is a transfer subject to all applicable provisions of the NFA and GCA, including completion of a Form 4473 and compliance with the Brady Law.

Any NFA firearms retained by the business that were imported under 26 U.S.C. 5844 for use as samples or for scientific or research purposes may only be transferred to government agencies for official use or to a Federal firearms licensee who has paid the special (occupational) tax to manufacture, import, or deal in NFA firearms.

CAUTION:

The mere possession of a license and a special (occupational) tax stamp as a dealer in NFA firearms does not qualify a person to receive firearms free of transfer tax. Where a person who possesses a license and tax stamp is not actually engaged in the business of selling NFA firearms and receives such firearms by a tax-free transfer on ATF Form 3, an unlawful transfer in violation of the NFA has occurred because the transfer tax was not paid. In such cases, the firearms involved are subject to seizure and forfeiture.

Also, see ATF Rul. 76-22, set out within the Rulings, Procedures, and Industry Circulars portion of this publication.

Should your special tax status lapse, your continued possession of certain firearms may place you in violation of various State laws and local ordinances. We urge you to carefully consider the consequences of possessing NFA firearms in your particular city, county, and State without being a Special (Occupational) Taxpayer.

(2) Disposition of Records

If someone is taking over the business, the licensee will underline the final entry in each bound book, note the date of transfer, and deliver all records and forms kept by the licensee to the successor (who must apply for and receive his own license before lawfully engaging in business) or deliver the records and forms to the ATF Out-of-Business Records Center. If there is no business successor, within 30 days of business discontinuance the licensee must ship the required records and forms to the ATF Out-of-Business Records Center, 244 Needy Road, Martinsburg, WV 25401.

h. If You Have Questions or Problems

In the event you have any inquiries relating to your NFA business activity, please contact: Bureau of Alcohol, Tobacco, Firearms and Explosives, National Firearms Act Branch.

11. MOVING NFA FIREARMS INTERSTATE

A person who desires to transport a machinegun, short-barreled rifle, short-barreled shotgun, or destructive device interstate must first apply to ATF for permission to do so. ATF Form 5320.20, Application to Transport Interstate or to Temporarily Export Certain National Firearms Act (NFA) Firearms, can be used for this purpose. Only after the person receives ATF approval can the firearm be taken into another State, even for a short period of time. This requirement does not apply to a licensee qualified under the National Firearms Act to engage in business with respect to the weapon or device to be transported or to a licensed collector if the device or weapon to be transported is a curio or relic.

Alternatively, the lawful owner of the firearm may write a letter, in duplicate, giving:

a. A complete description and identification of the device or weapon to be transported;

b. A statement whether such transportation involves a transfer of title;

c. The need for such transportation;

d. The approximate date such transportation is to take place;

e. The present location of such device or weapon, the place to which it is to be transported, and the transportation to be used (including, if by common carrier, the name and address of the carrier); and

f. Evidence that the transportation or possession of such device or weapon is not inconsistent with the laws at the place of destination.

An application will not be approved if possession of the firearm at the place of destination would place the possessor in violation of State or local law.

If you have any questions regarding this subject, please contact: Bureau of Alcohol, Tobacco, Firearms and Explosives, National Firearms Act Branch.

12. LISTS OF LICENSEES/ PERMITTEES

Current lists of Federal firearms licensees and Federal explosives licensees and permittees are available. Prices are quoted on request.

For a copy of the order form, contact: the Bureau of Alcohol, Tobacco, Firearms and Explosives, Disclosure Branch.

We must advise, however, that Federal law generally prohibits the disclosure (either affirmatively or negatively) of information concerning the registration of NFA firearms. Therefore, please do not ask us for the names of persons engaged in business with respect to machineguns or other NFA firearms.

13. IDENTIFYING FIREARMS (MARKINGS)

To ensure that firearms are properly identified, ATF wishes to remind licensees that it is their responsibility to ensure that firearms are properly identified in accordance with the law. The markings on the firearms are vital to our tracing program. Violations of the law and regulations may result in criminal and administrative action, including license revocation and denial of license renewal applications. For required markings see 27 CFR 478.92 and 479.104. See also ATF Ruling 2002-6, set out within the Rulings, Procedures, and Industry Circulars portion of this publication.

14. NFA FIREARMS IN DECEDENTS' ESTATES

Possession of an NFA firearm not registered to the possessor is a violation of Federal Law and the firearm is subject to seizure and forfeiture. However, a reasonable time is allowed for transfer of lawfully registered firearms in a decedent's estate.

It is the responsibility of the executor or the administrator of an estate to transfer firearms registered to a decedent. ATF Form 5, Application for Tax Exempt Transfer and Registration of a Firearm, is used to apply for a tax-exempt transfer to a lawful heir. A lawful heir is anyone named in the decedent's will or, in the absence of a will, anyone entitled to inherit under the laws of the State in which the decedent last resided. NFA firearms may be transferred directly interstate to a beneficiary of the estate. However, if any Federal, State or local law prohibits the heir from receiving or possessing the firearm, ATF will not approve the application. When a firearm is being transferred to an individual heir, his or her fingerprints on FBI Forms FD-258 must accompany the transfer application.

ATF Form 4 is used to apply for the taxpaid transfer of a serviceable NFA firearm to a person outside the estate (not a beneficiary). ATF Form 5 is also used to apply for the tax-exempt transfer of an unserviceable NFA firearm to a person outside the estate. As noted above, all requirements, such a fingerprint cards for transfers to individuals and compliance with State of local law, must be met before an application could be approved.

If the NFA firearm in the estate was imported for use as a "sales sample," this restriction on the firearm's possession remains. The NFA firearm may only be transferred to a Federal firearms licensee who has paid the special (occupational) tax to deal in NFA firearms or to a government agency.

For further information, contact: Bureau of Alcohol, Tobacco, Firearms and Explosives, National Firearms Branch.

15. STRAW PURCHASES

Questions have arisen concerning the lawfulness of firearms purchases from licensees by persons who use a "straw purchaser" (another person) to acquire the firearms. Specifically, the actual buyer uses the straw purchaser to execute the Form 4473 purporting to show that the straw purchaser is the actual purchaser of the firearm. In some instances, a straw purchaser is used because the actual purchaser is prohibited from acquiring the firearm. That is to say, the actual purchaser is a felon or is within one of the other prohibited categories of persons who may not lawfully acquire firearms or is a resident of a State other than that in which the licensee's business premises is located. Because of his or her disability, the person uses a straw purchaser who is not prohibited from purchasing a firearm from the licensee. In other instances, neither the straw purchaser nor the actual purchaser is prohibited from acquiring the firearm.

In both instances, the straw purchaser violates Federal law by making false statements on Form 4473 to the licensee with respect to the identity of the actual purchaser of the firearm, as well as the actual purchaser's residence address and date of birth. The actual purchaser who utilized the straw purchaser to acquire a firearm has unlawfully aided and abetted or caused the making of the false statements. The licensee selling the firearm under these circumstances also violates Federal law if the licensee is aware of the false statements on the form. It is immaterial that the actual purchaser and the straw purchaser are residents of the State in which the licensee's business premises is located, are not prohibited from receiving or possessing firearms, and could have lawfully purchased firearms from the licensee.

An example of an illegal straw purchase is as follows: Mr. Smith asks Mr. Jones to purchase a firearm for Mr. Smith. Mr. Smith gives Mr. Jones the money for the firearm. If Mr. Jones fills out Form 4473, he violates the law by falsely stating that he is the actual buyer of the firearm. Mr. Smith also violates the law because he has unlawfully aided and abetted or caused the making of false statements on the form.

Where a person purchases a firearm with the intent of making a gift of the firearm to another person, the person making the purchase is indeed the true purchaser. There is no straw purchaser in these instances. In the above example, if Mr. Jones had bought a firearm with his own money to give to Mr. Smith as a birthday present, Mr. Jones could lawfully have completed Form 4473. The use of gift certificates would also not fall within the category of straw purchases. The person redeeming the gift certificate would be the actual purchaser of the firearm and would be properly reflected as such in the dealer's records.

16. FEDERAL EXCISE TAX

A Federal excise tax is imposed by 26 U.S.C. 4181 on the sale by the manufacturer, importer or producer of firearms, shells, and cartridges. The tax is 10 percent of the sale price for pistols and revolvers, 11 percent of the sale price for firearms and 11 percent of the sale price for shells and cartridges. This tax was formerly administered by ATF. Now, the tax is administered by the Alcohol and Tobacco Tax and Trade Bureau (TTB), Department of the Treasury.

The excise tax attaches only to the sale of complete firearms and ammunition or firearms that, although in a knockdown condition, are complete as to all component parts.

The term **firearm** for excise tax purposes includes all portable weapons, such as rifles, carbines, machineguns, shotguns, and fowling pieces from which a shot, bullet, or projectile may be discharged by an explosive. The term firearm also includes pistols and revolvers. Antique firearms are also subject to the excise tax.

Shells and cartridges include any article consisting of a projectile, explosive, and container that is designed, assembled, and ready for use without further manufacture in firearms, pistols or revolvers.

Reloading of used shells or cartridges is considered manufacturing for purposes of excise tax. Sale of such shells by the reloader is subject to the excise tax. However, if the reloader merely reloads shells belonging to a customer and is paid for labor and materials, the reloading service is not a taxable sale, as long as the reloader returns the identical shells provided by the customer to that same customer. In such instances the customer is the manufacturer and would not be liable for tax if the shells are manufactured for personal use. If the customer sells reloaded shells or uses them in a business, e.g., shooting range, the customer would be liable for the tax.

Returns and Deposits

Regulations in 27 CFR Part 53 require that taxpayers incurring a tax liability on the sale or use of firearms and ammunition file excise tax returns quarterly on TTB Form 5300.26, Federal Firearms and Ammunition Excise Tax Return. In addition, taxpayers are required to make semimonthly deposits of tax on TTB Form 5300.27.

Further Information

For more detailed information regarding firearms and ammunition excise taxes, refer to the TTB web site, www.TTB.gov.

17. ARMOR PIERCING AMMUNITION

For purposes of the prohibitions imposed upon manufacture, importation, and transfer of armor piercing ammunition in 18 U.S.C. 922(a)(7)-(8) and

923(e), armor piercing ammunition includes the following:

KTW AMMUNITION, all calibers. Identified by a green coating on the projectile.

ARCANE AMMUNITION, all calibers. Identified by a pointed bronze or brass projectile.

THV AMMUNITION, all calibers. Identified by a brass or bronze projectile and a head stamp containing the letters SFM and THV.

CZECHOSLOVAKIAN manufactured 9mm Parabellum (Luger) ammunition having an iron or steel bullet core. Identified by a cupronickel jacket and a head stamp containing a triangle, star, and dates of 49, 50, 51, or 52. This bullet is attracted to a magnet.

GERMAN manufactured 9mm Parabellum (Luger) having an iron or steel bullet core. Original packaging is marked Pistolenpatronen 08 m.E. May have black colored bullet. This bullet is attracted to a magnet.

MSC AMMUNITION, caliber .25. Identified by a hollowpoint brass bullet. **NOTE:** MSC ammunition, caliber .25 identified by a hollowpoint copper bullet is **not** armor piercing.

BLACK STEEL ARMOR PIERCING AMMUNITION, all calibers, as produced by National Cartridge, Atlanta, Georgia.

BLACK STEEL METAL PIERCING AMMUNITION, all calibers, as produced by National Cartridge, Atlanta, Georgia.

7.62mm NATO AP, identified by black coloring in the bullet tip. This ammunition is used by various NATO countries. The U.S. military designation is M61 AP.

7.62mm NATO SLAP. Identified by projectile having a plastic sabot around a hard penetrator. The penetrator protrudes above the sabot and is similar in appearance to a Remington accelerator cartridge.

PMC ULTRAMAG, .38 Special caliber, constructed entirely of a brass type material, and a plastic pusher disc located at the base of the projectile. **NOTE:** PMC ULTRAMAG 38J late production made of copper with lead alloy projectile is **not** armor piercing.

OMNISHOCK. A .38 Special cartridge with a lead bullet containing a mild steel core with a flattened head resembling a wad cutter. **NOTE**: OMNISHOCK cartridges having a bullet with an aluminum core are **not** armor piercing.

7.62x39mm with steel core. These projectiles have a steel core. **NOTE:** Projectiles having a lead core with steel jacket or steel case are **not** armor piercing.

In addition, the Violent Crime Control and Law Enforcement Act of 1994 added to the definition of armor piercing ammunition the following:

> **"... a full jacketed projectile larger than .22 caliber designed and intended for use in a handgun and whose jacket has a weight of more than 25 percent of the total weight of the projectile."**

Exemptions: The following articles are exempted from the definition of armor piercing ammunition.

5.56 mm (.223) SS109 and M855 Ammunition, identified by a green coating on the projectile tip.

U.S. .30-06 M2AP, identified by a black coating on the projectile tip.

18. ASSEMBLY OF NONSPORTING SEMIAUTOMATIC RIFLES AND SHOTGUNS

Section 922(r), Title 18, U.S.C., makes it unlawful for any person to assemble from imported parts any semiautomatic rifle or any shotgun which is identical to any rifle or shotgun prohibited from importation under section 925(d)(3) of the GCA. Regulations implementing the law in 27 C.F.R. 478.39 provide that a violation of section 922(r) will result if a semiautomatic rifle or shotgun is assembled with more than 10 of the following imported parts:

(1) Frames, receivers, receiver castings, forgings, or stampings
(2) Barrels
(3) Barrel extensions
(4) Mounting blocks (trunnions)
(5) Muzzle attachments
(6) Bolts
(7) Bolt carriers
(8) Operating rods
(9) Gas pistons
(10) Trigger housings
(11) Triggers
(12) Hammers
(13) Sears
(14) Disconnectors
(15) Buttstocks
(16) Pistol grips
(17) Forearms, handguards
(18) Magazine bodies
(19) Followers
(20) Floorplates

Section 922(r) does not prohibit the importation, sale, or possession of parts which may be used to assemble a semiautomatic rifle or shotgun in violation of the statute. However, 18 U.S.C. § 2 provides that a person who aids or abets another person in the commission of an offense is also responsible for the offense. Therefore, a person who sells parts knowing that the purchaser intends to use the parts in assembling a firearm in violation of section 922(r) would also be responsible for the offense.

19. FIREARMS THEFT/LOSS REPORTING

Federal firearms licensees are required to report the theft or loss of firearms from their inventory or collection to local authorities as well as to ATF within 48 hours after the theft or loss is discovered.

The notification to ATF must be made through our toll free theft hotline number which is **1-800-930-9275**. The hotline is operational 24 hours a day, 7 days a week. The caller should indicate that he or she is a Federal firearms licensee and be ready to furnish his or her Federal firearms license number. The hotline representative will provide the licensee with an incident number which should be recorded in the licensee's bound book.

The verbal notification must be followed up by a written notification to ATF within the same 48-hour period. The written notification should be made on ATF F 3310.11, Federal Firearms Licensee Theft/Loss Report, in accordance with the instructions on the form. This form is available from the ATF Distribution Center.

Theft or loss of NFA firearms should be reported to the NFA Branch immediately upon discovery.

20. IMPORTATION OF SEMIAUTOMATIC ASSAULT RIFLES AND MODIFIED VERSIONS OF SUCH RIFLES

Section 925(d)(3), Title 18, U.S.C., states that the Attorney General shall authorize a firearm or ammunition to be imported or brought into the United States if it is of a type that does not fall within the definition of a firearm in section 5845(a) of the Internal Revenue

Code of 1986 and is generally recognized as particularly suitable for or readily adaptable to sporting purposes, excluding surplus military firearms.

In 1989, ATF identified a type of rifle known as a semiautomatic assault rifle and found that weapons of this type were not generally recognized as particularly suitable for or readily adaptable to sporting purposes under the sporting purposes test for importation in 18 U.S.C. 925(d). Accordingly, on July 6, 1989, ATF determined that rifles of this type were not importable into the United States. ATF's finding was based, in part, on the determination that these rifles have certain characteristics that are common to modern military assault rifles and that distinguish them from traditional sporting rifles. These characteristics include the ability to accept a detachable magazine, folding/telescoping stocks, separate pistol grips, ability to accept a bayonet, flash suppressors, bipods, grenade launchers, and night sights. It was decided that any of these military features, other than the ability to accept a detachable magazine, would make a semiautomatic assault rifle not importable. The particular rifles barred from importation were as follows:

AKS Variants
AK 47 type
AK47S type
AK74 type
AKS type
AKM type
AKMS type
ARM type
84S type
84S1 type
84S3 type
86S type
867S type
Galil type
Type 56 type
Type 56S type
Valmet M76 type
Valmet M78 type
M76 counter sniper type

FAL Variants
FAL type
L1A1A type
SAR 48

Other
AUG type
FNC type
Uzi carbine
Algimec AGM1 type
AR180 type
Australian Automatic Arms SAR type
Beretta AR70 type
Beretta BM59 type
CIS SR88 type
HK91 type
HK93 type
HK94 type
G3SA type
K1 type
K2 type
AR100 type
M14S type
MAS223 type
SIG 550SP type
SIG 551SP type
SKS with detachable magazine

Subsequent to the 1989 decision, certain semiautomatic assault rifles that failed the 1989 sporting purposes test for importation were modified to remove all of their military features other than the ability to accept a detachable magazine. They could still accept the large capacity magazines originally designed and produced for the military assault rifles from which they were derived. The modified weapons were permitted to be imported because they met the 1989 sporting purposes test for importation.

On November 14, 1997, the President and the Secretary of the Treasury ordered a review of the importation of modified versions of semiautomatic assault rifles into the United States. Consequently, it was decided on April 6, 1998, that the following modified versions of semiautomatic assault rifles could no longer be imported under the sporting purposes test.

AK47 Variants
MAK90
314
56V
89
EXP56A
SLG74
NHM90
NHM90-2
NHM91
SA85M
SA93
A93
AKS762
SA2000
ARM
MISR
MISTR
SA85M
Mini PSL
ROMAK 1
ROMAK 2
ROMAK 4
Hunter Rifle
386S
PS/K
Galil Sporter
Haddar
Haddar II
WUM 1
WUM 2
SLR95
SLR96
SLR97
SLG94
SLG95
SLG96

FN-FAL Variants
L1A1 Sporter
FAL Sporter
FZSA
SAR4800
XFAL
C3
C3A
LAR Sporter

HK Variants
BT96
Centurian 2000
SR9
PSG1
MSG90
G3SA
SAR8

Uzi Variants
Officers 9
320 carbine
Uzi Sporter

SIG SG550 Variants
SG550-1
SG550-2

21. IMPORTATION OF HANDGUNS

Section 925(d)(3), Title 18, U.S.C., states that the Attorney General shall authorize a firearm or ammunition to be imported or brought into the United States if it is of a type that does not fall within the definition of a firearm in section 5845(a) of the Internal Revenue Code of 1986 and is generally recognized as particularly suitable for or readily adaptable to sporting purposes, excluding surplus military firearms. The statute also prohibits the importation of any frame, receiver, or barrel of a firearm that would be prohibited if assembled.

In 1968, the Secretary of the Treasury established the Treasury Department Firearms Evaluation Panel. The purpose of this panel was to assist in the establishment of guidelines for use in determining if a particular firearm would be importable under the sporting purposes test prescribed by section 925(d)(3). This panel was composed of representatives from the firearms industry, law enforcement, and the military.

The panel developed objective numerical criteria with minimum qualifying scores to determine if a handgun is particularly suitable for or readily adaptable to sporting purposes. The criteria, ATF Form 4590, Factoring Criteria for Weapons, assigns point values to handguns based on dimensions, material used in construction, weight, caliber, safety features and miscellaneous equipment. The criteria also have prerequisite requirements concerning safeties and minimum dimensions.

Form 4590 is divided into two parts. The right side of the form is used to evaluate revolvers and the left side is used to evaluate pistols. The minimum qualifying score for a revolver is 45 points and the minimum qualifying score for a pistol is 75 points. The form also provides that revolvers must pass a safety test.

Any handgun being imported into the United States must pass these criteria. The fact that a particular weapon may be of domestic manufacture or classified as a curio or relic does not exempt it from the factoring criteria.

If you have any questions concerning Form 4590, or the score a particular handgun achieves, please contact the ATF Firearms Technology Branch.

22. IMPORTANT INFORMATION FOR CHIEF LAW ENFORCEMENT OFFICERS (CLEOs) RECEIVING COPIES OF ATF FORM 3310.4 (REPORT OF MULTIPLE SALE OR OTHER DISPOSITION OF PISTOLS AND REVOLVERS)

18 U.S.C. 923(g)(3)(B) provides that Chief Law Enforcement Officers (CLEOs) receiving copies of ATF Form 3310.4 from FFLs for sales to non-prohibited persons may not disclose the contents of the forms and must destroy the forms and any record of the content of the forms within 20 days of receipt. This provision also requires CLEOs to certify in writing to the United States Attorney General every six months that no such disclosures have been made and that the forms on non-prohibited persons have been destroyed as required by law. CLEOs making this certification should send the certification to:

U.S. DEPARTMENT OF JUSTICE
ATTN: MULTIPLE HANDGUN
SALE FORM CERTIFICATIONS
P.O. BOX 4278
CLARKSBURG, WV 26302

QUESTIONS & ANSWERS

TABLE OF CONTENTS

A. GENERAL QUESTIONS

(A1) Does the law regulate who can be in the business?

Yes. The Gun Control Act (GCA), administered by the Bureau of Alcohol, Tobacco, Firearms and Explosives (ATF) of the Department of Justice, contains Federal licensing standards for various firearms businesses (manufacturers, importers, and dealers). An example of these standards is that the applicant must have a business premises.

[18 U.S.C. 923(d), 27 CFR 478.47]

(A2) Does the Federal Government issue a license or permit to carry a concealed weapon?

No. Neither ATF nor any other Federal agency issues such a permit or license. Carrying permits may be issued by a State or local government.

(A3) Do antique firearms come within the purview of the GCA?

No.

[18 U.S.C. 921(a)(3) and (16), 27 CFR 478.11 and 478.141(d)]

(A4) What kinds of ammunition are covered by the GCA?

Ammunition includes cartridge cases, primers, bullets or propellant powder designed for use in any firearm other than an antique firearm. Items **NOT** covered include blank ammunition, tear gas ammunition, pellets and nonmetallic shotgun hulls without primers.

Generally, no records are required for ammunition transactions. However, information about the disposition of armor piercing ammunition is required to be entered into a record by importers, manufacturers, and collectors. A license is not required for dealers in ammunition only.

[18 U.S.C. 921(a)(17) and 922(b)(5), 27 CFR 478.11 and 478.125]

(A5) Does the GCA control the sale of firearms parts?

No, except that frames or receivers of firearms are "firearms" as defined in the law and subject to the same controls as complete firearms. Silencer parts are also firearms under the GCA, as well as under the National Firearms Act (NFA). Certain machinegun parts, such as conversion parts or kits, are also subject to the NFA.

[18 U.S.C. 921(a)(3) and (24), 26 U.S.C. 5845, 27 CFR 478.11 and 479.11]

(A6) Does the GCA prohibit anyone from making a handgun, shotgun or rifle?

With certain exceptions a firearm may be made by a nonlicensee provided it is not for sale and the maker is not prohibited from possessing firearms. However, a person is prohibited from assembling a nonsporting semi-automatic rifle or nonsporting shotgun from imported parts. In addition, the making of an NFA firearm requires a tax payment and approval by ATF. An application to make a machinegun will not be approved unless documentation is submitted showing that the firearm is being made for a Federal or State agency.

[18 U.S.C. 922(o) and (r), 26 U.S.C. 5822, 27 CFR 478.39, 479.62 and 479.105]

(A7) How can a person apply for relief from Federal firearms disabilities?

Under the provisions of the Gun Control Act of 1968 (GCA), convicted felons and certain other persons are prohibited from possessing or receiving firearms. The GCA provides the Attorney General with the authority to grant relief from this disability where the Attorney General determines that the person is not likely to act in a manner dangerous to the public safety and granting relief would not be contrary to the public interest. The Attorney General delegated this authority to ATF.

Since October 1992, however, ATF's annual appropriation has prohibited the expending of any funds to investigate or act upon applications for relief from Federal firearms disabilities submitted by individuals. As long as this provision is included in current ATF appropriations, the Bureau cannot act upon applications for relief from Federal firearms disabilities submitted by individuals.

[18 U.S.C. 922(g), 922(n) and 925(c)]

(A8) Are there any alternatives for relief from firearms disabilities?

A person is not considered convicted for Gun Control Act purposes if he has been pardoned, had his civil rights restored, or the conviction was expunged or set aside, unless the pardon, expungement, or restoration expressly provides the person may not ship, transport, possess, or receive firearms.

Persons convicted of a Federal offense may apply for a Presidential pardon. 28 CFR 1.1-1.10 specify the rules governing petitions for obtaining Presidential pardons. You may contact the Pardon Attorney's Office at the U.S. Department of Justice, 500 First Street, N.W., Washington, DC 20530, to inquire about the procedures for obtaining a Presidential pardon.

Persons convicted of a State offense may contact the State Attorney General's Office within the State in which they reside and the State of their conviction for information concerning any alternatives that may be available, such as pardons and civil rights restoration.

[18 U.S.C. 921(a)(20) and (a)(33)]

B. UNLICENSED PERSONS

(B1) To whom may an unlicensed person transfer firearms under the GCA?

A person may sell a firearm to an unlicensed resident of his State, if he does not know or have reasonable cause to believe the person is prohibited from receiving or possessing firearms under Federal law. A person may loan or rent a firearm to a resident of any State for temporary use for lawful sporting purposes, if he does not know or have reasonable cause to believe the person is prohibited from receiving or possessing firearms under Federal law. A person may sell or transfer a firearm to a licensee in any State. However, a firearm other than a curio or relic may not be transferred interstate to a licensed collector.

[18 U.S.C 922(a)(3) and (5), 922(d), 27 CFR 478.29 and 478.30]

(B2) From whom may an unlicensed person acquire a firearm under the GCA?

A person may only acquire a firearm within the person's own State, except that he or she may purchase or otherwise acquire a rifle or shotgun, in person, at a licensee's premises in any State, provided the sale complies with State laws applicable in the State of sale and the State where the purchaser resides. A person may borrow or rent a firearm in any State for temporary use for lawful sporting purposes.

[18 U.S.C 922(a)(3) and (5), 922(b)(3), 27 CFR 478.29 and 478.30]

(B3) May an unlicensed person obtain a firearm from an out-of-State source if the person arranges to obtain the firearm through a licensed dealer in the purchaser's own State?

A person not licensed under the GCA and not prohibited from acquiring firearms may purchase a firearm from an out-of-State source and obtain the firearm if an arrangement is made with a licensed dealer in the purchaser's State of residence for the purchaser to obtain the firearm from the dealer.

[18 U.S.C 922(a)(3) and 922(b)(3)]

(B4) May an unlicensed person obtain ammunition from an out-of-State source?

Yes, provided he or she is not a person prohibited from possessing or receiving ammunition.

[18 U.S.C. 922(g) and (n)]

(B5) Are there certain persons who cannot legally receive or possess firearms and/or ammunition?

Yes, a person who –

(1) Has been convicted in any court of a crime punishable by imprisonment for a term exceeding 1 year;

(2) Is a fugitive from justice;

(3) Is an unlawful user of or addicted to any controlled substance;

(4) Has been adjudicated as a mental defective or has been committed to a mental institution;

(5) Is an alien illegally or unlawfully in the United States or an alien admitted to the United States under a nonimmigrant visa;

(6) Has been discharged from the Armed Forces under dishonorable conditions;

(7) Having been a citizen of the United States, has renounced his or her citizenship;

(8) Is subject to a court order that restrains the person from harassing, stalking, or threatening an intimate partner or child of such intimate partner; or

(9) Has been convicted of a misdemeanor crime of domestic violence

cannot lawfully receive, possess, ship, or transport a firearm.

A person who is under indictment or information for a crime punishable by imprisonment for a term exceeding 1 year cannot lawfully receive a firearm. Such person may continue to lawfully possess firearms obtained prior to the indictment or information.

[18 U.S.C. 922(g) and (n), 27 CFR 478.32]

(B6) Do law enforcement officers who are subject to restraining orders and who receive and possess firearms for purposes of carrying out their official duties violate the law?

Not if the firearms are received and possessed for official use only.

The law prohibits persons subject to certain restraining orders from receiving, shipping, transporting or possessing firearms or ammunition. To be disabling, the restraining order must:

1. specifically restrain the person from harassing, stalking, or threatening an "intimate partner" of the person (e.g., spouse);

2. be issued after a hearing of which notice was given to the person and at which the person had an opportunity to participate; and

3. include a finding that the person subject to the order represents a credible threat to the "intimate partner" or child of the "intimate partner" OR explicitly prohibits the use, attempted use, or threatened use of force against the partner.

However, the GCA has an exception for the receipt and possession of firearms and ammunition on behalf of a Federal or State agency. Therefore, the GCA does not prohibit a law enforcement officer under a restraining order from receiving or possessing firearms or ammunition for use in performing official duties. Possession of the firearm for official purposes while off duty would be lawful if such possession is required or authorized by law or by official departmental policy. An officer subject to a disabling restraining order would violate the law if the officer received or possessed a firearm or ammunition for other than official use. (See Question **Q13** on officers' receipt and possession of firearms and ammunition after a conviction of a misdemeanor crime of domestic violence. The government exception does not apply to such convictions.)

[18 U.S.C. 921(a)(32), 922(g)(8) and 925(a)(1)]

(B7) May a nonlicensee ship a firearm through the U.S. Postal Service?

A nonlicensee may not transfer a firearm to a nonlicensed resident of another State. A nonlicensee may mail a shotgun or rifle to a resident of his or her own State or to a licensee in any State. The Postal Service recommends that long guns be sent by registered mail and that no marking of any kind which would indicate the nature of the contents be placed on the outside of any parcel containing firearms. Handguns are not mailable. A common or contract carrier must be used to ship a handgun.

[18 U.S.C. 1715, 922(a)(3), 922(a)(5) and 922 (a)(2)(A)]

(B8) May a nonlicensee ship a firearm by common or contract carrier?

A nonlicensee may ship a firearm by a common or contract carrier to a resident of his or her own State or to a licensee in any State. A common or contract carrier must be used to ship a handgun. In addition, Federal law requires that the carrier be notified that the shipment contains a firearm and prohibits common or contract carriers from requiring or causing any label to be placed on any package indicating that it contains a firearm.

[18 U.S.C. 922(a)(2)(A), 922(a) (3), 922(a)(5) and 922(e), 27 CFR 478.31 and 478.30]

(B9) May a nonlicensee ship firearms interstate for his or her use in hunting or other lawful activity?

Yes. A person may ship a firearm to himself or herself in care of another person in the State where he or she intends to hunt or engage in any other lawful activity. The package should be addressed to the owner. Persons other than the owner should not open the package and take possession of the firearm.

(B10) May a person who is relocating out-of-State move firearms with other household goods?

Yes. A person who lawfully possesses a firearm may transport or ship the firearm interstate when changing his or her State of residence.

Certain NFA firearms must have prior approval from the Bureau of ATF before they may be moved interstate. The person must notify the mover that firearms are being transported. He or she should also check State and local laws where relocating to ensure that movement of firearms into the new State does not violate any State law or local ordinance.

[18 U.S.C. 922(a)(4) and 922(e), 27 CFR 478.28 and 478.31]

(B11) What constitutes residency in a State?

The State of residence is the State in which an individual is present; the individual also must have an intention of making a home in that State. A member of the Armed Forces on active duty is a resident of the State in which his or her permanent duty station is located. If a member of the Armed Forces maintains a home in one State and the member's permanent duty station is in a nearby State to which he or she commutes each day, then the member has two States of residence and may purchase a firearm in either the State where the duty station is located or the State where the home is maintained. An alien who is legally in the United States is considered to be a resident of a State only if the alien is residing in that State and has resided in that State continuously for a period of at least 90 days prior to the date of sale of the firearm. See also **Item 5**, "Sales to Aliens in the United States", in the General Information section of this publication.

[18 U.S.C. 921(b), 922(a) (3), and 922(b)(3), 27 CFR 478.11]

(B12) May a person (who is not an alien) who resides in one State and owns property in another State purchase a handgun in either State?

If a person maintains a home in 2 States and resides in both States for certain periods of the year, he or she may, during the period of time the person actually resides in a particular State, purchase a handgun in that State. However, simply owning property in another State does not qualify the person to purchase a handgun in that State.

[27 CFR 478.11]

(B13) May aliens legally in the United States buy firearms?

An alien legally in the U.S. may acquire firearms if he has a State of residence. An alien has a State of residence only if he is residing in that State and has resided in a State continuously for at least 90 days prior to the purchase. An alien acquiring firearms from a licensee is required to prove both his identity, by presenting a government-issued photo identification, and his residency with substantiating documentation showing that he has resided in the State continuously for the 90-day period prior to the purchase. Examples of qualifying documentation to prove residency include: utility bills, lease agreements, credit card statements, and pay stubs from the purchaser's place of employment, if such documents include residential addresses.

See also **Item 5**, "Sales to Aliens in the United States," in the General Information section of this publication.

[18 U.S.C. 921, 922(b)(3), (d) and (g), 27 CFR 478.11 and 478.99(a)]

(B14) May a parent or guardian purchase firearms or ammunition as a gift for a juvenile (less than 18 years of age)?

Yes. However, possession of handguns by juveniles (less than 18 years of age) is generally unlawful. Juveniles generally may only receive and possess handguns with the written permission of a parent or guardian for limited purposes, e.g., employment, ranching, farming, target practice or hunting.

[18 U.S.C. 922(x)]

(B15) Are curio or relic firearms exempt from the provisions of the GCA?

No. Curios or relics are still firearms subject to the provisions of the GCA; however, curio or relic firearms may be transferred in interstate com merce to licensed collectors or other licensees.

(B16) What recordkeeping procedures should be followed when two private individuals want to engage in a firearms transaction?

When a transaction takes place between private (unlicensed) persons who reside in the same State, the Gun Control Act (GCA) does not require any record keeping. A private person may sell a firearm to another private individual in his or her State of residence and, similarly, a private individual may buy a firearm from another private person who resides in the same State. It is not necessary under Federal law for a Federal firearms licensee (FFL) to assist in the sale or transfer when the buyer and seller are "same-State" residents. Of course, the transferor/seller may not knowingly transfer a firearm to someone who falls within any of the categories of prohibited persons contained in the GCA. See 18 U.S. C. §§ 922(g) and (n). However, as stated above, there are no GCA-required records to be completed by either party to the transfer.

There may be State or local laws or regulations that govern this type of transaction. Contact State Police units or the office of your State Attorney General for information on any such requirements.

Please note that if a private person wants to obtain a firearm from a private person who resides in another State, the firearm will have to be shipped to an FFL in the buyer's State. The FFL will be responsible for record keeping. See also Question **B3**.

(B17) How do I obtain a classification from ATF for my "potato gun"?

Any person desiring a classification of a "potato gun," "spud gun" or similar device must submit a written request (not e-mail) to the Director and include a complete and accurate description of the device, the name and address of the manufacturer or importer, the purpose for which it is intended, and such photographs, diagrams, or drawings as may be necessary to make a classification. A final determination may require physical examination of the device. Such requests for classification should be submitted to: Bureau of ATF, Firearms Technology Branch.

C. LICENSING

(C1) Who can get a license?

ATF will approve the application if the applicant:

- Is 21 years or more of age;
- Is not prohibited from shipping, transporting, receiving or possessing firearms or ammunition;
- Has not willfully violated the GCA or its regulations;
- Has not willfully failed to disclose material information or willfully made false statements concerning material facts in connection with his application;
- Has premises for conducting business or collecting; and
- The applicant certifies that:

  (1) the business to be conducted under the license is not prohibited by State or local law in the place where the licensed premises is located;

  (2) within 30 days after the application is approved the business will comply with the requirements of State and local law applicable to the conduct of the business;

  (3) the business will not be conducted under the license until the requirements of State and local law applicable to the business have been met;

  (4) the applicant has sent or delivered a form to the chief law enforcement officer where the premises is located notifying the officer that the applicant intends to apply for a license; and

  (5) secure gun storage or safety devices will be available at any place in which firearms are sold under the license to persons who are not licensees ("secure gun storage or safety device" is defined in 18 U.S.C. 921(a)(34)).

[18 U.S.C. 923(d)(1), 27 CFR 478.47(b)]

(C2) How does one get a license?

Submit ATF Form 7, Application for License, or ATF Form 7CR, Application for License (Collector of Curios or Relics), with the appropriate fee in accordance with the instructions on the form to ATF. These forms may be obtained from the Federal Firearms Licensing Center or your local ATF office.

[18 U.S.C. 923, 27 CFR 478.44 and 478.45]

(C3) May one license cover several locations?

No. A separate license must be obtained for each location. However, storage facilities are not required to be covered by a separate license, although the records maintained on licensed premises must reflect all firearms held in the separate storage facility. Firearms may be shipped directly to separate storage facilities as long as they are properly recorded as an acquisition in the licensee's records.

[27 CFR 478.50]

(C4) Does an importer or manufacturer of firearms also need a dealer's license?

No, as long as the importer or manufacturer is engaged in the business of dealing in firearms at the licensed premises in the same type of firearms authorized by the importer's or manufacturer's license.

[27 CFR 478.41(b)]

(C5) If a person timely files an application for renewal of a license and the present license expires prior to receipt of the new license, may the person continue to conduct the business covered by the expired license?

Yes. A person who timely files an application for renewal of a license may continue operations authorized by the expired license until the application is finally acted upon. An application is timely filed when it is received accurate and completed at the P.O. Box listed on the application form with the appropriate renewal fee.

If a person does not timely file a license renewal application and the license expires, the person must file ATF Form 7, Application for License, or an ATF Form 7CR, Application for License (Collector of Curios or Relics), as required by 27 CFR 478.44, submit the application fee applicable to a new business, and obtain the required license before continuing business activity.

[27 CFR 478.45]

(C6) Must a licensed importer's, manufacturer's, or dealer's records be surrendered to ATF if the licensee discontinues business?

If the business is being discontinued completely, the licensed dealer, manufacturer or importer is required, within 30 days, to forward the business records to the following address:

BUREAU OF ATF
ATF OUT-OF-BUSINESS RECORDS CENTER
244 NEEDY ROAD
MARTINSBURG, WV 24501

Failure to surrender required records is a felony and could result in the licensee being fined up to $250,000, imprisoned up to 5 years, or both. A licensee discontinuing business also must notify the Federal Firearms Licensing Center within 30 days.

If someone is taking over the business, the original licensee should underline the final entry in each bound book, note the date of transfer, and forward all records and forms to the successor (who must apply for and receive his or her own license before lawfully engaging in business) or forward the records and forms to the ATF Out-of-Business Records Center. If the successor licensee receives records and forms from the original licensee, the successor licensee may choose to forward these records and forms to the ATF Out-of-Business Record Center.

[18 U. S. C. 923(g)(4), 22 CFR 478.57, 27 CFR 478.127]

(C7) What records am I required to forward to ATF upon discontinuance of my business?

The records consist of the licensee's bound acquisition/disposition (A/D) records, ATF Forms 4473, ATF Forms 3310.4 (Report of Multiple Sale or Other Disposition of Pistols and Revolvers), ATF Forms 3310.11 (Federal Firearms Licensee Theft/Loss Report), records of transactions in semiautomatic assault weapons, records of importation (ATF Forms 6 and 6A), and law enforcement certification letters. If the licensee was granted a variance to use a computerized recordkeeping system, the licensee is required to provide a complete printout of the entire A/D records.

[27 CFR 478.127]

(C8) May a successor owner of a business entity, other than one who is a successor under the provisions of 27 CFR 478.56 (for example, the surviving spouse or child, or a receiver or trustee in bankruptcy), commence a firearms business prior to receiving a Federal firearms license in the successor's name?

No. Each person intending to engage in business as a firearms dealer, importer or manufacturer or an ammunition importer or manufacturer must obtain the required Federal firearms license prior to commencing business.

[27 CFR 478.41]

(C9) Does a Federal firearms license allow the licensee to carry a firearm in the course of business?

No. A Federal firearms license confers no right or privilege to carry a firearm, concealed or otherwise. Permits to carry are issued by State or local authorities.

(C10) May a person obtain a dealer's license to engage in business only at gun shows?

No. A license may only be issued for a permanent premises at which the license applicant intends to do business. A person having such license may conduct business at gun shows located in the State in which the licensed premises is located and sell and deliver curio or relic firearms to other licensees at any location.

[18 U.S.C. 923(a) and (j)]

(C11) May a licensee change the location of the licensed business or activity?

To change your location, you must file an application for an amended license, ATF Form 5300.38, not less than 30 days prior to the move. You must obtain the amended license before commencing business at the new location. The application for an amended license includes a certification of compliance with State and local laws and notification of local law enforcement officials.

[27 CFR 478.52]

(C12) Are black powder dealers required to be licensed as ammunition dealers under the GCA?

No. However, black powder dealers are subject to the provisions of 27 CFR Part 555, Commerce in Explosives, which requires that a dealer in any quantity of black powder must have a license as a dealer.

[18 U.S.C. 842]

D. ATF FORM 4473 FIREARMS TRANSACTION RECORD

(D1) Where can a dealer get ATF Forms 4473?

They are available free of charge from the ATF Distribution Center. The current address is P.O. Box 5950, Springfield, VA 22150-5950. Please order a quantity of forms estimated for 6 months use.

(D2) Does an unlicensed person need an ATF Form 4473 to transfer a firearm?

No. ATF Form 4473 is required only for transfers by a licensee.

[27 CFR 478.124]

(D3) Does a dealer have to execute ATF Form 4473 to take a weapon out of the dealer's inventory for his or her own use?

No. However, the "bound book" must reflect the disposition of the firearm from business inventory to personal use.

However, if the business is a corporation, and the firearm is being transferred to a corporate officer or director for other than business purposes, then a Form 4473 must be executed.

[27 CFR 478.124 and 478.125a]

(D4) Who signs ATF Form 4473 for the seller?

ATF Form 4473 must be signed by the person who verified the identity of the buyer.

(D5) Is a Social Security card a proper means of identification for purchasing a firearm from an FFL?

No. A Social Security card, alien registration card, or military identification alone does not contain sufficient information to identify a firearms purchaser. However, a purchaser may be identified by any combination of government-issued documents which together establish all of the required information: Name, residence address, date of birth, and photograph of the holder.

[27 CFR 478.11 and 478.124(c)]

(D6) When must the ATF Form 4473 be signed?

Part I used for over-the-counter sales must be completed, signed and dated by the buyer prior to delivery of the firearm.

Part II (green form) used for intrastate non-over-the-counter sales must be completed, signed and dated in duplicate by the buyer at the time of sale.

[27 CFR 478.124(c) and 478.124(f)]

E. RECORDS REQUIRED – LICENSEES

(E1) What is a "bound book?"

A "bound book" is a permanently bound book or an orderly arrangement of loose-leaf pages which must be maintained on the business premises. The format must follow that prescribed in the regulations and the pages must be numbered consecutively.

[27 CFR 478.121 and 478.125]

(E2) May a dealer keep more than one "bound book" at the same time?

Yes. A dealer in firearms is not limited to using only one "bound book." It may be convenient for a dealer to account for different brands or types of firearms in separate "bound books."

(E3) Does the Government sell a record book for licensees to use in recording their receipts and dispositions of firearms?

No. Certain trade associations have them available at nominal cost. Your supplier should be able to tell you about this.

(E4) What is the dealer's responsibility where a variance from normal regulatory practice has been authorized?

The ATF letter authorizing the variance must be kept at the licensed premises and available for inspection. For businesses with more than a single licensed outlet, each outlet covered by the variance must have a copy of the letter authorizing the change.

[27 CFR 478.22 and 478.125(h)]

(E5) How much time does a dealer have to record acquisitions and dispositions of firearms in his or her "bound book?"

Generally, licensees have to enter the acquisition or purchase of a firearm by the close of the next business day after the acquisition or purchase and shall record sales or other dispositions within 7 days.

However, if commercial records containing the required information are available for inspection and are separate from other commercial documents, dealers have 7 days from the time of receipt to record the receipt in the "bound book."

If a disposition is made before the acquisition has been entered in the "bound book," the acquisition entry must be made at the same time as the disposition entry.

[27 CFR 478.125]

(E6) Are the ammunition recordkeeping requirements the same as for firearms?

No. No records are required for ammunition other than armor piercing ammunition. See 27 CFR 478.125 for details on recordkeeping requirements.

(E7) Are rental firearms subject to recordkeeping control?

Yes, if the firearms are taken off the premises of the licensee. However, the recordkeeping is not imposed on the loan or rental of firearms for use only on the premises of the licensee.

[27 CFR 478.97]

(E8) May a licensee who has firearms in his or her private collection sell any of these firearms without making firearms record entries?

A licensee may sell a firearm from his or her personal collection, subject only to the restrictions on firearm sales by unlicensed persons, provided the firearm was entered in the licensee's bound book and then transferred to the licensee's private collection at least 1 year prior to the sale. When the personal firearm is sold, the sale must be recorded in a "bound book" for dispositions of personal firearms, but no ATF Form 4473 is required.

[27 CFR 478.125a]

(E9) May a licensee maintain computer records in lieu of the "bound book?"

Yes. The Director of Industry Operations or other designated ATF official must approve a request for a recordkeeping variance before the licensee may use a computer system in lieu of the "bound book" record required by the regulations.

[27 CFR 478.22 and 478.125(h)]

F. CONDUCT OF BUSINESS - LICENSEES

(F1) Does the Federal firearms law require licensees to comply with State laws and local published ordinances when selling firearms?

Yes. It is unlawful for any licensed importer, licensed manufacturer, licensed dealer, or licensed collector to sell or deliver any firearm or ammunition to any person if the person's purchase or possession would be in violation of any State law or local published ordinance applicable at the place of sale or delivery.

[18 U.S.C. 922(b)(2), 27 CFR 478.99(b)(2)]

(F2) May a licensed dealer sell a firearm to a nonlicensee who is a resident of another State?

Generally, a firearm may not lawfully be sold by a licensed dealer to a nonlicensee who resides in a State other than the State in which the seller's licensed premises is located. However, the sale may be made if the firearm is shipped to a licensed dealer whose business is in the purchaser's State of residence and the purchaser takes delivery of the firearm from the dealer in his or her State of residence. In addition, a licensee may sell a rifle or shotgun to a person who is not a resident of the State where the licensee's business premises is located in an over-the-counter transaction, provided the transaction complies with State law in the State where the licensee is located and in the State where the purchaser resides.

[18 U.S.C. 922(b)(3)]

(F3) May a dealer sell firearms to law enforcement agencies and individual officers in another State?

Yes. Sales and deliveries of firearms to out-of-State police and sheriff departments are not prohibited by the GCA. A dealer may also sell or ship firearms, other than NFA firearms, to an individual law enforcement officer, regardless of age, if the dealer has a signed statement from the officer's agency, stating that the items are to be used in the buyer's official duties and that the officer has not been convicted of a misdemeanor crime of domestic violence. No ATF Form 4473 or NICS check is required; however, the bound book must be properly posted, and the signed statement included in the dealer's records. For further information on sales of firearms to law enforcement officers, see **Item 4**, "Sales to Law Enforcement Officers", in the General Information section of this publication. You should contact your State's Attorney General's Office to ensure there is no State prohibition on such sales.

[18 U.S.C. 925(a) (1), 27 CFR 478.134 and 478.141]

(F4) May an employee of a licensed dealer, such as a manager or clerk, who is under 21 years of age, sell handguns and ammunition suitable for use in handguns for the licensee?

Yes, if the employee is not a prohibited person (e.g., a felon). However, to sell handguns, a person less than 18 years of age must have the prior written consent of a parent or guardian and the written consent must be in the person's possession at all times. Also, the parent or guardian giving the written consent may not be prohibited by law from possessing a firearm. Moreover, State law must not prohibit the juvenile from possessing the handguns or ammunition.

[18 U.S.C. 922(x)]

(F5) As a licensed dealer, must I advise ATF if I sell more than one handgun to an individual?

If you sell or dispose of more than one handgun to any nonlicensee during a period of 5 consecutive business days, the sale must be reported on ATF Form 3310.4, Report of Multiple Sale or Other Disposition of Pistols and Revolvers, not later than the close of the business day on which you sold or disposed of the second handgun. The licensee must forward a copy of the Form 3310.4 to the ATF office specified thereon, and another copy must be forwarded to the State police or local law enforcement agency where the sale occurred. A copy of the Form 3310.4 also must be attached to the firearms transaction record, ATF Form 4473, documenting the sale or disposition of the second handgun.

A business day for purposes of reporting multiple sales of pistols or revolvers is a day that a licensee conducts business pursuant to the license, regardless of whether State offices are open. The application of the term "business day" is, therefore, distinguishable from the term "business day" as used in the NICS context. **Example**: A licensee conducts business only on Saturdays and Sundays, days on which State offices are not open. The licensee sells a pistol to an unlicensed person on a Saturday. If that same unlicensed person acquires another handgun the next day (Sunday), the following Saturday or Sunday, or the Saturday after the reporting requirement would be triggered, the subsequent acquisition of a handgun would have to be reported on a Form 3310.4 by the close of the day upon which the second or subsequent handgun was sold.

[18 U.S.C. 923(g)(3), 27 CFR 478.126a]

(F6) Does a customer have to be a certain age to buy firearms or ammunition from a licensee?

Yes. Under the GCA, long guns and long gun ammunition may be sold only to persons 18 years of age or older. Sales of handguns and ammunition for handguns are limited to persons 21 years of age and older. Although some State and local ordinances have lower age requirements, dealers are bound by the minimum age requirements established by the GCA. If State law or local ordinances establish a higher minimum age, the dealer must observe the higher age requirement.

[18 U.S.C. 922(b)(1), 27 CFR 478.99(b)]

(F7) May a licensee sell interchangeable ammunition such as .22 cal. rimfire to a person less than 21 years old?

Yes, provided the buyer is 18 years of age or older, and the dealer is satisfied that it is for use in a rifle. If the ammunition is intended for use in a handgun, the 21 year old minimum age requirement is applicable.

[18 U.S.C. 922(b)(1), 27 CFR 478.99(b)]

(F8) In transactions between licensees, how is the seller assured that a purchaser of a firearm is a licensed dealer?

Verification must be established by the transferee furnishing to the transferor a certified copy of the transferee's license and by any other means the transferor deems necessary (such as the FFL eZcheck).

[27 CFR 478.94]

(F9) Must a multi-licensed business submit a certified copy of each of its licenses when acquiring firearms?

No. It need only provide the seller a list, certified to be true, correct and complete, containing the name, address, and license number and expiration date for each location.

[27 CFR 478.94]

(F10) May a licensee continue to deliver to a business whose license has expired?

Yes, for a period of 45 days following the expiration date of the license. After the 45-day period, the transferor is required to verify the licensed status of the transferee with the Chief, Firearms Licensing Center. If the transferee's license renewal application is still pending, the transferor must obtain evidence from the Director of Industry Operations that a license renewal application has been timely filed by the transferee and is still pending.

[27 CFR 478.94]

(F11) Is a license required to engage in the business of selling small arms ammunition?

No. A license is not required for a dealer in ammunition only, but a manufacturer or an importer of ammunition must be licensed.

[18 U.S.C. 922 (a)(1)(B)]

(F12) May licensed dealers sell firearms at gun shows?

Generally, a licensee may sell firearms at a gun show located only in the same State as that specified on the seller's license. However, a licensee may sell curio or relic firearms to another licensee at any location.

[18 U.S.C. 923(j), 27 CFR 478.100]

(F13) What may a licensed dealer do at an out-of-State gun show?

A licensed dealer may sell and deliver curio or relic firearms to another licensee at an out-of-State gun show. With respect to other firearms transactions, a licensed dealer may only display and take orders for firearms at an out-of-State gun show. In filling any orders for firearms, the dealer must return the firearms to his or her licensed premises and deliver them from that location. Any firearm ordered by a nonlicensee must be delivered or shipped from the licensee's premises to a licensee in the purchaser's State of residence, and the purchaser must obtain the firearm from the licensee located in the purchaser's State. Except for sales of curio or relic firearms to other licensees, sales of firearms and simultaneous deliveries at the gun show, whether to other licensees or to nonlicensees, violate the law because the dealer would be unlawfully engaging in business at an unlicensed location.

[18 U.S.C. 922(a)(1), (b)(3), 923(a) and (j)]

(F14) Who may ship handguns through the U.S. Postal Service?

Federal firearm licensees may send an unloaded handgun in the mail to another FFL in customary trade shipments. Handguns also may be mailed to any officer, employee, agent, or watchman who is eligible under 18 U.S.C. 1715 to receive pistols, revolvers, and other firearms capable of being concealed on the person for use in connection with his or her official duties.

However, postal service regulations must be followed. Any person proposing to mail a handgun must file with the postmaster, at the time of mailing, an affidavit signed by the addressee stating that the addressee is qualified to receive the firearm, and the affidavit must bear a certificate stating that the firearm is for the official use of the addressee. See the current Postal Manual for details.

The Postal Service recommends that all firearms be sent by registered mail and that no marking of any kind which would indicate the nature of the contents be placed on the outside of any parcel containing firearms. (See also Questions **B7** and **B8**.)

(F15) Must a dealer record firearms received on consignment?

Yes. Firearms received for sale on consignment must be entered in the dealer's "bound book."

Sales of the firearms are handled in the same manner as other firearm sales. Return of the remaining firearms by the licensee to the consignor is entered in the dealer's disposition record. An ATF Form 4473 and a NICS check must be completed.

(F16) To whom does an FFL report stolen or lost firearms?

A theft or loss of firearms must be reported to your local police as well as to ATF within 48 hours after the discovery. Licensees should notify ATF on the 24-hour, 7 days a week toll free line at 1-800-800-3855 and by preparing and submitting ATF Form 3310.11, Federal Firearms Licensee Theft/Loss Report.

Theft or loss of NFA firearms should also be reported to the NFA Branch immediately upon discovery.

[18 U.S.C. 923(g)(6), 27 CFR 478.39 and 479.141]

(F17) If my firearms are stolen or lost, what do I do about my records?

Take an inventory of stock on hand and enter "stolen" or "lost" and the date in the disposition section of the "bound book" for those stolen or lost firearms. In addition, at the time a licensee reports the theft or loss on the ATF toll free line, the licensee will be provided a control number that should be placed in the records as well as on ATF Form 3310.11, Federal Firearms Licensee Theft/Loss Report.

(F18) How many copies of the ATF Form 3310.4, Report of Multiple Sale or Other Disposition of Pistols and Revolvers, must be completed and what becomes of each copy?

ATF Form 3310.4 must be completed in triplicate (3 copies). The original is sent to ATF's National Tracing Center by FAX at 1-877-283-0288 or by mail to Box 1061, Falling Waters, West Virginia 25419-1061. A copy is to be sent to the designated State police or the local law enforcement agency in the jurisdiction where the sale took place. The remaining copy is to be retained in the records of the dealer and held for not less than 5 years.

[27 CFR 478.126a and 478.129]

(F19) What is my responsibility to respond to a request to trace a firearm?

A licensee must provide the requested information immediately and in no event later than 24 hours after receipt of a request by ATF. Failure to respond to the request for trace information can result in monetary fines, imprisonment, and/or revocation of the licensee's Federal firearms license.

[18 U.S.C. 923(g)(7), 27 CFR 478.25a]

(F20) Does the requirement to give written notification to handgun transferees about juvenile handgun possession apply to a licensed dealer who returns firearms to their owners, for example, handguns that the dealer repaired?

Yes. The requirement to give written notification to nonlicensees applies to the return of handguns, as well as to their sale. It applies even if the licensee ships the repaired firearm to the customer.

[27 CFR 478.103]

(F21) Does the requirement to post a sign on the licensed premises about juvenile handgun possession apply to a licensed dealer who only disposes of handguns to nonlicensees who do not appear at the dealer's premises?

No. The sign posting requirement does not apply where the licensee only disposes of handguns to nonlicensees who do not appear at the licensed premises (for example, the licensee only ships repaired or replacement handguns to nonlicensees).

[27 CFR 478.103]

G. COLLECTORS

(G1) Is there a specific license which permits a collector to acquire firearms in interstate commerce?

Yes. A person may obtain a collector's license. However, this license applies only to transactions in curio or relic firearms.

[18 U.S.C. 923(b), 27 CFR 478.41(c), (d), 478.50(b) and 478.93]

(G2) Does a collector's license afford any privileges to the licensee with respect to acquiring or disposing of firearms other than curios or relics in interstate or foreign commerce?

No. A licensed collector has the same status under the GCA as a nonlicensee except for transactions in curio or relic firearms.
[27 CFR 478.93]

(G3) Does a license as a collector of curio or relic firearms authorize the collector to engage in the business of dealing in curios or relics?

No. A dealer's license must be obtained to engage in the business of dealing in any firearms, including curios or relics.

[18 U.S.C. 922(a) and 923(a), 27 CFR 478.41]

(G4) Since a licensed firearms dealer may legally deal in curio or relic firearms, is there any reason why a dealer would need both a dealer's license and collector's license?

No. A collector's license enables a collector to obtain curio or relic firearms interstate. A person holding a dealer's license may also acquire curio or relic firearms interstate, and so there is no need for a licensed dealer to obtain a collector's license.

(G5) Are licensed collectors required to execute ATF Form 4473 for transactions in curio or relic firearms?

No. However, licensed collectors are required to keep a "bound book" record.

[27 CFR 478.125(f)]

(G6) Are licensed collectors' transfers of curio or relic firearms subject to the Brady law, including the provision for making background checks on transferees?

No, but it is unlawful to transfer a firearm to any person knowing or having reasonable cause to believe that such person is a felon or is within any other category of person prohibited from receiving or possessing firearms. See also Questions **P12** and **P13**.

[18 U.S.C. 922(d), 27 CFR 478.32(d)]

(G7) Are licensed collectors required to comply with the requirements that written notification be given to handgun transferees and signs be posted on juvenile handgun possession?

The requirement that written notification concerning juvenile handgun possession be given by licensees to a nonlicensee to whom a handgun is delivered applies to curio or relic handguns transferred by licensed collectors. However, the sign posting requirement does not apply to licensed collectors.

[18 U.S.C. 922(x), 27 CFR 478.103]

(G8) Are licensed collectors required to turn in their acquisition/disposition records to ATF if their collector's license is not renewed or they discontinue their collecting activity?

No. The GCA requires the delivery of required records to the Government within 30 days after a firearms "business" is discontinued. A license as a collector of curios or relics does not authorize any business with respect to firearms. Therefore, the records required to be kept by licensed collectors under the law and regulations are not business records and are not required to be turned in to ATF when collector's licenses are not renewed or collecting activity under such licenses is discontinued.

[18 U.S.C. 923(g)(4), 27 CFR 478.127]

H. MANUFACTURERS

(H1) Must a person who engages in the business of manufacturing and importing firearms have a separate license to cover each type of business?

Yes. A separate license is required to cover each of these types of businesses.

[27 CFR 478.41]

(H2) May a person licensed as a manufacturer of ammunition also manufacture firearms?

No. A person licensed as a manufacturer of ammunition may not manufacture firearms unless he or she obtains a license as a firearms manufacturer.

(H3) May a person licensed as a manufacturer of firearms also manufacture ammunition?

Yes. The person may also manufacture ammunition (not including destructive device ammunition or armor piercing ammunition) without obtaining a separate license as a manufacturer of ammunition.

(H4) Is a person who reloads ammunition required to be licensed as a manufacturer?

Yes, if the person engages in the business of selling or distributing reloads for the purpose of livelihood and profit. No, if the person reloads only for personal use.

[18 U.S.C. 922(a) (i) and 923(a), 27 CFR 478.41]

(H5) Must a licensed manufacturer pay excise taxes?

Yes. Licensed manufacturers incur excise tax on the sale of firearms and ammunition manufactured. See Item 17, "Federal Excise Tax" in the General Information section of this publication.

I. GUNSMITHS

(I1) Is a license needed to engage in the business of engraving, customizing, refinishing or repairing firearms?

Yes. A person conducting such activities as a business is considered to be a gunsmith within the definition of a dealer. See **Item 16**, "Federal Excise Tax" in the General Information section of this publication.

[27 CFR 478.11]

(I2) Does a gunsmith need to enter in a permanent "bound book" record every firearm received for adjustment or repair?

If a firearm is brought in for repairs and the owner waits while it is being repaired or if the gunsmith is able to return the firearm to the owner during the same business day, it is not necessary to list the firearm in the "bound book" as an "acquisition." If the gunsmith has possession of the firearm from one business day to another or longer, the firearm must be recorded as an "acquisition" and a "disposition" in the permanent "bound book" record.

(I3) Is an ATF Form 4473 required when a gunsmith returns a repaired firearm?

No, provided the firearm is returned to the person from whom it was received.

[27 CFR 478.124(a)]

(I4) May a gunsmith make immediate repairs at locations other than his or her place of business?

Yes.

(I5) May a licensed gunsmith receive an NFA firearm for purposes of repair?

Yes, for the sole purpose of repair and subsequent return to its owner. It is suggested that the owner obtain permission from ATF for the transfer by completing and mailing ATF Form 5 to the NFA Branch and receive approval prior to the delivery. The gunsmith should do the same prior to returning the firearm.

Only the face of the form needs to be completed in each instance. ATF Forms 5 may be obtained from the Bureau of ATF, NFA Branch. ATF Form 5 is also available on the internet at www.ATF.gov.

(I6) Is a licensed gunsmith required to comply with the requirements to give written notification to handgun transferees and post signs on juvenile handgun possession?

The requirement that written notification on juvenile handgun possession be given to a nonlicensee to whom a handgun is delivered applies to all Federal firearms licensees. It applies to the return of handguns to their owners, as well as to their sale. Thus, a gunsmith who repairs or customizes a nonlicensee's handgun must provide the notification to the nonlicensee when the handgun is returned. The sign posting requirement also applies to gunsmiths, unless the gunsmith only disposes of handguns to nonlicensees who do not appear at the gunsmith's licensed premises, for example, when repaired handguns are shipped to nonlicensees.

[18 U.S.C. 922(x), 27 CFR 478.103]

(I7) Is a licensed gunsmith's return of repaired or customized firearms to their owners subject to the Brady law, including the provision for making background checks on transferees?

No, but it is unlawful to transfer a firearm to any person knowing or having reasonable cause to believe that such person is a felon or is within any other category of person prohibited from receiving or possessing firearms. (See also Question **P24**.)

[18 U.S.C. 922(d), 27 CFR 478.32(d)]

J. PAWNBROKERS

(J1) What disposition records must be kept by a pawnbroker upon the

redemption of a pawned firearm?

The redemption of a pawned firearm is a "disposition" of a firearm under Federal firearms law and is subject to all the recordkeeping requirements under the GCA. Disposition must be properly entered in the pawnbroker's "bound book," and ATF Form 4473 must be executed in connection with the redemption.

[27 CFR 478.124 and 478.125]

(J2) What is the procedure for a licensed pawnbroker to return a firearm?

The procedure varies, depending upon the firearm and the situation. ATF Form 4473 must be used in each situation. In addition, any State laws regarding pawn transactions must be followed.

Some Examples:

(1) Pawnbroker and nonlicensee are residents of the same State: The pawnbroker may return a handgun or long gun to either the person who pawned it or a holder of the pawn ticket who resides in the pawnbroker's State.

(2) Pawnbroker and nonlicensee are not residents of the same State:

a. The pawnbroker may return a handgun only to the person who pawned it.

b. The pawnbroker may return a rifle or shotgun to the person who pawned it.

c. The pawnbroker may transfer a rifle or shotgun to the holder of a pawn ticket who did not pawn it, provided the transaction complies with the law of the State where the pawnbroker's business is located and the law of the State where the pawn ticket holder resides.

[18 U.S.C. 922(a)(2), 922(a)(3) and 922(b)(3)]

(J3) Are there categories of persons to whom a pawnbroker cannot return firearms?

Yes. A pawnbroker cannot lawfully return a firearm to a person who is underage or within a prohibited category of persons to whom the sale or other disposition of the firearm would be unlawful. For example, a pawnbroker cannot lawfully return a pawned handgun to a person who is less than 21 years of age, nor can he or she return a firearm to a convicted felon or to anyone else who is prohibited from receiving the firearm under Federal or State law.

[18 U.S.C. 922(d), 27 CFR 478.99]

(J4) Are licensed pawnbrokers required to comply with the requirements that written notification be given to handgun transferees and signs be posted on juvenile handgun possession?

The requirement that written notification on juvenile handgun possession be given to a nonlicensee to whom a handgun is delivered applies to all types of Federal firearms licensees. It also applies to the return of handguns to their owners, as well as to their sale. Thus, a pawnbroker who returns a handgun to its owner, or to another person, upon its redemption from pawn must provide the notification. The sign posting requirement also applies to licensed pawnbrokers.

[18 U.S.C. 922(x), 27 CFR 478.103]

(J5) Are licensed pawnbrokers' firearms sales or return of firearms redeemed from pawn subject to the Brady law, including the provision for making background checks of transferees?

Yes. Moreover, as provided by Public Law 105-277, enacted on October 21, 1998, a licensed pawnbroker may also contact the National Instant Criminal Background Check System (NICS) for a background check on a person at the time the person offers to pawn a firearm. If NICS advises the pawnbroker that receipt or possession of the firearm by the person attempting to pawn the firearm would violate the law, the pawnbroker must advise local law enforcement within 48 hours after receipt of the information.

A pawnbroker who contacts NICS about a person prior to accepting the person's firearm in pawn must still comply with the requirements of the Brady law at the time of the firearm's redemption, i.e., the pawnbroker must contact NICS for another background check at the time of redemption. (See also Questions **P17** through **P23**.)

K. AUCTIONEERS

(K1) Does an auctioneer who is involved in firearms sales need a dealers license?

Generally speaking, there are two types of auctions: estate-type auctions and consignment auctions.

In estate-type auctions, the articles to be auctioned (including firearms) are being sold by the executor of the estate of an individual. The firearms belong to and are possessed by the executor. The firearms are controlled by the estate, and the sales of firearms are being made by the estate. The auctioneer is acting as an agent of the executor and assisting the executor in finding buyers for the firearms. In these cases, the auctioneer does not meet the definition of engaging in business as a dealer in firearms and would not need a license. An auctioneer who does have a license may perform this function away from his or her licensed premises.

In consignment-type auctions, an auctioneer often takes possession of firearms in advance of the auction. These firearms are generally inventoried, evaluated, and tagged for identification. The firearms belong to individuals who have entered into a consignment agreement with the auctioneer giving that auctioneer authority to sell the firearms. The auctioneer therefore has possession and control of the firearms. Under these circumstances, an auctioneer would generally need a license. If you are not sure if a license is needed in a particular consignment auction situation, contact your local ATF office.

See **ATF Ruling 96-2**.

(K2) If a licensed auctioneer is making sales of firearms, where may those sales be made?

In a consignment auction, firearms may be displayed at an auction site away from the auctioneer's licensed premises and sales of the firearms can be agreed upon at that location, but delivery may only be made to purchasers after the firearms have been returned to the auctioneer's licensed premises. The simultaneous sale and delivery of the auctioned firearms away from the licensed premises would violate the law, i.e., engaging in business at an unlicensed location.

However, if the auctioneer is assisting an estate in disposing of firearms, the estate is the seller of the firearms and the estate is in control and possession of the firearms. In this situation, the firearms may be delivered from the auction site. See also Question **K1** and **ATF Ruling 96-2**.

L. IMPORTING AND EXPORTING

(L1) May a licensed dealer who does not have an importer's license make an occasional importation?

Yes. A licensed dealer may make an occasional importation of a firearm for a nonlicensee or for the licensee's personal use (not for resale). The licensee must first submit an ATF Form 6, Part I to ATF for approval. The licensee may then present the approved Form 6 and completed ATF Form 6A to U.S. Customs and Border Protection. Contact the Bureau of ATF, Firearms and Explosives Imports Branch for forms, or download them from the ATF webpage at www.ATF.gov.

(L2) Does a licensee need an export license to export a firearm?

The GCA does not require export licenses. However, most firearms and ammunition must be exported in accordance with the provisions of the Arms Export Control Act of 1976. Regulations implementing this Act generally require a license to be obtained from the Directorate of Defense Trade Controls, Department of State, PM/DDTC, SA-1, Room 1200, 2401 E St., N.W., Washington, DC 20037; (202) 663-1282.

The export of sporting shotguns and ammunition for sporting shotguns is regulated by the U.S. Department of Commerce rather than the State Department. An export license is generally needed to export these shotguns and ammunition. For further information, contact them at their nearest district office or the Bureau of Industry and Security, Outreach and Educational Services Division, U.S. Department of Commerce, 14th St. & Pennsylvania Ave. N.W., Washington, DC 20230, (202) 482-4811.

When exporting NFA firearms, ATF Form 9 must be completed and approved by ATF prior to export.

[22 U.S.C. 2778, 27 CFR 479.114 and 479.116]

(L3) Does ATF regulate the importation of gas masks?

Yes. Gas masks are "defense articles" subject to regulation under the Arms Export Control Act. ATF regulates the importation of gas masks and generally requires an ATF Form 6 import permit for lawful importation of these items. The standard processing time is 4-6 weeks from the date ATF's Firearms and Explosives Imports Branch receives a complete Form 6 application. You can download the Form 6 at www.ATF.gov.

(L4) What must I do to import gas masks for resale?

A commercial importer must be registered in accordance with the Arms Export Control Act. You can download the application to register, ATF F 5330.4 (4587), Application to Register as an Importer of U.S. Munitions Import List Articles, at www.ATF.gov. There is a fee and the processing time is 2-4 weeks.

M. FIREARMS - NATIONAL FIREARMS ACT (NFA)

(M1) The types of firearms that must be registered in the National Firearm Registration and Transfer Record are defined in the NFA and in 27 CFR Part 479. What are some examples?

Some examples of the types of firearms that must be registered are:

Machineguns;

The frames or receivers of machineguns;

Any combination of parts designed and intended for use in converting weapons into machineguns;

Any part designed and intended solely and exclusively for converting a weapon into a machinegun;

Any combination of parts from which a machinegun can be assembled if the parts are in the possession or under the control of a person;

Silencers and any part designed and intended for fabricating a silencer;

Short-barreled rifles;

Short-barreled shotguns;

Destructive devices; and,

"Any other weapon."

A few examples of destructive devices are:

Molotov cocktails;

Anti-tank guns (over caliber .50); Bazookas; and,

Mortars.

A few examples of "any other weapon" are:

H&R Handyguns;

Ithaca Auto-Burglar guns;

Cane guns; and,

Gadget-type firearms and "pen" guns which fire a projectile by the action of an explosive.

[26 U.S.C. 5845]

(M2) How can an individual legally acquire NFA firearms?

Basically, there are 2 ways that an individual (who is not prohibited by Federal, State, or local law from receiving or possessing firearms) may legally acquire NFA firearms:

(1) By transfer after approval by ATF of a registered weapon from its lawful owner residing in the same State as the transferee.

(2) By obtaining prior approval from ATF to make NFA firearms.

[27 CFR 479.62-66 and 479.84-86]

(M3) What is the tax on making an NFA firearm?

The tax is $200 for making any NFA firearm, including "any other weapon."

(M4) How is this tax paid?

A money order or check made payable to the Bureau of ATF together with the application forms are to be mailed to the Bureau of ATF, NFA Branch.

(M5) What is an unserviceable firearm?

An unserviceable firearm is defined as one which is incapable of discharging a shot by means of an explosive and which is incapable of being readily re-stored to a firing condition.

An acceptable method of rendering most firearms unserviceable is to fusion weld the chamber closed and fusion weld the barrel solidly to the frame. Certain unusual firearms require other methods to render the firearms unserviceable.

An unserviceable NFA firearm is still subject to the controls of the NFA, but may be transferred tax free as a curio or ornament.

[26 U.S.C. 5845(h) and 5852, 27 CFR 479.11and 479.91]

(M6) What happens when a State acquires an unregistered NFA firearm through seizure or abandonment?

When a State wants to keep such NFA firearms for official use, they must be registered by filing ATF Form 10 with the Bureau of ATF, NFA Branch.

Since approval of the Form 10 is conditioned on an "official use only" basis, subsequent transfers will not be approved except if the transfer is to another government agency for official use.

[27 CFR 479.104]

(M7) May a private citizen who owns an NFA firearm which is not registered have the firearm registered?

No. The NFA permits only manufacturers, makers, and importers to register firearms. Mere possessors may not register firearms. An unregistered NFA firearm is a contraband firearm and it is unlawful to possess the weapon. The possessor should contact the nearest ATF office to arrange for its disposition.

[26 U.S.C. 5861(d)]

(M8) What can happen to someone who has an NFA firearm which is not registered to him?

Violators may be fined not more than $250,000, and imprisoned not more than 10 years, or both. In addition, any vessel, vehicle or aircraft used to transport, conceal or possess an unregistered NFA firearm is subject to seizure and forfeiture, as is the weapon itself.

[49 U.S.C. 781-788, 26 U.S.C. 5861 and 5872

(M9) What should a person do if he or she comes into possession of an unregistered NFA firearm?

Contact the nearest ATF office immediately.

(M10) Are there any exemptions from the making or transfer tax provisions of the NFA?

Yes. These are noted below, along with the required form number, if any, to apply for the exemption. Completed forms must be approved by the NFA Branch prior to the making or transfer:

(1) Tax-exempt transfer and registration of a firearm between special (occupational) taxpayers: ATF Form 3.

(2) Tax-exempt making of a firearm on behalf of a Federal or State agency: ATF Form 1. Tax-exempt transfer and registration of the firearm on behalf of a Federal or State agency: ATF Form 5.

(3) A licensed manufacturer under contract to make NFA firearms for the U.S. Government may be granted an exemption from payment of the special (occupational) tax as a manufacturer of NFA firearms and an exemption from all other NFA provisions (except importation) with respect to the weapons made to fulfill the contract. Exemptions are obtained by writing the NFA Branch, stating the contract number(s) and the anticipated date of termination. This exemption must be renewed each year prior to July 1.

(4) Tax-exempt transfer and registration of an unserviceable firearm which is being transferred as a curio or ornament: ATF Form 5.

(5) Tax exempt transfer of a firearm to a lawful heir: ATF Form 5.

(6) Tax-exempt transfer by operation of law (e.g., court order).

[26 U.S.C. 5851-5853, 27 CFR 479.69, 479.70 and 479.88–91]

(M11) How does a person qualify to import, manufacture, or deal in NFA firearms?

The person must be licensed under the GCA and pay the required special (occupational) tax imposed by the NFA. After becoming licensed under the GCA, he or she must file ATF Form 5630.7 with the appropriate tax payment in the entire amount with ATF. In addition, an importer (except importers of sporting shotguns and shotgun ammunition) must also be registered with ATF under the Arms Export Control Act of 1976.

[26 U.S.C. 5801, 18 U.S.C. 923, 27 CFR 447.31, 478.41 and 479.34]

(M12) When must firearms special (occupational) taxes be paid and how much are the taxes?

These taxes must be paid in full on first engaging in business and thereafter on or before the first day of July. The current taxes are set out in the following table.

SPECIAL (OCCUPATIONAL) TAX RATES UNDER THE NFA

CLASS OF TAXPAYER	ANNUAL FEE
1 Importer of Firearms (Including "Any Other Weapon")	**1000.00**
2 Manufacturer of Firearms (Including "Any Other Weapon")	**1000.00**
3 Dealer of Firearms (Including "Any Other Weapon")	**$500.00**
1 Importer of Firearms (Including "Any Other Weapon") REDUCED*	**$500.00**
2 Manufacturer of Firearms (Including "Any Other Weapon") REDUCED*	**$500.00**

* REDUCED = Rates which apply to certain taxpayers whose total gross receipts in the last taxable year are less than $500,000.

(M13) Does a single special (occupational) tax payment entitle a person or firm to import and manufacture firearms?

No. A separate special (occupational) tax payment must be made for each of these activities. However, Class 1 (importer) and Class 2 (manufacturer) special (occupational) taxpayers are qualified to deal in NFA firearms without also having to pay special (occupational) tax as a Class 3 dealer.

[27 CFR 479.39]

(M14) May a licensed collector obtain NFA firearms in interstate commerce?

Only if the firearms are classified as curios or relics, are registered, and are transferred in accordance with the provisions of the NFA. See Question **M15**.

(M15) What are the required transfer procedures for an individual who is not qualified as a manufacturer, importer, or dealer of NFA firearms?

ATF Form 4 (5320.4) must be completed, in duplicate. The transferor first completes the face of the form. The transferee completes the transferee's certification on the reverse of the form and must have the "Law Enforcement Certification" completed by the chief law enforcement officer.

The transferee is to place, on each copy of the form, a 2-inch by 2-inch photograph of the transferee taken within the past year (proofs, group photographs or photocopies are unacceptable). The transferee's address must be a street address, not a post office box. If there is no street address, specific directions to the residence must be included.

If State or local law requires a permit or license to purchase, possess, or receive NFA firearms, a copy of the transferee's permit or license must accompany the application. A check or money order for $200 ($5 for transfer of "any other weapon") shall be made payable to ATF by the transferor. All signatures on both copies must be in ink.

Fingerprints also must be submitted on FBI Form FD-258, in duplicate. Fingerprints must be taken by a person qualified to do so, and must be clear and classifiable. If wear or damage to the fingertips do not allow clear prints, and if the prints are taken by a law enforcement official, a statement on his or her official letterhead giving the reason why good prints are unobtainable should accompany the fingerprints.

Forward the completed application and appropriate tax payment to the Bureau of ATF, P.O. Box 73201, Chicago, IL 60673.

Transfer of the NFA firearm may be made only upon approval of the ATF Form 4 by the NFA Branch. If the application is approved, the original of the form with the cancelled stamp affixed showing approval will be returned to the applicant. If the tax application is denied, the tax will be refunded.

Upon approval of the ATF Form 4, the transferor should transfer the firearm as soon as possible, since the firearm is now registered to the transferee.

[26 U.S.C. 5812, 27 CFR 479.84-86]

(M16) How does an individual obtain authorization to make an NFA firearm?

Prior to making a firearm, the individual must submit ATF Form 1, Application to Make and Register a Firearm, to the Bureau of ATF, NFA Branch, and receive approval. The applicant must follow the procedures described in Question **M15** concerning photographs, fingerprints and certifications. The applicant must forward the original and a duplicate of the form along with a check or money order for $200 made payable to the Bureau of ATF. If the application is approved, the original of the form with the cancelled stamp affixed showing approval will be returned to the applicant. If the application is denied, the tax will be refunded.

Applications to make a firearm will not be approved if Federal, State, or local law prohibits possession of the firearm.

[26 U.S.C 5822, 27 CFR 479.61-65]

(M17) Are parts which would convert a firearm into an NFA firearm subject to registration?

Yes. Examples:

An M-2 conversion kit (See Question **M29**);

Any part designed and intended solely and exclusively to convert a weapon into a machinegun. (See Question **M1**.)

(M18) What law enforcement officials' certifications on an application to transfer or make an NFA weapon are acceptable to ATF?

As provided by regulations, certifications by the local chief of police, sheriff of the county, head of the State police, or State or local district attorney or prosecutor are acceptable. The regulations also provide that certifications of other officials are appropriate if found in a particular case to be acceptable to the Director. Examples of other officials who have been accepted in specific situations include State attorneys general and judges of State courts having authority to conduct jury trials in felony cases.

[27 CFR 479.63 and 479.85]

(M19) Is the chief law enforcement officer required to sign the law enforcement certification on ATF Form 1 or ATF Form 4?

No law enforcement officer can be compelled to sign the law enforcement certification under Federal or State law. However, ATF will not approve an application to make or transfer a firearm on ATF Forms 1 or 4 unless the law enforcement certification is completed by an acceptable law enforcement official who has signed the certification in the space indicated on the form. See Question **M18**.

(M20) If the chief law enforcement official whose jurisdiction includes the proposed transferee's residence refuses to sign the "law enforcement certification," will the signature of an official in another jurisdiction be acceptable?

No. (But see Question **M18** for the list of acceptable chief law enforcement officers.)

(M21) Does the registered owner of a destructive device, machinegun, short-barreled shotgun, or short-barreled rifle need authorization to lawfully transport such items interstate?

Yes, unless the owner is a qualified dealer, manufacturer or importer, or a licensed collector transporting only curios or relics. Prior approval must be obtained, even if the move is temporary. Approval is requested by either submitting a letter containing all necessary information, or by submitting ATF Form 5320.20 to the Bureau of ATF, NFA Branch. Possession of the firearms also must comply with all State and local laws.

[18 U.S.C. 922(a) (4), 27 CFR 478.28]

(M22) If an individual is changing his or her State of residence and the individual's application to transport the NFA firearm cannot be approved because of a prohibition in the new State, what options does a lawful possessor have?

NFA firearms may be left in a safe deposit box in his or her former State of residence. Also, the firearm could be left or stored in the former State of residence at the house of a friend or relative in a locked room or container to which only the registered owner has a key. The friend or relative should be supplied with a copy of the registration forms and a letter from the owner authorizing storage of the firearm at that location.

The firearms may also be transferred under the procedures referred to in Question **M15** or abandoned to ATF.

(M23) May a transferor submit an application to transfer an NFA firearm prior to the date on which the transferor receives the weapon?

No.

(M24) If a person has a pistol and an attachable shoulder stock, does this constitute possession of an NFA firearm?

Yes, unless the barrel of the pistol is at least 16 inches in length (and the overall length of the firearm with stock attached is at least 26 inches). However, certain stocked handguns, such as original semiautomatic Mauser "Broomhandles" and Lugers, have been removed from the purview of the NFA as collectors' items.

[26 U.S.C. 5845, 27 CFR 479.11]

(M25) Does the owner of a registered NFA firearm have to have any evidence to show it is registered lawfully to him or her?

Yes. The approved application received from ATF serves as evidence of registration of the NFA firearm in the owner's name. This document must be kept available for inspection by ATF officers. It is suggested that a photocopy of the approved application be carried by the owner when the weapon is being transported.

(M26) What is the status of unloaded or dummy grenades, artillery shell casings and similar devices?

Unloaded or dummy grenades, artillery shell casings, and similar devices, which are cut or drilled in an ATF approved manner so that they cannot be used as ammunition components for destructive devices are not considered NFA weapons.

(M27) Are muzzleloading cannons classified as destructive devices?

Generally, no. Muzzleloading cannons not capable of firing fixed ammunition and manufactured in or before 1898 and replicas thereof are antiques and not subject to the provisions of either the GCA or the NFA.

[26 U.S.C. 5845, 27 CFR 479.11]

(M28) Are grenade and rocket launcher attachments destructive devices?

Grenade and rocket launcher attachments for use on military type rifles generally do not come within the definition of destructive devices. However, the grenades and rockets used in these devices are generally within the definition.

[26 U.S.C. 5845, 27 CFR 479.11]

(M29) What is a "conversion kit?"

A conversion kit is any part or combination of parts designed and intended for use in converting a weapon into a machinegun. A conversion kit is a machinegun for purposes of the NFA. (See Question **M17**.)

[26 U.S.C. 5845, 27 CFR 479.11]

(M30) Are Paintball and/or Airgun Sound Suppressers NFA firearms?

The terms "firearm silencer" and "firearm muffler" mean any device for silencing, muffling, or diminishing the report of a portable firearm, including any combination of parts, designed or redesigned, and intended for use in assembling or fabricating a firearm silencer or firearm muffler, and any part intended only for use in such assembly or fabrication.

Numerous paintball and airgun silencers tested by ATF's Firearms Technology Branch have been determined to be, by nature of their design and function, firearm silencers. Because silencers are NFA weapons, an individual wishing to manufacture or transfer such a silencer must receive prior approval from ATF and pay the required tax. See Questions **M15** and **16** for application details.

If I have any further questions as to the classification of a paintball or airgun silencer, who should I contact?

Please send a written request to ATF's Firearms Technology Branch.

[18 U.S.C. 921(a)(24), 26 U.S.C. 5845(a), 27 CFR 479.11]

N. MACHINEGUNS - NATIONAL FIREARMS ACT (NFA)

(N1) May an unlicensed person make a machinegun?

Generally, no. However, if documentation can be provided, along with the Application to Make a Machinegun, which establishes that the weapon is being made for distribution to a Federal or State agency, an individual may be permitted to make the machinegun.

[18 U.S.C. 922(o)(2), 27 CFR 479.105(e)]

(N2) May machineguns be transferred from one registered owner to another?

Yes. If the machinegun was lawfully registered and possessed before May 19, 1986, it may be transferred pursuant to an approved ATF Form 4.

[18 U.S.C. 922(o)(2), 26 U.S.C. 5812]

O. SEMIAUTOMATIC ASSAULT WEAPONS AND LARGE CAPACITY AMMUNITION FEEDING DEVICES (SAWs and LCAFDs)

(O1) What was the semiautomatic assault weapon (SAW) ban?

The SAW ban was enacted on September 13, 1994, by PL 103-322, Title IX, Subtitle A, section 110105. The ban made it unlawful to manufacture, transfer, or possess SAWs. The law defined SAWs as 19 named firearms, as well as semiautomatic rifles, pistols, and shotguns that have certain named features. The ban was codified at 18 U.S.C. § 922(v). SAWs lawfully possessed on September 13, 1994 were not covered by the ban. There also were certain exceptions, such as possession by law enforcement.

(O2) Was the SAW ban permanent?

No. The law enacting the ban provided that it would expire 10 years from the date of enactment, which was September 13, 1994. Therefore, effective 12:01 a.m. on September 13, 2004, the provisions of the law ceased to apply.

(O3) What was the Large Capacity Ammunition Feeding Device (LCAFD) ban?

The LCAFD ban was enacted along with the SAW ban on September 13, 1994. The ban made it unlawful to transfer or possess LCAFDs. The law generally defined a LCAFD as a magazine, belt, drum, feed strip, or similar device manufactured after September 13, 1994 that has the capacity of, or can be readily restored or converted to accept, more than 10 rounds of ammunition. The ban was codified at 18 U.S.C. § 922(w). As with SAWs, there were certain exceptions to the ban, such as possession by law enforcement.

(O4) Was the LCAFD ban permanent?

No. The LCAFD ban was enacted by the same law as the SAW ban. Therefore, like the SAW ban, it expired 10 years from the date of enactment. Therefore, effective 12:01 a.m. on September 13, 2004, the provisions of the law ceased to apply.

(O5) Does expiration of the ban affect records maintained by licensed manufacturers, importers and dealers?

Yes. Federal firearms licensees are no longer required to collect special records regarding the sale or transfer of SAWs and LCAFDs for law enforcement or government sales. However, existing records on SAWs and LCAFDs must still be maintained for a period of 5 years. Moreover, records of importation and manufacture must be maintained permanently and licensees must maintain all other acquisition and disposition records for 20 years.

(O6) Are SAWs and LCAFDs marked "Restricted law enforcement/government use only" or "For export only" now legal to sell to civilians in the United States?

Yes. SAWs and LCAFDs are no longer prohibited. Therefore firearms with the restrictive markings are legal to transfer to civilians in the United States and it is legal for non-prohibited civilians to possess them. All civilians may possess LCAFDs.

(O7) Does the expiration of the SAW ban and the LCAFD ban affect importation?

LCAFDs are no longer prohibited from importation but they are still subject to the provisions of the Arms Export Control Act. An approved Form 6 import permit is still required

Non-sporting firearms are still prohibited from importation under sections 922(l) and 925(d)(3) of the GCA. Because the vast majority of SAWs are nonsporting, they generally cannot be imported.

Temporary importation of SAWs and LCAFDs is now lawful under the provisions of 27 CFR section 478.115(d) because firearms that are temporarily imported are not required to meet sporting purpose requirements.

(O8) Does the expiration of the SAW ban change laws regarding assembly of nonsporting shotguns and semiautomatic rifles from imported parts?

No. The provisions of section 922(r) of the GCA and the regulations in 27 CFR 478.39 regarding assembly of non-sporting shotguns and semiautomatic rifles from imported parts still apply.

(O9) Does the expiration of the SAW ban affect firearms under the National Firearms Act?

All provisions of the National Firearms Act (NFA) relating to registration and transfer of machineguns, short-barreled rifles, weapons made from rifles, short-barreled shotguns, weapons made from shotguns, any other weapons as defined in 26 USC section 5845(e), silencers, and destructive devices still apply. However, it is now lawful to possess NFA firearms that are also semiautomatic assault weapons, as long as all provisions of the NFA are satisfied.

For example, USAS-12 and Striker12/ Street-sweeper shotguns are still classified as destructive devices under ATF Rulings 94-1 and 94-2 and must be possessed and transferred in accordance with the NFA.

(O10) Can tribal law enforcement entities now possess SAWs and LCAFDs?

Yes.

(O11) Does the expiration of the ban affect State law?

Expiration of the federal law will not change any provisions of State law or local ordinances. Questions concerning State assault weapons restrictions should be referred to State and local authorities.

P. BRADY LAW

Editor's Note:

Unless otherwise noted, these questions and answers relate to the permanent provisions of the Brady law found in section 922(t) of the Gun Control Act. These provisions, including the requirement for licensees to initiate background checks of individuals to whom firearms are transferred by contacting the National Instant Criminal Background Check System (NICS), became effective on November 30, 1998. They replace the interim provisions of the Brady law that imposed a Federal 5-day waiting period on licensees' sales of handguns and required the sending of Brady forms to State or local officials.

(P1) Who must comply with the requirements of the Brady law?

Federally licensed firearms importers, manufacturers, and dealers must comply with the Brady law prior to the transfer of any firearm to a nonlicensed individual.

[18 U.S.C. 922(t), 27 CFR 478.102]

(P2) When did the provisions of the permanent Brady law take effect?

The permanent Brady law went into effect on November 30, 1998. Accordingly, any transfer occurring on or after November 30, 1998, is subject to the requirements of this law.

(P3) Is NICS operated by ATF?

No. NICS is operated by the Federal Bureau of Investigation (FBI).

(P4) Do all NICS checks go through the FBI's NICS Operations Center?

No. In many States, licensees initiate NICS checks through the State point of contact (POC).

(P5) If the State is acting as a point of contact (POC), does that mean that all NICS checks go through the POC rather than the FBI?

That depends on the State. In some States, the POC conducts background checks for all firearms transactions. In other States, licensees must contact the POC for handgun transactions and the FBI for long gun transactions. In some POC States, NICS checks for pawn redemptions are handled by the FBI.

(P6) How does a licensee know whether to contact the FBI or a State point of contact (POC) in order to initiate a NICS check?

Prior to November 30, 1998, ATF sent an open letter to licensees in each State, providing the licensees with instructions as to how to initiate a NICS check in their State. ATF has alerted FFLs if their State's procedures have changed since this time. Your local ATF office can advise you on the appropriate point of contact for NICS checks or you can check the ATF webpage at www.ATF.gov.

(P7) Is there a charge for NICS checks?

The FBI does not charge a fee for conducting NICS checks. However, States that act as points of contact for NICS checks may charge a fee consistent with State law.

(P8) Must licensees enroll with the FBI to get access to NICS?

Licensees must be enrolled with the FBI before they can initiate NICS checks through the FBI's NICS Operations Center. Licensees who have not received an enrollment package from the FBI should call the FBI NICS Operations Center at 1-877-444-6427 and ask that an enrollment package be sent to them. Licensees in States where a State agency is acting as a point of contact for NICS checks should contact the State for enrollment information.

(P9) Does the Brady law apply to the transfer of long guns as well as handguns?

Yes.

(P10) Does the Brady law apply to the transfer of antique firearms?

No. Licensees need not comply with the Brady law when transferring a weapon that meets the Gun Control Act's definition of an "antique firearm."

(P11) Does the Brady law apply to the transfer of firearms between two licensees?

No. The Brady law only applies when a licensed importer, manufacturer, or dealer is transferring a firearm to a nonlicensee.

(P12) Must licensed collectors comply with the Brady law prior to transferring a curio or relic firearm?

No. Transfers of curio or relic firearms by licensed collectors are not subject to the requirements of the Brady law.

(P13) Is the transfer of a firearm by a licensed dealer to a licensed collector subject to the Brady law?

The Brady law does not apply to the transfer of a curio or relic firearm to a licensed collector. However, a licensed collector who acquires a firearm other than a curio or relic from a licensee would be treated like a nonlicensee, and the transfer would be subject to Brady requirements.

(P14) Must a licensed importer, manufacturer or dealer comply with the Brady law when selling firearms from his or her own personal collection?

No, provided the licensee has maintained the firearm as part of his personal collection for at least 1 year from the date the firearm was transferred from his business inventory into his personal collection or otherwise acquired as a personal firearm and the licensee complies with the recordkeeping requirements in 27 CFR 478.125(a).

(P15) Do the provisions of the Brady law apply to a licensee's loan or rental of a firearm to a nonlicensee?

If the firearm is loaned or rented for use on the licensee's premises, the transaction is not subject to the Brady law. However, if the firearm is loaned or rented for use off the premises, the licensee must comply with the Brady law.

(P16) Must licensees conduct NICS checks for sales of firearms to nonlicensees at gun shows?

Yes. A licensed importer, manufacturer, or dealer may not transfer a firearm to a nonlicensee at a gun show without first complying with the requirements of the Brady law.

(P17) Is the redemption of a pawned firearm subject to the Brady law?

Yes. Unlike the interim Brady law, the permanent Brady law that went into effect on November 30, 1998 does not contain an exemption for the return of a firearm to the individual from whom it was received. Accordingly, the redemption of a pawned firearm is considered a transfer subject to the permanent Brady law.

(P18) What should a licensed pawnbroker do with a firearm he or she has in pawn when the NICS check results in a "denied" transaction?

The licensee cannot transfer the firearm to the transferee without violating the law and placing the transferee in violation of the law. Licensees with additional questions should contact their local ATF office.

[18 U.S.C. 922(d) and (g)]

(P19) If an individual repeatedly pawns the same firearm, is the FFL required to do a NICS check each time the firearm is redeemed?

Yes. The fact that the transferee has redeemed the firearm before does not excuse the pawnbroker from complying with the Brady law.

(P20) Can licensed pawnbrokers conduct NICS checks prior to accepting a firearm in pawn?

Yes. The law provides that a NICS check may be done, on an optional basis, prior to accepting a firearm in pawn. If the check results in a "denied" response, the licensee is required to notify local law enforcement officials within 48 hours after receipt of the "denied" response.

(P21) If a pawnbroker conducts an optional NICS check prior to receiving a firearm in pawn, should the owner of the firearm complete a Form 4473 before the pawnbroker contacts NICS?

ATF suggests that licensees have the owner complete Section A of the Form 4473, record the results of the NICS check on the form, and retain the form in their records for at least 5 years.

(P22) What should a licensed pawnbroker do when he or she gets a "denied" response from an optional NICS check conducted prior to the receipt of a firearm in pawn?

The licensee is required to notify local law enforcement officials within 48 hours after receipt of the "denied" response. If the licensee has taken possession of the firearm, he or she may not return it to the individual who offered it for pawn.

(P23) If a licensed pawnbroker conducts a NICS check prior to accepting a firearm in pawn, and gets an "approved" response from NICS, must the pawnbroker conduct another NICS check if the firearm is redeemed from pawn?

What if it is redeemed from pawn the same day?

The law provides that another NICS check must be done at the time of redemption, regardless of how recently the pre-pawn NICS check was conducted. Even if the firearm is redeemed the same day, a separate NICS check must be conducted at the time of redemption.

(P24) A firearm is delivered to a licensee by an unlicensed individual for the purpose of repair. Is the return of the repaired firearm subject to the requirements of the Brady law? Would the transfer of a replacement firearm from the licensee to the owner of the damaged firearm be subject to the requirements of the Brady law?

Neither the transfer of a repaired firearm nor the transfer of a replacement firearm would be subject to the requirements of the Brady law. Furthermore, the regulations provide that a Form 4473 is not required to cover these transactions. However, the licensee's permanent acquisition and disposition records should reflect the return of the firearm or the transfer of a replacement firearm.

[27 CFR 478.124-25]

(P25) Is a licensee's return of a consigned firearm to an unlicensed individual subject to permanent Brady?

Yes.

(P26) Do the requirements of the Brady law apply to sales of firearms to law enforcement officials for official use?

Transfers of firearms to law enforcement officials for their official use are exempt from the provisions of the Brady law when the transaction complies with the conditions set forth in the regulations at 27 CFR 478.134. In general, the purchaser must provide a certification on agency letterhead, signed by a person in authority within the agency (other than the officer purchasing the firearm), stating that the officer will use the firearm in official duties, and that a records check reveals that the purchasing officer has no convictions for misdemeanor crimes of domestic violence. If these conditions are met, the purchasing officer is not required to complete a Form 4473 or undergo a NICS check. However, the licensee must record the transaction in his or her permanent records and retain a copy of the certification letter.

[27 CFR 478.134]

(P27) Do the requirements of the Brady law apply to the sale of a firearm to a law enforcement official for his or her personal use?

Yes. In such transactions, the law enforcement official is treated no differently from any other unlicensed transferee, and a NICS check must be conducted.

[18 U.S.C. 922(t) and 925(a) (1)]

(P28) Are there exceptions to the Brady law's requirement for a NICS check prior to a licensee's transfer of a firearm to an unlicensed individual?

Firearm transfers are exempt from the requirement for a NICS check in 3 situations. These include transfers: (1) to buyers having a State permit that has been recognized by ATF as an alternative to a NICS check; (2) of National Firearms Act weapons approved by ATF; and (3) certified by ATF as exempt because compliance with the NICS check requirement is impracticable.

[18 U.S.C. 922(t), 27 CFR 478.102(d)]

(P29) What steps must be followed by an FFL prior to transferring a firearm subject to the requirements of the Brady law?

The following steps must be followed prior to transferring a firearm:

1. The licensee must have the transferee complete and sign ATF Form 4473, Firearms Transaction Record.

2. The licensee must verify the identity of the transferee through a government-issued photo identification.

3. The licensee must contact NICS through either the FBI or a State point of contact (POC). The licensee initially will get either a "proceed" or "delayed" response from NICS. If the licensee gets a "proceed" response, the firearm may be transferred if there is no additional State waiting period. If the licensee gets a "delayed response" it indicates the transaction is in "open" status and that more research is required prior to a NICS "proceed" or "denied" response. If the licensee gets a "delayed" response and there is no additional response from the FBI or POC, the licensee may transfer the firearm after 3 business days have elapsed. Of course, the licensee must still comply with any waiting period requirements under State law. FFLs contacting the FBI directly will receive information from the FBI indicating when the 3 business days time period elapses on "delayed" transactions.

If the licensee gets a "denied" response prior to the 3 business days elapsing, the firearm cannot be transferred.

In addition, after conducting additional research, the FBI may provide the FFL with a "cancelled" response. Transactions will be cancelled if the FBI discovers that the NICS check was not initiated in accordance with ATF or FBI regulations, e.g., information received from the Bureau of Immigration and Customs Enforcement demonstrates an alien cannot meet the requirement of 90 days of continuous state residence, or a NICS check was initiated for a non-authorized purpose or by a non-authorized individual. A licensee cannot transfer a firearm when he gets a "cancelled" response.

(P30) What form of identification must a dealer obtain from a purchaser under the Brady law?

The identification document presented by the purchaser must have a photograph of the purchaser, as well as the purchaser's name, address, and date of birth. The identification document must also have been issued by a governmental entity for the purpose of identification of individuals. An example of an acceptable identification document is a driver's license.

[18 U.S.C. 922(t), 27 CFR 478.124]

(P31) Under the Brady law, may a licensee transfer a firearm to a nonlicensed individual who does not appear in person at the licensed premises?

In any transaction that is subject to the requirement for a NICS check, the firearm may only be sold over-the-counter. Unless the purchaser appears in person at the licensed premises, the licensee cannot comply with the requirement in the Brady law that the identity of the purchaser be verified by means of a government-issued photo identification document.

(P32) If no NICS check is required, may a licensee transfer a firearm to

a nonlicensed individual who does not appear in person at the licensed premises?

Yes, assuming the transaction otherwise complies with Federal and State law. For example, a licensee may still ship firearms to out-of-state law enforcement officials for official use, as long as the transaction complies with the alternate procedure set forth in the regulations at 27 CFR 478.134. Furthermore, licensees may ship firearms intrastate to State residents who have valid permits that have been recognized as alternatives to the NICS check requirements, in compliance with the procedures set forth in 18 U.S.C. 922(c) and 27 CFR 478.96(b).

(P33) Is the transferee required to provide his or her social security number on the ATF Form 4473?

No. This information is solicited on an optional basis. However, providing this information will help ensure the lawfulness of the sale and avoid the possibility that the transferee will be incorrectly identified as a felon or other prohibited person.

[27 CFR 478.124]

(P34) Will NICS provide an instant response?

NICS will not always provide an instant response. Licensees will receive either a "proceed," "denied," "delayed," or "cancelled" response from NICS. If a "proceed" response is received, the transfer may proceed. If a "denied" response is received, the transfer may not proceed. If a "delayed" response is received, the transfer must be delayed until a "proceed" response is received from NICS or until the lapse of 3 business days, whichever occurs first. Of course, the licensee must still comply with any waiting period requirements under State law. See Question **P29** for a discussion of cancelled transactions.

(P35) For purposes of the Brady law, what is meant by a "business day?"

A business day is defined as any day on which State offices are open.

(P36) If a licensee contacts NICS on Thursday, December 2, and gets a "delayed" response, when may the licensee transfer the firearm if no further response is received from NICS?

The firearm may be transferred on Wednesday, December 8. Assuming State offices are open on Friday, Monday, and Tuesday, and closed on Saturday and Sunday, 3 business days would have elapsed at the end of Tuesday, December 7. Therefore, the licensee may transfer the firearm at the start of business on Wednesday, December 8. The 3 business days do not include the day the NICS check is initiated.

(P37) What should a licensee do if he or she gets a "denied" response from NICS or a State point of contact after 3 business days have elapsed, but prior to the transfer of the firearm?

If the licensee receives a "denied" response at any time prior to the transfer of the firearm, he or she may not transfer the firearm.

[18 U.S.C. 922(d)]

(P38) What should an FFL tell a transferee who is denied by NICS?

The FFL should inform the transferee that the NICS check indicates that the transfer of the firearm should not be made, but that it does not provide a reason for the denial. The FFL should provide the transferee with the NICS or State transaction number and an appeals brochure. The FBI has provided FFLs with brochures that outline the transferee's appeal rights and responsibilities. If you do not have any, the FBI can provide them.

(P39) If a transferee receives a "denied" response from NICS, can the transferee find out why he or she was denied?

Yes. Although the FFL will not know the reason for the denial, the transferee may contact the FBI or the State point of contact in writing to request the reason for denial.

(P40) What information do FFLs have to record on ATF Form 4473 once they hear back from NICS or the State point of contact?

FFLs must record any initial "proceed", "delayed", or "denied" response received, as well as any transaction number provided. If the initial NICS response was "delayed," FFLs also must record the date a later "proceed" or "denied" response was received, or that no resolution was provided within 3 business days if they transfer the firearm after the 3 days. In addition, if an FFL receives a response from NICS after a firearm has been transferred, he or she must record this information.

[27 CFR 478.124(c) (3) (iii)]

(P41) What should licensees do if no transaction number is provided?

The FBI's NICS Operations Center will always provide a transaction number at the time of the initial inquiry. Some State points of contact (POC) may not provide a transaction number for denied transactions. If a State POC does not provide a transaction number for a denied transaction, then the licensee should just record the response without a transaction number. Licensees should note, however, that a transaction number is required if a "proceed" response is issued and the firearm is being transferred within 3 business days of the initiation of a NICS check.

(P42) Do FFLs have to keep a copy of ATF Form 4473 if the transaction is denied or for some other reason is not completed?

FFLs must keep a copy of each ATF Form 4473 for which a NICS check has been initiated, regardless of whether the transfer of the firearm was made. If the transfer is not made, the FFL must keep the Form 4473 for 5 years after the date of the NICS inquiry. If the transfer is made, the FFL must keep the Form 4473 for 20 years after the date of the sale or disposition. Forms 4473 with respect to a transfer that did not take place must be separately maintained.

[27 CFR 478.129(b)]

(P43) When should FFLs contact NICS?

FFLs should contact NICS after the transferee has completed Section A of the ATF Form 4473.

(P44) For what period of time is a NICS check valid?

A NICS check is valid for 30 calendar days, as long as it applies to a single transaction. An FFL may not rely on a NICS check that was conducted more than 30 calendar days prior to the transfer of the firearm.

Example: A NICS check is initiated on December 15, 2004. The FFL receives a "proceed" from NICS. The purchaser does not return to pick up the firearm until January 22, 2005. The FFL must conduct another NICS check before transferring the firearm to the purchaser, and must record the results of the check on the Form 4473.

[27 CFR 478.124(c)]

(P45) Is a NICS check valid for 30 days from when the check was initiated, or from when a "proceed" is issued?

The NICS check is valid for 30 days from when the check was initiated.

Example: A NICS check is initiated on December 15, 2004. The FFL receives a "proceed" response from NICS on December 17, 2004. The purchaser does not return to pick up the firearm until January 16, 2005. The FFL must conduct another NICS check before transferring the firearm to the purchaser.

Example: A NICS check is initiated on December 15, 2004. The FFL receives a "delayed" response from NICS; no further response is received. The purchaser does not return to pick up the firearm until January 16, 2005. The FFL must conduct another NICS check before transferring the firearm to the purchaser.

[27 CFR 478.124(c)]

(P46) A purchaser places an order for a custom-made firearm, which will not be ready for at least 60 days. Should the NICS check be initiated on the date the order is placed or the day the firearm is ready for delivery?

The issue raised by this question also applies to lay-away purchases.

If the licensee knows that it will be more than 30 days before the firearm is ready, the licensee may want to wait until the firearm is ready for delivery (or in the case of a lay-away purchase, until full payment can be made) to have the purchaser complete Section A of the Form 4473 and contact NICS for a background check. This is because if the purchaser completes Section A of the Form 4473 on the date the order is placed, and a NICS check is initiated on that day, the licensee will have to conduct a second NICS check prior to transferring the firearm and the purchaser will have to complete section C of ATF Form 4473 at the time of delivery.

(P47) Can one NICS check cover two or more separate firearms transactions?

No. An FFL must initiate separate NICS checks for separate transactions. However, an individual may purchase several firearms in one transaction.

Example: A purchaser completes an ATF Form 4473 for a single firearm on February 15. The FFL receives a "proceed" from NICS that day. The FFL signs the form, and the firearm is transferred. On February 20, the purchaser returns to the FFL's premises and wishes to purchase a second firearm. The purchase of the second firearm is a separate transaction. Therefore, a new NICS check must be initiated by the FFL.

Example: A purchaser completes ATF Form 4473 for a single firearm on February 15. The FFL receives a "proceed" from NICS that day. The purchaser does not return to pick up the firearm until February 20. Before the FFL signs the Form 4473 for the first firearm, the purchaser decides to purchase an additional firearm. The second firearm may be recorded on the same Form 4473. The purchase of the 2 firearms is considered a single transaction. Therefore, the licensee is not required to conduct a new NICS check prior to transferring the second firearm.

Example: A purchaser wishes to purchase 1 rifle and 1 handgun. State law requires that a background check be conducted on the sale of all handguns. The State is acting as a point of contact (POC) for NICS checks for handgun sales, while the FBI is conducting NICS checks for long gun transactions in that State. To comply with State law, the dealer initiates a background check through the State POC. There is no need to initiate a separate background check through the FBI for the sale of the rifle, since the 2 firearms are being transferred in one transaction.

(P48) If no exceptions to the NICS check apply, must an FFL always wait 3 business days before transferring a firearm to a transferee?

No. An FFL may transfer a firearm to a transferee as soon as he or she receives a "proceed" from NICS (assuming that the transaction would be in compliance with State law). However, if the FFL does not receive a final "proceed" or "denied" response from NICS, he or she must wait until 3 business days have elapsed prior to transferring the firearm.

(P49) Will a State "instant check" or "point of sale check" system qualify as an alternative to a NICS check?

No. However, it should be noted that many States with their own background check requirements are also acting as points of contact for NICS checks.

(P50) What happens if the transferee successfully appeals the NICS denial but more than 30 calendar days have elapsed since the initial background check was initiated?

The FFL must initiate another NICS check before the firearm may be transferred.

(P51) How does a licensee know whether a permit may be accepted as an alternative to a NICS check?

Prior to November 30, 1998, ATF sent an open letter to licensees in each State, advising them what, if any, permits in that State qualified as alternatives to a NICS check at the time of transfer. Since that time, ATF has informed licensees when a State permit stopped qualifying and when a State permit became a qualifying NICS alternative. Licensees with questions about whether particular State permits are acceptable alternatives to NICS checks should contact their local ATF office, or check online at www.ATF.gov.

(P52) Does a permit qualify as an alternative to a NICS check if the purchaser is using it to purchase a type of firearm that is not covered by the permit?

Yes, assuming the transaction complies with State law.

Example: ATF recognizes the permit to purchase a handgun and the concealed weapons permit as alternatives to a NICS check in State A. Any purchaser who displays a permit to purchase a handgun or a concealed weapons permit in State A is not required to undergo a NICS check prior to purchasing a rifle, assuming the transaction complies with State law.

Example: In that same State, a person with a concealed weapons permit wants to purchase a handgun. State law prohibits the sale of any handgun to a person unless he or she has a permit to purchase a handgun. Accordingly, the licensee cannot lawfully sell the handgun (with or without a NICS check) unless the purchaser has a permit to purchase a handgun.

(P53) ATF has recognized the concealed weapons permits in State A and State B as valid alternatives to a NICS check. Can a resident of State A use a concealed weapons permit issued by State A to purchase a longgun in State B without undergoing a NICS check?

No. A permit qualifies as a NICS alternative only if it was issued by the State in which the transfer is to take place.

[18 U.S.C. 922(t), 27 CFR 478.102(d)]

(P54) If State law provides that permits are only valid for 2 years, can a licensee accept as a NICS alternative a permit that was issued 4 years ago?

No. The permit must be valid under State law in order to qualify as an alternative to a NICS check.

[18 U.S.C. 922(t), 27 CFR 478.102(d)]

(P55) If the State has its own instant check system, must the licensee comply with State requirements for a background check as well as the Brady law?

Yes. If the State is not acting as a point of contact for NICS checks, the licensee may have to initiate 2 background checks by contacting (1) the FBI's NICS Operations Center, and (2) the State system.

(P56) If a licensee gets a "proceed" response from NICS, does he or she still have to wait until the expiration of the State waiting period before transferring the firearm?

Yes. Compliance with the Brady law does not excuse a licensee from compliance with State law.

(P57) If a State is acting as a NICS point of contact (POC) and State law has requirements regarding the amount of time that a licensee must wait before transferring a firearm after contacting the State, should the licensee comply with the State requirements, the Federal requirements, or both?

The licensee must comply with both State and Federal requirements.

Example: State D is acting as a POC for NICS checks. State law requires a background check prior to the transfer of any firearm. State law also requires the licensee to wait 10 days to get a response from the State. The licensee must contact the State POC for a NICS check and a State background check. The licensee must comply with both Federal and State law by waiting 10 days for a response prior to transferring the firearm. If the licensee has not received a response from the State after 10 days, he or she may transfer the firearm.

Example: State E is acting as a POC for NICS checks. State law requires a background check prior to the transfer of any firearm. Under State law, the licensee may transfer the firearm if he or she gets no final response from the State by the next day. The licensee contacts the State POC for a NICS check, and gets a "delayed" response. Assuming that the licensee gets no further response from the State POC, the licensee must comply with both Federal and State law by waiting until 3 business days have elapsed prior to transferring the firearm.

(P58) Does an individual (not a corporation or partnership) licensee have to conduct a NICS check on himself or herself prior to transferring a firearm to his or her own personal collection?

No. The regulations do not require a licensee to complete a Form 4473 prior to transferring a firearm to his or her own personal collection. A NICS check is not required either. Such transfers must be recorded in the manner prescribed by the regulations at 27 CFR 478.125a.

(P59) Is a NICS check required for the sale of firearms registered under the National Firearms Act (NFA)?

No, assuming all NFA requirements have been satisfied.

[18 U.S.C. 922(t), 27 CFR 478.102(d)]

(P60) An organization without a firearms license wishes to acquire a firearm from a licensee for the purpose of raffling the firearm at an event. How does the licensee comply with the Brady law?

The licensee must comply with the Brady law by conducting a NICS check on the transferee. If the licensee wishes to transfer the firearm to the organization, a representative of the organization must complete a Form 4473 and a NICS check must be conducted on that representative prior to the transfer of the firearm. Alternatively, if the licensee transfers the firearm directly to the winner of the raffle, the winner must complete a Form 4473 and a NICS check must be conducted on the raffle winner prior to the transfer.

Please note, if the organization's practice of raffling firearms rises to the level of being engaged in the business of dealing in firearms, the organization must get its own Federal firearms license (and the examples below would not apply).

Example 1: A licensee transfers a firearm to the organization sponsoring the raffle. The licensee must comply with the Brady Law by requiring a representative of the organization to complete the Form 4473 and undergo a NICS check. As indicated in the instructions on the Form 4473, when the buyer of a firearm is a corporation, association, or other organization, an officer or other representative authorized to act on behalf of the organization must complete the form with his or her personal information and attach a written statement, executed under penalties of perjury, stating that the firearm is being acquired for the use of the organization and the name and address of the organization. Once the firearm had been transferred to the organization, the organization can subsequently transfer the firearm to the raffle winner without a Form 4473 being completed or a NICS check being conducted. This is because the organization is not an FFL. However, the organization cannot transfer the firearm to a person who is not a resident of the State where the raffle occurs and cannot knowingly transfer the firearm to a prohibited person.

Example 2: The licensee or his or her representative brings a firearm to the raffle so that the firearm can be displayed. After the raffle, the firearm is returned to the licensee's premises. The licensee must complete a Form 4473 for the transaction and must comply with the Brady Law prior to transferring the firearm to the winner of the raffle. If the firearm is a handgun, the winner of the raffle must be a resident of the State where the transfer takes place, or the firearm must be transferred through another FFL in the winner's State of residence. If the firearm is a rifle or shotgun, the FFL can lawfully transfer the firearm to the winner of the raffle as long as the transaction is over-the-counter and complies with the laws applicable at the place of sale and the State where the transferee resides.

Example 3: If the raffle meets the definition of an "event" at which the licensee is allowed to conduct business pursuant to 27 CFR 478.100, the licensee may attend the event and transfer the firearm at the event to the winner of the raffle. As in Example 2, the FFL must complete a Form 4473 and comply with the Brady law and the interstate controls in transferring the firearm.

Please note, procedures used in Examples 2 and 3 ensure that the winner is not a prohibited person and that there is a record of the final recipient of the firearm in the raffle.

[18 U.S.C. 922(t) and 922(a)(1)(A)]

Q. MISDEMEANOR CRIME OF DOMESTIC VIOLENCE

(Q1) What is a "misdemeanor crime of domestic violence"?

A "misdemeanor crime of domestic violence" means an offense that:

(1) is a misdemeanor under Federal or State law;

(2) has, as an element, the use or attempted use of physical force, or the threatened us of a deadly weapon; and

(3) was committed by a current or former spouse, parent, or guardian of the victim, by a person with whom the victim shares a child in common, by a person who is cohabiting with or has cohabited with the victim as a spouse, parent, or guardian, or by a person similarly situated to a spouse, parent, or guardian of the victim.

However, a person is not considered to have been convicted of a misdemeanor crime of domestic violence unless:

(1) the person was represented by counsel in the case, or knowingly and intelligently waived the right of counsel in the case; and

(2) in the case of a prosecution for which a person was entitled to a jury trial in the jurisdiction in which the case was tried, either –

(a) the case was tried by a jury, or

(b) the person knowingly and intelligently waived the right to have the case tried by a jury, by guilty plea or otherwise.

In addition, a conviction would not be disabling if it has been expunged or set aside, or is an offense for which the person has been pardoned or has had civil rights restored (if the law of the jurisdiction in which the proceedings were held provides for the loss of civil rights upon conviction for such an offense) unless the pardon, expunction, or restoration of civil rights expressly provides that the person may not ship, transport, possess, or receive firearms, and the person is not otherwise prohibited by the law of the jurisdiction in which the proceedings were held from receiving or possessing firearms.

[18 U.S.C. 921(a)(33), 27 CFR 478.11]

(Q2) What is the effective date of this disability?

The law was effective September 30, 1996. However, the prohibition applies to persons convicted of such misdemeanors at any time, even if the conviction occurred prior to the law's effective date.

(Q3) Does application of the law to persons convicted prior to the law's effective date violate constitutional rights?

No. This provision is not being applied retroactively or in violation of the Ex Post Facto clause of the Constitution. This is because the law does not impose additional punishment upon persons convicted prior to the effective date, but merely regulates the future possession and receipt of firearms on or after the effective date. The provision is not retroactive merely because the person's conviction occurred prior to the effective date.

(Q4) X was convicted of misdemeanor assault on October 10, 1996 for beating his wife. Assault has as an element the use of physical force, but is not specifically a domestic violence offense. May X lawfully possess firearms or ammunition?

No. X may not legally possess firearms or ammunition.

[18 U.S.C. 922(g)(9), 27 CFR 478.32(a)(9)]

(Q5) X was convicted of a misdemeanor crime of domestic violence on September 20, 1996, 10 days before the effective date of the statute. He possesses a firearm on October 10, 2004. Does X lawfully possess the firearm?

No. If a person was convicted of a misdemeanor crime of domestic violence at any time, he or she may not lawfully possess firearms or ammunition on or after September 30, 1996.

[18 U.S.C. 922(g)(9), 27 CFR 478.32(a)(9)]

(Q6) In determining whether a conviction in a State court is a "conviction" of a misdemeanor crime of domestic violence, does Federal or State law apply?

State law applies. Therefore, if the State does not consider the person to be convicted, the person would not have the Federal disability.

[18 U.S.C. 921(a)(33), 27 CFR 478.11]

(Q7) What State and local offenses are "misdemeanors" for purposes of 18 U.S.C. 922(d)(9) and (g)(9)?

The definition of misdemeanor crime of domestic violence in the GCA includes any offense classified as a "misdemeanor" under Federal or State law. In States that do not classify offenses as misdemeanors, the definition includes any State or local offense punishable by imprisonment for a term of 1 year or less or punishable by a fine. For example, if State A has an offense classified as a "domestic violence misdemeanor" that is punishable by up to 5 years imprisonment, it would be a misdemeanor crime of domestic violence. If State B does not characterize offenses as misdemeanors, but has a domestic violence offense that is punishable by no more than 1 year imprisonment, this offense would be a misdemeanor crime of domestic violence.

[18 U.S.C. 921(a)(33), 27 CFR 478.11]

(Q8) Are local criminal ordinances "misdemeanors under State law" for purposes of sections 922(d)(9) and (g)(9)?

Yes, assuming a violation of the ordinance meets the definition of "misdemeanor crime of domestic violence" in all other respects.

(Q9) In order for an offense to qualify as a "misdemeanor crime of domestic violence", does it have to have as an element the relationship part of the definition (e.g., committed by a spouse, parent, or guardian)?

No. The "as an element" language in the definition of "misdemeanor crime of domestic violence" only applies to the use of force provision of the statute and not the relationship provision. However, to be disabling, the offense must have been committed by one of the defined parties.

[18 U.S.C. 921(a)(33), 27 CFR 478.11]

(Q10) Is a person who received "probation before judgment" or some other type of deferred adjudication subject to the disability?

What is a conviction is determined by the law of the jurisdiction in which the proceedings were held. If the State law where the proceedings were held does not consider probation before judgment or deferred adjudication to be a conviction, the person would not be subject to the disability.

[18 U.S.C. 921(a)(33), 27 CFR 478.11]

(Q11) What should a licensee do if he or she has been convicted of a misdemeanor crime of domestic violence?

A licensee convicted of a disqualifying misdemeanor may not lawfully possess firearms or ammunition. In addition, a licensee who incurs firearms disabilities during the term of a license by reason of such a misdemeanor conviction may not continue operations under the license for more than 30 days after incurring the disability unless the licensee applies for relief from Federal firearms disabilities.

[18 U.S.C 922(g)(9) and 925(c), 27 CFR 478.144 (i)]

(Q12) What should an individual do if he or she has been convicted of a misdemeanor crime of domestic violence?

Individuals subject to this disability should immediately dispose of their firearms and ammunition. ATF recommends that such persons transfer their firearms and ammunition to a third party who may lawfully receive and possess them, such as their attorney, a local police agency, or a Federal firearms dealer. The continued possession of firearms and ammunition by persons under this disability is a violation of law and may subject the possessor to criminal penalties. In addition, such firearms and ammunition are subject to seizure and forfeiture.

[18 U.S.C. 922(g)(9) and 924(d)(1), 27 CFR 478.152]

(Q13) Does the disability apply to law enforcement officers?

Yes. The Gun Control Act was amended so that employees of government agencies convicted of misdemeanor crimes of domestic violence would not be exempt from disabilities with respect to their receipt or possession of firearms or ammunition. Thus, law enforcement officers and other government officials who have been convicted of a disqualifying misdemeanor may not lawfully possess or receive firearms or ammunition for any purpose, including performance of their official duties. The disability applies to firearms and ammunition issued by government agencies, purchased by government employees for use in performing their official duties, and personal firearms and ammunition possessed by such employees.

[18 U.S.C. 922(g)(9) and 925(a)(1), 27 CFR 478.32(a)(9) and 478.141]

(Q14) Is an individual who has been pardoned, or whose conviction was expunged or set aside, or whose civil rights have been restored, considered convicted of a misdemeanor crime of domestic violence?

No, as long as the pardon, expungement, or restoration does not expressly provide that the person may not ship, transport, possess, or receive firearms.

R. NONIMMIGRANT ALIENS

(R1) May nonimmigrant aliens legally in the United States purchase or possess firearms and ammunition while in the United States?

Nonimmigrant aliens generally are prohibited from possessing or receiving (purchasing) firearms and ammunition in the United States.

There are exceptions to this general prohibition. The exceptions are as follows:

1. nonimmigrant aliens who possess a valid (unexpired) hunting license or permit lawfully issued by a State in the United States;

2. nonimmigrant aliens entering the United States to participate in a competitive target shooting event or to display firearms at a sports or hunting trade show sponsored by a national, State, or local firearms trade organization devoted to the collection, competitive use or other sporting use of firearms;

3. certain diplomats, if the firearms are for official duties;

4. officials of foreign governments, if the firearms are for official duties, or distinguished foreign visitors so designated by the U.S. State Department;

5. foreign law enforcement officers of friendly foreign governments entering the United States on official law enforcement business; and

6. persons who have received a waiver from the prohibition from the U.S. Attorney General.

Significantly, even if a nonimmigrant alien falls within one of these exceptions, the nonimmigrant alien CANNOT purchase a firearm from a Federal firearms licensee (FFL) unless he or she (1) has an alien number or admission number from the Department of Homeland Security (formerly the Immigration and Naturalization Service) AND (2) can provide the FFL with documentation showing that he or she has resided in a State within the United States for 90 consecutive days immediately prior to the firearms transaction.

[18 U.S.C. 922(g)(5)(b) and 922(y), 27 CFR 478.124, ATF Rul. 2004-1]

(R2) Typically, who are "nonimmigrant aliens?"

In large part, nonimmigrant aliens

are persons traveling temporarily in the United States for business or pleasure, persons studying in the United States who maintain a foreign residence abroad, and certain foreign workers. Permanent resident aliens are NOT nonimmigrant aliens. Permanent resident aliens often are referred to as people with "green cards."

(R3) How do I obtain a waiver from the Attorney General?

You should contact ATF's Firearms Programs Division for information on that procedure. However, in order to apply for the waiver you must have resided in the United States continuously for at least 180 days prior to submitting your application.

(R4) I'm a nonimmigrant alien. I have a State concealed weapons permit. Does this exempt me from the prohibition on nonimmigrant aliens possessing or receiving firearms and ammunition?

No. A State concealed weapons license/permit does NOT satisfy the hunting license or permit exception to the prohibition.

(R5) What is an alien number or admission number?

These are 2 different types of numbers. An admission number is the number on an INS Form I-94 or INS Form I-94W, the arrival/departure form Customs and Border Protection (CBP) gives most nonimmigrant aliens when they arrive in the U.S. While most nonimmigrant aliens will automatically receive an admission number when they enter the U.S., Canadians will not. However, if a Canadian asks a CBP official for an admission number when he/she enters the United States, he/she will be given an admission number.

Most nonimmigrant aliens will not have an alien number. An alien number is a Department of Homeland Security file number. It is issued in a variety of limited situations, such as to nonimmigrant aliens with employment authorization documents.

(R6) I am a nonimmigrant alien. I purchased a firearm in Maine in early 1998 after providing the Federal firearms dealer with documentation showing I had resided in the State for more than 90 days. I was told that this transaction was legal then. Am I entitled to keep that firearm and any ammunition I have on hand? Is there a "grandfather" clause that would protect me from criminal liability?

Since October 21, 1998, when the Gun Control Act was amended to make nonimmigrant aliens a new category of prohibited persons, nonimmigrant aliens generally have not been able to possess firearms and ammunition in the United States. The law does not contain a "grandfather clause." Therefore, unless you obtain a valid State hunting license or permit (or fall within one of the other exceptions), your possession of the firearm and ammunition is NOT legal.

(R7) I have a "green card" and have lived in Texas for several years. Am I prohibited from purchasing firearms and ammunition from an FFL in Texas?

As long as you are not otherwise prohibited from purchasing or possessing firearms and ammunition (for example, a felon), Federal law does not prohibit you from purchasing or possessing firearms or ammunition. However, you will need to put your alien number or admission number on the Form 4473 and provide the FFL with documentation establishing you have resided in Texas for more than 90 consecutive days preceding the transaction. Moreover, you must make sure there are no State or local restrictions on such a purchase.

(R8) I am a nonimmigrant alien who has resided in Idaho for 1 year. I have a valid Montana hunting license. Can I use the Montana license as evidence that I fall within an exception to the nonimmigrant alien prohibition when I go to buy a gun from a dealer in Idaho?

Yes. A valid hunting license or permit from any State within the United States satisfies the hunting license exception to the nonimmigrant alien prohibition. The license does not have to be from the State where the nonimmigrant alien is purchasing the firearm. Please note, the transaction must comply with State and local laws.

(R9) I am a nonimmigrant alien who is on a month-long vacation in the United States. I have a hunting license and an admission number. Can I legally buy a firearm from a Federal firearms licensee (FFL) in the United States and take possession of it in the United States?

No. You cannot legally buy a firearm from an FFL and take possession of it in the U.S. because you have not resided in a State within the United States for 90 days.

(R10) I am a nonimmigrant alien. I'm coming to the United States for 2 weeks. I do not have a hunting license or any alien or admission number. Can I buy a firearm from a Federal firearms licensee (FFL) to take back to my home country?

You may not buy a firearm and take possession of it in the United States. However, the FFL may directly export the firearm to your home country. If the FFL directly exports it, you do not need a hunting license, alien number or admission number or 90 days of State residency. However, the FFL first must obtain an export license from the Department of State or, if the firearm is a sporting shotgun, from the Department of Commerce.

(R11) I am a nonimmigrant alien from Canada and am planning to reside in Florida for 6 months. Do I need an alien number or admission number if I plan to buy a gun after living in Florida for 90 days? If so, how do I get such a number?

All non-U.S. citizens need an alien number or admission number to purchase a firearm from a Federal firearms licensee (FFL). The FFL will not complete the sale if you do not have such a number. This is the case even if you have a State permit that ATF has determined qualifies as a "NICS alternative" and therefore do not need to have a National Instant Criminal Background Check System Check. Most nonimmigrant aliens will automatically receive an admission number when they enter the United States. However, Canadians will not automatically receive this number and therefore should specifically ask for this number when they enter the United States. These numbers only can be issued at a port of entry, so it is important that you request the number when you enter the United States. Please note, most nonimmigrant aliens will not receive alien numbers.

(R12) I am a nonimmigrant alien

and have been residing in Florida for 4 months. I do not have an alien number or admission number. Is there any way I can get a number?

Yes. However, because Customs and Border Protection (CBP) only issues admission numbers at ports of entry, you generally will have to leave the United States and return to the United States to get such a number. One way to accomplish this is to go to either the Canadian border or Mexican border, leave the United States, and then reenter the United States. Upon reentering the United States, you can ask a CBP inspector for an admission number. You do not have to stay in Canada or Mexico for any length of time to do this. You can simply drive over the border, turn around, and reenter at a Customs and Border Protection point of entry. You also could request an admission number at a U.S. airport or seaport if you left the United States and returned by plane or ship. Please note, however, that it will take several weeks for the admission number to be entered into the government records system. You likely will receive a "denied" response from NICS or the State POC if you try to purchase a firearm before the number is entered into the government records system. Most nonimmigrant aliens will not be able to obtain alien numbers.

(R13) ATF regulations give the Attorney General or his delegate (the ATF Director) the authority to require nonresidents temporarily bringing firearms and ammunition into the United States for hunting or other lawful sporting purposes to first obtain an approved import permit. Do all such importations now require an ATF-approved permit?

All nonimmigrant aliens (with a few exceptions which are listed below) must obtain an import permit from ATF to temporarily import firearms and ammunition for hunting or other lawful sporting purposes. Please note this requirement applies to all nonimmigrant aliens, not all nonresidents (e.g., it does not apply to U.S. citizens residing abroad). The exceptions to this permit requirement are for certain foreign military personnel, official representatives of foreign governments, distinguished foreign visitors, and foreign law enforcement officers of friendly foreign governments entering the U.S on official law enforcement business.

[27 CFR 478.115(d) and (e)]

(R14) What type of form do I, as a nonimmigrant alien, need to file with ATF to temporarily import a firearm or ammunition for hunting or other lawful sporting purposes?

You need to file ATF Form 6 NIA (Application and Permit for Importation of Firearms, Ammunition and Implements of War by Nonimmigrant Aliens). The Form is both the application and, once approved, the permit you present to the U.S. Customs and Border Protection when you enter the United States. The Form 6 NIA can be obtained by calling ATF's Firearms and Explosives Imports Branch or ATF's Distribution Center. It also can be downloaded from ATF's Web site at www.ATF.gov.

(R15). Why is the submission of a Form 6 NIA for every importation now necessary?

After the events of September 11th, it was determined that national security and public safety required ATF to know when nonimmigrant aliens are bringing firearms and ammunition into the country and the numbers and types of firearms and ammunition they are bringing in. Moreover, because nonimmigrant aliens generally cannot possess firearms and ammunition in the United States, the permit process is necessary to ensure any nonimmigrant alien bringing firearms or ammunition into the country falls within an exception to the prohibition.

(R16) Do I need to attach any particular documentation to the Form 6 NIA application I submit to ATF?

Yes. When you file your Form 6 NIA application, you must provide ATF with appropriate documentation demonstrating you fall within an exception to the nonimmigrant alien prohibition.

Appropriate documentation includes a valid hunting license/permit from a State of the United States. In order to be valid, the license cannot have expired at the time of submission. An application for a hunting license/permit does not qualify for the exception. A Canadian hunting license/permit also does not qualify.

Appropriate documentation also includes an invitation/registration to attend a target shooting competition or sports or hunting trade show sponsored by a national, State, or local firearms trade organization devoted to the collection, competitive use, or other sporting use of firearms. Invitations/registrations only will qualify for the exception if they are addressed to/filled out by the Form 6 NIA applicant.

If a hunting license/permit or an invitation/registration is faxed to ATF, all information must be legible on the faxed copy.

(R17) Do I need to show anything other than the approved Form 6 NIA permit to U. S. Customs and Border Protection (CBP) when I enter the United States?

When you enter the United States, you must show CBP both your approved Form 6 NIA permit and appropriate documentation demonstrating you fall within an exception to the nonimmigrant alien prohibition.

(R18) I'm a nonimmigrant alien. I provided ATF with a copy of a hunting license when I filed my Form 6 NIA import application. Can I provide Customs and Border Protection (CBP) with a copy of the same hunting license when I enter the United States?

It depends. If the hunting license is still valid (meaning it has not expired) at the time you enter the United States, you may present CBP with a copy of the same license. However, if the hunting license has expired, you must get a new hunting license to present to CBP, or CBP will not allow you to import the firearm or ammunition. Please note, you must have a valid hunting license the entire time you possess firearms or ammunition in the United States or your possession will be illegal.

Similarly, if you provided ATF with a copy of an invitation/registration to attend a target shooting competition or sports or hunting trade show, you may present CBP with a copy of the same invitation/registration, as long as it is for a future event.

(R19) How long does it usually take for ATF to approve a Form 6 NIA permit application?

6 to 8 weeks. Please note, this time

frame begins to run once ATF receives a correctly completed application. The time frame is the same whether the application is submitted to ATF by U.S. mail, FedEx, fax, or hand delivery.

(R20) Can I list more than one firearm on a Form 6 NIA application? Can I also list ammunition on the same form?

You may list more than one firearm on a Form 6 NIA application and may include ammunition on the same form.

(R21) I am a nonimmigrant alien who wants to bring firearms into the U.S. for a hunting trip. Can I fill out and submit the Form 6 NIA myself? Or must I have a Federally licensed importer or dealer complete it and submit it for me?

You may complete the form yourself and submit it yourself.

(R22) May I submit my Form 6 NIA application by fax?

Yes. You may fax your application documents to the Firearms and Explosives Imports Branch. If you would like the permit faxed back to you once it is processed, please include a return fax number, as well as a daytime telephone number so that we can follow-up if necessary.

(R23) I am a nonimmigrant alien. I have obtained an approved ATF Form 6 NIA import permit. Someone mentioned that I have to give Customs and Border Protection (CBP) both the approved Form 6 NIA permit and something called "ATF Form 6A." Is this correct?

No. Because you are only importing the firearms temporarily, you do not need to complete ATF Form 6A or provide it to CBP.

(R24) I am a nonimmigrant alien and come to the United States at least 6 times a year to go hunting. Do I need to file a new Form 6 NIA application for each trip?

No. The import permit you receive authorizes you to bring in the firearms and ammunition listed on the permit repeatedly for 12 months after the date the permit is approved, as long as you have a valid State hunting license to present to the Customs and Border Protection Inspector at the time of entry.

(R25) I am a nonimmigrant alien and come to the United States for multiple competitive target shooting events each year. Do I need to file a new Form 6 NIA application for each event?

No. When you apply for your import permit, you should attach your invitations/registrations to all the events you will be attending during the next 12 months. These will be attached to your approved permit. The import permit you receive will authorize you to bring in the firearms and ammunition listed on the permit repeatedly for 12 months after the date the permit is approved, as long as the invitations/registrations attached to the permit are for future events. If you do not expect to have invitations/registrations for future competitions until shortly before the competitions, you may want to rely on the hunting license exception to the nonimmigrant alien prohibition. If you attach a hunting license to your application, you will be able to bring in the firearms and ammunition listed on the permit repeatedly for 12 months after the date the permit is approved, as long as you have a valid State hunting license to present to the Customs and Border Protection Inspector at the time of entry. You may rely on this exception even though you are coming to the United States for competitive shooting events, and not to hunt.

(R26) I am a nonimmigrant alien coming to the U.S. for a two-week hunting trip with 10 other nonimmigrant aliens. Do we all have to file separate Form 6 NIA applications?

Yes. Each person must file a separate Form 6 NIA application. Each applicant also must provide his or her own hunting license/permit as part of his/her application. Similarly, if a group of nonimmigrant aliens is coming to the U.S. to compete in a qualifying competitive target shooting event or to attend a sports or hunting trade show, each person must file a separate Form 6 NIA application and provide his/her own invitation/registration as part of the application.

(R27) I'm attending a shooting event in the United States and am not sure if it qualifies for the exception. What should I do?

We suggest that before filing your Form 6 NIA application, you call the Firearms and Explosives Imports Branch and ask if the event qualifies. If it does not qualify, you may obtain a hunting license as an alternative way of meeting one of the exceptions. Checking if the event qualifies before you submit your application will speed up the processing time of your application.

(R28) I am a nonimmigrant alien. Do I need a Form 6 NIA import permit to import a muzzle loading gun that is considered an antique firearm under the Gun Control Act?

No. Because antique firearms are not considered firearms for purposes of the Gun Control Act, none of the import regulations apply to the importation of antique firearms. Moreover, a nonimmigrant alien may possess antique firearms, even if the alien does not fall within an exception to the nonimmigrant alien prohibition. If you are not sure if your firearm is an antique firearm as defined by the Gun Control Act, contact ATF's Firearms Technology Branch.

(R29) I am a nonimmigrant alien. Are there any restrictions on the types of guns I can temporarily bring into the United States for hunting or for a shooting competition?

Yes. Unregistered National Firearms Act weapons (which include machineguns, short-barreled rifles and shotguns, and silencers); U.S. government origin firearms or firearms that contain U.S. government origin manufactured parts or components; and firearms from certain proscribed countries may not be temporarily imported. If you think the firearm(s) you wish to import may be affected by these restrictions, please contact the Firearms and Explosives Imports Branch.

[26 U.S.C. 5844, 27 CFR 447.52 and 447.57]

(R30) If I import a firearm temporarily for hunting or match shooting, do I have to take it out of the United States at all? My brother-in-law is a United States citizen and a U.S. resident. Can I just import the firearm, and leave it in the United States with him until I come back

to the U.S. and need the gun again?

You may not leave the firearm with your brother-in-law. The regulations only allow you to temporarily import firearms. Therefore, you must take the firearms you import back out with you when you complete your sporting activity. In fact, the import permit you receive will have a stamp on it stating that the firearm must be taken out of the United States at the conclusion of the hunting or sporting event, as well as a stamp saying the firearm may not be transferred to another person. Therefore leaving the firearm in the United States with your brother will result in an unlawful importation.

(R31) If I can't leave the gun with a friend or relative, could I leave it in the custody of a Federal Firearms Licensee in the United States?

No. Because the regulations require you to take the firearms you import back out with you when you complete the sporting activity, you cannot leave them with anyone in the United States, even a Federal firearms licensee. Also, leaving it with an FFL will violate the stamps on your permit that state the firearm must be taken out of the United States at the conclusion of the hunting or sporting event and may not be transferred to another person.

(R32) If I receive a Form 6 NIA import permit, do I need to obtain an export permit to take the guns and/or any remaining ammunition back out of the United States?

No.

(R33) I'm a nonimmigrant alien who is coming to the United States for two weeks to go hunting. Can I rent a firearm in the United States to use on this trip? What about if I want to go to a shooting range one day - can I rent a firearm there as well?

As long as you possess a valid hunting license from a State within the United States, you may rent firearms to hunt and to use at a shooting range. If you do not have the hunting license, your possession of the firearms and ammunition will be unlawful. The hunting license does NOT have to be from the State where you will be possessing the guns and ammunition.

[18 U.S.C. 922(a)(5) and (9), 922(g)(5)(B) and 922(y)]

(R34) I am a U.S. citizen living in Canada with no State of residency in the United States. Do I need an import permit to temporarily bring my gun to the U.S. to hunt or to attend target shooting events?

No. You are not required to obtain an import permit. The import permit requirement only applies to nonimmigrant aliens.

(R35) I'm Canadian and am going to be driving through the United States as a short cut to get from one part of Canada to another part of Canada. I'm going to have a rifle with me. What documentation will I need at the U.S. borders?

You do not need a Form 6 NIA import permit from ATF because you are not temporarily bringing your gun in for hunting or other lawful sporting purposes. You also do not need a DSP-61 import license from the State Department because there is an exception to their license requirement that applies in this situation. However, you do need to have a valid hunting license from a State within the United States to make your possession of firearm legal while you are in the United States.

(R36) I'm a nonimmigrant alien and have been living in the United States for several years. I have a firearm that I legally purchased in the United States and legally possess (I have a valid State hunting license). I'm going to bring my gun with me on a hunting trip to Canada. Will I need to obtain a Form 6 NIA permit to bring the gun back into the United States with me?

You will not need to obtain a Form 6 NIA permit if, when you return to the U.S., you can satisfy the U.S. Customs and Border Protection (CBP) official that you previously took the firearm out of the United States with you. The easiest way to do this is to complete Customs Form 4457 when you leave the U.S. with the gun. Please note that when you return to the U.S. you will have to present the CBP official with documentation (for example, a hunting license/permit) demonstrating you are exempt from the general non-immigrant alien prohibition on possessing firearms.

(R37) I am a Canadian law enforcement officer planning to attend a qualifying competitive target shooting event in the United States. Do I need an import permit?

If you are attending the event in your official capacity as a law enforcement officer and all the firearms you are bringing are for use in that event, you do not need a permit. This is because of a regulatory exception in 27 CFR 478.115(d)(5). You will need to show Customs and Border Protection documentation (such as a letter on agency letterhead) demonstrating that you are a law enforcement officer and need to use these firearms for official purposes.

If, however, you are attending the event as a private individual, the nonimmigrant alien import permit requirements will apply to you.

Please note that even if you are exempt from the permit requirements as a law enforcement officer on official business, you may not bring in unregistered National Firearms Act (NFA) weapons (e.g., short-barreled rifles and shotguns, machineguns, silencers).

ATF POINTS OF CONTACT

Report Lost or Stolen Firearms, including National Firearms Act Weapons .... **888-930-9275 or 800-ATF-GUNS**

Report Unlawful Firearms Activity **800-ATF-GUNS**

Report Multiple Sales of Handguns

Please make sure to send a completed ATF Form 3310.4 to:

ATF National Tracing Center (NTC)
244 Needy Road, Martinsburg WV 25401.. **304-260-1500**

Rather than mailing this form, you may **FAX** it to the NTC at **877-283-0288**

(You also must send a copy of the Form 3310.4 to the Chief Law Enforcement Officer where your business is located and attach a copy to the Form 4473 for the second handgun.)

Federal Firearms License Application or Renewal

Federal Firearms Licensing Center
P.O. Box 409567
Atlanta, GA 30384-9567 ....**404-417-2750**
Toll-free **866-662-2750**

ATF Forms and Publications

ATF Distribution Center
P.O. Box 5950
Springfield, VA 22150-5950 ....**703-455-7801**
Most forms and publications also are available at the ATF website at.. **www.ATF.gov**

Firearms Technology Branch

244 Needy Road
Martinsburg, WV 24501 .... **304-260-1700**

Going Out-of-Business

Records should be shipped to:
ATF Out-of-Business Records Center
244 Needy Road, Martinsburg, WV 25401 .... **800-788-7133**

Questions About Federal Firearms Laws and Regulations

Contact your nearest ATF Field Office..**See listing starting on page 202**

FOR QUESTIONS CONCERNING FEDERAL FIREARMS LAWS, REGULATIONS, PROCEDURES OR POLICES CONTACT AN ATF INDUSTRY OPERATIONS FIELD OFFICE BELOW

ATF Web Site: http://www.atf.gov/contact/field.htm

Atlanta Field Division:	404-417-2600
States: Georgia	
Atlanta, GA (Group I)	404-417-1300
Atlanta, GA (Group II Arson)	404-417-1300
Atlanta, GA (Group III)	404-417-1300
Atlanta, GA (Group IV)	404-815-4400
Atlanta, GA (Group V - Industry Operations)	404-417-2670
Atlanta, GA (Group VI)	404-417-2600
Atlanta, GA (Group VII)	404-417-1300
Augusta, GA (Satellite Office)	706-724-9983
Columbus, GA (Satellite Office)	706-653-3545
Macon, GA	478-474-0477
Macon, GA (Group II - Industry Operations)	478-474-0477
Savannah, GA	912-790-8326

TOP

Baltimore Field Division:	410-779-1700
States: Delaware, Maryland	
Baltimore, MD (Group I Arson)	410-779-1710
Baltimore, MD (Group II)	410-579-5011
Baltimore, MD (Group III)	410-779-1730
Baltimore, MD (Group IV)	410-779-1740
Baltimore, MD (Group V - Industry Operations)	410-779-1750
Hyattsville, MD	301-397-2640
Wilmington, DE (Criminal Enforcement)	302-252-0110
Wilmington, DE (Industry Operations)	302-252-0130

TOP

Boston Field Division:	617-557-1200
States: Connecticut, Maine Massachusetts, New Hampshire, New York, Rhode Island, Vermont	
Boston, MA (Group I Arson)	617-557-1210
Boston, MA (Group II)	617-557-1220
Boston, MA (Group III)	617-557-1326
Boston, MA (Group IV)	617-557-1240
Boston, MA (Group V - Industry Operations)	617-557-1250
Burlington,VT	802-951-6593
Hartford, CT (Industry Operations)	860-240-3400
Manchester, NH	603-471-1283
New Haven, CT	203-773-2060
Portland, ME	207-780-3324
Providence, RI	401-528-4366

Providence, RI (Satellite Office - Industry Operations)	401-528-4366
Springfield, MA (Satelite Office)	413-785-0007
Worcester, MA	508-793-0240

TOP

Charlotte Field Division:	704-716-1800
States: North Carolina, South Carolina	
Asheville, NC (Satellite Office)	828-271-4075/4076
Charleston, SC	843-763-3683
Charlotte, NC (Group I)	704-716-1810
Charlotte, NC (Group II)	704-716-1820
Charlotte, NC (Group III - Industry Operations)	704-716-1830
Charlotte, NC (Group IV)	704-716-1840
Charlotte, NC (Violent Crime Task Force)	704-716-1850
Columbia, SC	803-765-5723
Columbia, SC (Satellite Office - Industry Operations)	803-765-5722
Fayetteville, NC	910-483-3030
Fayetteville, NC (Satellite Office - Industry Operations)	910-483-3030
Florence, SC (Satellite Office)	843-667-3985
Greensboro, NC (Group I)	336-547-4224
Greensboro, NC (Group II - Industry Operations)	336-547-4150
Greenville, SC	864-282-2937
Greenville, SC (Satellite Office - Industry Operations)	864-282-2937
Raleigh, NC	919-856-4366
Wilmington, NC	910-343-6801

TOP

Chicago Field Division:	312-846-7200
States: Illinois	
Chicago, IL (Group I)	312-846-7230
Chicago, IL (Group II)	312-846-7250
Chicago, IL (Group III)	312-846-7270
Chicago, IL (Group IV)	312-846-8850
Chicago, IL (Group V)	312-846-8870
Fairview Heights, IL	618-632-9380
Fairview Heights, IL (Satellite Office - Industry Operations)	618-632-0704
Downers Grove, IL (Group I)	630-725-5220
Downers Grove, IL (Group II -Arson & Explosives)	630-725-5230
Downers Grove, IL (Group III - Industry Operations)	630-725-5290
Peoria, IL (Satellite Office - Industry Operations)	309-671-7108
Rockford, IL (Satellite Office)	309-732-0636
Springfield, IL (Group I - Criminal Enforcement)	217-547-3650
Springfield, IL (Group II - Industry Operations)	217-547-3675

TOP

Columbus Field Division:	614-827-8400
States: Indiana, Ohio	
Cincinnati Field Office	513-684-3354
Cincinnati II IO Office	513-684-3351
Cleveland I Field Office	216-522-3080
Cleveland II Field Office	216-522-3786
Cleveland III IO Office	216-522-3374
Columbus I Field Office	614-827-8450
Columbus II Intel Group	614-827-8430
Columbus IO Satellite	614-827-8470
Fort Wayne Field Office	260-424-4440
Indianapolis I Field Office	317-226-7464
Indianapolis II IO Office	317-248-4002
Merrillville Field Office	219-755-6310
Toledo Field Office	419-259-7520
Youngstown Field Office	330-707-2300

TOP

Dallas Field Division:	469-227-4300
States: Oklahoma, Texas	
Dallas, TX (Group I)	469-227-4350
Dallas, TX (Group II Arson)	469-227-4370
Dallas, TX (Group III)	469-227-4395
Dallas, TX (Group IV)	972-915-9570
Dallas, TX (Group V - Industry Operations)	469-227-4415
El Paso, TX	915-534-6449
El Paso, TX (Satellite Office - Industry Operations)	915-534-6475
Fort Worth, TX	817-862-2800
Fort Worth, TX - Industry Operations	817-862-2850
Lubbock, TX (Group I)	806-798-1030
Lubbock, TX (Satellite Office - Industry Operations)	806-798-1030
Oklahoma City, OK (Group I - Industry Operations)	405-297-5073
Oklahoma City, OK (Group II)	405-297-5060
Tulsa, OK	918-581-7731
Tyler, TX	903-590-1475

TOP

Detroit Field Division:	313-259-8050
States: Michigan	
Ann Arbor	734-741-2456
Detroit, MI (Group I)	313-259-8110
Detroit, MI (Group II)	313-259-8120
Detroit, MI (Group III Arson)	313-259-8140
Detroit, MI (Group IV)	313-259-8320
Detroit, MI (Group V - Industry Operations)	313-259-8390

Detroit, MI (Group VI)	313-259-8760
Flint, MI	810 341-5710
Flint, MI (Satellite Office - Industry Operations)	810 341-5730
Grand Rapids, MI (Group I)	616-301-6100
Lansing, MI (Satellite Office)	517-337-6645
Grand Rapids, MI (Group II - Industry Operations)	616-301-6100

TOP

Houston Field Division:	281-372-2900
States: Texas	
Austin, TX	512-349-4545
Beaumont, TX	409-835-0062
Beaumont, TX (Satellite Office - Industry Operations)	409-835-0062
Corpus Christi, TX	361-888-3392
Houston, TX (Group I)	281-372-2990
Houston, TX (Group II)	281-372-2960
Houston, TX (Group III Arson)	281-372-2930
Houston, TX (Group IV)	281-372-2980
Houston, TX (Group V)	281-372-3010
Houston, TX (Group VI - Industry Operations)	281-372-2950
McAllen, TX	956-687-5207
San Antonio, TX (Group I)	210-805-2727
San Antonio, TX (Group II - Industry Operations)	210-805-2777
Waco, TX (Satellite Office)	254-741-9900

TOP

Kansas City Field Division:	816-559-0700
States: Iowa, Kansas, Missouri, Nebraska	
Cape Girardeau, MO (Satellite Office)	573-331-7300
Des Moines, IA	515-284-4372
Des Moines, IA (Satellite Office - Industry Operations)	515-284-4857
Kansas City, MO (Group I)	816-559-0710
Kansas City, MO (Group II)	816-559-0720
Kansas City, MO (Group III - Industry Operations)	816-559-0730
Kansas City, MO (Group IV)	816-746-4962
Omaha, NE	402-493-3651
Omaha, NE (Satellite Office - Industry Operations)	402-493-4183
Springfield, MO	417-575-8015
St. Louis, MO (Group I)	314-269-2200
St. Louis, MO (Group II)	314-269-2200
St. Louis, MO (Group III - Industry Operations)	314-269-2250
Wichita, KS	316-269-6229

TOP

Los Angeles Field Division:	213-534-2450
States: California	

Los Angeles , CA (Group I - Metro)	213-534-1050
Los Angeles, CA (Group II)	213-534-1070
Los Angeles, CA (Group III Arson)	213-534-6480
Los Angeles, CA (Group IV - Industry Operations)	213-534-2430
Los Angeles, CA (Group V)	213-534-5050
Riverside, CA	909-276-6031
San Diego, CA (Group I)	619-446-0700
San Diego, CA (Group II)	619-446-0720
San Diego, CA (Group III - Industry Operations)	619-446-0740
Santa Ana, CA (Group I)	714-246-8210
Santa Ana, CA (Group II - Industry Operations)	714-246-8252
Santa Maria, CA (Enforcement)	805-348-1820
Santa Maria, CA (Industry Operations)	805-348-0027
Van Nuys, CA	818-756-4350
Van Nuys, CA (Satellite Office - Industry Operations)	818-756-4364

TOP

Louisville Field Division:	502-753-3400
States: Kentucky, West Virginia	
Ashland, KY	606-329-8092
Bowling Green, KY	270-781-7090
Bowling Green, KY (Satellite Office - Industry Operations)	270-781-1757
Charleston, WV (Charleston I)	304-347-5249
Charleston, WV (Satellite Office - Industry Operations)	304-347-5172
Lexington, KY (Lexington I Field Office)	859-219-4500
Lexington, KY (Group II - Industry Operations)	859-219-4508
London, KY (Satellite Office)	606-878-3011/3012
Louisville, KY (Group I)	502-753-3450
Louisville, KY (Group II - Industry Operations)	502-753-3500
Louisville, KY (Group III)	502-753-3550
Wheeling, WV	304-232-4170
Wheeling, WV (Satellite Office - Industry Operations)	304-232-4170

TOP

Miami Field Division:	305-597-4800
States: Florida, Puerto Rico, Virgin Island	
Fort Lauderdale, FL	954-453-6001
Fort Lauderdale, FL HIDTA	954-888-1661
Fort Pierce, FL (Satellite Office)	561-835-8878
San Juan, PR (Group I - HIDTA)	787-766-5084
San Juan, PR (Group II)	787-766-5084
San Juan, PR (Group III - Industry Operations)	787-766-5584
Mayaguez, PR (Satellite Office - Industry Operations)	787-344-8636

Miami, FL (Group I)	305-597-4910
Miami, FL (Group II)	305-597-4920
Miami, FL (Group III)	305-597-4930
Miami, FL (Group IV)	305-597-4940
Miami, FL (Group V - HIDTA)	305-597-2056
Miami, FL (Group VI - Industry Operations)	305-597-4960
St. Croix, VI (Satellite Office)	340-719-4799
St. Thomas, VI	340-774-2398
West Palm Beach, FL	561-835-8878

TOP

Nashville Field Division:	615-565-1400
States: Alabama, Tennessee	
Birmingham, AL (Group I)	205-583-5920
Birmingham, AL (Group II - Industry Operations)	205-583-5950
Birmingham, AL (Group III)	205-583-5970
Chattanooga, TN	423-855-6422
Huntsville, AL (Satellite Office)	256-539-0623
Jackson, TN (Satellite Office)	731-265-4258
Johnson City, TN (Satellite Office)	423-283-7262/7104
Knoxville, TN	865-545-4505
Memphis, TN	901-544-0321
Mobile, AL	251-405-5000
Mobile, AL (Satellite Office - Industry Operations)	251-405-5000
Montgomery, AL	334-206-6050
Nashville, TN (Group I)	615-565-1400
Nashville, TN (Group II - Industry Operations)	615-565-1420
Nashville, TN (Group III)	615-565-1430

TOP

New Orleans Field Divison:	504-841-7000
States: Arkansa, Louisiana, Mississippi	
Baton Rouge, LA	225-819-4314
Biloxi, MS	228-388-5092
Fort Smith, AR	501-709-0872
Jackson, MS	601-292-4000
Jackson, MS (Group II – Industry Operations)	601-292-4025
Little Rock, AR	501-324-6181
Little Rock, AR (Satellite Office - Industry Operations)	501-324-6457
New Orleans, LA (Group I)	504-841-7040
New Orleans, LA (Group II)	504-841-7080
New Orleans, LA (Group III - Industry Operations)	Temporary number 985-893-8333
New Orleans, LA (Group IV)	504-841-7160
Oxford, MS (Group I)	662-234-3751
Shreveport, LA	318-424-6850

Shreveport, LA (Satellite Office - Industry Operations)	318-424-6861

TOP

New York Field Division:	718-650-4000
States: New Jersey, New York City	
Albany, NY (Criminal)	518-431-4182
Albany, NY (Satellite Office - Industry Operations)	518-431-4188
Bath, NY (Satellite Office - Industry Operations)	607-776-4549
Buffalo, NY (Group I - Criminal)	716-853-5070
Buffalo, NY (Group II - Industry Operations)	716-853-5160
Melville, NY	631-694-8372
Melville, NY (Satellite Office - Industry Operations)	631-694-1366
New York, NY (Group I)	718-552-1610
New York, NY (Group II)	718-552-1620
New York, NY (Group III Arson)	718-896-6400
New York, NY (Group IV)	718-650-4040
New York, NY (Group V)	718-650-4050
New York, NY (Group VI - Industry Operations)	718-650-4060
New York, NY (Group VII)	718-650-4070
Rochester, NY (Satellite Office)	716-263-5720
Syracuse, NY (Criminal)	315-448-0889
Syracuse, NY (Satellite Office - Industry Operations)	315-448-0898
West Patterson, NJ (NJ Group I)	973-247-3010
West Patterson, NJ (NJ Group II Arson)	973-247-3020
West Patterson, NJ (NJ Group III- Industry Operations)	973-247-3030
White Plains, NY	914-682-6164
White Plains, NY (Satellite Office - Industry Operations)	914-682-6164

TOP

Philadelphia Field Division:	215-717-4700
States: Pennsylvania, New Jersey	
Atlantic City, NJ (Satellite Office)	609-487-2110
Camden, NJ	856-488-2520
Harrisburg, PA	717-221-3402
Lansdale, PA (Industry Operations)	215-362-1840
Philadelphia, PA (Group I - Violent Crime Impact Team)	215-717-4710
Philadelphia, PA (Group II Arson & Explosives)	215-597-9080
Philadelphia, PA (Group III)	215-560-1631
Philadelphia, PA (Group IV - Industry Operations)	215-597-2203
Philadelphia, PA (Group V - Intelligence Group)	215-717-4750
Philadelphia, PA (Group VI - Ceasefire)	215-717-4760
Philadelphia, PA (Group VII)	215-717-4710
Pittsburgh, PA (Group I and Group II Arson)	412-395-0540
Pittsburgh, PA (Group III - Industry Operations)	412-395-0600

Reading, PA	610-320-5222
Trenton, NJ (Arson and Explosives)	609-989-2155
Trenton, NJ (Satellite Office - Industry Operations)	609-989-2142
Wilkes-Barre, PA (Industry Operations)	570-826-6551

TOP

Phoenix Field Division:	602-776-5400
States: Arizona, Colorado, New Mexico, Utah, Wyoming	
Albuquerque, NM	505-346-6914
Albuquerque, NM (Satellite Office - Industry Operations)	505-346-6910
Cheyenne, WY	307-772-2346
Colorado Springs, CO	719-473-0166
Denver, CO (Group I)	303-844-7540
Denver, CO (Group II Arson)	303-844-7570
Denver, CO (Group III- Industry Operations)	303-844-7545
Phoenix, AZ (Group I)	602-776-5440
Phoenix AZ (Group II)	602-776-5460
Phoenix, AZ (Group III - Industry Operations)	602-776-5480
Phoenix AZ (Group IV)	602-776-5500
Salt Lake City, UT	801-524-7000
Salt Lake City, UT (Satellite Office - Industry Operations)	801-524-7012
Tucson, AZ (Group I)	520-670-4725
Tucson, AZ (Satellite Office - Industry Operations)	520-670-4804

TOP

San Francisco Field Division:	925-479-7500
States: California, Nevada	
Bakersfield, CA (Satellite Office)	661-861-4420
Fresno, CA (Group I)	559-487-5393
Fresno, CA (Group II - Industry Operations)	559-487-5093
Las Vegas, NV	702-387-4600
Stockton, CA (Stockton I Satellite Office and Stockton II Satellite Office)	209-321-8878
Napa, CA (Satellite Office - Industry Operations)	707-224-7801
Oakland, CA (Group I)	510-267-2200
Oakland, CA (Satellite Office - Industry Operations)	925-479-7500
Redding, CA (Satellite Office)	530-224-1862
Reno, NV	775-784-5251
Sacramento, CA (Group I)	916-498-5100
Sacramento, CA (Group II - Industry Operations)	916-498-5095
San Francisco, CA (Group I)	415-436-8020
San Francisco, CA (Group II Arson)	925-479-7520
San Francisco, CA (Group III - Industry Operations)	925-479-7530
San Francisco, CA (Group IV)	925-479-7540

San Jose, CA (Group I)	408-535-5015
San Jose, CA (Group II - Industry Operations)	408-535-5538
Santa Rosa, CA (Industry Operations)	707-576-0184

TOP

Seattle Field Division:	206-389-5800
States: Alaska, Guam, Hawaii, Idaho, Oregon, Washington	
Anchorage, AK	907-271-5701
Anchorage, AK (Satellite Office - Industry Operations)	907-271-5701
Boise, ID	208-334-1160
Boise, ID (Satellite Office - Industry Operations)	208-334-1164
Hawaii County, HI	809-933-8139
Honolulu, HI	808-541-2670
Honolulu, HI (Satellite Office - Industry Operations)	808-541-2670
Mongmong, Guam	671-472-7129
Portland, OR (Group I)	503-331-7810
Portland, OR (Group II)	503-331-7820
Portland, OR (Group III)	503-331-7830
Seattle, WA (Group I)	206-389-6860
Seattle, WA (Group II - Industry Operations)	206-389-6800
Seattle, WA (Group III - Arson & Explosives)	206-389-6830
Seattle, WA (Group IV)	206-389-5870
Spokane, WA (Group I)	509-324-7866
Spokane, WA (Group II - Industry Operations)	509-324-7881
Yakima, WA	509-454-4403

TOP

St. Paul Field Division:	651-726-0200
States: Minnesota, Montana, North Dakota, South Dakota, Wisconsin	
Billings, MT (Group I)	406-657-6886
Fargo, ND (Group I)	701-293-2860
Fargo, ND (Group II - Satellite Office, Industry Operations)	701-293-2880
Helena, MT (Group I)	406-441-1100
Helena, MT (Group II - Satellite Office, Industry Operations)	406-441-1100
Madison, WI (Group I Field Office)	608-441-5050
Milwaukee, WI (Group I)	414-727-6170
Milwaukee, WI (Group II - Industry Operations)	414-727-6200
Missoula, MT (Satellite Office)	406-721-2611
Rapid City, SD (Satellite Office)	605-343-3288
Sioux Falls, SD (Group I)	605-330-4368
St. Paul, MN (Group I)	651-726-0300
St Paul, MN (Group II- Industry Operations)	651-726-0220
St. Paul, MN (Group III)	651-726-0230

TOP

Tampa Field Division:	813-202-7300
States: Florida	

Fort Myers, FL (Group I Satellite Office)	239-334-8086
Fort Myers, FL (Group II Satellite Office - Industry Operations)	239-334-8086
Gainesville, FL (Satellite Office)	352-374-9503
Jacksonville, FL	904-232-3468
Jacksonville, FL (Satellite Office - Industry Operations)	904-232-2868
Orlando FL	407-384-2411
Orlando FL II (Industry Operations)	407-384-2420
Panama City , FL (Satellite Office)	850-769-0234
Pensacola, FL	850-435-8485
Tallahassee, FL	850-942-9660
Tampa, FL (Group I)	813-202-7310
Tampa, FL (Group II - Industry Operations)	813-202-7320
Tampa, FL (Group III)	813-202-7330

TOP

Washington Field Division:	202-927-8810
States: Virginia, West Virginia, District of Columbia	
Bristol, VA	276-466-2727
Charlottesville, VA (Satellite Office)	434-970-3872
Falls Church, VA (Group I Arson)	703-287-1110
Falls Church, VA (Group II)	703-287-1120
Falls Church, VA (Group III - Industry Operations)	703-287-1130
Martinsburg, WV, LE	304-260-3400
Martinsburg, WV (Satellite Office - Industry Operations)	304-260-3400
Norfolk, VA	757-616-7400
Norfolk, VA (Satellite Office - Industry Operations)	757-441-3192
Richmond, VA (Group I)	804-200-4200
Richmond, VA (Group II - Industry Operations)	804-200-4141
Richmond, VA (Group III)	804-775-4200
Roanoke, VA	540-857-2300
Roanoke, VA (Satellite Office - Industry Operations)	540-857-2304
Washington, DC (Group I)	202-305-8189
Washington, DC (Group II)	202-927-7105
Washington, DC (Group III)	703-658-7842
Washington, DC (Group IV Arson Task Force)	202-927-0890
Washington, DC (Group V)	202-927-7100

NON-ATF POINTS OF CONTACT

Directorate of Defense Trade Controls

Department of State
PM/DDTC, SA-1, Room 1200
2401 E St., N.W.
Washington, DC 20037

Telephone **202-663-1282**
Web site .... **www.pmdtc.org**

Bureau of Industry and Security

Outreach and Educational Services Division
14th Street & Pennsylvania Ave., N.W.
Washington, DC 20230

Telephone **202-482-4811**
Web site .... **www.bis.doc.gov**

Federal Bureau of Investigation
NICS Operations Center

Appeals Services Unit
P.O. Box 4278
Clarksburg, WV 26302-9922

Telephone **877-444-6427**
Web site .... **www.fbi.gov/hq/cjisd/nics/index.htm**

Canada Firearms Centre

Ottawa, ON
Canada K1A 1M6

Telephone **800-731-4000**
Web site .... **www.cfc-cafc.gc.ca**
E-mail .... **cfc-cafc@cfc-cafc.gc.ca**

Questions about State Laws or Local Ordinances

Contact your State Police, local law enforcement authority or State Attorney General's Office
See listing of State Attorney Generals starting on page 206

STATE ATTORNEY GENERALS

If any of this contact information changes over time, the website for the National Association of Attorney Generals (www.naag.org) should contain updated information.

Alabama
Office of the Attorney General
State House
11 South Union Street
Montgomery, AL 36130
(334) 242-7300

Alaska
Office of the Attorney General
Post Office Box 110300
Diamond Courthouse, 4th Floor
Juneau, AK 99811-0300
(907) 465-3600

American Samoa
Office of the Attorney General
Post Office Box 7
Pago Pago, AS 96799
011 (684) 633-4163

Arizona
Office of the Attorney General
1275 West Washington Street
Phoenix, AZ 85007
(602) 542-4266

Arkansas
Office of the Attorney General
200 Tower Building
323 Center Street
Little Rock, AR 72201-2610
(800) 482-8982

California
Office of the Attorney General
1300 I Street, Suite 1740
Sacramento, CA 95814
(916) 445-9555

Colorado
Office of the Attorney General
Department of Law
1525 Sherman Street, 5th Floor
Denver, CO 80203
(303) 866-4500

Connecticut
Office of the Attorney General
55 Elm Street
Hartford, CT 06106
(860) 808-5318

Delaware
Office of the Attorney General
Carvel State Office Building
820 North French Street
Wilmington, DE 19801
(302) 577-8500

District of Columbia
Office of the Corporation Counsel
1350 Pennsylvania Ave., NW
Suite 409
Washington, DC 20004
(202) 727-3400

Florida
Office of the Attorney General
The Capitol
PL 01
Tallahassee, FL 32399-1050
(850) 414-3990

Georgia
Office of the Attorney General
40 Capitol Square
Atlanta, GA 30334-1300
(404) 656-3300

Guam
Office of the Attorney General
Suite 2-200E
Judicial Center Building
120 West O'Brien Drive
Hagatna, Guam 96910
(671) 475-3324

Hawaii
Office of the Attorney General
425 Queen Street
Honolulu, HI 96813
(808) 586-1282

Idaho
Office of the Attorney General
P.O. Box 83720
Boise, ID 83720-0010
(208) 334-2400

Illinois
Office of the Attorney General
James R. Thompson Center
100 West Randolph Street
12th Floor
Chicago, IL 60601
(312) 814-3000

Indiana
Office of the Attorney General
Indiana Government Center
302 West Washington Street
5th Floor
Indianapolis, IN 46204
(317) 232-6201

Iowa
Office of the Attorney General
Hoover State Office Building
1305 East Walnut
Des Moines, IA 50319
(515) 281-5164

Kansas
Office of the Attorney General
120 S.W. 10th Avenue, 2nd Floor
Topeka, KS 66612-1597
(785) 296-2215

Kentucky
Office of the Attorney General
State Capitol, Room 118
700 Capitol Avenue
Frankfort, KY 40601
(502) 696-5300

Louisiana
Office of the Attorney General
Department of Justice
Post Office Box 94005
Baton Rouge, LA 70804
(225) 326-6000

Maine
Office of the Attorney General
Six State House Station
Augusta, ME 04333-0006
(207) 626-8800

Maryland
Office of the Attorney General
200 Saint Paul Place
Baltimore, MD 21202-2202
(410) 576-6300

Massachusetts
Office of the Attorney General
One Ashburton Place
Boston, MA 02108-1698
(617) 727-2200

Michigan
Office of the Attorney General
Post Office Box 30212
525 West Ottawa Street
Lansing, MI 48909-0212
(517) 373-1110

Minnesota
Office of the Attorney General
State Capitol
Suite 102
St. Paul, MN 55155
(651) 296-3353

Mississippi
Office of the Attorney General
Department of Justice
Post Office Box 220
Jackson, MS 39205-0220
(601) 359-4279

Missouri
Office of the Attorney General
Supreme Court Building
207 West High Street
Jefferson City, MO 65101
(573) 751-3321

Montana
Office of the Attorney General
Justice Building
215 North Sanders
Helena, MT 59620-1401
(406) 444-2026

Nebraska
Office of the Attorney General
State Capitol
Post Office Box 98920
Lincoln, NE 68509-8920
(402) 471-2682

Nevada
Office of the Attorney General
Old Supreme Court Building
100 North Carson Street
Carson City, NV 89701
(775) 684-1100

New Hampshire
Office of the Attorney General
State House Annex
33 Capitol Street
Concord, NH 03301-6397
(603) 271-3658

New Jersey
Office of the Attorney General
Department of Law and
Public Safety
Box 080
Trenton, NJ 08625
(609) 292-8740

New Mexico
Office of the Attorney General
Post Office Drawer 1508
Santa Fe, NM 87504-1508
(505) 827-6000

New York
Office of the Attorney General
Department of Law -The Capitol
Room 220
Albany, NY 12224
(518) 474-7330

North Carolina
Office of the Attorney General
Department of Justice
9001 Mail Service Center
Raleigh, NC 27699
(919) 716-6400

North Dakota
Office of the Attorney General
State Capitol
600 East Boulevard Avenue
Bismarck, ND 58505-0040
(701) 328-2210

No. Mariana Islands
Office of the Attorney General
2nd Floor, Juan A Saplan Bldg.
Capitol Hill
Caller Box 10007
Saipan, MP 96950
(670) 664-2366/2341

Ohio
Office of the Attorney General
State Office Tower
30 East Broad Street, 17th Floor
Columbus, OH 43215-3428
(614) 466-4320

Oklahoma
Office of the Attorney General
State Capitol, Room 112
2300 North Lincoln Boulevard
Oklahoma City, OK 73105
(405) 521-3921

Oregon
Office of the Attorney General
Justice Building
1162 Court Street NE
Salem, OR 97301-4096
(503) 378-4732

Pennsylvania
Office of the Attorney General
1600 Strawberry Square
Harrisburg, PA 17120
(717) 787-3391

Puerto Rico
Office of the Attorney General
Post Office Box 9020192
San Juan, PR 00902-0192
(787) 721-7700

Rhode Island
Office of the Attorney General
150 South Main Street
Providence, RI 02903
(401) 274-4400

South Carolina
Office of the Attorney General
Rembert C. Dennis Office Bldg
Post Office Box 11549
Columbia, SC 29211-1549
(803) 734-3970

South Dakota
Office of the Attorney General
500 East Capitol
Pierre, SD 57501-5070
(605) 773-3215

Tennessee
Office of the Attorney General
2nd Floor
Cordell Hull Bldg.
Nashville, Tennessee 37243
(615) 741-3491

Texas
Office of the Attorney General
Capitol Station
Post Office Box 12548
Austin, TX 78711-2548
(512) 463-2100

U.S. Virgin Islands
Office of the Attorney General
Department of Justice
G.E.R.S. Complex 2nd Floor
48B-50C Kronprinsdens Gade
St. Thomas, VI 00802
(340) 774-5666

Utah
Office of the Attorney General
P.O. Box 142320
Salt Lake City, Utah 84114-2320
(801) 538-9600

Vermont
Office of the Attorney General
109 State Street
Montpelier, VT 05609-1001
(802) 828-3171

Virginia
Office of the Attorney General
900 East Main Street
Richmond, VA 23219
(804) 786-2071

Washington
Office of the Attorney General
P.O. Box 40100
1125 Washington Street, SE
Olympia, WA 98504-0100
(360) 753-6200

West Virginia
Office of the Attorney General
State Capitol, Room E26
1900 Kanawha Boulevard East
Charleston, WV 25305
(304) 558-2021

Wisconsin
Office of the Attorney General
114 East State Capitol
Post Office Box 7857
Madison, WI 53707-7857
(608) 266-1221

Wyoming
Office of the Attorney General
123 State Capitol Building
Cheyenne, WY 82002
(307) 777-7841

Federal Bureau of Investigation OMB NO. 1110-0026

National Instant Criminal Background Check System (NICS)

Federal Firearms Licensee (FFL) Enrollment / E-Check Enrollment

Please TYPE or PRINT neatly in BLACK INK using UPPERCASE letters.

1. FFL NUMBER *Note: 03 Licensees need not enroll.*

☐ - ☐☐ - ☐☐☐ - ☐☐ - ☐☐ - ☐☐☐☐☐

2. CODE WORD (Must be six to ten characters - NO PROFANITY) 3. BUSINESS PHONE NUMBER

☐☐☐☐☐☐☐☐☐☐ ☐☐☐ - ☐☐☐ - ☐☐☐☐

NAME OF FFL: (Name that appears on FFL License. If company name, place in LAST Name block and place overflow in FIRST and MI blocks-if necessary.)

4. LAST NAME

FIRST NAME MI CADENCE

5. NAME OF LICENSEE BUSINESS (STORE NAME):

6. POST OFFICE BOX OR STREET ADDRESS OF LICENSEE STORE PREMISE:

CITY STATE ZIP CODE

7. POINT OF CONTACT PERSON (If different than Item 4 above):

LAST NAME FIRST NAME MI CADENCE

8. POINT OF CONTACT PHONE NUMBER: 9. BUSINESS FAX NUMBER (optional):

☐☐☐ - ☐☐☐ - ☐☐☐☐ ☐☐☐ - ☐☐☐ - ☐☐☐☐

(FBI NICS E-Check users only) **FBI NICS E-Check Users Complete this Section**

Every FFL & employee wanting to use the FBI NICS E-Check must complete and submit, by mail, this entire form and provide the following additional information then access our web site at www.nicsezcheckfbi.gov and request a digital certificate:

10. LAST NAME FIRST NAME MI CADENCE

11. MOTHER'S MAIDEN NAME (Last name only)

12. E-MAIL ADDRESS: ______________________

Note: If there is a change in FFL ownership, the FBI NICS Section must be notified and a new acknowledgment must be signed.

By executing this document and/or by the use of the above code word, the FFL acknowledges understanding of its obligations and responsibilities under the NICS (as detailed in the Gun Control Act of 1968, as amended and the Responsibilities of a Federal Firearms Licensee (FFL) under the National Instant Criminal Background Check System, dated October 7, 2002) and intent to honor those obligations and responsibilities. Intending to be legally bound, I hereby execute this acknowledgment on behalf of the above-mentioned FFL and certify under penalty of perjury that I have full authority from the FFL to make a legally binding commitment on its behalf.

13. User/Applicant Signature: ______________________ **Date executed:** ______________________

14. FFL Witness: ______________________ **Date executed:** ______________________

FEDERAL FIREARMS LICENSEE (FFL) ENROLLMENT FORM ITEMS OMB NO. 1110-0026

1. **FFL NUMBER** - This is the 15 digit number assigned by the Bureau of Alcohol, Tobacco, Firearms and Explosives.

2. **CODE WORD** - This is a code word of your choice to be used as a verification of identity when you contact the NICS Section. The code word MUST be between 6 and 10 characters in length, and can include both numbers and letters. NO OBSCENE WORDS OR PHRASES PLEASE.

3. **BUSINESS PHONE NUMBER** - Please write the complete business telephone number including area code.

4. **NAME OF FEDERAL FIREARM LICENSEE** - The name of the individual or corporation license by the Bureau of Alcohol, Tobacco, Firearms and Explosives to sell firearms. f the name is a corporation name, and additional space is required, you may use the First Name spaces.

5. **NAME OF LICENSEE BUSINESS** - Store name.

6. **STREET ADDRESS** - Please fill in the entire mailing address section - complete with city, state, and zip code.

7. **POINT OF CONTACT PERSON/FBI NICS E-CHECK APPLICANT** - This is the person the FBI will call first when contacting your store. It is only needed if different from the name of the licensee. This may be a store manager or other responsible individual chosen by the licensee to represent the licensee for NICS matters, including codeword changes, validation of the FBI NICS E-Check users, etc.

8. **POINT OF CONTACT PHONE NUMBER** - This is the telephone number the FBI will call first when contacting your store. It is only needed if different from the business phone number. Please include area code.

9. **BUSINESS FAX NUMBER** (optional)- Please fill in the complete number including area code.

For FBI NICS E-Check Enrollment only. **(FBI NICS E-Check users only)**

10. **NAME of FFL or Employee who will be accessing the FBI NICS E-Check** - Every user of the FBI NICS E-Check must complete, sign, and mail in the enrollment form to the NICS Section as well as submitting a digital request on the internet.

11. **MOTHER'S MAIDEN NAME** (last name only) - Enter your mother's maiden name. This will be used to help identify users of the FBI NICS E-Check should the need arise.

12. **E-MAIL ADDRESS** - If you provide your e-mail address you will be notified by e-mail when your digital certificate is ready for downloading otherwise you will be contacted by telephone.

---

13. **SIGNATURE** - Under the authority of Brady Handgun Violence Prevention Act (Brady Act), 18 U.S.C. Chapter 44, as implemented by 28 C.F.R. Part 25, the FBI requires completion of this acknowledgment by all FFLs as a condition of being granted NICS inquiry privileges. The NICS has been established within the FBI's Criminal Justice Information Services Division (CJIS) of the FBI for the purpose of performing instant background checks on prospective firearm transferees. The primary purpose of this acknowledgment is to ensure that FFLs accessing and using the NICS understand and accept the attendant obligations and responsibilities. This acknowledgment will be used to identify and validate those FFLs who may be granted NICS inquiry privileges, to legally obligate the FFL to comply with these obligations and responsibilities, and as evidence of an FFL's knowledge and acceptance of these obligations and responsibilities whenever such matters may be in issue. Completion of this acknowledgment by an FFL is voluntary, but an FFL which does not complete this acknowledgment will not be able to make a NICS inquiry. It is a criminal violation of federal law for an FFL to transfer a firearm to a non-FFL without making a NICS inquiry.

14. **FFL WITNESS** - This is the signature of an individual witnessing the aforementioned signature.

    DATE EXECUTED - This is the date that the enrollment document was signed by each person.

A person is not required to respond to any collection of information unless it displays a currently valid OMB control number. The FBI has created the NICS Enrollment Form so that it is easily understood and requires the least possible burden on you to provide us with information. The reporting burden for collection of information on the NICS Enrollment Form is computed as: 1) learning about the documents, 2 minutes; 2) completing the NICS Enrollment Form, 3 minutes; 3) assembling, mailing (to enroll in the FBI NICS E-Check the enrollment form must be mailed), faxing or e-mailing the form to the FBI, 3 minutes for an estimated average of 8 minutes per response. If you have comments regarding the accuracy of the estimates, or suggestions for making this form simpler, you can write the Federal Bureau of Investigation, NICS Section, P. O. Box 4278, Clarksburg, West Virginia 26302-4278.

Privacy Act Statement: Pursuant to the Privacy Act of 1974, 5 U.S.C. 552a, we are providing the following information regarding this collection of information. The authority under which this information is being collected is the Brady Handgun Violence Prevention Act (Public Law 103-159) and title 28 CFR part 25. The principal purposes for which the information will be used are to enroll an FFL into the NICS, log a user onto the FBI NICS E-Check, and identify FFLs that submit NICS checks. The information collected may be shared with other government agencies for authorized purposes and with certain other persons and entities for other purposes as provided for in the most recently published routine uses for the NICS (Justice/FBI-018). The form requests both mandatory and optional information. If you omit mandatory information we may not be able to process your request.

A NICS DELAY . . .

A Federal Firearms Licensee (FFL) will receive the following instructions when a call is transferred from the FBI National Instant Criminal Background Check System (NICS) call center to the FBI NICS Section in an open transaction resulting in a delay:

"--NTN-- will be delayed while the NICS continues its research of potentially prohibiting information on this open transaction and will advise you if it reaches a final determination of proceed or denied. If you do not receive a response from us, the Brady Law does not prohibit the transfer of the firearm on ___day/date___."

The following table specifies the day after a delay response on which a firearm may be lawfully transferred under federal law if a final determination has not been received from the NICS (assuming there are no intervening state holidays or closures):

Delay Response On	*Can Legally Transfer Under Federal Law On*
Monday	Friday
Tuesday	Saturday
Wednesday	Tuesday
Thursday	Wednesday
Friday	Thursday
Saturday	Thursday
Sunday	Thursday

If the FFL has not received from the NICS a final determination after 3 business days have elapsed since the delay response, it is within the FFL's discretion whether or not to transfer the firearm (if state law permits the transfer). If the FFL transfers the firearm, the FFL must note "no resolution was provided within 3 business days" on line 21d of the ATF Form 4473. ***(Please refer to pages 24 and 25 of the FBI NICS FFL User Manual.)***

***Best practice recommendation: In open transactions, the FFLs should record on the ATF Form 4473 the date provided in the delay response on which the firearm may be lawfully transferred under federal law if a final determination of proceed or denied is not received from the NICS.**

Applicable Federal Regulations

28 CFR Part 25 - The National Instant Criminal Background Check System

Section 25.6(c)(1)(iv)(B) – Delayed response provided to FFL:

(B) "Delayed" response, if the NICS search finds a record that requires more research to determine whether the prospective transferee is disqualified from processing a firearm by Federal or state law. A "Delayed" response to the FFL indicates that the firearm transfer should not proceed pending receipt of a follow-up "Proceed" response from the NICS or the expiration of three business days (exclusive of the day on which the query is made), whichever occurs first. (Example: An FFL requests a NICS check on a prospective firearm transferee at 9:00 a.m. on Friday and shortly thereafter receives a "Delayed" response from the NICS. If state offices in the state in which the FFL is located are closed on Saturday and Sunday and open the following Monday, Tuesday, and Wednesday, and the NICS has not yet responded with a "Proceed" or "Denied" response, the FFL may transfer the firearm at 12:01 a.m. Thursday.)

Section 25.2 – Definition of "Open" transaction:

"Open" means those non-canceled transactions where the FFL has not been notified of the final determination. In cases of "open" responses, the NICS continues researching potentially prohibiting records regarding the transferee and, if definitive information is obtained, communicates to the FFL the final determination that the check resulted in a proceed or a deny. An "open" response does not prohibit an FFL from transferring a firearm after three business days have elapsed since the FFL provided to the system the identifying information about the prospective transferee.

What is a Business Day?

A business day is any 24-hour day **beginning at 12:01 a.m. the day *after* the check was initiated,** in which state offices are open. A business day **does not include Saturday, Sunday, or holidays.**

The table below advises when a firearm can be transferred if no response is received:

NICS Contacted On	Can Legally Transfer On*
Sunday	Thursday
Monday	Friday
Tuesday	Saturday
Wednesday	Tuesday
Thursday	Wednesday
Friday	Thursday
Saturday	Thursday

*** The transfer day may change depending on holidays.**

Reference

Hours of Operation

8 a.m. - 1 a.m.
Eastern Standard Time
7 days a week
(Except Christmas Day)

What is a Business Day?

A business day is any 24-hour day **beginning at 12:01 a.m. the day after the check was initiated,** in which state offices are open. A business day **do snot include Saturday, Sunday, or holidays.**

The table below advises when a firearm can be transferred if no response is received:

NICS Contacted On	Can Legally Transfer On*
Sunday	Thursday
Monday	Friday
Tuesday	Saturday
Wednesday	Tuesday
Thursday	Wednesday
Friday	Thursday
Saturday	Thursday

*The transfer day may change depending on holidays.

FFL LIAISON

The NICS has dedicated a Program Analyst to assist you in the following areas:

- Enrollment
- Code word changes/modifications
- Activation and deactivation of NICS privileges
- Troubleshooting system access
- Registration of gun shows
- Providing written correspondence concerning program advancement
- Educating the FFLs and their employees on the NICS via telephone or by representing the NICS Section at various conventions/seminars

Contact
Giget Stover – (304) 625-7387

Fax on Demand

Have documents faxed directly to you!!

✓ Call 1-877-444-6427
✓ Enter Option 4
✓ Enter the document number you want faxed

☆ Appeal Brochure

☆ E-Check/NICS FFL Enrollment Form

✓ Enter your fax number
✓ The documents will automatically be faxed to you.
✓ Other documents will be available soon.

Phone Numbers

NICS Customer Service	**1-877-444-NICS (6427)**
NICS Call Center	**1-877-FBI-NICS (324-6427)**
To Request a NICS Background Check	1-877-FBI-NICS *(Press 1)*
To Request the Status of a Delayed NICS Background Check	1-877-444-NICS *(Press 2)*
To Change Enrollment Information	1-877-444-NICS *(Press 2)*
To Request Information About the NICS Section	1-877-444-NICS *(Press 2)*
Fax on Demand	1-877-444-NICS *(Press 4)*

NICS

National Instant Criminal Background Check System

Benefits of Using the NICS E-Check

- A reduction in the NICS Call Center traffic.
- A more accurate search facilitated based on the direct entry of descriptive data by the transaction originator, thereby increasing data integrity.
- The ability to retrieve NICS background check results 24 hours per day, 7 days per week.
- The ability to retrieve all checks initiated at the NICS Call Centers or via the NICS E-Check.
- The ability to print completed NICS background check search requests.
- Spanish translation capabilities (a future enhancement under development).
- Increased usability for the hearing and speech impaired.
- The availability of messages regarding the NICS operational status.

Additional Information

For additional information pertaining to the NICS E-Check or the system's availability in your state, you may contact the FBI NICS Section via 1-877-444-6427 (select option three) or access the E-Check web site.

NICS Section Information:

Customer Service:
1-877-444-NICS (6427)
Select Option 3 for NICS E-Check

Facsimile:
1-888-550-6427

Telecommunications Device for the Deaf (TDD):
1-877-NICS-TTY

NICS Web Site:
www.fbi.gov/hq/cjisd/nics/index.htm

NICS E-Check E-mail Address:
echeck@leo.gov

NICS E-Check Web Site:
www.nicsezcheckfbi.gov

NICS E-mail Address:
a_nics@leo.gov

[Rev. August 2003]

U.S. Department of Justice
Federal Bureau of Investigation
Criminal Justice Information Services Division

NICS E-Check

NATIONAL INSTANT CRIMINAL BACKGROUND CHECK SYSTEM

NICS E-Check

Brady Act Requirements

In November 1993, the Brady Handgun Violence Prevention Act of 1993 (Brady Act), Public Law 103-159, was signed into law requiring Federal Firearms Licensees (FFLs) to request background checks on prospective firearm transferees. The permanent provisions of the Brady Act, which went into effect on November 30, 1998, required the U.S. Attorney General to establish the National Instant Criminal Background Check System (NICS) so that any FFL may contact by telephone, or by other electronic means, for information to be supplied immediately, on whether the transfer of a firearm would violate Section 922 (g) or (n) of Title 18, United States Code, *or* state law.

NICS Operations and the E-Check

Depending upon the level of each state's participation with the NICS, every FFL is provided access to the NICS via one of the following three ways:

- Through a designated state point of contact (POC) for those states that have chosen to implement and maintain their own Brady NICS Program;

- Through the FBI NICS Section for those states that have declined to serve as a POC for the system (non-POC states); or

- Through the designated state POC for handguns and the FBI NICS Section for long guns.

In the non-POC states, the FFLs contact the FBI NICS Section using a toll-free telephone number to provide the requisite information to a customer service representative who initiates the check on their behalf. However, 28 Code of Federal Regulations (CFR) Part 25, NICS Regulations, allowed for the development of other electronic means of contact as alternatives in addition to the telephone.

Therefore, the FBI NICS Section, in a joint effort with the FBI Information Technology Management Section, Lockheed Martin Energy Systems, Science Applications International Corporation and an FFL focus group, developed the NICS E-Check. This function enables the FFLs to conduct an *unassisted* NICS background check for firearm transfers via the Internet. The FFLs, via electronic communication, data enter the prospective firearm transferee's descriptive information directly into the NICS and initiate the transaction search process.

The NICS E-Check is very easy to use once the registration process has been completed. However, to utilize the NICS E-Check capability, certain restrictions apply:

- You must be a registered FFL;
- You must have Internet access; and
- You must use a web browser with 128-bit encryption technology.

Currently, the NICS E-Check is only available in those states whose FFLs are fully serviced by the FBI NICS Section inclusive of those states whose FFLs contact the FBI NICS Section for long gun transactions only.

Security

Access to the NICS E-Check is restricted through computer software and certification authority, thereby providing secure and restricted access. The NICS E-Check is monitored 24 hours a day, 7 days per week, for misuse, etc. In addition, the NICS E-Check denies access to any individual whose identification is not known to the system.

Reasons Why You May Receive a Delayed Response

If you have ever been arrested, (juvenile offenses, old arrest/conviction, misdemeanor arrest/conviction, non-convictions, investigation arrest, and/or current cases), charged, and/or fingerprinted for a criminal investigation.

- **A Criminal History with an offense or conviction that could possibly fall under one of nine federal or various state prohibiting standards.**
  - Often the National Instant Criminal Background Check System (NICS) receives criminal history records that are incomplete. This requires extensive research by a NICS Legal Instrument Examiner (NICS Examiner) to obtain information that updates any criminal charges listed on the record.
  - Any arrest/conviction that you have could cause a delayed response from the NICS. There is no limit on age of arrest/conviction. If you feel the arrest/conviction is not disqualifying, it may have been reported to the FBI differently, and would require additional research by the NICS.
  - The NICS does not always receive complete disposition information from the courts and may need to research your criminal history record to determine if a specific offense and/or conviction is no longer disqualifying or has been cleared from your record.
- **Stolen, Misplaced or Similar Identity:**
  - If you have a common name, you may experience a short delay every time you have a background check initiated.
  - Someone with a criminal history may have a similar or altered name that causes an incorrect match on your descriptive data. The NICS checks are based on name and descriptive data provided on the Alcohol, Tobacco, Firearm, and Explosives (ATF) Form 4473 (firearm application) form.
  - Stolen or misplaced identity occurs when someone has used your key descriptive data (e.g.) full name, social security number, date of birth, and place of birth). An individual may have used this data to identify themselves at the time of arrest for the commission of a crime.

A Delayed Response

When a delayed response is received from the NICS, this indicates that information you have supplied on the ATF Form 4473 has been matched with information contained in one or more of the three National Criminal Computer Databases. Complete arrest and/or judicial information are not always provided on the criminal history record. When complete information is not provided, the NICS Examiners attempt to obtain complete record information by contacting law enforcement agencies, i.e., local, state, federal courts, arresting and judicial agencies to obtain dispositions, court records and police reports. The NICS updates criminal history records with information received, resolving many delayed transactions.

- Often the various judicial and law enforcement agencies are unable to meet the demand the NICS places on their resources. These agencies are often small and do not have the manpower to support the NICS requests or their court records are not maintained for extended periods of time. The information these agencies maintain is generally public record and can be easily obtained by you, the subject of the record.
- If you are able to obtain a certified copy of your court records, you may send this information along with your fingerprints to the FBI Special Correspondence Unit to be updated on criminal history files. Updating this information may resolve your repeated delayed response to an immediate proceed.

Action You Can Take to Resolve Delayed Responses

- If you have any criminal history arrests/convictions, you should obtain the court certified documentation of the **final** outcome of your offense and forward the information to the FBI Special Correspondence Unit to update your FBI criminal history record.
- The FBI Special Correspondence Unit will then send the information you provided to the state that holds your record.
- Updating your FBI criminal history record can include having cases expunged, pardoned, conviction level changed or rights restored depending upon the legal process in the state of conviction and/or state of residence.
- State policy on restoration of rights varies from state to state. Contact your state Office of the Attorney General for clarification.
- If you are unsure of what may be on your criminal record, you may request a copy of your FBI record by contacting the FBI Special Correspondence Unit.
- The FBI does not maintain all criminal history records. You may need to contact the state repository for criminal history information.

① A state repository is responsible for maintaining criminal history records that are reported to the FBI. This repository, not the FBI NICS Section, is responsible for maintaining and updating the information accessed on a criminal background check.

② If you are unsure how to contact the state repository, please contact the state Office of the Attorney General for additional information.

FEDERAL PROHIBITORS

1. Currently under formal charges for a felony or convicted of a felony that can receive more than a one-year sentence or convicted of a misdemeanor that could receive more than a two-year sentence.

2. A fugitive from justice or subject of an active criminal warrant. This includes misdemeanor warrants.

3. Illegal drug possession, current use, or a conviction of controlled substance within the past year.

4. In a court proceeding, formally determined to be a mental defective, involuntarily committed to a mental institution or deemed incompetent to handle your own affairs. This includes final dispositions to criminal charges of "found not guilty by reason of insanity" or "found incompetent to stand trial."

5. An alien illegally/unlawfully in the United States or a non immigrant who does not qualify for the exceptions under Title 18, United States Code, Section 922(y) (i.e., not having a valid hunting license).

6. Dishonorable discharge from United States Armed Forces.

7. Renounced citizenship of the United States.

8. The subject of a protection order issued after a hearing of which the accused had the opportunity to participate. The protection order restrains the subject from harassing, stalking, or threatening an intimate partner or child of such partner.

9. Persons convicted of a misdemeanor crime of domestic violence including offenses that contain the element of use or attempted use of physical force or threatened use of a deadly weapon in which the victim was a spouse, former spouse, parent, guardian, a person with whom the victim shares a child in common, a person who is cohabitating with or has cohabitated with the victim.

How to contact the FBI

To obtain a copy of your FBI Record, you must send a written request to:

Federal Bureau of Investigation
Special Correspondence Unit
1000 Custer Hollow Road
Clarksburg, WV 26302

You are required to provide the following with your written request:

- An $18 money order payable to the U.S. Treasury;
- A ten-print fingerprint card bearing your fingerprints from a local law enforcement office;
- A return mailing address for your response; and
- A court-certified copy of a final disposition regarding your criminal offense in order to update your FBI Criminal History Record.

Updating this information may resolve your repeated delayed response into an immediate proceed.

The FBI Special Correspondence Unit can only provide information on criminal history records. They have no access to NICS background check records.

Locate information on the FBI Special Correspondence Unit at:

www.fbi.gov/hq/cjisd/fprequest.html

NICS Customer Service

1 (877) 444-NICS (6427)

Due to the Privacy Act of 1974 the FBI NICS Section cannot provide criminal history information over the telephone.

U.S. Department of Justice

Federal Bureau of Investigation

Criminal Justice Information Services Division

Gun Buyer's Resolution Guide

Explanation and Information on a Delayed NICS Response

What Prohibits an Individual From the Transfer or Possession of a Firearm?

A delay or deny message from the NICS indicates that either you or another individual with a similar name and/or similar descriptive features has been matched with one or more of the following federally prohibitive criteria:

- Persons convicted of/under indictment (or information) for a crime punishable by imprisonment for a term exceeding one year, whether or not sentence was imposed. This includes misdemeanor offenses with a potential term of imprisonment in excess of two years, whether or not sentence was imposed.
- Persons who are fugitives from justice (the subject of an active warrant).
- An unlawful user and/or an addict of any controlled substance.
- Persons adjudicated as a mental defective or involuntarily committed to a mental institution or incompetent to handle their own affairs.
- An alien illegally/unlawfully in the United States, with the exception of non-immigrants pursuant to Subsection (y)(2).
- Persons dishonorably discharged from the United States Armed Forces.
- A renouncer of United States citizenship.
- The subject of a protective order.
- Persons convicted of a misdemeanor crime of domestic violence.

NICS Section Information:

NICS Section Facsimile
1-888-550-6427

NICS Appeal Facsimile
1-304-625-0535

Telecommunications Device for the Deaf (TDD)
1-877-NICS-TTY

NICS Web Site
www.fbi.gov/hq/cjisd/nics/index.htm

NICS E-mail Address
a nics@leo.gov

NICS Appeals E-Mail Address
nicsappeals@leo.gov

NICS Customer Service
1-877-444-NICS (6427)

(Rev. October 2003)

U.S. Department of Justice
Federal Bureau of Investigation
Criminal Justice Information Services Division

NICS

Guide For Appealing A Firearm Transfer DENIAL

Your Rights and Responsibilities

To be provided by the FFL
NTN: ____________

If you have been **denied** by a Federal Firearms Licensee (FFL) from receiving a firearm/firearm permit because of a record in the FBI's National Instant Criminal Background Check System (NICS), you may submit a request to appeal your denial decision. The provisions for appeals are outlined by the NICS Regulations at 28 Code of Federal Regulations, Part 25.10 and Subsection 103 (f) and (g) and Section 104 of the Brady Handgun Violence Prevention Act (Brady Act) of 1993.

Requesting The Reason For Your Denial

You may request the reason for your denial by **writing** to:

Federal Bureau of Investigation
NICS Section
Appeal Services Team, Module A-1
Post Office Box 4278
Clarksburg, WV 26302-9922

- **You must include** your complete mailing address in the request. You may also submit your request by facsimile or by e-mail. (See reverse side for the NICS facsimile number and e-mail address.)

- **You must include** your NICS Transaction Number (NTN) with your written request. The NTN can be obtained from the FFL.

- The NICS Appeal Services Team (AST) **cannot** initiate an appeal for you at the request of another individual without your written *and* signed authorization.

- You may submit your fingerprints, which **must** be rolled by a law enforcement agency, with your initial written correspondence. *The submission of fingerprints with your initial written request may hasten the appeal process.*

- The NICS AST will respond to your written request by providing the reason for your denial within five business days **after** receiving your correspondence.

Appealing your Denial

The following information outlines the steps you must take to either challenge your record or make a claim that the record used as the basis for your **denial** pertains to someone other than you.

Questions of Identity

In cases involving criminal history records, if fingerprints are not submitted with your initial request, you may be required to submit your fingerprints to establish *positive proof* of your identity. If your fingerprints are required by the NICS Section and you wish to further the appeal process, you must have your fingerprints rolled by a local law enforcement agency. The law enforcement agency rolling your fingerprints **must stamp** its agency name, address and telephone number on the fingerprint card *and* the reason fingerprinted must be marked "For NICS Purposes."

Record Challenges

You may challenge the accuracy of the record used in the evaluation of your denial or declare that your rights to obtain a firearm have been restored, etc. If you have any additional information (e.g., court documentation) that may assist the NICS AST in correcting or updating the record, you should attach the information to your written correspondence. The NICS AST will evaluate your information and provide you with its decision on your appeal **in writing**.

- If the NICS AST is unable to resolve your appeal, **you** will be provided information to contact the agency that created the record. For correction of the record, you must follow procedures established by the state or federal agency that maintains the original record. The FBI, as custodian of arrest information that has been submitted voluntarily by local, state and federal law enforcement agencies, does not have the authority to change such records **unless** notified to do so by an authorized criminal justice agency.

You may submit *any* information to the originating agency that would assist with the correcting and/or updating of your record. *(This may also hasten the appeal process.)*

- If the **originating agency** corrects your record, the NICS AST must be notified and provided documentation indicating such. The NICS AST will verify and evaluate the information and provide you with its decision on your appeal **in writing**.

Appeal Inquiry

Any inquiry concerning your appeal should be directed to the NICS AST **in writing**. Due to the Privacy Act of 1974, the NICS Section cannot disseminate *specific* information to you via the telephone.

Successful Appeal

If your appeal is successful, you will be notified by the NICS AST that your denial has been overturned and that you are eligible to receive a firearm. You will be issued a letter, which **must** be presented to the FFL who initiated your background check.

INDEX

SUBJECT	LAW	REGULATIONS	RULINGS, PROCEDURES, INDUSTRY CIRCULARS	QUESTIONS AND ANSWERS
Auctioneer			Rul. 96-2	K
Auto sear			Rul. 81-4	
Bequest				
Acquiring firearm interstate by	922(a)(3)	478.29(a)		
Transfer of firearm interstate to carry out	922(a)(5)	478.30(a)		
Body armor				
Convicted violent felons, prohibition	931(a)			
Affirmative defenses	931(b)			
Bound book (see "Records")				
Brady law (National Instant Criminal Background Check System, "NICS")				G, J, P
Accessing NICS	922(t)(1)(A)	478.102		
For purpose other than NICS check		25.6(j)		
Through FBI		25.6(a), (b)		
Through State point of contact		25.6(d)		
Civil fine	922(t)(5)			
Hearing		478.74		
Notice of		478.73		
Correction of erroneous information	925A	25.10		
Voluntary appeal by individual		25.10(g)		
Definitions				
NICS		478.11; 25.2		
Other terms		25.2		
Enrollment with FBI				P
Establishment of NICS	922(t)(1)	25.3		
Exceptions				
Certain geographical areas	922(t)(3)(C)	478.102(d)(3); 478.150		
NFA transfer	922(t)(3)(B)	478.102(d)(2)		
Permit	922(t)(3)(A)	478.102(d)(1)		
Government documents as ID			Rul. 2001-5	
Identification				
Document	922(s)(3)(A)	478.11; 478.102(a)(3); 478.124(c)(3)(i)		
Transaction identification number	922(t)(2)(A); 922 (4)	478.102(b); 478.124(c)(3)(iv) 25.6(c)		
Verifying identity of transferee	922(t)(1)(C)	478.102(a)(3); 478.124(c)(3)(i)		
Liability in civil action	922(t)(6)			
Loan or rental of firearm		478.97		
Out-of-State, mail-order sales		478.96(b)		
Pawnbroker				J
Optional NICS check before pawn				J
Penalties				
Prohibited activities		25.11		
Violation of Brady Law	924(a)(5)			
Prohibited transferees	922(g); 922(n); 922(t)			
Records				
Audit Log				
Audit Log contents		25.9 (b)(1)		
Use of information		25.9 (b)(1)(iii)		
Limitation on use		25.9 (b)(3)		
Destruction	922(t)(2)(C)	25.9		
NICS checks where transfer not made		478.129(b)		
Response to NICS check		478.124(c)(3)(iv)		
Retention periods		478.102(e); 478.129(b); 478.131		
Transaction identification number	922(t)(4)	478.124(c)(3)(iv)		
Transactions not subject to NICS		478.131		
Transfer of firearm		478.124(c)		
Remedy for erroneous denial	925A	25.10		
Requirement to contact NICS	922(t)(1)	478.102(a)		
Exceptions	922(t)(3)	478.102(d)		
Responses to NICS checks				
"Delayed"		25.6(c)(1)(iv)(B)		
"Denied"		25.6(c)(1)(iv)(C)		
"Proceed"		25.6(c)(1)(iv)(A)		
Revocation of license	922(t)(5)			
Hearing		478.74		
Notice of		478.73		

SUBJECT L	AW	REGULATIONS	RULINGS, PROCEDURES, INDUSTRY CIRCULARS	QUESTIONS AND ANSWERS
Suspension of license	922(t)(5)			
Hearing		478.74		
Notice of		478.73		
System safeguards		25.8		
Time limitation on NICS check		478.102(c)		
Transfer of firearm	922(t)(1)	478.102(a)-(c); 25.6(a)		
Validation of records in system		25.5		
Business				
Changes				
Bankruptcy		479.43		
Death of owner		479.42		
In corporation		479.45		
In partnership		479.44		
New license required for change of address		479.52		
Notification of change of control		478.54		
Notification of new trade name		478.53		
Continuance by certain successors		478.56		
Continuing operations after indictment, conviction	925(b); 925(C)	478.143		
Delivery to business with expired license		478.94		
Discontinuance of	923(g)(4)	478.57; 478.127; 479.105(f)		
Hours				
Gunsmith			Rul. 73-13	
Inspection during	923(g)(1)(A)	478.23		
License to engage in	922(a)(1); 923(a)	478.41(a)		
Premises				
Defined		478.11		
Dwelling as business premises		478.11		
Separate license for each location		478.41(b)		
Business premises (see "Business")				
Canal Zone				
Exclusion from "Interstate or foreign commerce"	921(a)(2)	478.11		
Certified copy of license	926(a)	478.94; 478.95		F
Certified list of licenses		478.94		F
Collector's item				
Obtaining determination of Director	5845(a)	479.25		
Collector of curios or relics				G
Application for license	923(b)	478.41(c); 478.44(b)		
Authorized operations	923(c)	478.41(d); 478.93		
Curios or relics defined		478.11		
Discontinuance of collecting activity				G
Fee for license	923(b)	478.42(d)		
Inspection	923(g)(1)(C)	478.23(a); 478.23(c)		
License denial	923(f)	478.71		
License revocation	923(e)	478.73		
Licensing standards	923(d)(1)	478.47(b)		
Location of firearm acquisition/disposition		478.50(b)		
Notice to handgun purchaser		478.103(a)-(c)		
Obtaining curio or relic determination		478.26		
Records (see "Records")				
Reports (see "Reports")				
Combination of parts				
Machinegun	921(a)(23); 5845(b)	478.11; 479.11		
Commerce				
Defined		478.11		
Committed to a mental institution				B
Defined		478.11		
Possession, receipt of firearm/ammunition by committed person	922(g)(4)	478.32(a)(4)		
Sale of firearm, ammunition to such person	922(d)(4)	478.32(d)(4)		
Common or contract carrier				
Delivery of firearm without receipt	922(f)(2)	478.31(d)		
Interstate transportation of firearm/ammunition	922(f)(1)	478.31(c)		
Labeling firearm container	922(e)	478.31(b)		
Notification that package contains firearm	922(e)	478.31(a)		

SUBJECT	LAW	REGULATIONS	RULINGS, PROCEDURES, INDUSTRY CIRCULARS	QUESTIONS AND ANSWERS
Concealed Firearms				
Firearms - exceptions	926B(e); 926C(e)			
Identification required to carry	926B(d); 926C(d)			
Qualified law enforcement officer	926B(a), (c)			
Qualified retired law enforcement officer	926C(a), (c)			
Continuance of business (see "Business")				
Crime punishable by imprisonment for term exceeding 1 year				B
Continuing licensed operations after conviction	925(b); 925(c)	478.143		
Defined	921(a)(20)	478.11		
Pardon, expunction	921(a)(20); 921(33)	478.142		
Possession, receipt by convicted person	922(g)(1)	478.32(a)(1)		
Sale to convicted person	922(d)(1)	478.32(d)(1)		
Curio or relic firearm (see also "Collector of curios or relics")				
Defined		478.11		
Importation	925(e)	478.118		
From proscribed countries		447.52		
Surplus military	22 USC 2778(b)		Rul. 85-10	
U.S. military defense articles		447.57		
Obtaining curio or relic determination		478.26		
Off premises sale	923(j)	478.50(d); 478.100(a)(2) 478.100(c)		
Dealer in firearms (see also "Brady law", "License", "NFA controls")				A, C, F, I, J
Application for license	923(a)			
Revocation, suspension, civil fine	922(t)(5)	478.73(a)		
Consultant, expert			Rul. 73-19	
Defined	921(a)(11); 5845(k)	478.11; 479.11		
Engaged in business as dealer				
Defined	921(a)(21)(C)	478.11		
Gunsmith	921(a)(21)(D)	478.11	Rul. 73-13; Rul. 77-1	I
License denial	923(d)(2); 923(f)(1); 923(f)(2)	478.71; 478.72		
License requirement	922(a)(1); 923(a)	478.41(a)		
License revocation	923(e)	478.73		
License suspension, civil fine	922(t)(5)	478.73(a)		
Licensing standards	923(d)	478.47(b)		
Notice to handgun purchaser		478.103(a)-(c)		
Pawnbroker	921(a)(11); 921(12)	478.11	Rul. 76-15	J
Posting sign on juvenile handgun possession		478.103(d)-(f)		
Records (see "Records")				
Registration under NFA	5802	479.34		
Reports (see "Reports")				
Sale (see "Sale or delivery by licensee")				
Defense article				
Importation	22 U.S.C. 2778	Part 447		
U.S. Munitions Import List		447.21		
Definitions of terms	921; 5845	447.11; 478.11; 479.11		
Destructive device (see also "NFA controls")				
Defined	921(a)(4); 5845(f)	478.11; 479.11		
Gas/flare gun			Rul. 95-3	
Striker-12/Streetsweeper shotgun			Rul. 94-2; Rul. 2001-1	
Removal from definition	5845(f)	479.24		
Sale to nonlicensee	922(b)(4)	478.98		
Sale to research organization		478.145		
Transportation	922(a)(4)	478.28		
USAS-12 shotgun			Rul. 94-1; Rul. 2001-1	
Discharged under dishonorable conditions				B
Defined		478.11		
Possession, receipt of firearm/ammunition by discharged person	922(g)(6)	478.32(a)(6)		
Sale of firearm, ammunition to discharged person	922(d)(6)	478.32(d)(6)		
District of Columbia				
State	921(a)(2)	447.11; 478.11; 479.11		

SUBJECT	LAW	REGULATIONS	RULINGS, PROCEDURES, INDUSTRY CIRCULARS	QUESTIONS AND ANSWERS
Drug user, addict (see "Unlawful drug user, addict")				
Dwelling as business premises		478.11		
Engaged in the business				
Defined				
Dealer in firearms	921(a)(21)(C)	478.11		
Gunsmith	921(a)(21)(D)	478.11		
Importer of firearms/ammunition	921(a)(21)(E); 921(a)(21)(F)	478.11		
Manufacturer of firearms/ammunition	921(a)(21)(A); 921(a)(21)(B)	478.11		
Emergency variations from requirements		478.22(b); 479.26(b)		
Exportation (see also "NFA controls")				L
Application to export NFA firearm, exemption	5854	479.114		
Proof of	5854	479.118		
Tax, exemptions, records	5843	478.171; 479.131		
Firearm				
Defined				
Antique	921(a)(16); 5845(g)	478.11; 479.11		
Arms Export Control Act		447.11		
GCA	921(a)(3)	478.11		
NFA	5845(a)	479.11		
Identification of firearms		479.102		
Information to be marked		479.102(a)(2)		
Depth of markings		479.102(b)		
Destructive devices	921(a); 5845(f)	479.102(d)		
Machine guns, silencers, mufflers	921(a)(23); 921(a)(24); 5845(b)	479.102 (f)		
Manner of markings		479.102(a)(1), (2)	Rul. 2002-6	
Other means of identification		479.102(f)(2)		
Muzzle loading firearm	921(a)(16)			
National Firearm Registry		479.101(a)		
Taser			Rul. 76-6; Rul. 80-20	
Tear gas weapons			Rul. 75-7	
Transfer to nonimmigrant aliens			Rul. 2004-1	
Undetectable firearm	922(p)			
False statement, records				
Arms Export Control Act	2778(c)	447.62		
GCA	922(a)(6); 924(a)(1)(A); 924(a)(3)	478.128		
NFA	5861(l)	479.181		
Frame or receiver				
Firearm	921(a)(3)(B)	478.11		
Machinegun	5845(b)	479.11		
Fugitive from justice				B
Defined		478.11		
Possession, receipt of firearm/ammunition by	922(g)(2)	478.32(a)(2)		
Sale of firearm/ammunition to	922(d)(2)	478.32(d)(2)		
Gun free school zone				
Defined	921(a)(25); 921(a)(26)			
Discharging firearm in	922(q)(3)(A)			
Exceptions to prohibitions	922(q)(2)(B); 922(q)(3)(B)			
Possession of firearm in	922(q)(2)(A)			
Gun show				F
Inspection	923(j)	478.23		
Defined	923(j)	478.100(b)		
Posting license at		478.100(a)		
Posting sign on juvenile handgun possession		478.103(d)-(f)		
Record of firearm sale at	923(j)	478.100(c)		
Sale of firearm at show in licensee's State	923(j)	478.100(a)		
Sale of firearm at show outside licensee's State			Rul. 69-59	
Gunsmith (see "Dealer in firearms", "License")				

SUBJECT	LAW	REGULATIONS	RULINGS, PROCEDURES, INDUSTRY CIRCULARS	QUESTIONS AND ANSWERS
Handgun				
Defined	921(a)(29)	478.11		
Notice to handgun purchaser		478.103(a)-(c)		
Possession by juvenile	922(x)(2)			
Posting sign on juvenile handgun possession		478.103(d)-(f)		
Sale to juvenile	922(x)(1)			
Sale to underage person	922(b)(1)	478.99(b)		
Hearing				
After civil fine		478.74		
After license application denial		478.72		
After notice of revocation		478.74		
After notice of suspension		478.74		
Notice of		478.73		
Place of		478.77		
Service of notices		478.75		
Identification (see also "Brady law")				D, F, O
Adoption of foreign serial number			Rul. 75-28	
Alternate means of		478.92(a)(4)(i); 478.92(c)(3)(ii); 479.102(c)		
Armor piercing ammunition	923(k)	478.92(b)(1)		
Document				
Defined	922(s)(3)(A)	478.11		
Employer identification number		479.35	Proc. 90-1	
Firearm	923(i)	478.92; 479.102		
Destructive device	923(i); 5842(a); 5842(c)	478.92(a)(4)(ii); 479.102(d)		
Imported	923(i); 5842(a); 5842(c)	478.92(a)(1); 479.102(a)(1)		
Machineguns, silencers, parts	923(i); 5842(a)	478.92(a)(3)(iii); 479.102		
Manner of markings		478.92(a)		
Manufactured	923(i); 5842(a); 5842(c)	478.92(a); 479.102		
Measurements		478.92(a)(5)		
Personal firearm on licensed premises			IC 72-30	
Semiautomatic assault weapon	923(i)	478.92(a)(2)		
Identifying firearm purchaser		478.124(c)	Rul. 79-7	
Large capacity ammunition feeding device	923(i)	478.92(c)		
Links for belted ammunition	923(i)	478.92(c)(3)(i)		
Verifying identity of licensee	926(a)	478.94-478.95	IC 74-13	
Importation				L
Alternate method for temporary importation			Rul. 2004-2	
Ammunition feeding device		478.119		
Approved application as permit		447.42(a)(2)(i)		
Barrel, receiver for prohibited firearm	925(d)(3)	478.111		
Brokering activities		447.2 (d)		
By licensed dealer		478.113		
By licensed importer		478.112		
By licensed manufacturer		478.113		
By nonlicensee	922(a)(3)	478.111; 478.113a; 478.119		
By nonresident		478.115(d)(1)	Rul. 81-3	
By U.S. Armed Forces member	925(a)(4)	478.114	Rul. 69-309 (revoked); Rul. 74-13	
Conditional importation	925(d); 5844	478.116; 478.119(g); 479.113		
Curio or relic firearm	2778(b); 925(e)	447.52(d); 447.52(e); 447.57(b); 478.118	Rul. 85-2; Rul. 85-10	
Defense article	2778	Part 447		
Defense article from prohibited country		447.52		
Defined		447.11; 478.11; 479.11		
Delegation of authority to ATF officers		447.58		
Exempt importation		478.115		
Export license for Category I and II		447.45(a)		
Firearm, ammunition taken out of U.S.	925(d)(4)	478.115(a); 479.111(c)		
For government agency	925(a)(1); 5844(1)	447.53(a)(1); 478.115(b); 479.111(a)(1)	Rul. 80-8	
For scientific, research purpose	925(d)(1); 5844(2)	479.111(a)(2)		
For testing, use as model	5844(3)	479.111(a)(3)		
Gas masks				L

SUBJECT	LAW	REGULATIONS	RULINGS, PPROCEDURES, INDUSTRY CIRCULARS	QUESTIONS AND ANSWERS
Identification				
Firearm	923(i)	478.92(a); 479.102		
Large capacity ammunition feeding device	923(i)	478.92(c)		
Form 6		478.112(b)		
Need for export license		478.112(c)		
Licensee other than importer		478.113(b)		
Semiautomatic assault weapon	923(i)	478.92(a)(2)		
Large capacity ammunition feeding device	922(w)	478.40a(c); 478.119		
Machinegun		479.105(c)		
Modified semiautomatic assault rifle	925(d)(3)			
NFA firearm	925(d)(2); 925(d)(3); 5844	479.111(a)		
Nonimmigrant aliens				
Documentation		478.120(a)		
Waiver		478.120(b)		
Nonsporting firearm/ammunition	925(d)(3)	478.111(a)	Rul. 80-8	
Permit		447.41-.46; 478.111-.119; 479.111		
Denial, revocation, suspension		447.44		
Sales sample of NFA firearm	5844(3)	479.111(a)(3)	Rul. 85-2	
Semiautomatic assault rifle	925(d)(3)			
Surplus military firearm	925(d)(3)	478.111	Rul. 80-8; Rul. 85-10	
Surplus military curios or relics			Rul. 2001-3	
Unserviceable firearm	925(d)(2)			
U.S. military defense articles	2778(b)	447.57		
U.S. Munitions Import List		447.21		
War souvenir	925(a)(4)	478.114(c)		
Imported parts (see also "Assembly")				
Assembly of semiautomatic rifle, shotgun	922(r)	478.39		
Defined		478.39(c)		
Importer of firearms/ammunition (see also "Brady law", "License", "NFA controls")				A, C
Application for license	923(a)	478.41(b); 478.44(a)		
Civil fine	922(t)(5)	478.73(a)		
Defined	921(a)(9)	478.11; 479.11		
Engaged in business as importer				
Defined	921(a)(21)(E); 921(a)(21)(F)	478.11; 479.11		
License denial	923(d)(2); 923(f)(1); 923(f)(2)	478.71; 478.72		
License requirement	922(a)(1); 923(a)	478.41(a)		
License revocation	923(e)	478.73		
License suspension	922(t)(5)	478.73(a)		
Licensing standards	923(d)(1)	478.47(b)		
Notice to handgun purchaser		478.103(a)-(c)		
Posting of sign on juvenile handgun possession		478.103(d)-(f)		
Records (see "Records")				
Registration				
Arms Export Control Act	22 U.S.C. 2778(b)	447.31		
NFA	5802	479.34		
Reports (see "Reports")				
Sales (see "Sale or delivery by licensee")				
Indictment				
Continuing licensed operations after	925(b); 925(c)			
Defined		478.11		
Receipt of firearm, ammunition by person under	922(n)	478.32(b)		
Sale of firearm, ammunition to person under	922(d)(1)	478.32(d)(1)		
Inspection		478.23		
Gun show	923(j)			
Importers under Arms Export Control Act		447.34(b)		
Importers, manufacturers, dealers	923(g)(1)(A); 923(g)(1)(B)	478.23; 478.121(b)		
Collectors	923(g)(1)(C); 923(g)(1)(D)	478.23; 478.121(b)		
Special (occupational) taxpayers under NFA		479.22		
Insular possessions				
State	921(a)(2)			
Interstate or foreign commerce				
Defined	921(a)(2)	478.11		

SUBJECT	LAW	REGULATIONS	RULINGS, PROCEDURES, INDUSTRY CIRCULARS	QUESTIONS AND ANSWERS
Interstate transactions				
Licensees				F
Loan, rental of firearm for sporting purposes	922(b)(3)(B)	478.97; 478.99(a)(2)		
Return, replacement of firearm	922(a)(2)(A)	478.147		
Sale of curio or relic	923(j)	478.100(a)(2)		
Sale of firearm to nonresident	922(b)(3)	478.99(a)		
Sale of rifle or shotgun	922(b)(3)(A)	478.99(a)(1)		
Transfer of firearm to writer, consultant			IC 72-23	
Transfer of firearm to agent, employee			IC 72-23	
Nonlicensees				B
Acquisition of firearm generally	922(a)(3)	478.29		
Acquisition of rifle/shotgun from licensee	922(a)(3)(B)	478.29(b)		
Interstate shipment of firearm to licensee		478.147		
Loan, rental of firearm	922(a)(5)(B)	478.30(b)		
Transfer of firearm to nonlicensee generally	922(a)(5); 922(b)(3)	478.30		
Prohibited persons	922(g)	478.32(a)-(c)		
Transfer, receipt of firearm to carry out				
Bequest	922(a)(3)(A); 922(a)(5)(A)	478.29(a); 478.30(a)		
Intestate succession	922(a)(3)(A); 922(a)(5)(A)	478.29(a); 478.30(a)		
Transportation	922(a)(4); 926A	478.28		
Intestate succession				
Acquiring firearm interstate by	922(a)(3)(A)	478.29(a)		
Transfer of firearm interstate to carry out	922(a)(5)(A)	478.30(a)		
Juvenile				F
Defined	922(x)(5)			
Exceptions to prohibitions	922(x)(3)			
Notice to handgun purchaser		478.103(a)-(c)		
Possession of handgun, ammunition by	922(x)(2)			
Posting sign on juvenile handgun possession		478.103(d)-(f)		
Sale of handgun, ammunition to	922(x)(1)			
Large capacity ammunition feeding device				O
Defined	921(a)(31)	478.11		
Discontinuance of business		478.57(c)		
Identification	923(i)	478.92(c)		
Alternate means of		478.92(c)(3)(ii)		
Metallic links for belted ammunition		478.92(c)(3)(i)		
Importation		478.119		
Possession, transfer prohibited	922(w)(1)	478.40a(a)		
Exceptions	922(w)(2), (3)	478.40a(b), (c)		
Sale to law enforcement officer	922(w)(3)(A)	478.40a(b)(2); 478.132		
License				C, G, H, I, K
Abandoned application		478.59		
Application	923(a); 923(b)	478.41(b); 478.41(c)		
Certified copy	926(a)(2)	478.94; 478.95		
Certified list of licenses		478.94		
Correction of error		478.48		
Delivery to business with expired license		478.94		
Denial of application	923(d)(2); 923(f)(1); 923(f)(2)	478.71		
Duration		478.49		
Fees	923(a); 923(b)	478.42		
Refund		478.43		
Hearing (see "Hearing")				
Identifying licensees to local authorities	923(l)			
Location covered	923(a)	478.50		
Multiple locations			Rul. 73-9	
Not transferable		478.51		
New business or previously unlicensed		478.44(a)(1)		
Nonimmigrant alien as collector		478.44(b)		
Posting	923(h)	478.91		
Gun show	923(h)	478.100(a)		
Renewal				
Generally	923(a)	478.45	Rul. 75-27	
Nonimmigrant alien		478.45		
Requirement	922(a)(1); 923(a)	478.41(a)		
Revocation	923(e)	478.73(a)		
Suspension	922(t)(5)	478.73(a)		

SUBJECT L	AW	REGULATIONS	RULINGS, PROCEDURES, INDUSTRY CIRCULARS	QUESTIONS AND ANSWERS
Standards				
Business premises	923(d)(1)(E)	478.47(b)(5)(i)		
Compliance with local law	923(d)(1)(F)	478.47(b)(6)		
Minimum age	923(d)(1)(A)	478.47(b)(1)		
Notification of local authorities	923(d)(1)(F)(iii)	478.47(b)(6)(iv)		
Prohibited persons	923(d)(1)(B)	478.47(b)(2)		
Secure gun storage, safety device	923(d)(1)(G)			
Loan or rental of firearm	922(a)(5)(B); 922(b)(3)(B)	478.30(b); 478.97; 478.99(a)(2)		
Machinegun (see also "NFA controls")				N
Anti-aircraft 7.62mm			Rul. 2004-5	
Auto sear			Rul. 81-4	
Defined	921(a)(23); 5845(b)	478.11; 479.11		
Interstate transport of	922(a)(4)	478.28		
KG-9 pistol			Rul. 82-2	
Possession, transfer prohibited	922(o)	478.36		
Exceptions	922(o)(2)	478.36(a); 478.36(b); 479.105		
SAC carbine			Rul. 82-8	
Sale to nonlicensee	922(b)(4)	478.98		
Sale to research organization		478.145		
SM10, SM11A1 pistol			Rul. 82-8	
YAC STEN MK II carbine			Rul. 83-5	
Transfer as sales samples			Rul. 2002-5	
Mailing concealable firearms	18 U.S.C. 1715			B, F
Manufacturer of firearms/ammunition (see also "Brady law", "License", "NFA controls")				A, C, H
Application for license	923(a)	478.41(b); 478.44(a)		
Civil fine	922(t)(5)	478.73(a)		
Defined	921(a)(10)	478.11; 479.11		
Engaged in business as a manufacturer				
Defined	921(a)(21)(A); 921(a)(21)(B)	478.11		
License denial	923(d)(2); 923(f)(1); 923(f)(2)	478.71; 478.72		
License requirement	922(a)(1); 923(a)	478.41(a)		
License revocation	923(e)	478.73		
License suspension	922(t)(5)	478.73(a)		
Licensing standards	923(d)	478.47(b)		
Notice to handgun purchaser		478.103(a)-(c)		
Posting sign on juvenile handgun possession		478.103(d)-(f)		
Records (see "Records")				
Registration				
Arms Export Control Act	22 U.S.C. 2778(b)	447.31		
NFA	5802	479.34		
Reports (see "Reports")				
Sales (see "Sale or delivery by licensee")				
Measuring certain NFA firearms		479.11		
Misdemeanor crime of domestic violence				B, Q
Certification for law enforcement sale		478.134(a)		
Defined	921(a)(33)	478.11		
Possession, receipt by convicted person	922(g)(9)	478.32(a)(9)		
Sale of firearm/ammunition to convicted person	922(d)(9)	478.32(d)(9); 478.99(c)(9)		
Multiple handgun sales report	923(g)(3)(A)	478.126(a)		F
Muzzle loading firearm	921(a)(16)(C)			
National Instant Criminal Background Check System (see "Brady law")				
NFA controls				M, N
Alternate methods, procedures		479.26(a)		
Antique firearm				
Defined	5845(g)	479.11		
Change of business ownership		479.42-479.45		
Collector's item				
Determination by Director	5845(a)	479.25		
Dealer				
Defined	5845(k)	479.11		
Registration	5802	479.34		
Special (occupational) tax	5801(a)	479.31		
Destructive device determination	5845(f)	479.24		
Discontinuance of NFA business		479.105(f)		
Emergency variations		479.26(b), (c)		

SUBJECT	LAW	REGULATIONS	RULINGS, PROCEDURES, INDUSTRY CIRCULARS	QUESTIONS AND ANSWERS
Employer identification number		479.35	Proc. 90-1	
Exemptions				
Business with U.S.	5851(a)	479.33		
Exportation	5854	479.114		
Making by qualified manufacturer	5852(c)	479.68		
Making for government entities	5852(b); 5853(b)	479.69; 479.70		
Transfer between special (occupational) taxpayers	5852(d)	479.88		
Transfer to State	5853(a)	479.90(a)		
Transfer to U.S.	5852(a)	479.89		
Unserviceable firearm	5852(e)	479.91		
Exportation				
Application	5812	479.114		
Defined		479.11		
Exemption from tax	5854	479.114		
Exporter defined		479.11		
Permit		479.114		
Proof of		479.118		
Refund of transfer tax		479.120		
Transfer to insular possession		479.121		
Transportation to effect		479.119		
Firearm				
Antique				
Defined	5845(g)	479.11		
Excluded from definition	5845(a)	479.11		
Any other weapon				
Defined	5845(e)			
Taser			Rul. 76-6	
Auto sear			Rul. 81-4	
Defined	5845	479.11		
Destructive device				
Defined	5845(f)	479.11		
Gas/flare gun			Rul. 95-3	
Removal from definition	5845(f)	479.24		
Striker-12/Streetsweeper shotgun			Rul. 94-2	
USAS-12 shotgun			Rul. 94-1	
Machinegun				N
Auto sear			Rul. 81-4	
Combination of parts for	5845(b)	479.11		
Dealer sales sample	5844(3)	479.105(d)		
Defined	5845(b)	479.11		
Frame or receiver	5845(b)	479.11		
Importation	5844	479.105(c)		
KG-9 pistol			Rul. 82-2	
Making after May 18, 1986		479.105(e)		
Manufacture		479.105(c)		
Part to convert weapon into	5845(b)	479.11		
Possessed prior to May 19, 1986		479.105(b)		
Possession, transfer prohibited		479.105(a)		
SAC carbine			Rul. 82-8	
SM10, SM11A1 pistols			Rul. 82-8	
YAC STEN MK11 carbine			Rul. 83-5	
Removal from definition	5845(a); 5845(f)	479.24; 479.25		
Rifle	5845(a)(3); 5845(c)	479.11		
Shotgun	5845(a)(1); 5845(d)	479.11		
Silencer (Muffler)	5845(a)(7)			
Unserviceable firearm	5845(h)	479.11; 479.91		
Weapon made from shotgun, rifle	5845(a)(2); 5845(a)(4)			
Identification				
Alternate means of identifying firearm		479.102		
Destructive device	5842(c)	479.102		
Employer identification number		479.35	Proc. 90-1	
Firearm	5842	479.102		
Firearm without serial number	5842(b)			
Maker of firearm	5822(d)	479.63		
Transferee of firearm	5812(a)(3)	479.85		
Importation				
Conditional	5844	479.113		
Criteria for	5844	479.111		
Dealer sales sample	5844(3)	479.105(d); 479.111(a)(3)	Rul. 85-2	
Defined		479.11		

SUBJECT	LAW	REGULATIONS	RULINGS, PROCEDURES, INDUSTRY CIRCULARS	QUESTIONS AND ANSWERS
For scientific, research purpose	5844(2)	479.111(a)(2)		
For testing, use as model	5844(3)	479.111(a)(3)		
For government	5844(1)	479.111(a)(1)		
Machinegun	5844	479.105(c), (d)		
Permit		479.111(a)		
Records	5843	479.131		
Registration of firearm	5841(b)	479.112		
Return of firearm to U.S.		479.111(c)		
Importer				
Defined	5845(l)	479.11		
Identification of firearm	5842(a)	479.102		
Registration	5802	479.34		
Registration of firearm	5841(b)	479.101; 479.112		
Special (occupational) tax	5801(a)	479.31; 479.32		
Small business	5801(b)	479.32(a)		
Inspection		479.22; 479.131		
Making firearm				
Application	5822	479.62		
Defined	5845(i)	479.11		
Identification of firearm	5842(a)	479.102		
Identification of maker	5822(d)	479.63		
Machinegun		479.105(a), (e)		
Proof of registration of firearm	5841(e)	479.71		
Registration of firearm	5822; 5841(b)	479.62		
State law		479.65; 479.101(b)		
Tax	5821	479.61		
Manufacturer				
Defined	5845(m)	479.11		
Identification of firearm	5842(a); 5842(c)	479.102		
Registration	5802	479.34		
Registration of firearm	5841(b)	479.103		
Special (occupational) tax	5801(a)	479.31- 479.32		
Small business	5801(b)	479.32(a)		
Measuring certain NFA firearms		479.11		
Penalties and interest	5871	479.181		
Delinquency		479.50		
Failure to pay special (occupational) tax		479.48		
Failure to register change, removal		479.49		
Fraudulent return		479.51		
Person				
Defined		479.11		
Pistol		479.11		
Records and returns				
Failure to make		479.151		
Penalties		479.152		
Records required	5843	479.131		
Refund of tax		479.172		
Registration				
By government agency for official use		479.104	Rul. 74-8	
Firearm				
Importation	5841(b); 5841(c)	479.101(b); 479.112		
Making	5822; 5841(b); 5841(c)	479.62; 479.101(b)		
Manufacture	5841(b); 5841(c)	479.101(b); 479.103		
Transfer	5812; 5841(c)	479.84; 479.101(b)		
Proof of	5841(e)	479.71; 479.101(e)		
Special (occupational) taxpayer	5802			
Repair of firearm			Rul. 77-1	
Reports				
Stolen, lost registration document		479.142		
Stolen, lost firearm		479.141		
Revolver		479.11		
Rifle	5845(c)	479.11		
Seizure and forfeiture	5872(a)	479.182		
Shotgun	5845(d)	479.11		
Special (occupational) tax/taxpayer				
Business at more than 1 location		479.34(c); 479.38		
Business, more than 1 at same location		479.39		
Certificate in lieu of stamp lost/destroyed		479.37		
Change of business location		479.38		
Change of ownership		479.42-479.45		

SUBJECT	LAW	REGULATIONS	RULINGS, PROCEDURES, INDUSTRY CIRCULARS	QUESTIONS AND ANSWERS
Change of trade name		479.47		
Employer identification number		479.35	Proc. 90-1	
Liability for tax	5801	479.31		
Partnership liability		479.40		
Penalties (see "Penalties, criminal")				
Rate of tax	5801(a)	479.32		
Small business	5801(b)	479.32a		
Registration, return, payment	5802	479.34		
Single sale of firearm		479.41		
State law	5812(a); 5822	479.52; 479.65		
Stolen, lost firearm/document		479.141; 479.142		
Tax				
Making firearm	5821	479.61		
Special (occupational) tax	5801(a)	479.31; 479.32		
Small business	5801(b)	479.32a		
Stamps		479.161-.163		
Transfer of firearm	5811	479.82		
Transfer				
Application	5812(a)	479.84		
Defined	5845(j)	479.11		
Identification of transferee	5812(a)	479.85		
Insular possessions		479.121		
Machinegun		479.105		
Registration	5812(a)	479.84		
State law	5812(a)	479.86		
Tax	5811	479.82		
To special taxpayer not engaged in business			Rul. 76-22	
Unserviceable firearm	5852(e)	479.91		
Panama Canal Zone				
State	921(a)(2)	478.11		
Pardon, expunction	921(a)(20); 921(33)(B)(ii)	478.142		
Pawnbroker (see "Dealer in firearms," "License")				J
Penalties, criminal	22 U.S.C. 2778(c); 924; 5871			
Person				
Defined	921(a)(1)	447.11; 478.11; 479.11		
Pistol				
Any other weapon	5845(e)	479.11		
Defined		447.11; 478.11; 479.11		
Multiple handgun sales reports	923(g)(3)	478.126(a)		
Possession by juvenile	922(x)(2)			
Sale to juvenile	922(x)(1)			
Sale to underage person	922(b)(1)	478.99(b)		
Puerto Rico				
State	921(a)(2)	478.11		
Records				C, D, E, F
Alternate		478.122(c); 478.123(c); 478.125(h)		
Armor piercing ammunition	922(b)(5)	478.121(c); 478.122(b); 478.125(a)-(d)		
Bound Book				
Collector's receipt/sale of firearms	923(g)(2)	478.125(f)		
Collector's sale of armor piercing ammunition	922(b)(5)	478.125(a)		
Dealer's receipt/sale of firearms	923(g)(1)(A)	478.125(e)		
Importer's firearms sales to nonlicensees	923(g)(1)(A)	478.122(d)		
Manufacturer's firearms sales to nonlicensees	923(g)(1)(A)	478.123(d)		
Sales of personal firearms	923(c)	478.125(a)		
Certification for law enforcement sales		478.132; 478.134		
Collector of curios or relics	923(g)(2)	478.121(c); 478.125(a), (b), (f)		
Discontinuance of collecting activity				G
Commercial		478.125(d), (g)		
Dealer in firearms	923(g)(1)(A); 5843	478.124; 478.125(e); 479.131		
Exportation	923(g)(1)(A); 5843	478.171		
Firearms receipt and disposition	923(g)(1)(A); 5843	478.121-.125(a); 479.131		
Firearms transaction record, Form 4473	923(g)(1)(A); 5843	478.124		D
Firearms transaction record, Form 4473(LV)	923(g)(1)(A); 5843	478.124(a), (b), (c)		
Gun show transactions	923(j)	478.100(c)		
Importer of defense articles		447.34		
Importer of firearms/ammunition	923(g)(1)(A); 5843	478.122		
Large capacity ammunition feeding device		478.40a(c); 478.132		

SUBJECT	LAW	REGULATIONS	RULINGS, PROCEDURES, INDUSTRY CIRCULARS	QUESTIONS AND ANSWERS
Manufacturer of firearms/ammunition	923(g)(1)(A); 5843	478.123		
Multiple handgun sales reports	923(g)(3)	478.126(a)		
NFA records	5843	479.131		
Off premises sale of curio or relic	923(j)	478.100(c)		
Out-of-business records	923(g)(4)	478.127		
Nonimmigrants		478.124(c)(3)(iii)		
Personal firearms	923(c)	478.125(a)		
POC determination messages		25.6		
Private transactions				B
Retention periods				
Collector		478.125(a); 478.129(a), (e)		
Dealer		478.125(c); 478.129(a), (b), (e)		
Form 4473, 4473(LV)		478.129(b)		
Importation		478.129(a), (d)		
Importer of defense articles		447.34(b)		
Manufacture		478.129(a), (d)		
Multiple handgun sales reports		478.129(c)		
Semiautomatic assault weapon		478.129(f)		
Statement of intent to obtain handgun		478.129(c)		
Theft, loss report		478.129(c)		
Redemption of pawned firearm			Rul. 76-15	
Repair of firearm			Rul. 76-25; Rul. 77-1	
Replacement of firearm			Rul. 74-20; Rul. 76-25	
Salvaged firearm			Rul. 76-25	
Semiautomatic assault weapon		478.40(c); 478.133		
Theft, loss report		478.39(a)		
Relief from Disabilities	925(c)	478.143; 478.144		A
Renunciation of citizenship				B
Defined		478.11		
Possession, receipt of firearm/ammunition by renunciate	922(g)(7)	478.32(a)(7)		
Sale of firearm/ammunition to renunciate	922(d)(7)	478.32(d)(7)		
Reports				F
Multiple handgun sales	923(g)(3)	478.126(a)		
On demand	923(g)(5)	478.126		
Out-of-business records	923(g)(4)	478.127		
Theft, loss of firearm	923(g)(6)	478.39(a); 479.141		
Theft, loss of NFA registration document		479.142		
Response to trace request	923(g)(7)	478.25(a)		
Restraining order				B
Defined	922(d)(8); 922(g)(8)	478.32(a)(8)		
Possession, receipt of firearm/ammunition by person under order	922(g)(8)	478.32(a)(8)		
Sale of firearm/ammunition to person under order	922(d)(8)	478.32(d)(8)		
Revolver				
Any other weapon	5845(e)	479.11		
Defined		447.11; 478.11; 479.11		
Multiple handgun sales report	923(g)(3)	478.126(a)		
Possession by juvenile	922(x)(2)			
Sale to juvenile	922(x)(1)			
Sale to underage person	922(b)(1)	478.99(b)		
Rifle				
Defined	921(a); 5845(c)	447.11; 478.11; 479.11		
Sale to underage person	922(b)(1)	478.99(b)		
Short-barreled rifle (see also "NFA controls")				
Defined	921(a)(8)	478.11		
Interstate transport of	922(a)(4)	478.28		
Sale to nonlicensee	922(b)(4)	478.98		
Sale to research organization		478.145		
Transport of short-barreled rifle	922(a)(4)	478.28		
Weapon made from	5845(a)(4)			
Safety device	921(a)(34); 923(d)(1)(G)			
Sale or delivery by licensee (see also "NFA controls")				B, D, E, F
Armor piercing ammunition	922(a)(8); 923(e)	478.99(d), (e)		
Between licensees				
Certified copy of transferee's license	926(a)(1)	478.94; 478.95		
Certified list of transferee's licenses		478.94		
Off premises sale of curio or relic	923(j)	478.50(d); 478.100(a)(2)		
Verifying identity of transferee		478.94	IC 74-13	
By mail		478.146		

SUBJECT	LAW	REGULATIONS	RULINGS, PROCEDURES, INDUSTRY CIRCULARS	QUESTIONS AND ANSWERS
Brady law (see "Brady law")				
Compliance with State, local law	922(b)(2)	478.99(b)(2)		
Delivery to agents, employees			IC 72-23; Rul. 69-248	
Delivery to business with expired license		478.94		
Delivery to writers, evaluators			IC 72-23	
Drop shipment			Proc. 75-3	
Gun shows	923(j)	478.100	Rul. 69-59	
Large capacity ammunition feeding device		478.40(a)		
Machinegun	922(o)	478.36; 479.105		
Non-over-the-counter	922(c)	478.96(b), (c)		
Notice to handgun purchaser		478.103(a)-(c)		
Off licensed premises	922(a)(1); 923(a); 923(j)	478.100	Rul. 69-59	
Personal firearm	923(c)	478.125(a)		
Posting sign on juvenile handgun possession		478.103(d)-(f)		
Sales to aliens, nonimmigrant aliens		478.99(c)		
Semiautomatic assault weapon		478.40		
To juvenile	922(x)(1)			
To law enforcement officer		478.132; 478.134		
To licensee after license expiration		478.94		
To nonresident	922(b)(3)	478.99(a)		
To prohibited person	922(d)	478.32(d); 478.99(c)		
To underage person	922(b)(1)	478.99(b)		
Undetectable firearm	922(p)			
Secure gun storage	921(a)(34); 923(d)(1)(G)			
Seizure and forfeiture	924(d); 5872(a)	447.63; 478.152; 479.182		
Semiautomatic assault rifle				
Importation	925(d)(3)			
Semiautomatic assault weapon				O
Defined		478.11		
Discontinuance of business		478.57(b)		
Identification	923(i)	478.92(a)(2)		
Alternate means of		478.92(a)(3)(i)		
Special markings		478.92(a)(3)		
Possession, transfer prohibited		478.40(a)		
Exceptions		478.40(b), (c)		
Sale to law enforcement officer		478.40(b)(6); 478.132; 478.134		
Serial number				O
Adoption of foreign numbers			Rul. 75-28	
Duplication of numbers			IC 77-20	
Identification of firearm by	923(i); 5842(a); 5842(c)	478.92(a); 479.102	Rul. 75-28	
Large capacity ammunition feeding device	923(i)	478.92(c)		
Possessing firearm with altered number	922(k)	478.34		
Semiautomatic assault weapon	923(i)	478.92(a)(2)		
Shotgun				
Defined	921(a)(5); 5845(d)	478.11; 479.11		
Sale to underage person	922(b)(1)	478.99(b)		
Short-barreled shotgun (see "NFA controls")				
Defined	921(a)(6)	478.11		
Interstate transport of	922(a)(4)	478.28		
Sale to nonlicensee	922(b)(4)	478.98		
Sale to research organization		478.145		
Weapon made from	5845(a)(2)			
Silencer (Muffler) (see also "NFA controls")	921(a)(3)(C); 5845(a)(7)			
Defined	921(a)(24)	479.11		
Skeet, trap, similar activity		478.35		
State of residence				B
Alien legally in U.S.		478.11		
College student			Rul. 80-21	
Defined		478.11		
Member of Armed Forces	921(b)	478.11		
Stolen firearm				F
Possession, receipt	922(j)	478.33		
Report of theft, loss	923(g)(6)	478.39a		
Theft from licensee	922(u); 924(l)	478.33a		
Theft from any person	924(k)			
Transportation, shipment	922(i)	478.33		
Storage, guns	921(a)(34); 923(d)(1)(G)			
Theft of firearm (see "Stolen firearm")				
Transportation				
By prohibited person	922(g)	478.32(a)-(c)		

SUBJECT	LAW	REGULATIONS	RULINGS, PROCEDURES, INDUSTRY CIRCULARS	QUESTIONS AND ANSWERS
Interstate transport of				
Destructive device	922(a)(4)	478.28		
Machinegun	922(a)(4)	478.28		
Short-barreled rifle	922(a)(4)	478.28		
Short-barreled shotgun	922(a)(4)	478.28		
Personal firearm	926A	478.38		
Undetectable firearm	922(p)			
Unlawful drug user, addict				B
Defined		478.11		
Possession, receipt of firearm/ammunition by	922(g)(3)	478.32(a)(3)		
Sale of firearm/ammunition to	922(d)(3)	478.32(d)(3)		
United States				
Defined	921(a)(2)	447.11; 479.11		
U.S. Munitions Import List		447.21		
Advance certification	22 U.S.C. 2778 (j)(3)			
Bilateral agreements				
Requirements	22 U.S.C. 2778 (j)(2)			
Watchlist	22 U.S.C. 2778 (j)(2)(A)(iii)			
Foreign country exemptions	22 U.S.C. 2778 (f)(2)(3)			
Foreign country policies	22 U.S.C. 2778 (j)(2)(B)			
Identification of consignees and freight forwarders	22 U.S.C. 2778 (g)(2)			
Electronic filing of Form 6			Rul. 2003-6	
Periodic review of List	22 U.S.C. 2778 (f)			
30-day notice before removal from List	22 U.S.C. 2778 (f)(1)			
U.S. military defense articles		447.57		
Curios or relics		447.57(b)(2)		
Defined		447.57(c)		
Dept. of Defense authorization		447.57(b)		
Foreign assistance or military sales		447.57(a)		
Verification of identity of licensee			IC 74-13	F
Certified copy of license	926(a)(1)	478.95		
Certified list of licenses		478.94		
Waiver by Attorney General	922(y)(3)			
Weapons of mass destruction				
Export licenses for articles related to	2778(a)(2)			
Weapon made from rifle, shotgun (see "NFA controls")				

Made in United States
Troutdale, OR
11/03/2026

56844490R00190